Pythium

Diagnosis, Diseases and Management

Editors

Mahendra Rai

Department of Biotechnology, SGB Amravati University
Amravati, Maharashtra, India

Kamel A. Abd-Elsalam

Unit of Excellence in Nano-Molecular Plant Pathology
Plant Pathology Research Institute, Giza, Egypt

Avinash P. Ingle

Department of Biotechnology, Engineering School of Lorena,
University of Sao Paulo, Lorena, SP, Brazil

CRC Press is an imprint of the
Taylor & Francis Group, an **informa** business

A SCIENCE PUBLISHERS BOOK

Cover figures have been taken from Chapter 11 of this book. Reproduced by kind permission of the authors, Reza Mostowfizadeh-Ghalamfarsa and Fatemeh Salmaninezhad, Department of Plant Protection, Shiraz University, Shiraz, Iran.

CRC Press
Taylor & Francis Group
6000 Broken Sound Parkway NW, Suite 300
Boca Raton, FL 33487-2742

First issued in paperback 2021

© 2020 by Taylor & Francis Group, LLC
CRC Press is an imprint of Taylor & Francis Group, an Informa business

No claim to original U.S. Government works

ISBN-13: 978-0-367-25941-9 (hbk)
ISBN-13: 978-1-03-217555-3 (pbk)
DOI: 10.1201/9780429296406

Library of Congress Cataloging-in-Publication Data

Names: Rai, Mahendra, editor. | Abd-Elsalam, Kamel A., 1969- editor. | Ingle, Avinash P., editor.
Title: Pythium : diagnosis, diseases and management / editors: Mahendra Rai, Kamel A. Abd-Elsalam, Avinash P. Ingle.
Description: Boca Raton, FL : CRC Press, [2019] | Includes bibliographical references and index.
Identifiers: LCCN 2019040770 | ISBN 9780367259419 (hardcover)
Subjects: LCSH: Pythium. | Phytopathogenic fungi.
Classification: LCC QK621.P9 P98 2019 | DDC 579.5--dc23
LC record available at https://lccn.loc.gov/2019040770

**Visit the Taylor & Francis Web site at
http://www.taylorandfrancis.com**

**and the CRC Press Web site at
http://www.crcpress.com**

Preface

The genus *Pythium* is one of the most important fungi of class Oomycota. It is a well-known soil-borne phytopathogenic fungus causing significant damage in agriculture, forest, nurseries, etc. and is ubiquitously found across the world. It also affects the seed germination and attacks the seedlings both at pre- and post-emergence stage. It is an unseen enemy of the root zone of various plants and is hence considered as "hidden terror" for a number of plants. The accurate diagnosis of *Pythium* causing root rot in plants is very important because it is often confused with root rots caused by various other fungi such as *Phytophthora, Rhizoctonia, Chalara, Cylindrocladium, Fusarium* and *Aphanomyces*. Taxonomic identification of *Pythium* species is also important as they vary in their host range and temperature requirements. It is proposed that *Pythium* root rot is difficult to control once rot has begun; therefore, its effective and eco-friendly management is a major concern. In addition, *Pythium* is responsible for causing infections in different animals including horses, dogs, human being, etc. and such infections are known as Pythiosis.

Considering these facts, the present book is focused on all the important aspects related to *Pythium* biology. Broadly, this book is divided into four parts: **Part I** deals with an overview, host range and plant diseases, **Part II** presents various challenges in the taxonomy and diagnosis, and also current technological developments in detection and diagnosis of *Pythium*, **Part III** is devoted to the role of *Pythium* as human pathogen and **Part IV** is focused on the management of *Pythium*. The text in each chapter is supported by numerous clear, informative tables and figures. Each chapter contains relevant references of published articles, which offers a potentially large amount of primary information and further links to a nexus of data and ideas.

All the chapters included in the present book have been written by the specialists, experts in the concerned topic and these chapters are highly informative and detailed. Therefore, we believe that this book will serve as a rich guide for undergraduate or graduate students of various disciplines like agriculture, plant pathology, plant physiology, mycology, molecular biology, biotechnology, and allied subjects. In addition, the book will be useful for researchers in these fields and the people working in various agro-based industries, regulatory bodies, food and agriculture organizations.

The editors are highly thankful to all the contributors for their outstanding efforts to provide state-of-the-art information on the subject matter for their respective chapters. The collective efforts taken by all the authors will help to enhance and update the knowledge of the readers, particularly about *Pythium* and its pathogenesis in both plants and animals. We express our sincere thanks to the publisher and the authors of the chapters, whose research work have been cited in the book. We are also thankful to the entire team at CRC Press for their generous cooperation and efforts in producing this book. MR wishes to thank University Grants Commission, New Delhi for the award of BSR Faculty Fellowship.

We hope that the book will be useful for all the readers to find the required information on the latest research and advances in the field of *Pythium* biology.

Mahendra Rai
Kamel A. Abd-Elsalam
Avinash P. Ingle

Contents

PART I
Incidence, Host-range and Diseases

The Genus *Pythium*: An Overview

Mahendra Rai[1,*], Kamel A. Abd-Elsalam[2], Avinash P. Ingle[3], Priti Paralikar[1] and Pramod Ingle[1]

[1] Nanobiotechnology Lab, Department of Biotechnology, SGB Amravati University,
Amravati - 444602, Maharashtra, India
[2] Unit of Excellence in Nano-Molecular Plant Pathology, Plant Pathology Research Institute,
9 Gamaa St., 12619 Giza, Egypt
[3] Department of Biotechnology, Engineering School of Lorena, University of Sao Paulo,
Estrada municipal do Campinho, sn, 12602-810 Lorena, SP, Brazil

Introduction

Pythium is a soil-borne pathogen which contains more than 300 species and the majority of them are plant pathogenic. *Pythium* is classified into 10 different clades based on morphological and genetic characteristics (Rossman et al. 2017). It is commonly found in soil, sand, various water sources and dead and decaying part of plants. It is ubiquitously distributed across the globe including America, Asia, Africa and Australia. The old taxonomic criteria proposed was confusing and there were some difficulties in the validation of various *Pythium* species. Pringsheim discovered the genus *Pythium* for the first time in 1858, and placed it in the family Saprolegniaceae (Pringsheim 1858). Thereafter, a number of attempts were made to propose different taxonomic systems for the classification of *Pythium* but all such systems were rejected time to time by various taxonomists (Ho 2018). According to current taxonomic system, *Pythium* is placed in family Pythiaceae, order Pythiales, class Oomycetes, phylum Oomycota, and kingdom Chromista (Kirk et al. 2008).

As discussed earlier, *Pythium* spp. are mainly pathogenic to a wide variety of crop plant families and is a major problem in greenhouses and nurseries. Root rot and damping-off are the most important diseases caused by *Pythium* (Rai et al. 2018). It is reported that *Pythium* can infect some important stages of plant's growth which mainly include infection to the seed before germination or during germination. It may attack the young seedlings before or just after emergence. The causative agents generally feed on the root system causing damping-off, which ultimately results in poor germination and spindly plants (McKellar and Nelson 2003). The infection also leads to shortened or distorted leaves, fewer tillers and smaller heads; collectively, it leads to a great loss in crop yield and economic loss (http://www.syngenta-us.com/prodrender/imagehandler.ashx?ImID=37907769-8aca-45f1-9ed2-71ad09069464&fTy=0&et=8).

In addition, it is proposed that, among different species of *Pythium*, *P. insidiosum* is the most prevalent etiologic agent responsible for pythiosis in mammals (Krajaejun et al. 2018). However, among mammals, pythiosis is commonly reported in dogs, horses and humans. Apart from these, sometimes it is also observed in other animals, such as calves (Perez et al. 2005), cats (Rakich et al. 2005), sheep (Santurio et al. 2008), a bird (Pesavento et al. 2008), etc. It is also responsible for rare, non-transmissible disease generally found in tropical, subtropical and temperate regions (Mendoza

*Corresponding author: mahendrarai7@gmail.com; pmkrai@hotmail.com

2005). However, recently it was reported that the infections are not only restricted to these climatic conditions, but also observed in regions like California and Arizona, where the climate does not fit this description. It was clearly indicated that the environmental niche for *P. insidiosum* is expanding, possibly due to outcome of environmental changes like unnecessary flooding of rice fields or irrigated landscape development (Gaastra et al. 2010).

Apart from the pathogenesis, there are two important concerns about *Pythium* which are always raised in many available reports, one is about the confirm identifications of *Pythium* species and another is about its management (Sutton et al. 2006, Tambong et al. 2006). The overlapping mycelial and sporangial characters mainly hinder the morphological identification of *Pythium* spp. Some conventional identification methods are available which include identification of species on the basis of morphology of antheridia, oogonia and sporangia, but it greatly varies under different cultural conditions. In addition, these approaches are time-consuming and require highly expert hands (Kumar et al. 2008). It has been proved that management of *Pythium* is very difficult once the infection is established. There are some traditional approaches involving the use of chemical fungicides (antifungal agents), phytochemicals and biological agents which are routinely in practice for the management of *Pythium*; however, still there is need to develop more sensitive, eco-friendly and economically viable methods for the efficient management of *Pythium* (Rai et al. 2018).

Considering all these aspects, in the present chapter we have focused on overview of *Pythium*, which mainly includes its recent taxonomic status, worldwide distribution, pathogenesis and management.

Taxonomic challenges

Traditionally, the identification of *Pythium* species is based on morphological characteristics (Dick 1990, Van der Plaats-Niterink 1981) such as sporangia, shape and size of oogonia, antheridia and, sporangia, oospores, and rate of growth on culture medium. However, being biologically and ecologically diverse species, the morphological characteristics are highly variable and hence determination of identity is a major challenge (Ghalamfarsa 2015). Considering these difficulties, morphological characteristics can not be used for identification of *Pythium* species (Lévesque and de Cock 2004).

Mycelia of *Pythium* species branch out apically at right angles; hyphae are hyaline, with mostly 5-7 μm wide; cross septa are generally present in old culture and not in new ones (Vander PlaatsNiterink 1981). Streaming of protoplast is often conspicuous in newly formed hyphae. The hyphal walls of *Pythium* are mainly composed of polysaccharides (80-90%) like ß1-6 linked glucans and ß1-3 and ß1-4 (cellulose) (Postma et al. 2009). Interestingly, hyphal wall of *Pythium* spp. does not contain chitin or chitosan, whereas it consists of varying concentration of protein (3-8%) and lipid (1-3%) (Postma et al. 2009). These characteristics are variable under different culture conditions, and therefore, many species are morphologically similar. Some of these characteristics can also change or be acquired or lost readily (Lévesque and De Cock 2004). In addition, the morphological characteristics used for species differentiation has not always correlated with the major clades in *Pythium* determined by molecular methods (e.g. Lévesque and De Cock 2004). Therefore, molecular markers have been found to be essential tools for determining identity of *Pythium* spp.

Molecular markers are important tools for confirming the identity of the fungi being fast, authentic, specific and sensitive as compared to the use of morphological markers. The molecular techniques are promising alternatives to determine the identity of fungi even without the knowledge of taxonomy. Many researchers used Internal transcribed spacer (ITS) region for determination of identity of a particular pathogen (Robideau et al. 2011, Rai et al. 2014, Salmaninezhad and Mostowfizadeh-Ghalamfarsa 2019). In spite of much variation in sequences in ITS region, the availability of primers that provide sequence data is also essential (Lévesque and de Cock 2004).

Based on phylogenetic study, polyphyletic nature of *Pythium* has been suggested because its species have originated from two or more independent ancestors. Considering these facts, it is felt that taxonomic revisions of the genus *Pythium* are necessary, while some authors have recommended creation of five new genera (Uzuhashi et al. 2010). Lévesque and de Cock (2004) studied 116 species and varieties of *Pythium* on the basis of ITS rDNA sequencing and further, based on phylogeny, classified these species in 11 major clades (i.e. A to K). The authors further confirmed that the *Pythium* species in clade K are genetically different from the rest of the genus. Interestingly, the members of this clade demonstrated morphology intermediary between *Phytophthora* and *Pythium* and therefore, this group has been designated as a new genus termed as *Phytopythium*. The morphological and molecular studies of clade K together with improved taxon sampling, led to its reassignment as genus *Phytopythium* (Bala et al. 2010, Marano et al. 2014, De Cock et al. 2015, Jesus et al. 2016). *Phytopythium* generally showed morphological characteristics similar to both *Pythium* and *Phytophthora* as they proliferate internally similar to some species of *Phytophthora;* however, development and release of zoospores is external (De Cock et al. 2015) or partly internal and partly external to sporangia (Marano et al. 2014, Jesus et al. 2016). The important morphological difference between these clades is due to the sporangial morphology (ovoid, globose, elongated or filamentous) (Levesque and de Cock 2004, Uzuhashi et al. 2010)

Pathogenicity of *Pythium*

Many members of the genus *Pythium* cause infections and diseases in plants, animals and human beings. The pathogenicity of *Pythium* to plants, animals and humans has been briefly discussed below.

Diseases in plants

As discussed above, the genus *Pythium* is a readily recognized plant pathogen with a very wide host range and distribution (Van Buyten and Höfte 2013). Certainly, all the members are not pathogenic but most of them cause serious loss to crops under favorable conditions like susceptible host, environment and geographical range. *Pythium* spp. primarily causes infection to the juvenile or succulent tissues, limiting their damage to seedlings or feeder roots. In non-seedling plant hosts like grass, tomato transplants, peanuts, and chrysanthemums, most affected parts are stems and foliage leaves. A fruit rot was also seen in crops like beans, squash, and watermelon (Hendrix and Campbell 1973). *Pythium* spp. involve in the destruction of the fine roots and root tips of trees (Lorio 1966), curtailing the inability of roots to absorb sufficient nitrogen from the soil (Campbell and Otis 1954). Peach and citrus decline also hampers the production which is associated with *Pythium* spp. (Sleeth 1953, Hendrix et al. 1966).

Michigan is the third largest producer of floriculture in the USA. The million dollar business is mainly affected by various diseases caused by *Pythium* which mainly includes damping-off, crown and root rot. Del Castilo Munera and Hausbeck (2016) worked on the isolation of pathogenic fungi from various flowering plants and identified them on the basis of morphology and ITS sequencing. Among the various isolates, 287 isolates were obtained from poinsettias, 726 from geranium and other greenhouse floral cultures. *P. aphanidermatum, P. ultimum, P. cylindrosporum,* and *P. irregulare* were the most commonly reported species (Del Castilo Munera and Hausbeck 2016). Seed rot, root rot, seedling damping off, flower rot and black leg in ornamental plants are profoundly caused by *Pythium* spp. (Martin and Loper 1999, Moorman et al. 2002). Stephen and Powell (1982) reported different *Pythium* spp. such as *P. aphanidermatum, P. ultimum, P. spinosum* and *P. debaryanum* associated with damping-off of impatiens, vinca and celosia. In case of crop plants, *Pythium* infects both underground and aerial parts. *P. aphanidermatum* and *P. myriotylum* are the major crop pathogenic *Pythium* spp. (Agrios 2005). *Pythium* infections are confined to the meristematic tissues of mature plants. They cause necrotic lesions on root tips and less commonly affect the tap roots. But the deeper invasion may cause infection of vascular parts (Watanabe et al. 2008).

Studies on seedling diseases of corn (*Zea mays* L.) and soybean (*Glycine max* (L.) Merr.) in Ohio showed the predominance of *Pythium* spp. Broders et al. (2007) reported eleven *Pythium* spp.: *P. attrantheridium*, *P. dissotocum*, *P. echinulatum* Matthews, *P. graminicola*, *P. inflatum*, *P. irregulare*, *P. helicoides* Drechs., *P. sylvaticum*, *P. torulosum*, *P. ultimum* var. *ultimum*, and *P. ultimum* Trow var. *sporangiiferum* Drechs. associated with corn and soybean seeds rot. Out of these, six species were pathogenic to corn and nine species were pathogenic to soybean seeds. According to van der Plaats-Niterink (1981), *Pythium* spp. commonly infects seedlings, tap roots, root tips or feeder roots and also mature plants which leads to death of respective plant. *P. aphanidermatum* is reported to cause damping-off of seedlings, root and crown rots of mature cucumber (*Cucumis sativus* L.) plants (Zitter et al. 1996, Al-Sa'di et al. 2007). Pegg and Manners (2014) reported the association of *Pythium* spp. with the nursery plants like blueberries, causing cutting and stem rot, and aerial rot. Rai et al. (2018) have summarized the *Pythium* spp. and *Fusarium* spp. responsible for the soft rot of ginger (*Zingiber officinale* Rosc.). *P. aphanidermatum* and *P. myriotylum* are the majorly reported *Pythium* spp. responsible for soft rot in ginger rhizome (Dohroo 2005, Le et al. 2014, 2016).

Diseases in animals and humans

P. insidiosum is an oomycete pathogen in mammals in the tropical and subtropical region. The disease is reported throughout the world. Mostly pythiosis is reported in horses and dogs. Hasika and colleagues (2019) studied the demographic distribution of *Pythium* keratitis in South India and reported the large series of patients with *Pythium* keratitis. This indicated the devastating need of diagnosis, treatment and awareness in clinicians (Mendoza and Newton 2005, Wilson 2012, Hasika et al. 2019) and rarely in rabbits (Gaastra et al. 2010, Botton et al. 2011). The systemic infections are observed in horses involving ulcerative, proliferative, pyogranulomatous lesion (Reis et al. 2003). As mentioned earlier, *Pythium* species are present in soil and water resources, when animals like horses with open wound come in contact with these *Pythium* species get infected through such wounds. Similarly, *Pythium* causes infections in humans also in the same way through severely injured tissues. Apart from these, *Pythium* was also reported to infect the gastrointestinal tract of cats (Prasertwitayakij et al. 2003, Rakich et al. 2005, Bosco et al. 2005, Badenoch et al. 2009, Fortin et al. 2017). In case of dogs, cutaneous/subcutaneous and gastrointestinal lesions are seen (Miller 1985, Grooters 2003).

Pythiosis has also been reported in other domestic animals such as cats (nasal cavity, retrobulbar space, skin and subcutaneous tissues) (Foil et al. 1984, Bissonnette et al. 1991, Thomas and Lewis 1998), a dromedary camel (skin, gastric) (Wellehan et al. 2004), cattle (skin/subcutaneous) (Miller et al. 1985), and sheep (skin/subcutaneous, disseminated) (Tabosa et al. 2004). The first case of human pythiosis was reported from patients in Thailand in 1985 (Imwidthaya 1994).

Commonly, three forms of pythiosis have been reported in humans: (i) granulomatous and ulcerative lesions of skin and subcutaneous tissues of face and limbs (ii) Ophthalmic pythiosis responsible for keratitis (Pal and Mahendra 2014, Garg 2019), and (iii) systemic pythiosis responsible for vasculitis, thrombosis and aneurysms (Tanphaichitra 1989, Thianprasit et al. 1996, Prasertwitayakij et al. 2003). Pupaibool et al. (2006) reported *P. insidiosum* as a predominant causative agent in humans, leading to human pythiosis and ultimately death.

The animals from tropical, subtropical and temperate regions are mostly susceptible to *Pythium* infections; moreover, these region mainly include countries like Australia, Argentina, Brazil, Costa Rica, Colombia, Egypt, Greece, Haiti, Indonesia, India, Japan, Thailand, Papua New Guinea and the USA (Mendoza et al. 1996). First case of *Pythium* infection in animals (pythiosis) was reported in late 1800s for the first time. A controversy was reported when the similarity between lesions due to pythiosis and equine cutaneous habronemiasis was explained by Miller (1983). The infection caused by *P. insidiosum* triggers the T helper 2 [Th2] subset present in the host which further leads to inflammatory reaction which occurs mainly in eosinophils and mast cells. Later, these cells degranulate around the hyphal elements of *P. insidiosum* where a Splendore-Hoeppli-like reaction occurs (Mendoza and Newton 2005). Periorbital cellulitis caused by *P. insidiosum* was reported in a two-year child in the US, which extended to the nasopharynx and compromised airway, leading to gastrostomy (Shenep et al. 1998).

Different approaches for management of *Pythium* spp.

Chemical and physical agents in *Pythium* management

The genus *Pythium* is one of the most important plant pathogenic genera responsible for causing infections in broad host range of plants. Due to its ubiquitous nature, it is very difficult to control using chemical and physical methods. As it becomes resistant to chemical antifungals, researchers are encouraged to find new alternatives for its management. *Pythium* spp. are the ubiquitous water molds with habitats ranging from terrestrial to aquatic. It causes multibillion dollar losses of the crop yield worldwide every year. *P. aphanidermatum* is one of the most destructive member among the various *Pythium* species which causes many economically important diseases (van West et al. 2003, Parveen and Sharma 2015). Most widely used control method by farmers is a soil solarization, i.e. hydrothermal heating of soil during hot months, when soil is exposed to direct sunlight and change in soil environment can help in controlling *Pythium* progression (Katan 2000). Chemical methods usually used for the control of *Pythium* involves the use of different fungicides like mancozeb, carbendazim, ridomil, mancozeb, and topsin, etc. (Elliott 2003, Anomynous 2005, ANSAB 2011, Poduyal 2011). These synthetic fungicides are more effective because they control pathogen by either destroying cell membrane or increasing its permeability or by inhibiting their metabolic processes (Osman and Al-Rehiayam 2003).

Phytochemicals in *Pythium* management

The use of organic materials can also be a good option for control of *Pythium* infections because of its unique composition and microbial activity. There are many reports available claiming the management of *P. aphanidermatum* and some other *Pythium* species using certain waste materials like sugarcane residues, poultry slurry and municipal bio-solids. This supplementary organic matter contributes higher nutrient content to soil, ultimately increasing the yield (Parveen and Sharma 2015). Antimicrobial components of plant extracts can be used as biocontrol agents as they bring about the change in proton flux across the membrane, thus changing the cell environment and ultimately death of cell (Omidbeygi et al. 2007, Pane et al. 2011). Plant extracts rich in hydrophobic lipids can be deleterious to fungus as they partition the fungal cell wall and mitochondria, disturbing their structure, rendering the leakage of cellular contents and finally death of cell (Burt 2004, Gonçalves et al. 2017). The extracts of *Zygophyllum fabago, Azadirachta indica, Allium sativum* and *Curcuma longa* were reported to have significant inhibition of *P. aphanidermatum* in *in vitro* study (Dana et al. 2010, Singh et al. 2010). However, the leaf extract of *Zimmu* (*Allium cepa* × *Allium sativum*) showed potential antifungal efficacy against *Pythium in vivo* (Muthukumar et al. 2010). Nowadays, it is possible to use extracts of various plants such as: *Thymus vulgaris* and *Zingiber officinale* instead of chemical fungicides for the control of tomato damping-off diseases caused by *Pythium* and *Fusarium* species. Vinayaka et al. (2014) reported antifungal activity of aqueous extract of *Usnea pictoides* against *P. aphanidermatum* isolated from rotten ginger rhizome.

In addition, methanolic extract of *Vitex agnus-castus* showed potential antifungal activity against *P. ultimum* causing infection in tomato under both *in vitro* and *in vivo* conditions. The involvement of pathogenesis-related (PR) genes in delayed infection was demonstrated by treating tomato plants with *Vitex agnus-castus* methanolic extract and/or *P. ultimum.* The expression of PR genes, i.e. PR-1, PR-2, PR-5 and PR-6 involved in plant defense mechanism was monitored (Svecová et al. 2013). More et al. (2017) reported the effect of different medicinal plant waste extracts in various solvents against the soil borne pathogen *P. debaryanum,* which was isolated from infected tomato on Vaartaza's medium. Various solvent-extract system used were include acetone, ethanol, methanol and chloroform extracts of *A. marmelos, S. cumini* and *P. pinnata.* One of the study includes evaluation of pathogenicity of *P. debaryanum* in some varieties of tomato by the soil inoculation technique. After inoculation, the initial symptoms were observed between 7-21 days. Pre-damping off seed decay and post-damping off stem lesions were observed in addition to top rot and root rot. Further, authors tested the efficacy of methanolic, ethanolic and chloroform extracts of leaves and fruits of *A. marmelos* against this pathogen. Maximum growth inhibition of pathogen was reported in case of methanolic extract of leaves and fruits of *A. marmelos* followed by ethanolic extract; however, minimum antifungal activity was reported in chloroform extract (More et al. 2017).

Biological methods of *Pythium* management

Biological control of *Pythium* involves the accumulation of antagonistic metabolites, spatial competition and nutrients diversification, hyphal interactions, mycoparasitism, enzymes secretion and microbes feeding on *Pythium* propagules. Use of *Trichoderma harzianum* against *Pythium* was reported by Elad (1982) and it was proposed that it can be used in disease reduction and increased seed germination (Shanmugam et al. 1999). Lumsden and Locke (1989) demonstrated significant control of damping-off caused by *P. ultimum* in various crops like cotton, cabbage and zinnia using *Gliocladium virens.* In addition, different species of *Trichoderma* are known for their antagonistic activity and are hence commonly used as potential biocontrol agents. Manoranjitham and Prakasam (2000) demonstrated the significant efficacy of *T. hamatum,T. harzianum, T. reesai, T. viridae* and *Psudomonas fluorescens* against *P. aphanidermatum.* Similarly, in another study, Ram (2000) evaluated potential of *T. harzianum, T. aureoviride* and *T. virens* against *Pythium* spp. and reported significant reduction in the disease and increased the yield. *P. fluorescens* controlled the damping-off in sugar beet (Bardin et al. 2004). Management of pre- and post-emergence of *P. aphanidermatum* mediated damping-off in chilli was reported by Haritha et al. (2010). *T. hamatum* and *T. harzianum* were also reported to be antagonistic against *P. aphanidermatum* and *P. ultimum* in chilli (Kamala and Indira 2011, Sharma et al. 2014). Biological control methods cannot be considered as curative methods of disease control because they are efficient only when pathogen load is low to moderate. Biocontrol methods may become ineffective in case the crop is already infected with the pathogen.

Nanotechnological approaches in *Pythium* management

In present days, nanotechnology is expected to turn into the innovative frontier in agriculture practices by contributing novel application. Nanoparticles have benefits depending upon type, concentration, size and mode of synthesis in agriculture application (Singh et al. 2016, 2017, Tripathi et al. 2017, Yang et al. 2017, Elmer et al. 2018). Among the different bioactivities possessed by nanoparticles, antimicrobial activity is important because there is a harmful impact of plant pathogens in agriculture and to tackle this problem nanotechnology is extensively used. Increased application of bactericides and fungicides for plant pathogens management caused harmful effect by contaminating vulnerable ecosystem. Furthermore, the use of nanomaterial offers many additional properties like improved solubility, long lasting residual activity of agrichemical, and delivery can be exploited in plant health management (Fig. 1.1). The application of nanoparticles in agriculture not only suppresses the growth of pathogen but also promotes the growth of plant (Pourkhaloee et

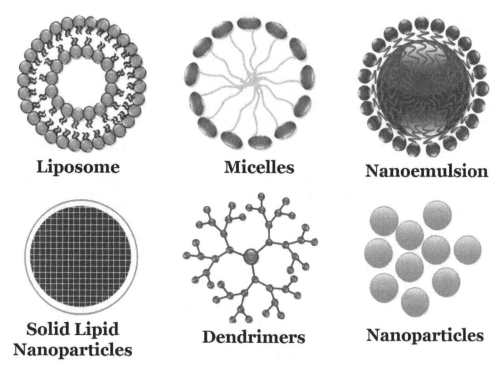

Figure 1.1. Nanomaterials for delivery of fungicides

Color version at the end of the book

al. 2011, Hu et al. 2014, Shinde et al. 2018, Mohamed et al. 2018). Different classes of nanoparticles are used in agriculture depending upon their mode of action. For the antimicrobial application in plant pathogen management, nanoparticles such as silver, copper, sulfur, zinc, carbon nanotubes, etc. are widely used (Fosso-Kankeu et al. 2016, Athawale et al. 2018, Rai et al. 2018). Nanoparticles were not only reported to show strong antifungal activity but they also help to maintain soil nutrients status (Ponmurugan et al. 2016, Prasad et al. 2017).

Conclusion

Pythium is ubiquitous in distribution, basically a soil-borne fungus with a large number of pathogenic species and polyphyletic. It is responsible for damping-off, soft-rot, and blight; consequently, there is a huge economic loss of crops in agriculture. *Pythium* also causes pythiosis in mammals, particularly horses, dogs and humans. The identification and differentiation of different *Pythium* spp. is an arduous task based on morphological characters, such as antheridia, oogonia and sporangia, which are not stable in different culture media. Therefore, the use of molecular markers for confirmation of species are essentially required. Moreover, these markers are rapid, accurate, sensitive and specific as compared to the morphological markers. Another important problem with *Pythium* spp. is its management. Although various chemicals/fungicides, phytochemicals are commonly used for the management of different *Pythium* spp., none of them is effective for complete management of the disease. The biological methods or green methods have advantage over the chemical methods because these are eco-friendly and economically viable. However, these methods are useful when the pathogen load is low and their efficacy reduces when pathogen is already present in crops or soils. There is a greater need to apply nanotechnological strategies which are emerging with remarkable efficacy.

References

Agrios, G.N. 2005. Plant Pathology. Elsevier Academic Press, New York, 922 p.

Al-Sa'di, A.M., Drenth, A., Deadman, M.L., de Cock, A.W.A. and Aitken, E.A.B. 2007. Molecular characterization and pathogenicity of *Pythium* species associated with damping-off in greenhouse cucumber (*Cucumis sativus*) in Oman. Plant Pathol. 56: 140-149.

Anomynous. 2005. Experiences in Collaboration – Ginger Pests and Diseases. Indo-Swiss Project Sikkim. Intercooperation India Programme Series 1, Intercooperation Delegation, Hyderabad, India. 57.

ANSAB (Asia Network for Sustainable Agriculture and Bioresources). 2011. A report on value chain analysis of Ginger in Nepal. Report Submitted to The Netherland Development Organization (SNV), Nepal (Unpublished material).

Athawale, V., Paralikar, P., Ingle, A.P. and Rai, M. 2018. Biogenically engineered nanoparticles inhibit *Fusarium oxysporum* causing soft-rot of ginger. IET Nanobiotech. 12(8): 1084-1089.

Badenoch, P.R., Mills, R.A.D., Chang, J.H., Sadlon, T.A., Klebe, S. and Coster, D.J. 2009. *Pythium insidiosum* keratitis in an Australian child. Clin. Exp. Ophthalmo. 37: 806-809.

Bahraminejad, S. 2012. *In vitro* and *In vivo* antifungal activities of Iranian plant species against *Pythium aphanidermatum*. Annals Biol. Res. 3(5): 2134-2143.

Bala, K., Robideau, G., de Cock, A.W.A.M., Abad, Z.G., Lodhi, A.M., Shahzad, S., Ghaffar, A., Coffey, M.D. and Lévesque, C.A. 2010. *Phytopythium* Abad, de Cock, Bala, Robideau, Lodhi & Lévesque. Gen. Nov. Persoonia 24: 136-137.

Bardin, S.D., Huang, H.C. and Moyer, J.R. 2004. Control of *Pythium* damping-off of sugar beet by seed treatment with crop straw powders and a biocontrol agent. Biol Control. 29(3): 453-460.

Bissonnette, K.W., Sharp, N.J.H., Dykstra, M.H., Robertson, I.R., Davis, B., Padhye, A.A. and Kaufman, L. 1991. Nasal and retrobulbar mass in a cat caused by *Pythium insidiosum*. J. Med. Vet. Mycol. 29: 39-44.

Bosco, S.D.G., Bagagli, E., Araujo, J.P., Candeias, J.M.G., de Franco, M.F., Marques, M.E.A., Mendoza, L., de Camargo, R.P. and Marques, S.A. 2005. Human pythiosis, Brazil. Emerging Infect. Dis. 11: 715-718.

Botton, S.A., Pereira, D.I.B., Costa, M.M., Azevedo, M.I., Argenta, J.S., Jesus, F.P.K., Alves, S.H. and Santurio, J.M. 2011. Identification of *Pythium insidiosum* by Nested PCR in cutaneous lesions of brazilian horses and rabbits. Curr. Microbiol. 62: 1225-1229.

Broders, K.D., Lipps, P.E., Paul, P.A. and Dorrance, A.E. 2007. Characterization of *Pythium* spp. associated with corn and soybean seed and seedling disease in Ohio. Plant Dis. 91: 727-735.

Burt, S. 2004. Essential oils: Their antimicrobial properties and potential applications in foods – A review. Int. J. Food Microbiol. 94: 223-253.

Campbell, W.A. and Copeland, Otis L. 1954. Littleleaf disease of shortleaf and loblolly pines. USDA Circ. 940.

Dana, E.D., Delomas, J.G. and Sanchez, J. 2010. Effects of the aqueous extracts of *Zygophyllum fabago* on the growth of *Fusarium oxyosporum f. sp. melonis* and *Pythium aphanidermatum*. Weed Biol. Manag. 10: 170-175.

Del Castilo Munera, J. and Hausbeck, M.K. 2016. Characterization of *Pythium* species associated with greenhouse floriculture crops in Michigan. Plant Dis. 100: 569-576.

Dohroo, N.P. 2005. Diseases of ginger. pp. 305-340. *In*: Ravindran, P.N. and Babu, K.N. (eds.). Ginger, the genus Zingiber. CRC Press, Boca Raton.

Elad, Y., Kalfon, A. and Chet, I. 1982. Control of *Rhizoctonia splani* in cotton by seed-coating with *Trichoderma* spp. spores. Plant Soil. 66: 279-281.

Elliott, S.M. 2003. Rhizome Rot Disease of Ginger. Ministry of Agriculture Research and Development Division. St. Catherine, Jamaica.

Elmer, W.H., Ma, C. and White, J.C. 2018. Nanoparticles for plant disease management. Curr. Opinion Env. Sci. Health. doi: 10.1016/j.coesh.2018.08.002.

Foil, C.S.O., Short, B.G., Fadok, V.A. and Kunkle, G.A. 1984. A report of subcutaneous pythiosis in five dogs and a review of the etiologic agent *Pythium* spp. J. Am. Anim. Hosp. Assoc. 20: 959-966.

Fortin, J.S., Calcutt, M.J. and Kim, D.Y. 2017. Sublingual pythiosis in cats. Acta. Vet. Scand. 59(1): 63. https://dx.doi.org/10.1186%2Fs13028-017-0330-z

Fosso-Kankeu, E., De Klerk, C.M., Botha, T.A., Waanders, F., Phoku, J. and Pandey, S. 2016. The antifungal activities of multi-walled carbon nanotubes decorated with silver, copper and zinc oxide particles. pp. 55-59. *In*: International Conference on Advances in Science, Engineering, Technology and Natural Resources (ICASETNR-16), Parys, South Africa, 24-25 November 2016.

Gaastra, W., Lipman, L.J.A., De Cock, W.A.M., Exel, T.K., Pegge, R.B.G., Scheurwater, J., Vilela, R. and Mendoza, L. 2010. *Pythium insidiosum*: A overview. Vet. Microbio. 146: 1-16.

Garg, P. 2019. Commentary: *Pythium insidiosum* keratitis. Indian J. Ophthalmol. 67(1): 47-48. doi: 10.4103/ijo.IJO_1491_18

Ghalamfarsa, R.M. 2015. The current status of *Pythium* species in Iran: Challenges in taxonomy. Mycologia Iranica 2(2): 79-87.

Gonçalves, A.P., Heller, J., Daskalov, A., Videira, A. and Glass, N.L. 2017. Regulated forms of cell death in fungi. Front. Microbiol. 8: 1837. doi:10.3389/fmicb.2017.01837

Grooters, A.M. 2003. Pythiosis, lagenidiosis and zygomycosis in small animals. (Emerging and re-emerging infectious diseases). *In*: Veterinary Clinics of North America, Small Animal Practice. Philadelphia: W.B. Sanders, 33: 695-720.

Haritha, V., Gopal, K., Madhusudhan, P., Vishwanath, K. and Rao, S.V.R.K. 2010. Integrated management of damping off disease incited by *Pythium aphanidermathum* (Edson) pitzpin tobacco nursery. J. Plant Dis. Sci. 5(1): 41-47.

Hasika, R., Lalitha, P., Radhakrishnan, N., Rameshkumar, G., Prajna, N.V. and Srinivasan, M. 2019. *Pythium* keratitis in South India: Incidence, clinical profile, management, and treatment recommendation. Indian J. Ophthalmol. 67: 42-47.

Hendrix, F.F. Jr., Powell, W.M. and Owen, J.H. 1966. Relation of root necrosis caused by *Pythium* species to peach tree decline. Phytopath. 56: 1229-1232.

Hendrix, F.F. Jr. and Campbell, W.A. 1973. *Pythium* as plant pathogens. Annu. Rev. Phytopathol. 11: 77-98.

Ho, H.H. 2018. The taxonomy and biology of *Phytophthora* and *Pythium*. J. Bacteriol. Mycol. 6(1): 40-45.

Hu, X. and Zhou, Q. 2014. Novel hydrated graphene ribbon unexpectedly promotes aged seed germination and root differentiation. Sci Rep. 4: 3782.

Imwidthaya, P. 1994. Human pythiosis in Thailand. Postgrad. Med. J. 70: 558-560.

Kalu, N.N., Sutton, J.C. and Vaartaja, O. 1976. *Pythium* spp. associated with root dieback of carrot in Ontario. Can. J. Plant Sci. 56: 555-561.

Kamala, T.H. and Indira, S. 2011. Evaluation of indigenous *Trichoderma* isolates from Manipur as biocontrol agent against *Pythium aphanidermatum* on common beans. 3Biotech. 1(4): 217-225.

Katan, J. 2000. Physical and culture methods for the management of soil borne pathogens. Crop Prot. 19: 725-731.

Kirk, P.M., Cannon, P.F., Minter, D.W. and Stalpers, J.A. 2008. Ainsworth & Bisby's Dictionary of the Fungi, 10th edn. CAB International, Wallingford.

Krajaejun, T., Rujirawat, T., Kanpanleuk, T., Santanirand, P., Lohnoo, T., Yingyong, W., Kumsang, Y., Sae-Chew, P., Kittichotirat, W. and Patumcharoenpol, P. 2018. Biochemical and genetic analyses of the oomycete *Pythium insidiosum* provide new insights into clinical identification and urease-based evolution of metabolism related traits. Peer J. 6: e4821, doi: 10.7717/peerj.4821.

Kumar, A., Reeja, S.T., Bhai, S.S. and Shiva, K.N. 2008. Distribution of *Pythium myriotylum* Drechsler causing soft rot of ginger. J. Spices Arom. Crops 17(1): 5-10.

Le, D.P., Smith, M., Hudler, G.W. and Aitken, E. 2014. *Pythium* soft rot of ginger: Detection and identification of the causal pathogens and their control. Crop Prot. 65: 153-167.

Le, D.P., Smith, M.K. and Aitken, E.A.B. 2016. An assessment of *Pythium* spp. associated with soft rot disease of ginger (*Zingiber officinale*) in Queensland, Australia. Australasian Plant Pathol. 45(4): 377-387.

Lévesque, C.A. and De Cock, W.A.M. 2004. Molecular phylogeny and taxonomy of the genus *Pythium*. Mycol Res. 108: 1363-1383, doi: 10.1017/S0953756204001431.

Lorio, P.L. 1966. *Phytophthora cinnamomi* and *Pythium* species associated with loblolly pine decline in Louisiana. Plant Dis. Reptr. 50: 596-597.

Lumsden, R.D. and Locke, J.C. 1989. Biological control of damping-off caused by *Pythium ultimum* and *Rhizoctonia solani* with *Gliocladium virens* in soilless mix. Phytopath. 79: 361-366.

Manoranjitham, S.K. and Prakasam, V. 2000. Management of chilli damping off using biocontrol agents. Capsicum Eggplant News Lett. 19: 101-104.

Marano, A.V., Jesus, A.L., De Souza, J.I., Leãno, E.M., James, T.Y., Jerônimo, G.H., De Cock, A. and Pires-Zottarelli, C.L.A. 2014. A new combination in Phytopythium: P. kandeliae (Oomycetes, Straminipila). J. Fungal Biol. 5: 510-522.

Martin, F.N. and Loper, J.E. 1999. Soil-borne plant diseases caused by Pythium spp.: Ecology, epidemiology, and prospects for biological control. Crit. Rev. Plant Sci. Boca Raton 18(2): 111-181.

McKellar, M.E. and Nelson, E.B. 2003. Compost-induced suppression of Pythium damping-off is mediated by fatty-acid-metabolizing seed-colonizing microbial communities. Appl. Environ. Microbiol. 69(1): 452-460.

Mendoza, L. 2005. Pythium insidiosum. pp. 617-630. In: Merz, W.G. and Hay, R.J. (eds.). Topley and Wilson's Microbiology and Microbial Infections. 10th ed. Medical Mycology. John Wiley & Sons, Inc., UK.

Mendoza, L., Ajello, L. and McGinnis, M.R. 1996. Infections caused by the oomycetous pathogen Pythium insidiosum. J. Mycol. Med. 6: 151-164.

Mendoza, L. and Newton, J.C. 2005. Immunology and immunotherapy of the infections caused by Pythium insidiosum. Med. Mycol. 43: 477-486.

Miller, R.I. 1983. Granulomatous and neoplastic diseases of the equine skin: A review. Vet Rev Monograph No. 6. Townsville, Australia. Dept Trop Vet Sci, James Cook University. 1-52.

Miller, R.I. 1985. Gastrointestinal phycomycosis in 63 dogs. J. Am. Vet. Assoc. 186: 473-478.

Miller, R.I., Olcott, B.M. and Archer, M. 1985. Cutaneous pythiosis in beef calves. J. Am. Vet. Med. Assoc. 186: 984-986.

Mohamed, M.A., Hashim, A.F., Alghuthaymi, M.A. and Abd-Elsalam, K.A. 2018. Nano-carbon: Plant growth promotion and protection. In: Abd-Elsalam, K. and Prasad, R. (eds.). Nanobiotechnology Applications in Plant Protection. Nanotechnology in the Life Sciences. Springer, Cham.

Moorman, G., Kang, S., Geiser, D. and Kim, S. 2002. Identification and characterization of Pythium species associated with greenhouse floral crops in Pennsylvania. Plant Disease 86: 1227-1231.

More, Y.D., Gade, R.M. and Shitole, A.V. 2017. Evaluation of antifungal activities of extracts of Aegle marmelos, Syzygium cumini and Pongamia pinnata against Pythium debaryanum. Indian J. Pharm. Sci. 79(3): 377-384.

Muthukumar, A., Eswaran, A., Nakkeeran, S. and Sangeetha, G. 2010. Efficacy of plant extracts and biocontrol agents against Pythium aphanidermatum inciting chilli damping-off. Crop Protect. 29: 1483-1488.

Omidbeygi, M., Barzegar, M., Hamidi, Z. and Naghdibadi, H. 2007. Antifungal activity of thyme, summer savory and clove essential oils against Aspergillus flavus in liquid medium and tomato paste. Food Control. 18: 1518-1523.

Osman, K.A. and Al-Rehiayam, S. 2003. Risk assessment of pesticide to human and the environment. Saudi J. Biol. Sci. 10: 81-106.

Pal, M. and Mahendra, R. 2014. Pythiosis: An emerging oomycetic disease of humans and animals. Int. J. Livestock Res. 4(6), ISSN 2277-1964 ONLINE.

Pane, C., Spaccini, R., Piccolo, A., Scala, F. and Bonanomi, G. 2011. Compost amendments enhance peat suppressiveness to Pythium ultimum, Rhizoctonia solani and Sclerotinia minor. Biol. Control. 56: 115-124.

Parveen, T. and Sharma, K. 2015. Pythium diseases, control and management strategies: A review. Int. J. Plant Animal Env. Sci. 5(1): 244-257.

Pegg, K. and Manners, A. 2014. Pythium species: A constant threat to nursery production. Agri-science Queensland, Department of Agriculture, Fisheries and Forestry (DAFF), as part of NY11001 Plant health biosecurity, risk management and capacity building for the nursery industry in 2014.

Perez, R., Luis-Leon, J.J., Vivas, J.L. and Mendoza, L. 2005. Epizootic cutaneous pythiosis in beef calves. Vet Microbiol. 109: 121-128.

Pesavento, P.A., Barr, B., Riggs, S.M., Eigenheer, A.L., Pamma, R. and Walker, R.L. 2008. Cutaneous pythiosis in a nestling white faced ibis. Vet. Pathol. 45: 538-541.

Poduyal, B.K. 2011. The Control of Soft Rot of Ginger by Jeevatu based Organic Liquid Manure. Central Vegetable Seed Production Centre Khumaltar, Lalitpur, Nepal.

Ponmurugan, P., Manjukarunambika, K., Elango, N. and Gnanamangai, B.M. 2016. Antifungal activity of biosynthesised copper nanoparticles evaluated against red root-rot disease in tea plants. J. Exp. Nanosci. DOI: 10.1080/17458080.2016.1184766

Postma, J., Stevens, L.H., Wiegers, G.L., Davelaar, E. and Nijhuis, E.H. 2009. Biological control of *Pythium aphanidermatum* in cucumber with a combined application of Lysobacter enzymogenes strain 3.1 T8 and chitosan. Bio. Control 48(3): 301-309.

Pourkhaloee, A., Haghighi, M., Saharkhiz, M.J., Jouzi, H. and Doroodmand, M.M. 2011. Carbon nanotubes can promote seed germination via seed coat penetration. J. Seed Technol. 33(2): 155-169.

Prasad, R., Bhattacharyya, A. and Nguyen, Q.D. 2017. Nanotechnology in sustainable agriculture: Recent developments, challenges, and perspectives. Front Microbiol. 8: 1014, doi:10.3389/fmicb.2017.01014.

Prasertwitayakij, N., Louthrenoo, W., Kasitanon, N., Thamprasert, K. and Vanittanakom, N. 2003. Human pythiosis, a rare cause of arteritis: Case report and literature review. Sem Arthr Rheumat. 33: 204-214.

Pringsheim, N. 1858. Beitraege zur morphologie und systematic algae. Die Saprolegnieen. Jb. Wiss Bot. 1: 284-306.

Pupaibool, J., Chindamporn, A., Patarakul, K., Suankratay, C., Sindhuphak, W. and Kulwichit, W. 2006. Human Pythiosis. Emerg. Infec. Dis. 12(3): 517-518.

Rai, M., Ingle, A.P., Paralikar, P., Anasane, N., Gade, R. and Ingle, P. 2018. Effective management of soft rot of ginger caused by *Pythium* spp. and *Fusarium* spp.: Emerging role of nanotechnology. App. Micro. Biotech. 102: 6827-6839.

Rai, M., Tiwari, V.V., Irinyi L. and Kövics, G.J. 2014. Advances in taxonomy of genus *Phoma*: Polyphyletic nature and role of phenotypic traits and molecular systematics. Indian J. Microbiol. 54(2): 123-128.

Rakich, P.M., Grooters, A.M. and Tang, K.N. 2005. Gastrointestinal pythiosis in two cats. J. Vet. Diagn. Invest. 17: 262-269.

Ram, D., Mathur, K., Lodha, B.C. and Webster, J. 2000. Evaluation of resident biocontrol agents against ginger rhizome rot. Indian Phytopath. 53: 451-454.

Reis Jr., J.L., Carvalho, E.C.Q. de, Nogueira, R.H.G., Lemos, L.S. and Mendoza, L. 2003. Disseminated pythiosis in three horses. Vet. Micro. 96: 289-295.

Robideau, G.P., de Cock, A.W., Coffey, M.D., Voglmayr, H., Brouwer, H., Bala, K., Chitty, D.W., Désaulniers, N., Eggertson, Q.A., Gachon, C.M., Hu, C.H., Küpper, F.C., Rintoul, T.L., Sarhan, E., Verstappen, E.C., Zhang, Y., Bonants, P.J., Ristaino, J.B. and LéVesque, C.A. 2011. DNA barcoding of oomycetes with cytochrome c oxidase subunit I and internal transcribed spacer. Mol. Ecol. Resour. 11(6): 1002-1011.

Rossman, D.R., Rojas, A., Jacobs, J.L., Mukankusi, C., Kelly, J.D. and Chilvers, M.I. 2017. Pathogenicity and virulence of soil-borne oomycetes on *Phaseolus vulgaris*. Plant Diseases 101: 1851-1859.

Salmaninezhad, F. and Mostowfizadeh-Ghalamfarsa, R. 2019. Three new *Pythium* species from rice paddy fields. Mycologia. 111(2): 274-290.

Santurio, J.M., Argenta, J.A., Schwendler, S.E., Cavalheiro, A.S., Pereira, D.I.B., Zanette, R.A., Alves, S.H., Dutra, V., Silva, M.C., Arruda, L.P., Nakazata, L. and Colodel, E.M. 2008. Granulomatous rhinitis associated with *Pythium insidiosum* infection in sheep. Vet. Rec. 163: 276-277.

Shanmugam, V., Varma, A.S. and Surendran, M. 1999. Management of rhizome rot of ginger by antagonistic microorganisms. Madras Agric. J. 86: 339-341.

Sharma, S., Kaur, M. and Prashad, D. 2014. Isolation of fluorescent *Pseudomonas* strain from temperate zone of Himachal Pradesh and their evaluation as plant growth promoting rhizobacteria (PGPR). The Bioscan. 9(1): 323-328.

Shenep, J.L., English, B.K., Kaufman, L., Pearson, T.A., Thompson, J.W., Kaufman, R.A., Frisch, G. and Rinaldi, M.G. 1998. Successful medical therapy for deeply invasive facial infection due to *Pythium insidiosum* in a child. Clin. Infec. Dis. 27(6): 1388-1393.

Singh, A.K., Singh, V.K. and Shukla, D.N. 2010. Effect of plant extracts against *Pythium aphanidermatum* – The incitant of fruit rot of muskmelon (*Cucumis melo*). Indian J. Agr. Sci. 80(1): 51-53.

Singh, S., Tripathi, D.K., Dubey, N.K. and Chauhan, D.K. 2016. Effects of nano-materials on seed germination and seedling growth: Striking the slight balance between the concepts and controversies. Mater. Focus. 5(3): 195-201.

Singh, S., Vishwakarma, K., Singh, S., Sharma, S., Dubey, N.K., Singh, V.K., Liu, S., Tripathi, D.K. and Chauhan, D.K. 2017. Understanding the plant and nanoparticle interface at transcriptomic and proteomic level: A concentric overview. Plant Gene. https://doi.org/10.1016/j.plgene.2017.03.006.

Sleeth, B. 1953. Winter Haven decline of citrus. Plant Dis. Reptr. 37: 425-426.

Stephen, C. and Powell, C. 1982. *Pythium* species causing damping-off of seedling bedding plants in Ohio greenhouses. Plant Dis. 66: 731-733.

Sutton, J.C., Sopher, C.R., Owen-Going, T.N., Liu, W., Grodzinski, B., Hall, J.C. and Benchimol, R.L. 2006. Etiology and epidemiology of *Pythium* root rot in hydroponic crops: Current knowledge and perspectives. Summa Phytopathol. 32(4): 307-321.

Svecová, E., Proietti, S., Caruso, C., Colla, G. and Crinò, P. 2013. Antifungal activity of Vitex agnus-castus extract against *Pythium ultimum* in Tomato. Crop Protec. 43: 223-230.

Tabosa, I.M., Riet-Correa, F., Nobre, V.M., Azevedo, E.O., Reis-Júnior, J.L. and Medeiros, R.M. 2004. Outbreaks of pythiosis in two flocks of sheep in Northeastern Brazil. Vet. Pathol. 41: 412-415.

Tambong, J.T., de Cock, A.W.A.M., Tinker, N.A. and Lévesque, C.A. 2006. Oligonucleotide array for identification and detection of *Pythium* species. Appl. Environ. Microbiol. 72(4): 2691-2706.

Tanphaichitra, D. 1989. Tropical disease in the immunocompromised host. Rev. Infect. Dis. 11: S1629-S1643.

Thianprasit, M., Chaiprasert, A. and Inwidthaya, P. 1996. Human pythiosis. Curr. Trop. Med. Mycol. 7: 43-54.

Thomas, R.C. and Lewis, D.T. 1998. Pythiosis in dogs and cats. Compend Contin Ed. 20: 63-74.

Tripathi, D.K., Tripathi, A., Guar, S., Singh, S., Singh, Y., Vishwakarma, K., Yadav, G., Sharma, S., Singh, V.K., Mishra, R.K., Upadhyay, R.G., Dubey, N.K., Lee, Y. and Chauhan, D.K. 2017. Uptake, accumulation and toxicity of silver nanoparticle in autotrophic plants, and heterotrophic microbes: A concentric review. Front Microbiol. 8: 7. https://doi.org/10.3389/fmicb.2017.00007.

Uzuhashi, S., Kakishima, M. and Tojo, M. 2010. Phylogeny of the genus Pythium and description of new genera. Mycoscience 51(5): 337-365.

Van Buyten, E. and Höfte, M. 2013. *Pythium* species from rice roots differ in virulence, host colonization and nutritional profile. BMC Plant Bio. 13: 203.

van der Plaats-Niterink, A.J. 1981. Monograph of the genus *Pythium*. Baarn: Centraalbureau voor Schimmelcultures. Studies in Mycol. 21: 1-242.

Van West, P., Appiah, A.A. and Gow, N.A.R. 2003. Advances in research on oomycete root pathogens. Physiol Mol. Plant Pathol. 62: 99-113.

Vinayaka, K.S., Prashitha Kekuda, T.R., Noor Nawaz, A.S., Junaid, S., Dileep, N. and Rakesh, K.N. 2014. Inhibitory Activity of *Usnea pictoides* G. Awasthi (Parmeliaceae) against *Fusarium Oxysporum* F. Sp. *Zingiberi* and *Pythium aphanidermatum* isolated from rhizome rot of ginger. Life Sci. Leaflets. 49: 17-22.

Watanabe, H., Kageyama, K., Taguchi, Y., Horinouchi, H. and Hyakumachi, M. 2008. Bait method to detect *Pythium* species that grow at high temperatures in hydroponic solutions. J. General Plant Pathol. 74: 417-424.

Wellehan, J.F.X., Farina, L.L., Keoughan, C.G., Lafortune, M., Grooters, A.M., Mendoza, L., Brown, M., Terrell, S.P., Jacobson, E.R. and Heard, D.J. 2004. Pythiosis in a dromedary camel (*Camelus dromedaries*). J. Zoo Wildlife Med. 35: 564-568.

Wilson, D.A. 2012. Pythiosis. pp. 485-487. *In*: Clinical Veterinary Advisor: The Horse. https://doi.org/10.1016/B978-1-4160-9979-6.00587-0

Yang, J., Cao, W. and Rui, Y. 2017. Interactions between nanoparticles and plants: Phytotoxicity and defense mechanisms. J. Plant Interact. 12(1): 158-169.

Zitter, T.A., Hopkins, D.L. and Thomas, C.E. 1996. Compendium of Cucurbit Diseases. St. Paul, MN: American Phytopathological Society Press.

The Genus *Pythium* in Three Different Continents

Hani Mohamed Awad Abdelzaher[1,2*], Shaima Mohamed Nabil Moustafa[1,2] and Hashem Al-Sheikh[3]

[1] Biology Department, College of Science, Jouf University, P.O. Box: 2014, Sakaka, Saudi Arabia
[2] Department of Botany & Microbiology, Faculty of Science, Minia University, El-Minia City, Egypt
[3] Department of Biology, College of Science, King Faisal University, Al-Hassa, Saudi Arabia

Introduction

Pythium species belong to Oomycetes which are fungal like microorganisms located under the kingdom Straminipila (Uzuhashi 2015). So far, more than 150 species of *Pythium* have been reported (Uzuhashi 2015). *Pythium* is a unique fungus in many respects. The members of this genus can be terrestrial and aquatic, parasitic, phytopathogenic and biocontroller, beneficial and harmful, psychrophilic and thermophiles, and ultimately friend and foe. The study of *Pythium* began after its definition by the biologist Pringsheim in 1858. Serial studies and amendments in the classification and definition have been performed until we reached the current situation by dividing the genus into four (*Ovatisporangium*, *Elongisporangium*, *Globisporangium* and *Pilasporangium*) distinct genera (Abdelzaher et al. 1995, Abdelzaher 1999, Rahman et al. 2015, Uzuhashi et al. 2015).

Hyphae of *Pythium* spp. are characterized by the absence of cross-sectional walls (Coenocytic) only at the limits of sexual structure (antheridia and oogonia) and the boundaries of zoosporangia and hyphal swellings. Incidental cross walls may be formed at the edge of the colony. These fungi have multiple forms of zoosporangia: filamentous, lobulated, spherical, oval and internal to external proliferated ones. Some species do not produce zoospores. Swimming biflagellated zoospores develop in a transparent thin vesicle, from where many zoospores release after maturity in appropriate water medium. Sexual reproduction takes place in *Pythium* by means of antheridia and oogonia. There are many forms of antheridia, including terminal and intercalary, and small and large. Oogonia are of various shapes, which mainly include spherical and oval, having smooth and rough surface, some are spiny and some are without spines. The antheridia are intertwined with oogonia in several ways: front and side, broad fusion and fine fusion, single antheridium with one oogonium, and several antheridia with one oogonium. As a result of the fusion of the two gametes, zygote is formed, and subsequently one or more oospores are produced inside each oogonium. These oospores are either thin or thick walled, and may fill or not fill the vicinity of the oogonia (Plaats-Niterink 1981).

Pythium is composed of many morphological groups. Modern molecular analyses have revealed that the genus *Pythium* exists as a polyphyletic group that contains several monophyletic groups. Uzuhashi et al. (2010) limited the genus *Pythium* to those species that possess filamentous zoosporangia and generated four new genera to represent fungi with non- filamentous zoosporangia. These genera were (1) *Ovatisporangium*, (2) *Elongisporangium*, (3) *Globisporangium* and (4)

---

*Corresponding author: hmdaher@ju.edu.sa

Pilasporangium. Similarly, Bala (2010) reported a new genus *Phytopythium*, the members of which are with globose to ovoid, often papillate and internally proliferating zoosporangia. Lately, de Cock et al. (2015) published molecular-based proof that members of *Pythium* clade K as described by Lévesque and de Cock (2004) belong to the genus *Phytopythium* while spotting the genus status of remaining species of *Pythium*.

The purpose of this chapter is to provide historical account of the evolution of the study of the genus *Pythium* as a result of research work during nearly three decades in three different continents.

Pythium species in Japan

During the period of the doctoral study in Japan, many species of *Pythium* were isolated and identified, including those that were isolated for the first time outside of Ireland, and new recordings in Japan as well as other fungi registered in the past (Abdelzaher 1994). From 1992 to the end of 1994, the use of molecular methods to assist in the identification of species of *Pythium* was not known. Therefore, identification of *Pythium* species was only by means of morphological identification, which was, in my personal opinion, sufficient at that time. What is noteworthy is that the molecular identification is not sufficient in itself, but it is useful in confirming the morphological identification and the separation of very similar species (Abdelzaher et al. 1994a, b, c, d, e, f, 1995, Uzuhashi et al. 2015, Chenari et al. 2015).

Methods for isolation of *Pythium* species from aquatic habitats

Many methods have been proposed for the isolation of aquatic *Pythium* spp. However, the best way is to use baits from various parts of plants. It was found that the best baits for isolation of these fungi were parts of the leaf blade of the seedlings of maize plant, as well as the internal pectin rind of orange or mandarin fruits. What is important for the great potential for isolation of aquatic pythia is the selection of stagnant water such as pond water, lakes, streams and small water pools, and not fast runoff. However, there are some *Pythium* spp. which can be found in flowing river water, most of which are species of filamentous zoosporangia with oospores that are difficult to detect. It is preferable to study the presence of *Pythium* spp. in soil adjacent to water basins. For isolation of pythia, 15 ml pond water or 5 gm soil plus 10 ml pond water is placed in deep Petri dishes together with 10 sterile baits as mentioned above, and then incubated at temperatures between 20-25°C, for nearly 5 days (Abdelzaher et al. 1995). For the use of different baits than that mentioned, filter paper discs were used to isolate specific types of Pythia such as *P. fluminum* var. *fluminum* which can utilize complex cellulose (Abdelzaher et al. 1994d).

Suitable selective medium for isolation and purification of *Pythium* spp.

It is very important to use a selective medium to isolate and cultivate *Pythium* spp. Several antibiotics can be used to eradicate and inhibit growth of bacteria and other non-pythiaceous fungi. Over time, vancomycin, ampicillin, streptomycin, nystatin, pimarcin, rifampicin, pentachloronitrobenzene and miconazole were used as antibacterial and antifungal (rather than *Pythium*) agents. Recently, some of the above-mentioned antibiotics have been banned because they cause serious diseases. The problem with the commercial type of the fungicide like Pancreatoronetropenazine is to contain hexachlorobenzene (HCB), which possesses carcinogenic activity. The problem with Pimarcin is the difficulty of getting it in many countries of the world because of the food sanitation law (Tojo 2017). For this reason, it became necessary to use a selective medium that did not contain the previously mentioned banned antibiotics. Therefore, the use of selective media of NARF (nystatin+ampicillin+rifampicin+fluazinam) or NARM (nystatin+ampicillin+rifampicin+miconazole) is recommended. NARF selective medium was prepared by mixing antibiotics of 50 mg of nystatin dissolved in 1 ml C_2H_5OH, 250 mg of ampicillin dissolved in 1 ml distilled H_2O, 10 mg of rifampicin dissolved in 1 ml dimethylsulphoxide, 0.5 mg of fluazinam dissolved in 1 ml of autoclaved water agar (0.1%), all then thoroughly added to 1 liter sterilized corn meal agar medium that is cooled to 50 °C. NARM selective medium was prepared by mixing antibiotics of 10 mg of nystatin dissolved in 1ml C_2H_5OH, 250 mg of ampicillin dissolved in 1 ml distilled H_2O, 10 mg of rifampicin dissolved in 1 ml dimethylsulphoxide, 1 mg of miconazole dissolved in 1 ml dimethylsulphoxide, all then thoroughly added to 1 liter sterilized corn meal agar medium that is cooled to 50 °C. Researchers who wish to isolate *Pythium* spp. can choose one of the previous mentioned selective media based on the availability of these antibiotics.

Pythium spp. from aquatic habitats in Osaka

From 1942 to 1994, 24 *Pythium* species and groups of fungi were isolated from aquatic environments in Osaka (Abdelzaher et al. 1995) (Fig. 2.1). All of these fungi were identified on the basis of morphological criteria, including shape and thickness of mycelia, shape of zoosporangia, male (antheridia) and female (oogonia) structures, and profile of oospores (Abdelzaher et al. 1994a, b, f, 1995). During 1992-1994, the identification was only based on morphological characteristics because groups of *Pythium* were known for lack of sexual reproduction. Different *Pythium* spp. like *P. carolinianum* Matthews, *P. catenulatum* Matthews, *P. coloratum* Vaataja, *P. deliense* Meurs, *P. diclinum* Tokunaga, *P. dissotocum* Drechsler, *P. fluminum* Park var. *fluminum*, *P. irregular* Buisman, *P. marsipium* Drechsler, *P. middletonii* Sparrow, *P. monospermum* Pringsh., *P. myriotylum* Drechsler, *P. papillatum* Matthews, *P. pleroticum* T. Ito, *P. spinosum* Sawada, *P. sylvaticum* Campbell et Hendrix, *P. torulosum* Coker et Patterson, *P. ultimum* Trow var. *ultimum*, *P. undulatum* H. E. Petersen, *P. vexans* de Bary, *Pythium* 'group F', *Pythium* 'group HS', *Pythium* 'group P' and *Pythium* 'group T' were recorded and identified from aquatic habitats in Osaka, Japan. It is worth mentioning that after application of molecular criteria in identification process, the identification

Figure 2.1. One of the ponds used in isolation of *Pythium* spp. in Osaka, Japan, in 1993, is a pond used to irrigate crop plants on adjacent farms

of the majority of these fungi was confirmed, as well as renaming of a few, especially *Pythium* groups. Some of these fungi were previously isolated in Japan and others were isolated for the first time in Japan, one of which was isolated for the first time outside its native habitat and another was reassigned as a new species of a new genus (Abdelzaher et al. 1994a, b, c, d, e, f, 1995, Uzuhashi et al. 2019).

P. fluminum var. *fluminum* from pond water in Osaka

Pythium fluminum var. *fluminum* was isolated from cellulose in fresh water habitats in Northern Ireland and registered as a new species in 1977 (Park 1977). This species of *Pythium* was then considered as type locality of North Ireland. No one has been able to isolate this fungus since its first isolation until 1994. It was again isolated from a fresh water lake named "Tatsumi" located in Sakai district of Osaka city, Japan (Abdelzaher et al. 1994d). However, after 1994 there is no report regarding the isolation of this fungus till today. This might be due to the following reasons:

- Vast majority of *Pythium* spp. are not capable of degrading complex cellulose, such as filter papers and the like, but they can degrade cellulose compounds of relatively small molecular weight, and soften cellulose and pectin walls to facilitate penetration with their penetration pegs.
- This species of *Pythium* can easily degrade cellulose as a result of secretion of C1 and Cx cellulases and can therefore be isolated using filter paper discs, which most researchers do not use as a bait.
- This taxon is one of the major cellulose decomposers in the water bodies, so it was found in a lake next to a field of rice that produced large quantities of rice straw during harvesting time. Therefore, we advise researchers to try to isolate this fungus from places similar to that mentioned previously.

P. marsipium from pond water in Osaka

During the study of occurrence of pythiacous fungi in three ponds in Osaka, from 1992 to 1994, *P. Marsipim* was isolated from one of the ponds using baits of mandarin internal rind (Abdelzaher et al. 1994e). At that time, the fungus was considered a rare occurrence in Japan, isolated a few times since it was first isolated by Ito in 1936 from a pond water in Kyoto, Japan. This fungus showed unique structures of the genus *Pythium* as it possesses swimming zoospores differentiating in a distinct transparent vesicle, characteristic of pythiaceous fungi. This fungus is rarely present or rather difficult to isolate, since it has been isolated only a few times, and may be due to the use of certain baits of pectinaceous layer of internal mandarin rind.

Moreover, due to the discovery of the molecular biology methods for identification, this *P. marsipium* has been converted into a new species of the genus *Globisporangium*. At the time of writing this chapter, the process of publishing a scientific paper on the transformation of *P. marsipium* to *Globisporangium lacustre* sp. nov. is now proceeding. This was done after the use of molecular phylogenetic analysis based on the internal transcribed spacer regions (ITS) of the nuclear ribosomal RNA as well as cytochrome c oxidase subunit 1 genes of fungal mitochondria (our new paper). Since I first isolated this fungus in 1992, I (the first author) found some morphological structures of *P. marsipium* that differed, in part, from characters of the genus *Pythium*, such as the bladder-shaped zoosporangium, which is similar to that in the genus *Pythiogeton* (Uzuhashi et al. 2019). However, a transparent vesicle emerges from the zoosporangia through evacuation tubes and remains for a period of time in which zoospores are differentiated. The vesicles explode and zoospores are released. Differentiation of zoospores inside vesicles are an inherent characteristic of this genus. I recently contributed to a research on converting *P. marsipium* to *Globisporangium lacustre* using the above-mentioned molecular criteria.

Groups of *Pythium*

Pythium spp. from water bodies are particularly those fungi which do not produce sexual reproductive structures (Abdelzaher et al. 1995). Such fungi are therefore defined on the basis of zoosporangial forms. The group of filamentous zoosporangia was defined as *Pythium* 'group of F', the group of lobulated zoosporangia was defined as *Pythium* 'group of T', the group of proliferated zoosporangia was defined as *Pythium* 'group of P', the group of globose zoosporangia was defined as *Pythium* 'group of G', and the group of fungi with no zoosporangia and only hyphal swellings was defined as *Pythium* 'group of HS'.

Differences have been found in the physiology and virulence between isolates of *Pythium* 'group F', which led us to try to distinguish between 10 isolates within this group (Abdelzaher et al. 1994f). Total soluble protein and ioszymes were used as tools available at the time to differentiate these isolates. Subsequently, it has been proved that there were differences between these isolates on the basis of total soluble protein and ioszymes. It should be noted that subsequent recent studies on individuals belonging to those groups proved that these fungi belong to already known and identified *Pythium* species. These fungi have lost their ability to reproduce sexually. This may be partly due to the availability of appropriate conditions for fungi during their presence in aquatic environments, which does not make it necessary for sexual reproduction. To name a few, one of the isolates of *Pythium* 'group F' was converted to *P. dissotocum* after following the methods of molecular biology known in the genetic identification of pythiaceous fungi.

One of the most important advices for those interested in identification of *Pythium* species is to concentrate well on the forms and shapes of zoosporangia. Researchers must make sure that there are transparent vesicles emerging from zoosporangia, whereas zoospores differentiate inside these vesicles and then release.

Pathogenicity of *Pythium* species

It was found that isolated *Pythium* spp. caused serious diseases to many crop plants. These diseases ranged from damping-off of seedling to root-rot of adult plants. It was also found that many of the crop plants that produce a weak production are due to attack on their feeder roots by *Pythium* species. Therefore, the ability of these plants to absorb water and salts from the soil is reduced that leads to weak plant and low yield. Because *Pythium* spp. are facultative parasites, they are already saprophytes, but when conditions become suitable for the fungus and unfavorable for the host plant, they become pathogenic and virulent, causing serious diseases of germinating seeds, seedling and even roots of adult plants. Of the conditions not suitable for the plant, but suitable for fungus are: increase in the proportion of water in the soil with bad drainage, inappropriate climatic conditions for plant growth and abundance of insects and worms in the vicinity of the plant. Any plant in the stage of germinating seeds and seedling is susceptible to infection by *Pythium*. It has been studied that diseases of bush okra, cauliflower, cucumber, maize, melon, soybean, spinach, tomato and wheat (Ichitani et al. 1993, Abdelzaher et al. 1994c, 1997a, d, Abdelzaher and Elnaghy 1998, Abdelzaher et al. 2000, Abdelzaher 2003, 2004, 2006, Elnaghy et al. 2003, 2014, Abdelzaher et al. 2004a, Feng et al. 2019).

Many chemicals are used to control the infection of crop plants by *Pythium*. One of the most famous of these chemicals is metalaxyl. Studies have found that this fungicide kills fungi that are harmful and also beneficial, causing an imbalance in the environment (Abdelzaher et al. 2004c). Therefore, studies have sought to find a safe alternative way of biological control. It has been found that there are many natural enemies of the pathogenic pythia, including one from the same genus which is *Pythium oligandrum*. Several biological control applications using *P. oligandrum* to overcome infection by plant pathogens have been successful. However, application of *P. oligandrum* to plants must be confirmed prior to infection in the expected infested fields (Elnaghy et al. 2014).

Pythium spp. from different habitats in Gifu and Rishri, Japan, during 2008

Now-a-days, there is a trend to use molecular tools for identification as a means of confirming the morphological identification and separation of closely related *Pythium* spp. (Kageyama et al. 2007, Tojo et al. 2012, Chenari et al. 2015, Uzuhashi et al. 2015, Shiba et al. 2018).

    Pythium adhaerens, *P. aquatile*, *P. diclinum*, *P. dissotocum*, *P. pachycaule* and *P. torulosum* as well as asexual isolate of *P. dissotocum* (*Pythium* "group F" and *Pythium* "group P") were isolated from six rivers and a lake in Gifu, Japan during spring and summer of 2007 (Fig. 2.2)). All of the isolated species have been previously recorded from aquatic habitats except of *P. pachycaule*. Sequencing of the internal transcribed spacer regions of ribosomal DNA (rDNA-ITS) including the 5.8SrDNA of these pythia confirmed identification based on morphological characteristics (Kageyama et al. 2011).

    A scientific mission to study biodiversity was conducted on the island of Rishiri in northern part of Japan in the summer of 2007. The first author collaborated with Professor Koji Kageyama, for isolation and identification of *Pythium* spp. in water and soil of this island.

Figure 2.2. Collection of water samples for isolation of aquatic *Pythium* spp. from rivers and a lake of the city of Gifu, Japan, during spring and summer of 2007

Aquatic pythia from Rishiri Island

Some *Pythium* were isolated from aquatic habitats of the island of Rishiri (Rahman et al. 2015) as in the following table (Table 2.1 and Fig. 2.3).

Table 2.1. Distribution of different *Pythium* spp. in Hime-numa pond, Otatomari-numa pond, Minamihara-shitsugen pond, Tanetomi-shitugen marsh, Yamunai-sawa river, Fureai Land Shukei pond and Loge Yukiguni river, located in Rishiri Island, Japan (24-27 July, 2007)

Sampling site	Bait (grass)	Bait (filter paper disks)
Hime-numa pond	*"Group T" (100%)[a]	-
Otatomari-numa pond	*P. undulatum* (75%) *Saprolegnia* sp. (25%)	-
Minamihara-shitsugen	*New species (later named: *Pythium rishiriense* Rahman MZ, Abdelzaher HMA and Kageyama K. sp. nov.) (Rahman et al. 2015)	*P. myriotylum* (25%)
Tanetomi-shitugen marsh	*P. dissotocum* (50%) *P. torulosum* (50%)	-
Yamunai-sawa river	*"Group F" (100%)	*"Group F" (25%)
Fureai Land Shukei pond	*"Group F" (100%)	*"Group F" (25%)
Loge Yukiguni river	*"Group F" (100%)	*"Group F" (25%)

[a] Each calculation entry represents the percentage (%) of colonies obtained on NARM selective medium, using 4 baits for each sample.

Figure 2.3. The first author collects water samples for isolation of aquatic *Pythium* spp. in Hime-numa pond, Otatomari-numa pond, Minamihara-shitsugen pond, Tanetomi-shitugen marsh, Yamunai-sawa river, Fureai Land Shukei pond and Loge Yukiguni river, located in Rishiri Island, Japan (24-27 July, 2007)

Terrestrial pythia from Rishiri Island

The following *Pythium* spp. were isolated from soils of the island of Rishiri (Table 2.2).

Table 2.2. Distribution of *Pythium* spp. in soils of different locations in Rishiri Island on July (24-27, 2007)

Sampling site	Bait (grass)	Bait (hemp-seed)	Bait (mandarin internal rind)	Surface soil dilution method
Beside Hime-numa pond	-	-	*"Group HS" (60%)[a)]	
Beside Otatomari-numa pond	-	-	-	
Minamihara-shitsugen	-	-	-	
Mikaeridai Picnic Site	*Saprolegnia* sp. (100%)	-	-	
Mt. Rishiri Kutsugata Course, 5-goume	-	-	-	
Mt. Rishiri Kutsugata Course, Goyou-no-saka, 6-goume	-	-	-	
Mt. Rishiri Kutsugata Course, Rebun-iwa, 6-goume	-	-	-	
Hokuroku Camping Site	**P. ultimum* var. *sporangiiferum* (40%) *"group HS" (60%)	*"Group HS" (60%)	*"Group HS" (100%)	New species (later named: *Pythium alternatum* Rahman MZ, Abdelzaher HMA and Kageyama K sp. nov.) (Rahman et al. 2015)
Pon-yama	-	*"Group HS" (100%)	-	
Tanetomi-shitugen		**P. dissotocum* (100%)		

[a)] Each calculation entry represents the percentage (%) of colonies obtained on NARM selective medium, using 5 baits for each sample.

Heterothalic species isolated from Rishiri soil

Some heterothalic species of *Pythium* were isolated. The compatible strains were found alone, and when they meet, sexual reproduction occurs and the ovarian microbes are formed. These species are abundant in forests and natural vegetation. Many have been found in the forests and jungles of the island of Rishiri (Fig. 2.4). *P. intermedium* and *P. sylvaticum* were isolated during study in Rishiri Island.

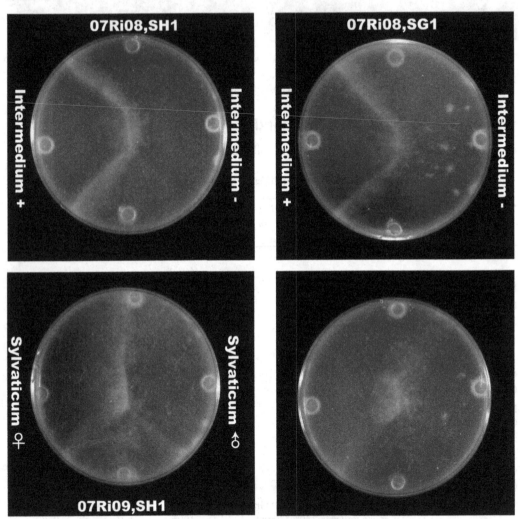

Figure 2.4. Reaction between compatible stains of heterothallic *Pythium* spp.

Isolation of two new species of *Pythium* from water and soil of Rishiri Island

During the search of *Pythium* species in the natural environments of Rishiri Island, two new species of *Pythium rishiriense* and *P. alternatum*, were isolated and identified based on morphological and molecular criteria (Rahman et al. 2015). It should be noted that they were initially recognized as *Pythium* but we were unable to identify them morphologically at the species level. For this reason, we expected them to be new types. The first new species of *Pythium rishiriense* was completely different from other similar *Pythium* species on morphological basis by its shape and structure of oogonia which was occasionally organized in chains. The second new species of *Pythium alternatum* varied morphologically from related *Pythium* species by its oogonia that alternates with antheridia in chains. Molecular identification using phylogenetic analyses based on ITS region and cytochrome *c* oxidase 1 gene sequences indicated that these two new *Pythium* spp. are obviously separate from related similar species that are based on morphology.

Pythium species from Saudi Arabia (2010-2018)

Saudi Arabia is a desert country, with rainfall concentrated in the south on the border with Yemen and adjacent to the Sarawat Mountains. The rest of the kingdom is desert, arid, and with very little rainfall. The majority of agriculture is concentrated in certain places of Qassim and Besita, which are the Saudi food basket where irrigation methods depend on sprinkling, drip irrigation and very little irrigation. The main source of water in these places is groundwater from some rather large underground reservoirs. Therefore, the presence of *Pythium* spp. in some places and farms with a history in agricultural production has been studied (Al-Enazi et al. 2018, Bibi et al. 2018). Research has been carried out on the presence of Pythiaceous fungi in Al Jouf area, which is characterized by abundant groundwater and from Al-Ahssa Oasis in the Kingdom of Saudi Arabia (Al-Sheikh and Abdelzaher 2012a, b).

Many saprophytic, pathogenic, biocontrol agents of pythia have been isolated from aquatic and soil habitats of rhizosphere and non-rhizosphere zones. From those fungi that have been isolated and identified on the basis of morphological and molecular basis, include: *P. aphanidermatum*, *P. diclinum*, *P. dissotocum*, *P. oligandrum*, and *P. ultimum* var. *ultimum* (Al-Sheikh 2010). More than one study has been conducted on the use of *P. oligandrum* in biological control of some fungal plant diseases such as *P. diclinum* in aquatic farms. This is in addition to the use of *P. oligandrum* in production of natural fertilizers to increase soil fertility in the northern region of Saudi Arabia (Abdelzaher et al. 2018).

Pythium species from Egypt (1996-2010)

Egypt has a variety of environments, including the desert, the humid environment in the north of the country, the only forests in the Mountain Olba, the Nile Valley environment, and the Mediterranean environment. It is an arid country with almost no rainfall except the northern coast adjacent to the Mediterranean Sea. The agricultural land in the Nile Valley in Egypt, which extends more than 1500 km inside the Egyptian territory, represents more than 90% of the total agricultural land in the Egyptian country (http://www.eroshen.com/site/?p=16664 and https://en.wikipedia.org/wiki/Nile#cite_ref-OED_8-0). Nile Valley is characterized by agriculture and irrigation in the form of permanent irrigation system through agricultural channels emanating from the Nile River. One of the obvious phenomena in the agricultural lands in Nile Valley is the increase in moisture in the morning in the form of dense fog, which makes the environment very suitable for the prosperity of pythiaceous fungi and the possibility of the increase of diseases of germinating seeds and seedlings. For these reasons, many of *Pythium* spp. have been isolated from the waterways and even from the agricultural soil. Few species of the rhizosphere region were isolated from naturally occurring plants in the desert lands (Figs 2.5 and 2.6).

The species that were isolated and identified from different habitats in Egypt have been listed as follows: *P. aphanidermatum*, *P. aquatile*, *P. arrhenomanes*, *P. carolinianum*, *P. catenulatum*, *P. deliense*, *P. diclinum*, *P. dissotocum*, *P. graminicola*, *P. irregular*, *P. longisporangium*, *P. monospermum*, *P. myriotylum*, *P. oligandrum*, *P. spinosum*, *P. ultimum* var. *sporangiferum*, *P. perillium*, *P. ultimum* var. *ultimum*, *P. undulatum*, *P. vexans*, *P. violae*, *Pythium* 'group F', *Pythium* 'group HS', *Pythium* 'group P' and *Pythium* 'group T' (Abdelzaher 1994, 1995, Abdelzaher et al. 1997a, b, c, d, Abdelzaher and Elnaghy 1998, Abdelzaher 1999, Abdelzaher et al. 2000, Abdelzaher 2003, 2004, 2006, 2009, Abdelzaher et al. 2004a, b, c, Ali and Abdel-Raheem 2003, El-Hissy et al.

Figure 2.5. Isolation from water of the river Nile in Assiut city, middle south of Egypt

Figure 2.6. Isolation from water of irrigation canal near agricultural field, Fayoum, Egypt

1989, El-Nagdy and Abdel-Hafez 1990, Elnaghy et al. 2003, El-Komy et al. 2004, El-Katatny et al. 2006, Elnaghy et al. 2002, 2003, 2009, El-Sharouny and Badran 1995, Fadl Allah et al. 2000, Gherbawy and Abdelzaher 2002, Gherbawy et al. 2005, Moustafa et al. 2009).

Many researchers studied a number of plant diseases caused by *Pythium* spp. as follows:

Damping-off of bush okra,
Damping-off of maize,
Damping-off of soybean,
Damping-off of spinach,
Damping-off of *Trifolium alexandrinum*,
Damping-off of wheat,
Root-rot of cauliflower,
Root-rot of cotton, and Root-rot of soybean.

Most of these diseases have been controlled by *P. oligandrum*, some terrestrial fungi or saprophytic bacteria.

A significant contribution has been made by Abdelzaher and his co-workers between 1995 and 2010 in taxonomic, environmental, physiological, pathological, and biological-control aspects of the genus *Pythium* in Egypt (Abdelzaher 1995, Abdelzaher et al. 1997a, b, c, d, Abdelzaher and Elnaghy 1998, Abdelzaher 1999, Abdelzaher et al. 2000, Abdelzaher 2003, 2004, 2006, 2009, Abdelzaher et al. 2004a, b, c, Elnaghy et al. 2003, El-Komy et al. 2004, El-Katatny et al. 2006, Elnaghy et al. 2002, 2003, 2009, Gherbawy and Abdelzaher 2002, Gherbawy et al. 2005, Moustafa et al. 2009).

Pythium species from Germany

Many species of *Pythium* were isolated and identified from Germany on the basis of morphological and molecular criteria (Nechwatal et al. 2005, Nechwatal and Mendgen 2006, Heine et al. 2007). These studies addressed the isolation and identification of some new species of *Pythium*, as well as some ways to increase the resistance of some plants to the attack by the phtopathogenic *Pythium aphanidermatum*.

Research studies were carried out on pythiaceous fungi in Berlin, Germany in 2009 during my collaboration with Prof. Dr. Franken Phillip, when I was a visiting professor to Institute for Vegetables and Ornamental plants, Grossbeeren, Germany. Some *Pythium* spp. were isolated from some waterways, soil and some diseased plants (Al-Sheikh and Abdelzaher 2010, Abdelzaher 2013).

P. aphanidermatum, P. aquatile, P. dissotocum, P. ultimum var. *sporangiferum, P. ultimum* var. *ultimum* were identified on morphological and molecular criteria. The majority of the practical study in Germany focused on the use of *P. oligandrum* to control tomato disease caused by a pathogenic species of *Pythium* as well as to increase production of fruits in aquaculture (Moustafa and Abdelzaher 2018).

Conclusion

The study on the genus *Pythium* has undergone many changes since the establishment of this fungus by Pringsheim in 1858. The principal characteristics of this genus were the aseptic mycelia, the antheridium as a male structure, the oogonium as a female structure, the oospores as sexual spores, and the zoospores which differentiate within vesicles outside different shapes zoosporangia.

After the revolution of molecular biology and its role in the recognition of fungi, the identification of *Pythium* was exposed to many permutations and combinations until we reached two distinct genera. So much so that the genus was transferred from the kingdom fungi to another Kingdom Straminipila, outside the traditional five kingdoms known.

As a result of the studies of the presence of these fungi in three continents in nearly three decades, we found a diversity between the similarities of the presence of these fungi in many places and between their presences in certain places. Here we review the stages of the transformation of a species of the genus of *Pythium* to another genus, as well as the emergence of new species isolated from new places in an island which was far from the scientists.

Useful and harmful role of some *Pythium* spp. in their ability to infect crop plants and their beneficial role in the biological control of some fungal plant diseases as well as its catalytic role in increasing the productivity of some plant crops were discussed.

References

Abdelzaher, H.M.A. 1994. Studies on isolation, identification and pathogenicity of aquatic *Pythium* spp. from pond water. D. Agr. Thesis, University of Osaka Prefecture, Japan.

Abdelzaher, H.M.A. 1995. Ecological and physiological studies on *Pythium* spp. from aquatic habitats. Ph.D. Thesis, Faculty of Science, Minia University, Egypt.

Abdelzaher, H.M.A. 1998. Separation of *Pythium spinosum* Sawada into two varieties: *P. s.* var. *spinosum* and *P. s.* var. *sporangiiferum*. Sixth Egyptian Botanical Conference, Egyptian Botanical Society, Proceeding Sixth Egyptian Botanical Conference, Cairo University, Giza, 393-404.

Abdelzaher, H.M.A. 1999. The Genus *Pythium* in Egypt. The second international conference on Fungi: Hopes & Challenges, Al-Azhar University. Afr. J. Mycol. Biotechnol. 1-31.

Abdelzaher, H.M.A. 2003. Biological control of root rot of cauliflower caused by *Pythium ultimum* var. *ultimum* using selected antagonistic rhizospheric strains of *Bacillus subtilis*. New Zeal. J. Crop. Hort. 31(3): 209-220.

Abdelzaher, H.M.A. 2004. Occurrence of damping-off of wheat caused by *Pythium diclinum* Tokunaga in El-Minia, Egypt and its possible control by *Gliocladium roseum* and *Trichoderma harzianum*. Arch. Phytopathol. Pl. Prot. 37(2): 147-159.

Abdelzaher, H.M.A. 2006. Biological Control of Damping-off and Root Rot Diseases of Soybean Caused by *Pythium spinosum* Sawada var. *spinosum* using Rhizosphere Species of *Aspergillus sulphureus*, *Penicillium islandicum* and *Paecilomyces variotii*. Annual Meeting of Phytopathological Society of Japan, Phytopathological Society of Japan. Jpn. J. Phytopathol. 247-248.

Abdelzaher, H.M.A. 2013. Negative interaction between tomato growth-promoting *Pythium oligandrum* and the damping-off pathogen *Pythium aphanidermatum*. Annual Meeting of Phytopathological Society of Japan, Gifu, March, 27-29, Japan.

Abdelzaher, H.M.A. and Elnaghy, M.A. 1998. *Pythium carolinianum* causing "Root Rot" of cotton and its possible biological control by *Pseudomonas fluorescens*. Mycopathologia 142: 143-151.

Abdelzaher, H.M.A., Elnaghy, M.A. and Fadl-Allah, E.M. 1997a. Isolation of *Pythium oligandrum* from Egyptian soil and its mycoparasitic effect on *Pythium ultimum* var. *ultimum* the damping-off organism of wheat. Mycopathologia 139(2): 97-106.

Abdelzaher, H.M.A., Elnaghy, M.A., FadlAllah, E.M. and Zohri, S.S. 1997b. Some physical and chemical factors affecting a sexual reproduction of 3 *Pythium* spp. Cryptogamie. Mycol. 18(3C): 267-277.

Abdelzaher, H.M.A., Shaban, G.M., Fadl-Allah, E.M. and Elnaghy, M.A. 1997c. Seasonal fluctuations of *Pythium oligandrum* in cultivated field of El-Minia, Egypt. Afr. J. Mycol. Biotechnol. 5(2): 39-50.

Abdelzaher, H.M.A., Shoulkamy, M.A. and Elnaghy, M.A. 1997d. Occurrence and pathogenicity of *Pythium catenulatum*, *Pythium ultimum* var. *ultimum* and *Pythium violae* in Egyptian soil. Afr. J. Mycol. Biotechnol. 5(2): 51-61.

Abdelzaher, H.M.A., Gherbawy, Y.A.M.H. and Elnaghy, M.A. 2000. Damping-off disease of Maize caused by *Pythium deliense* Meurs in El-Minia, Egypt and its possible control by some antagonistic soil fungi. Egypt. J. Microbio. 35(1): 21-45

Abdelzaher, H.M.A., Haifa Abdulaziz, S., Alhaithloul, H.A.S. and Moustafa, S.M.N. 2018. Olive-pressed solid residues as a medium for growing mushrooms and increasing soil fertility. *In*: Roland Nuhu Issaka and Mohammed Moro Buri (eds.). Soil Productivity Enhancement. IntechOpen. DOI: 10.5772/intechopen.78562. Available from: https://www.intechopen.com/books/soil-productivity-enhancement/olive-pressed-solid-residues-as-a-medium-for-growing-mushrooms-and-increasing-soil-fertility.

Abdelzaher, H.M.A., Ichitani, T. and Elnaghy, M.A. 1994a. Effect of temperature, hydrogen ion concentration and osmotic potential on oospore germination of five *Pythium* spp. isolated from pond water. Mycosci. 35(4): 315-318.

Abdelzaher, H.M.A., Ichitani, T. and Elnaghy, M.A. 1994b. Effect of temperature, hydrogen ion concentration and osmotic potential on zoospore production by three *Pythium* species isolated from pond water. Mycosci. 35(4): 377-382.

Abdelzaher, H.M.A., Ichitani, T. and Elnaghy, M.A. 1994c. Virulence of *Pythium* spp. isolated from pond water. Mycosci. 35(4): 429-432.

Abdelzaher, H.M.A., Ichitani, T. and Elnaghy, M.A. 1994d. *Pythium fluminum* var. *fluminum* from pond water in Osaka. Mycol. Res. 98: 982-984.

Abdelzaher, H.M.A., Ichitani, T. and Elnaghy, M.A. 1994e. *Pythium marsipium* from pond water in Osaka. Mycol. Res. 98: 920-922.

Abdelzaher, H.M.A., Morikawa, T., Ichitani, T. and Elnaghy, M.A. 1994f. Classification of *Pythium* 'group F' based on mycelial protein and isozyme pattern. Mycosci. 36(1): 45-49.

Abdelzaher, H.M.A., Ichitani, T., Elnaghy, M.A., Hassan, S.K.M. and Fadl-Allah, E.M. 1995. Materials for *Pythium* flora of Japan. X. Occurrence, identification and seasonality of *Pythium* spp. in three pond waters and mud soils in Osaka. Mycosci. 36: 71-85.

Abdelzaher, H.M.A., Imam, M.M., Shoulkamy, M.A. and Gherbawy, Y.M.A. 2004a. Biological control of *Pythium* damping-off of bush okra using rhizosphere strains of *Pseudomonas fluorescens*. Mycobio. 32(3): 139-147.

Abdelzaher, H.M.A., Shoulkamy, M.A. and Yaser, M.M. 2004b. Kind, abundance and pathogenicity of *Pythium* species isolated from maize rhizosphere of various habitats in El-Minia governorate, Egypt. Mycobio. 32(1): 35-41.

Abdelzaher, H.M.A., Shoulkamy, M.A. and Yasser, M.M. 2004c. Effect of benomyl and metalaxyl on reproduction of the plant parasite (*Pythium deliense*) and the mycoparasite (*P. oligandrum*). Arch. Phytopathol. Pl. Prot. 37(4): 307-317.

Abdelzaher, H.M.A., Shoulkamy, M.A., Gherbawy, Y.A.M.H. and Fadl-Allah, E.M. 1998. Possible involvement of cell wall degrading enzymes in pathogenesis of *Pythium spinosum* Sawada, the damping-off fungus of *Trifolium alexandrinum*. Sixth Egyptian Botanical Conference, Egyptian Botanical Society, Giza, Egypt, Proceeding.

Al-Enazi, N.M., Awaad, A.S., Al-Othman, M.R., Al-Anazi, N.K. and Alqasoumi, S.I. 2018. Isolation, identification and anti-candidal activity of filamentous fungi from Saudi Arabia soil. Saudi Pharm. J. 26(2): 253-257. doi: 10.1016/j.jsps.2017.12.003.

Ali, E.H. and Abdel-Raheem A. 2003. Distribution of zoosporic fungi in the mud of major Egyptian lakes. J. Basic Microbiol. 43(3): 175-184.

Al-Sheikh, H. 2010. Two pathogenic species of *Pythium*: *P. aphanidermatum* and *P. diclinum* from a wheat field. Saudi J. Biol. Sci. 17: 347-352.

Al-Sheikh, H. and Abdelzaher, H.M.A. 2010. Differentiation between two isolates of *Pythium ultimum* var. *ultimum* isolated from diseased plants in two different continents. J Biol. Sci. 10: 306-315.

Al-Sheikh, H. and Abdelzaher, H.M.A. 2012a. Materials for *Pythium* flora of Saudi Arabia (I) occurrence, pathogenicity and physiology of reproduction of *Pythium aphanidermatum* (Edson) Fitzp. isolated from north and east regions of Saudi Arabia. Res. J. Microbiol. 7: 82-100.

Al-Sheikh, H. and Abdelzaher, H.M.A. 2012b. Occurrence, identification and pathogenicity of *Pythium aphanidermatum*, *P. diclinum*, *P. dissotocum* and *Pythium* "Group P" Isolated from Dawmat Al-Jandal Lake, Saudi Arabia. Res. J. Environ. Sci. 6: 196-209.

Bibi, F., Strobel, G.A., Naseer, M.I., Yasir, M., Khalaf Al-Ghamdi, A.A. and Azhar, E.I. 2018. Microbial flora associated with the halophyte – *Salsola imbricate* and its biotechnical potential. Front Microbiol. 31: 9-65. doi: 10.3389/fmicb.2018.00065.

Chenari, B.A., Arzanlou, M., Tojo, M. and Babai-Ahari, A. 2015. *Pythium kandovanense* sp. nov., a fungus-like eukaryotic micro-organism (Stramenopila, Pythiales) isolated from snow-covered ryegrass leaves. Int. J. Syst. Evol. Microbiol. 65(8): 2500-2506. doi: 10.1099/ijs.0.000291.

de Cock, A., Lodhi, A.M., Rintoul, T.L., Bala, K., Robideau, G.P. and Gloria, Abad Z. 2015. *Phytopythium*: Molecular phylogeny and systematics. Persoonia. 34: 25-39. http://dx.doi.org/10.3767/003158515X685382.

El-Hissy, F.T., Khallil, A.R. and El-Nagdy, M.A. 1989. Aquatic fungi associated with seven species of Nile fishes (Egypt). Zentralbl. Mikrobiol. 144(5): 305-314.

El-Katatny, M.H., Abdelzaher, H.M.A. and Shoulkamy, M.A. 2006. Antagonistic actions of *Pythium oligandrum* and *Trichoderma harzianum* against phytopathogenic fungi (*Fusarium oxysporum* and *Pythium ultimum* var. *ultimum*). Arch. Phytopathol. Pl. Prot. 39(4): 289-301.

El-Komy, H.M., Attia, A.M.A., Abdelzaher, H.M.A. and Shoulkamy, M.A. 2004. Does *Azospirillum* protect wheat seedlings against infection by *Pythium ultimum* var. ultimum. El-Minia Sci. Bull. 15(1): 257-282.

El-Nagdy, M.A. and Abdel-Hafez, S.I. 1990. Occurrence of zoosporic and terrestrial fungi in some ponds of Kharga Oases, Egypt. J. Basic Microbiol. 30(4): 233-240.

Elnaghy, M.A., Abdelzaher, H.M.A. and Mohamed, S.R. 2003. Occurrence of *Pythium oligandrum* in Egypt, and its antagonistic behavior to *Pythium ultimum* var. *ultimum* the damping-off organism of spinach. Fac. Sci. Assiut Univ. 32(2-D): 371-382.

Elnaghy, M.A., Abdelzaher, H.M.A., Shoulkamy, M.A. and Sayed, S.R. 2014. Bean (*Phaseolus vulgaris* L.) damping-off caused by *Pythium ultimum* var. *ultimum* and its possible control by *Pythium oligandrum*. J. Pure Appl. Microbio. 8(Spl. Edn.): 161-169.

Elnaghy, M.A., Fadl-Allah, E.M., Abdelzaher, H.M.A. and Moharam, S.A. 2002. Rhizosphere pythia of some plants grown in Egypt. Bull. Fac. Sci. Assiut Univ. 31(2-D): 149-157.

Elnaghy, M., Shaban, G.M., Abdelzaher, H.M.A. and Yaser, M.M. 1999. Isolation, occurrence and seasonality of *Trichoderma* in Egyptian soil. El-Minia Sci Bull 12(1): 67-81.

El-Sharouny, H.M. and Badran, R.A. 1995. Experimental transmission and pathogenicity of some zoosporic fungi to Tilapia fish. Mycopath. 132(2): 95-103.

Fadl Allah, E.M., Elnaghy, M.A., Hassan, S.K.M., Abdelzaher, H.M.A. and Ichitani, T. 2000. Cellulolytic and pectinolytic abilities of some aquatic *Pythium* species isolated from pond water. Egyptian J. Microbiol. 35(1): 73-91.

Feng, W., Hieno, A., Kusunoki, M., Suga, H. and Kageyama, K. 2019. Lamp detection of four plant-pathogenic oomycetes and its application in lettuce fields. Plant Dis. 4: PDIS05180858RE. doi: 10.1094/PDIS-05-18-0858-RE.

Gherbawy, Y.A.M.H. and Abdelzaher, H.M.A. 1999. Isolation of fungi from tomato rhizosphere and evaluation of the effect of some fungicides and biological agents on the production of cellulase enzymes by *Nectria haematococca* and *Pythium ultimum* var. *ultimum*. Czech. Mycology 51: 157-170.

Gherbawy, Y.A.M.H. and Abdelzaher, H.M.A. 1999. Studies on the mycoflora of cheese in Egypt and their enzymatic activities. J. Food Mycol. 2(1): 281-289.

Gherbawy, Y.A.M.H. and Abdelzaher, H.M.A. 2002. Using of RAPD-PCR for separation of *Pythium spinosum* Sawada into two varieties: var. *spinosum* and var. *sporangiferum*. Cytologia. 67: 83-94.

Gherbawy, Y.A.M.H., Abdelzaher, H.M.A., Meens, J. and El-Hariry, H. 2005. Morphological and molecular identification of some closely related *Pythium* species in Egypt. Arch. Phytopathol. Pl. Protec. 38: 193-208.

Heine, G., Tikum, G. and Horst, W.J. 2007. The effect of silicon on the infection by and spread of *Pythium aphanidermatum* in single roots of tomato and bitter gourd. J. Exp. Bot. 58(3): 569-577.

Ichitani, T., Abdelzaher, H.M.A., Kanamori, M., Matsufuji, H. and Fukunishi, T. 1993. Occurrence of damping-off of melon caused by *Pythium spinosum* in Kyoto Prefecture. Proc. Kansai. Pl. Prot. Soc. 35: 95-96.

Kageyama, K., Senda, M., Asano, T., Suga, H. and Ishiguro, K. 2007. Intra-isolate heterogeneity of the ITS region of rDNA in *Pythium helicoides*. Mycol Res. 111(Pt 4): 416-423.

Kageyama, K., Motohashi, K., Abdelzaher, H.M.A., Li, M. and Suga, H. 2011. *Pythium* flora diversity in cool temperate and subtropical areas of Japan. Conference, Incheon, Korea, 2011/08.

Lévesque, C.A. and de Cock, A.W. 2004. Molecular phylogeny and taxonomy of the genus *Pythium*. Mycol Res. 108: 1363-1383.

Moustafa, S.M.N. and Abdelzaher, H.M.A. 2018. Increasing of tomato yield grown in hydroponic system using *Pythium oligandrum* isolated from Khoaa, Aljouf, Saudi Arabia. Egypt. J. Micro. 53: 1-8.

Moustafa, S.M.N., Abdelzaher, H.M.A. and Hassan, S.K.M. 2009. Effect of some environmental factors and quality of irrigation waters on mycelia growth and zoospore production of five *Pythium* spp. El-Minia Sci. Bull. 20(2): 1-18.

Nechwatal, J. and Mendgen, K. 2006. *Pythium litorale* sp. nov., a new species from the littoral of Lake Constance, Germany. FEMS Microbiol. Lett. 255(1): 96-101.

Nechwatal, J., Wielgoss, A. and Mendgen K. 2005. *Pythium phragmitis* sp. nov., a new species close to P. arrhenomanes as a pathogen of common reed (*Phragmites australis*). Mycol Res. 109(Pt 12): 1337-1346.

Plaats-Niterink, A.J. Van der 1981. Monograph of the genus *Pythium*. Stud. Mycol. 21: 1-244.

Rahman, M.Z., Abdelzaher, H.M.A., Mingzhu, L., Motohashi, K., Suga, H. and Kageyama, K. 2015. *Pythium rishiriense* sp. nov. from water and *P. alternatum* sp. nov. from soil, two new species from Japan. FEMS Microbiol. Lett. 362(13). DOI: http://dx.doi.org/10.1093/femsle/fnv086.

Shiba, K., Hatta, C., Sasai, S., Tojo, M., Ohki, S. and Mochizuki, T. 2018. Genome sequence of a novel partitivirus identified from the oomycete *Pythium nunn*. Arch. Virol. doi: 10.1007/s00705-018-3880-0.

Tojo, M. 2017. Selective media for practical isolations of *Pythium* spp. from natural and agricultural environments. Agri Res & Tech: Open Access J. 7(5): 555723. DOI: 10.19080/ARTOAJ. 2017.07.555723.

Tojo, M., Van West, P., Hoshino, T., Kida, K., Fujii, H., Hakoda, A., Kawaguchi, Y., Mühlhauser, H.A., Van Den Berg, A.H., Küpper, F.C., Herrero, M.L., Klemsdal, S.S., Tronsmo, A.M. and Kanda, H. 2012. *Pythium polare*, a new heterothallic oomycete causing brown discolouration of Sanionia uncinata in the Arctic and Antarctic. Fungal Biol. 116(7): 756-768. doi: 10.1016/j.funbio.2012.04.005.

Uzuhashi, S. 2015. Taxonomy and phylogenic studies of the genus *Pythium*. Jpn. J. Mycol. 56(1): 15-22.

Uzuhashi, S., Tojo, M. and Kakishima, M. 2010. Phylogeny of the genus *Pythium* and description of new genera. Mycosci. 51: 337-365.

Uzuhashi, S., Okada, G. and Ohkuma, M. 2015. Four new *Pythium* species from aquatic environments in Japan. Antonie Van Leeuwenhoek, 107(2): 375-391. doi: 10.1007/s10482-014-0336-8. Epub 2014 Nov 20.

Uzuhashi, S., Nakagawa, S., Abdelzaher, H.M.A. and Tojo, M. 2019. Phylogeny and morphology of new species of *Globisporangium*. Fung. Syst. Evol. 3: 15-20. doi.org/10.3114/fuse.2019.03.02.

3

Pythium: Diseases and Their Management

Patrycja Golińska* and Magdalena Świecimska

Department of Microbiology, Nicolaus Copernicus University, Lwowska 1, Torun 87 100, Poland

Introduction

The genus *Pythium*, with *Pythium monospermum* as a type species, was described for the first time by Pringsheim (Pringsheim 1858, Fischer 1892, Schröter 1893, Edson 1915). The genus *Pythium* belongs to the family *Pythiaceae*, order *Pythiales*, class *Oomycetes*, Phylum *Oomycota* and Kingdom *Chromista* (Kirk et al. 2008). Oomycetes are distinct from the true Fungi (Kingdom *Eumycota*) in several ways. They contain cellulose and β-glucans in their cell walls, have coenocytic hyphae, produce biflagellate zoospores, with tinsel and whiplash flagella, and have a diploid vegetative state, with meiosis occurring in gametangia (oogonia and antheridia) (Schroeder et al. 2013). According to MycoBank (http://www.mycobank.org), this fungal genus currently encompasses 355 described species, although many of these are not valid or are of doubtful identity (Schroeder et al. 2013, Ho 2018). It was reported that about half of the *Pythium* species were validly named (Bala et al. 2010, Robideau et al. 2011) and many of them came from Japan (Senda et al. 2009, Katsumoto 2010).

The genus *Pythium* is cosmopolitan, common in terrestrial habitats, found in soils, i.e. both natural (grassland, forest) and agricultural ecosystems (Senda et al. 2009, Schroeder et al. 2013, Kageyama 2014). Some species occupy the aquatic environments, namely seawater (Kurokawa and Tojo 2010, Kageyama 2014) rivers, lakes and ponds (Nechtawal et al. 2008), and irrigation systems (Hong and Moorman 2005). But its occurrence in the aquatic environments is still poorly understood when compared to terrestrial habitats (Uzuhashi et al. 2015). *Pythium* is mainly soil saprophytes, but many species are pathogenic to plants (Martin and Loper 1999, Schroeder et al. 2013). *Pythium* causes diseases of crop plants, flowers, weeds and marine plants (Uzuhashi et al. 2008, 2010). Most plant-pathogenic *Pythium* species have wide host ranges (e.g. *Pythium ultimum* with 719 host records), but some are restricted to certain taxa (e.g. *P. graminicola* and *P. arrhenomanes* on *Poaceae*) (Farr and Rossman 2012, Schroeder et al. 2013). Some species also attack mammals (*P. insidiosumis*, Phillips et al. 2008), fish (*P. flevoense*, Miura et al. 2010), and algae (*P. porphyrae*, Kawamura et al. 2005). *Pythium insidiosumis* causes pythiosis in humans and horses in subtropical and tropical regions (Gaastra et al. 2010, Ribeiro et al. 2017). Human pythiosis occurs mainly in Thailand (Krajaejun et al. 2006b), while equine pythiosis in Brazil (Santurio et al. 2006a). The infection leads to high morbidity and mortality in most affected patients (Krajaejun et al. 2006b, Gaastra et al. 2010, Ribeiro et al. 2017). Some *Pythium* species are mycoparasites on other *Oomycetes* and fungi (*P. oligandrum*, Rey et al. 2005). Pathogenic *Pythium* spp. may produce

*Corresponding author: golinska@umk.pl

hyphae with swollen digitate regions, called appressoria, which enable the fungus to attach and penetrate the host cells (Lévesque and de Cock 2004).

The aim of this chapter is to present the state-of-the-art on the plant and animal *Pythium* diseases and the strategies developed for their management.

Plant diseases

Pythium is a serious, worldwide pathogen of plants which causes both pre- and post-emergence damping-off of seeds and seedlings, and root rot of more mature plants as well as soft-rot of mature fruits and vegetables (Weiland et al. 2013, Yellareddygari et al. 2015). *Pythium* occurs most often in humid conditions, thus infections of plant are more common in nurseries of garden and forest plants, outdoor gardens and greenhouses and less in cereal crops (Weiland et al. 2013, Yellareddygari et al. 2015). Damping-off and root diseases in vegetable crops, beddings, foliage and woody plants as well as cut flowers caused by *Pythium* are responsible for many economic losses, especially in the nursery industry. These economic losses are the result of decreased plant production and quality, and plant death as well as increased management costs to minimize losses (Yellareddygari et al. 2015, Wagner et al. 2018).

Many species attack a variety of host plants while others are restricted to one host species (Farr and Rossman 2012, Schroeder et al. 2013). *Pythium ultimum* and *Pythium aphanidermatum* can infect many different species of hosts, while *Pythium arrhenomanes* and *Pythium graminicola* attack only monocot grass (Martin and Loper 1999). Numerous *Pythium* spp. are known to be pathogenic on soybean causing decrease in yield, mainly in USA (Broders et al. 2007, Jiang et al. 2012, Matthiesen and Robertson 2013). However, many *Pythium* spp. have been isolated from a single soybean plant, suggesting that a *Pythium* spp. complex is involved in some diseases (Wei et al. 2011, Zitnick-Anderson and Nelson 2015). A *Pythium* spp. complex on a single plant creates difficulty in assessing which *Pythium* sp. is the primary pathogen (Zitnick-Anderson and Nelson 2015). The difference in host numbers of various *Pythium* species was not explained (Martin and Loper 1999). It is believed that these differences may be related to the ability to infect host cells (Martin 1995), zoospore motility (Morris and Gow 1993, van West et al. 2002), various root exudates affecting zoospore attraction (Martin and Loper 1999) or diverse reaction of the host to colonization by pathogen (Glassgen et al. 1998, Koch et al. 1998).

Pythium disease of many plants such as cucurbit plants including pumpkin, melon, watermelon, cucumber, summer and winter squash, cantaloupe, and soybean, corn or petunia, chilli, etc. (Table 3.1) were noticed worldwide, e.g. in United States, Canada, Iran, Israel, China, and South, Eastern

Table 3.1. Review of most common diseases of seedlings caused by various *Pythium* species

Pythium species	Disease	Organism	Symptoms	Country of study	References
P. aphanidermatum, P. spinosum, P. splendens, P. oligandrum	Damping-off, Root and Crown rots	Cucumber	Damage	Oman	Al-Sa'di et al. 2007
P. aphanidermatum, P. ultimum, P. irregulare	Damping-off and root	Cucumber, Melon, Watermelon, Pumpkin, Summer and Winter squash	Decline or rapid wilting	Production areas in the United States, Israel, Iran, Canada and China.	Deadman 2017

(Contd.)

Table 3.1. (*Contd.*)

Pythium species	Disease	Organism	Symptoms	Country of study	References
P. aphanidermatum, P. irregulare, P. ultimum	Root rot	Geranium, Snapdragon, Hibiscus, Lantana	Stunting and wilting	Michigan, United States	Del Castillo Munera and Hausbeck 2016
P. aphanidermatum, P. dissotocum, P. heterothallicum, P. irregulare, P. myriotylum, P. ultimum	Not defined	Pelargonium	Not defined	Pennsylvania, United States	Moorman et al. 2002
P. conidiophorum, P. intermedium, P. recalcitrans, P. ultimum, P. heterothallicum, P. sylvaticum, P. acrogynum, P. oopapillum	Root rot, Damping-off	Corn, Soybean	Stunting and water soaked lesions on hypocolyls and cotyledons	Michigan, United States	Radmer et al. 2017
P. aphanidermatum, P. ultimum var. *ultimum, P. deliense*	Root rot, Damping-off	Cantaloupe	Damping-off, yellowing and wilting	Iran	Teymoori et al. 2012
P. heterothalicum, P. irregulare, P. sylvaticum, P. ultimum, P. debaryanum	Root rot	Soybean	Pre- and post-emergence damping-off and disintegration of the seed and discoloration of the hypocotyl and roots	Midwest United States	Wagner et al. 2018, Zitnick-Anderson and Nelson 2015
P. dissotocum, P. irregulare, P. aff. macrosporum P. mamillatum, P. aff. oopapillum P. rostratifingens, P. sylvaticum	Damping-off	Douglas-fir	Root lesions	United States	Weiland et al. 2013
P. aphanidermatum	Root rot, Damping-off	Petunia	Browning steam tissues, wilting	United States	Yellareddygari et al. 2015
P. ultimum	Damping-off	Chilli	Pre- and post-emergence damping-off	India	Zagade et al. 2013
26 different *Pythium* species	Root rot	Soybean	Stunting, discoloration of the hypocotyl and roots, root lesions	North Dakota United States	Zitnick-Anderson et al. 2018

and Central Africa, such as Burundi, the Democratic Republic of Congo, Kenya and Uganda and many more (Otsyula et al. 2003, Nzungize et al. 2012, Teymoori et al. 2012, Zagade et al. 2013, Weiland et al. 2013, Yellareddygari et al. 2015, Zitnick-Anderson and Nelson 2015, Deadman 2017, Radmer et al. 2017). Cucurbit damping-off and seedling root rot have been ascribed to several *Pythium* spp., most frequently *P. aphanidermatum, P. ultimum,* and *P. irregulare. P. ultimum* and *P. aphanidermatum* also have been shown to cause root rot of mature plants, resulting in decline or rapid wilting (Deadman 2017). Overall, the most common plant pathogen of *Pythium* in temperate climates is the *P. ultimum* while in tropical climate is the *P. aphanidermatum* (Bouhot 1988, Campbell 1989, Dick 1990, Deadman 2017).

Symptoms of diseases caused by *Pythium* spp.

Species of *Pythium* such as *P. dissotocum* Drechsler, *P. acanthicum* Drechsler, *P. torulosum* Coker and Patterson, and *P. rostratum* E.J. Butler reduce root system length whereas, others like *P. ultimum, P. irregulare* Buisman and *P. sylvaticum* Campbell and Hendrix can cause pre- and post-emergence damping-off (Mihail et al. 2002, Paulitz and Adams 2003). Symptoms of damping-off caused by this pathogen vary with the age and the stage of development of the infected plant (Martin and Loper 1999). In pre-emergence damping-off, susceptible seeds fail to germinate in the soil due to rotting of diseased tissues. Germinating seedlings which have not yet emerged from the soil level are also infected. The pathogens induce brownish, water-soaked lesions on radicals or young hypocotyls, causing seedlings to die (Adandonon et al. 2003, Moorman 2004). In post-emergence damping-off, susceptible seedlings are attacked at or below the soil level. The fungus penetrates the tissues of the lower steam, causing water-soaked and brown lesions. Due to the rotted basal part of the plant, the upper part cannot support and the seedlings collapse on the soil surface. Further rotting of the tissues leads to seedling death (Adandonon et al. 2003, Moorman 2004).

Mature plants show symptoms of root and crown rot. Secondary wall thickening occurs in the cells of stems and main roots and after its occurrence infection is restricted to the root tips or feeder roots, limiting plant vigour and yield which can result in plant death (Olsen and Young 1998). Initially, feeder roots are destroyed soon after brown lesions develop on lateral roots. Roots may have one to several lesions. Subsequent infections occur on the taproot or in the hypocotyl area. In severe infections, there is complete decay of the root cortex while the stele remains intact. As the severity and size of the lesions increase, the plant may show varying degrees of stress. The crown leaves often become chlorotic, necrosis soon follows as the symptoms gradually move outward towards the vine tips (Deadmen 2017), fruit can be sunburned, and soluble solids are reduced (Deadmen 2017). Sudden wilt is another symptom of *Pythium* infection, in which healthy-appearing plants suddenly collapse during the heat of the day. Although some plants recover turgidity at night, wilting reoccurs the next day, and the plants die in few days. The suddenness of collapse is directly proportional to the degree and speed of root infection (Deadman 2017). Infection with *P. ultimum* var. *ultimum, P. aphanidermatum,* and *P. deliense* are capable of causing reductions in emergence and maturation of cantaloupes (Teymoori et al. 2012). However, only a few *Pythium* species are very virulent, most of them are "minor pathogens" that limit the growth of the plant without the occurrence of disease symptoms (Cherif et al. 1997).

Factors affecting *Pythium* species activity in the environment

The activity of *Pythium* species in the soil depends on environmental factors, namely moisture, temperature, pH and nutrition (Martin and Loper 1999). Soil moisture is one of the most important factors influencing the growth and survival of soil-borne plant pathogens (Green and Jansen 2000).

Both highly dried and water-saturated soils inhibit the growth of *Pythium* spp. (Martin and Loper 1999). Water availability in soil affects conversion of thick-walled oospores to thin-walled oospores as well as pathogenic activity of species like *P. ultimum*. Similarly, temperature influences survival, germination, dispersal and ability to infect host cells of *Pythium* species (Green and Jansen 2000) but vary between species. For example, *Pythium irregulare, P. torulosum, P. sylvaticum* and *P. macrosporum* are the most virulent at temperatures below 20°C, *P. ultimum* has the ability to infect different hosts in a temperature range from 12 to 25°C (Wei et al. 2011), whereas *P. aphanidermatum,* which is a soybean pathogen, is most virulent in the range of temperature from 20-25°C (Rojas et al. 2017). Green and Jansen (2000) observed less pre-emergence damping-off of sugar beet and watermelon caused by *P. ultimum* at 30 to 35ºC, whereas *P. irregulare* causes damping-off at 5°C only. The highest conversion of thick- to thin-walled oospores of *P. ultimum* was observed at 25°C (Martin and Loper 1999).

Similarly, pH of soil affects growth and germination of *Pythium* species. The effect of pH on *Pythium* varies among and within species (Green and Jansen 2000). Previously published studies showed that saprotrophic and pathogenic activities of *P. ultimum* were favoured by acidic soils (Simon and Sivasithampapram 1988a, b). Maximum infection was observed at pH levels between 4.8 and 6.9, with a decrease in infection at pH 7.6. Conversion of thick-walled to thin-walled oospores of *P. ultimum* may be inhibited at acidic and alkalic soils (Martin and Loper 1999, Green and Jansen 2000). Soluble and volatile host exudates stimulate conversion of thick-walled to thin-walled oospores and germination of sporangia of *P. ultimum* leading to increase in its biomass and infectious capacity (Nelson 1990). The exact chemical nature of the stimulatory molecules is not yet known. However, it has been suggested that sugars and amino acids are the compounds most likely to be responsible for stimulating propagule germination in the spermosphere and rhizosphere (Nelson 1990).

Animal and human diseases

Although most of *Pythium* species cause plant diseases, these fungus-like organisms can also be a pathogen of humans and animals (Krajaejun et al. 2006a, b, Gaastra et al. 2010, Miura et al. 2010, Konradt et al. 2016, Ribeiro et al. 2017).

Pythiosis is an emerging life-threatening disease, caused by *Pythium insidiosum*, and occurs in tropical and subtropical areas of the world (Krajaejun et al. 2006b, Gaastra et al. 2010, Beakes et al. 2012, Ribeiro et al. 2017, Chitasombat et al. 2018a, b). *Pythium insidiosum* may affect humans and animals (Table 3.2). Pythiosis in humans was reported as an endemic disease in Thailand (Gaastra et al. 2010, Supabandhu et al. 2008, Chitasombat et al. 2018a, b), while animal pythiosis appears all over the world, among horses in Brazil (Santurio et al. 2006a), and also in cattle, sheep, cats, dogs, birds and animals kept in captivity (Chaiprasert et al. 2009, Dos Santos et al. 2011, Konradt et al. 2016). Most cases among humans were recorded as ocular or vascular forms of pythiosis, whereas skin forms most often occur in animals, and gastrointestinal forms mainly affect dogs (Miller 1985, Fischer et al. 1994, Mendoza et al. 1996).

Pythiosis inhabits aquatic area, and motile spores act as infective unit which attach to skin (mainly with pathologically changed or damaged tissue) and penetrate to deeper tissues, resulting in several forms, e.g. skin, subcutaneous tissue, cornea, vascular, disseminated form (Krajaejun et al. 2006b, Gaastra et al. 2010, Chitasombat et al. 2018a, b). The known risk factors for vascular pythiosis include thalassemia, hemoglobinopathy, paroxysmal nocturnal hemoglobinuria, aplastic anemia, and leukemia (Krajaejun et al. 2006b Chitasombat et al. 2018a, b). However, vascular pythiosis has been found in patients without risk factors, who have presented with unusual features, e.g. necrotizing cellulitis and brain abscesses (Narkwiboonwong et al. 2011, Kirzhner et al. 2015, Chitasombat et al. 2018b). There were also reported cases of keratitis caused by infection of *P. inosidiosum* in India, Australia and South America. In most cases, kerannoplasty was unable to save

Table 3.2. Review of human and animal pythioses caused by *Pythium insidiosum*

Diseases	Organism	Symptoms	Country	References
Vascular pythiosis of carotid artery with meningitis and cerebral septic emboli	Human	Pain and swelling at the left side of the neck, low-grade fever, narrowing of upper airway causing the difficulty in swallowing and hoarseness, bulging of left posterior pharyngeal wall and tonsil enlargement	Thailand	Chitasombat et al. 2018b
Keratitis	Human	Necrotizing infections	Costa Rica	Neufeld et al. 2018
Vascular infection treated with limb-sparing surgery	Human	Small bullae with a black center and later developed severely painful left inguinal swelling	Jamaica	Pan et al. 2014
Pythiosis	Horse	Massive muzzle pain, many coral-like necrotic tissues were drained from the ulcerative nodule	Thailand	Tonpitak et al. 2018
Gastrointestinal pythiosis	Dogs	Weight loss, chronic anorexia, vomiting, diarrhea, multifocal pyogranulomatous inflammation in the submucosa, mucosa, and/or muscularis of the stomach, small intestine and pylorus, with rare involvement of the mesentery, mesenteric lymph nodes, and pancreas	United States	Fischer et al. 1994
Cutaneous pythiosis	Calves	Ulcerated cutaneous lesions in thoracic and pelvic limbs, sometimes extending to the ventral thoracic region	Brazil	Konradt et al. 2016
Pythiosis	Cattle	Fast weight loss, swelling and ulceration	Brazil	Dos Santos et al. 2011

the eye and it was necessary to enucleate (Kobayashi et al. 2003, Azuara-Blanco et al. 1999, Liang et al. 2012).

In the case of animal pythiosis, contact of animals with the pathogen is probably caused by drinking water containing zoospores (Tonpitak et al. 2018). Phytosis in horses is very often located on skin and subcutaneous tissue of limbs and in the ventral portion of thoracoabdominal wall. However, there was a case of pythiosis in the nasal cavity of horses in Thailand and Brazil (Santurio et al. 2008, Ubiali et al. 2013, Souto et al. 2016, Tonpitak et al. 2018), and also of sheep in Brazil (Souto et al. 2016, Santurio et al. 2008, Ubiali et al. 2013). During the course of the disease in the ulcerative tissue appears granulomatous nodule or fistula tract with coral-like necrotic tissue (Souto et al. 2016).

Management and treatment of *Pythium* diseases

Plant *Pythium* diseases

Proper identification of the species causing infection and disease of plants, determination of how many different species are present in the field and their relative abundance as well as their sensitivity to fungicides are important for developing effective and long-term management strategies (Moorman et al. 2002, Teymoori et al. 2012).

Presently, numerous strategies have been employed to prevent plant infection from different oomycetes, including *Pythium* spp. (Wagner et al. 2018). The disease of plants in nurseries or fields is managed through the application of chemical fungicides and certain cultural control methods as well as biopesticides (Yellareddygari et al. 2015). Nevertheless, the greenhouse production of *Petunias* is still affected largely by the disease (Yellareddygari et al. 2015). *Pythium* spp. can be easily transferred from plant to plant. Therefore, in greenhouses, it is necessary to minimize contact with possible routes of contamination such as contaminated soil, potted mixtures, irrigation equipment and other tools that may have been in contact with the pathogen. Root rot can be managed by planting on raised beds to allow for maximum water drainage after each irrigation (Deadman 2017).

Management of seedlings diseases caused by *Pythium* spp. has relied on the use of seeds (e.g. corn or soybean) treated with fungicides such as metalaxyl and mefenoxam (Radmer et al. 2017). The common practice that limits the growth of *Pythium* is the use of mixtures of pesticides containing mefenoxam, metalaxyl and fosetyl-Al which provide protection against many different broad host-range *Pythium* pathogens during seed germination (Esker and Conley 2011).

Pre- and post-transplant applications of phenylamide (FRAC group 4) fungicides are common for management of damping-off in different parts of the world (Deadman 2017). *Pythium* spp. vary in their sensitivity to different fungicides (Broders et al. 2007, Weiland et al. 2014). There are some reports on the resistance of *Pythium* spp. to fungicides e.g. to propamocarb and mefenoxam (Moorman and Kim 2004). Therefore, all *Pythium* spp. cannot be managed with the same fungicide active ingredient (Zitnick-Anderson et al. 2015) and hence it is necessary to develop other effective methods of controlling this pathogen (Wagner et al. 2018).

The promising management tool was found to prevent soybean from disease caused by other plant pathogen, *Phytophthora sojae*, that was the deployment of soybean varieties containing single genes which confer resistance (Tyler et al. 2006, Dorrance et al. 2007). Nevertheless, continuant use of resistant soybean lines resulted in the selection for oomycete strains that were capable of overcoming these soybean resistance genes and lead to increased soybean yield losses due to selection for evolved *P. sojae* pathotypes (Wrather and Koenning 2006). More than 200 *P. sojae* pathotypes have already been isolated from soils, and selection for new ones seems to be most rapid when there is partial resistance to the deployed soybean varieties (Stewart et al. 2014). In case of *Pythium* species, major soybean resistance genes have not been identified, so seeds treatments with pesticides remain the primary strategy management tool to reduce seedling rot and damping-off (Dorrance et al. 2009, Vossenkemper et al. 2015).

Biological control of soil-borne pathogens, including *Pythium* spp., is one of the available ways to limit the development of pathogens (Nzungize et al. 2012, Wagner et al. 2018). However, this method is particularly complex because the pathogens occur in a dynamic environment of the rhizosphere which is abundant with a high population of other microorganisms and exposed to changeable conditions of pH, salinity, and osmotic and water potential (Handelsman and Stabb 1996). Microorganisms autochthonic to the rhizosphere can protect the plant from fungal attacks through the production of antifungal metabolites, competition with the pathogen for nutrients, niche exclusion, parasitism or lysis of the pathogen, or through induction of plant resistance mechanisms (Whipps 2001). Therefore, they are considered to be ideal for biological control since the rhizosphere provides a first-line defense for roots against attacks by plant pathogens (Weller 1988). Beneficial microorganisms used for biological control of plant pathogenic *Pythium* spp. have been identified among both fungi and bacteria. Fungal isolates of *Trichoderma* spp. and *Gliocladium* spp. are antagonists of *Pythium* caused soil-borne diseases (Howell et al. 1993, Martin and Loper 1999, Fravel 2005).

Among bacteria, various filamentous actinobacterial species including *Streptomyces*, *Actinoplanes*, and *Micromonospora* have been considered as the potential biological agents to inhibit *Pythium* species such as *P. coloratum* causing cavity-spot on carrots (El-Tarabily et al. 1997).

There are several commercial bioproducts from actinobacteria available in the market against soil-borne fungal diseases, including those caused by *Pythium* spp., e.g. Actinovate (Novozymes BioAg Inc., USA) containing *Streptomyces lydicus* WYEC 108 and Mycostop (Verdera Oy, Finland) with *Streptomyces* sp. strain K61 (Kabaluk et al. 2010, Vurukonda et al. 2018). Other effective bacterial antagonists against *Pythium* were found in *Enterobacter* sp., *Erwinia* sp., *Bacillus* sp., *Burkholderia* sp., *Stenotrophomonas* sp., and *Rhizobium* sp., but the most extensively studied bacterial biological control agents are *Pseudomonas* spp. (Chin-A-Woeng et al. 2003, Bardin et al. 2004, Wagner et al. 2018). Wagner and co-workers (2018) studied 330 *Pseudomonas* strains isolated from soil and water and found that 118 of tested strains showed antagonistic properties against various *Pythium* pathogens. The proposed mechanism of biological control of *Pythium* spp. is the competition for organic carbon and iron (Hoitink and Boehm 1999, Martin and Loper 1999). However, it is also known that seed or root exudates rapidly stimulate germination of *Pythium* spp. propagules which quickly infect seeds or roots (Whipps and Lumsden 1991).

Human and animal pythiosis

The rare occurrence of the disease leads to under-recognition, under-diagnosis, and delays in diagnosis, and this causes an advanced form of disease, which significantly reduces the effectiveness of treatment and affects survival (Permpalung et al. 2015, Reanpang et al. 2015, Sermsathanasawadi et al. 2016, Tonpitak et al. 2018). There has been no dependency of *Pythium* infection on gender and age of patients. There is also no information about the possibility of transmission of this infection from animals to humans and vice versa (Mendoza et al. 1996, Gaastra et al. 2010).

The gold standard for diagnosis of pythiosis in humans is the histopathological investigation of fungal tissue obtained after the surgical procedure (Krajaejun et al. 2006b). However, this method requires fungal culture, which is a time-consuming procedure (Chaiprasert et al. 1990, Lacaz et al. 2002, Krajaejun et al. 2006a, 2009, Intaramat et al. 2016, Chitasombat et al. 2018a, b). Advances in diagnostic methods made over the past decades include serodiagnosis, e.g. immunodiffusion (Pracharktam et al. 1991), ELISA (Krajaejun et al. 2002, Lacaz et al. 2002, Santurio et al. 2006b, Dial 2007), Western blot analysis, the hemagglutination test (Jindayok et al. 2009), the immunochromatographic test (Chareonsirisuthigul et al. 2013, Chitasombat et al. 2018b) and PCR (Rodrigues et al. 2006, Dial 2007, Gabriel et al. 2008, Ribeiro et al. 2017). However, pythiosis is often detected only during surgery (Krajaejun et al. 2006a).

Early recognition of pythiosis and its confirmation by rapid serodiagnosis could enhance the rapid clinical management for such life-threatening disease (Chitasombat et al. 2018a, b). Most antimicrobial drugs (e.g. itraconazole, terbinafine), designed to control human pathogenic fungi, are ineffective against *Pythium insidiosum* (Krajaejun et al. 2006b, Beakes et al. 2012). Resistance of *P. insidiosum* to antifungal drugs results from the lack of ergosterol, a component of the cell membrane, which is the target in traditional antifungal therapy with azole drugs (Lerksuthirat et al. 2017). A more effective pharmacological treatment seems to be the use of a caspofungin beta-glycan synthesis inhibitor or inhibitor of RNA synthesis, namely mefenoxam (Neufeld et al. 2018). An alternative treatment is *Pythium insidiosum* antigen (PIA) vaccine; however, the efficacy of this method of treatment has not been fully confirmed (Thitithanyanont et al. 1998, Wanachiwanawin et al. 2004, Permpalung et al. 2015). After penetration of *Pythium* into host cells, the exoantigens are released that are recognized by antigen presenting cells, which consequently stimulate the Th2 response (Miller 1985, Shipton 1987), thereby releasing IL-4 and mobilizing mast cells and eosinophiles. Immunotherapy is based on changing the Th2 response of the body to Th1. Th1 response leads to the release of IL-2 and INF-γ as well as mobilization of macrophages and T lymphocytes that destroy *P. insidiosum* (Mendoza et al. 2003). The use of immunotherapy is an effective method of treatment of *Pythiosis* in horses and in 80% of cases it leads to cure (Mendoza and Newton 2005). In contrast, in humans, it is an alternative method of therapy, used when the other treatment methods such as pharmacological agents or surgical removal of changed tissues fail

(Pan et al. 2014, Wanachiwanawin et al. 2004). However, as reported previously, in 50% of cases of human pythiosis immunotherapy leads to cure (Krajaejun et al. 2006a).

Successful treatment of vascular pythiosis still relies on surgery, which is a mainstay to achieve organism-free surgical margins and often leads to amputation (Sermsathanasawadi et al. 2016). Patients with unresectable disease, e.g. those with suprainguinal lesions or aortic involvement, often die (Krajaejun et al. 2006b, Permpalung et al. 2015, Sermsathanasawadi et al. 2016).

Conclusion

Pythium diseases are a serious problem in both animals and plants. Soil-borne diseases of plants are widespread and are a limitation to plant production, as they significantly reduce their yield and are responsible for many economic losses worldwide. Various disease control options are available, including different methods such as cropping practices, chemical and biological control, and genetic resistance methods. Similarly, pythiosis in human and animals, caused by *Pythium insidiosum,* is a life-threatening disease with rapid progression. Early diagnosis and surgery to achieve organism-free margins, along with adjunctive antifungal therapy and use of an immunotherapeutic vaccine, remain the primary treatment for the disease, which is associated with high morbidity and mortality. However, further research is required to understand the biology and virulence of *Pythium* spp. and to better control diseases caused by this pathogen.

References

Adandonon, A., Aveling, T.A.S., Labuschagne, N. and Ahohuendo, B.C. 2003. Epidemiology and biological control of the causal agent of damping-off and stem rot of cowpea in the Ouémé valley, Bénin. Agron. Sci. 6: 1-2.

Al-Sa'di, A.M., Drenth, A., Deadman, M., de Cock, A.W.A.M. and Aitken, E.A.B. 2007. Molecular characterization and pathogenicity of *Pythium* species associated with damping-off in greenhouse cucumber (*Cucumis sativus* L.) in Oman. Plant Path. 56: 140-149.

Azuara-Blanco, A., Pillai, C.T. and Dua, H.S. 1999. Amniotic membrane transplantation for ocular surface reconstruction. Br. J. Ophthalmol. 83: 399-402.

Bala, K., Robideau, G.P., Désaulniers, N., de Cock, A.W.A.M. and Lévesque, C.A. 2010. Taxonomy, DNA barcoding and phylogeny of three new species of *Pythium* from Canada. Persoonia 25: 22-31.

Bardin, S.D., Huang, H.-C., Pinto, J., Amundsen, E.J. and Erickson, R.S. 2004. Biological control of *Pythium* damping-off of pea and sugar beet by *Rhizobium leguminosarum* bv. *viceae*. Can. J. Bot. 82: 291-296.

Beakes, G.W., Glockling, S.L. and Sekimoto, S. 2012. The evolutionary phylogeny of the oomycete fungi. Protoplasma 249: 3-19.

Bouhot, D. 1988. Introductory remarks about *Pythium*. pp. 1-2. *In*: International *Pythium* group Pankhurts, C.E., Bouhot, D. and Ichitani, T. (eds.). 5th International Congress of Plant Pathology and the 1st International *Pythium* workshop, Kyoto, Japan.

Broders, K.D., Lipps, P.E., Paul, P.A. and Dorrance, A.E. 2007. Characterization of *Pythium* spp. associated with corn and soybean seed and seedling disease in Ohio. Plant Dis. 91: 727-735.

Campbell, R. 1989. Biological Control of Microbial Plants Pathogen. Cambridge University Press, Cambridge.

Chaiprasert, A., Krajaejun, T., Pannanusorn, S., Prariyachatigul, C., Wanachiwanawin, W., Sathapatayavongs, B., Juthayothin, T., Smittipat, N., Vanittanakom, N. and Chindamporn, A. 2009. *Pythium insidiosum* Thai isolates: Molecular phylogenetic analysis. Asian Biomed. 6: 623-633.

Chaiprasert, A., Samerpitak, K., Wanachiwanawin, W. and Thasnakorn, P. 1990. Induction of zoospore formation in Thai isolates of *Pythium insidiosum*. Mycoses 33: 317-323.

Chareonsirisuthigul, T., Khositnithikul, R., Intaramat, A., Inkomlue, R., Sriwanichrak, K., Piromsontikorn, S., Kitiwanwanich, S., Lowhnoo, T., Yingyong, W., Chaiprasert, A., Banyong, R., Ratanabanangkoon,

K., Brandhorst, T.T. and Krajaejun, T. 2013. Performance comparison of immunodiffusion, enzyme-linked immunosorbent assay immunochromatography and hemagglutination for serodiagnosis of human pythiosis. Diagn. Microbiol. Infect. Dis. 76: 42-45.

Cherif, M., Tirilly, Y. and Belanger, R.R. 1997. Effect of oxygen concentration on plant growth, lipidperoxidation, and receptivity of tomato roots to *Pythium* F under hydroponic conditions. Eur. J. Plant Pathol. 103: 255-264.

Chin-A-Woeng, T.F., Bloemberg, G.V. and Lugtenberg, B.J. 2003. Phenazines and their role in biocontrol by *Pseudomonas* bacteria. New Phytol. 157: 503-523.

Chitasombat, M.N., Chindamporn, N.L.A. and Krajaejun, T. 2018a. Clinicopathological features and outcomes of pythiosis. Int. J. Infect. Dis. 71: 33-41.

Chitasombat, M.N., Petchkum, P., Horsirimanont, S., Sornmayura, P., Chindamporn, A. and Krajaejun, T. 2018b. Vascular pythiosis of carotid artery with meningitis and cerebral septic emboli: A case report and literature review. Med. Mycol. Case Rep. 21: 57-62.

Deadman, M. 2017. *Pythium* damping off and root-rot. pp. 48-50. *In*: Keinath, A.P., Wintermantel, W.M. and Zitter, T.A. (eds.). Compendium of Cucurbit Diseases and Pests. 2nd edition. American Phytopathological Society, Saint Paul, Minnesota.

Del Castillo Munera, J. and Hausbeck, M.K. 2016. Characterization of *Pythium* species associated with greenhouse floriculture crops in Michigan. Plant Dis. 100: 569-576.

Dial, S.M. 2007. Fungal diagnostics: Current techniques and future trends. Vet. Clin. North Am. Small Anim. Pract. 37: 373-392.

Dick, M.W. 1990. Phylum *Oomycota. In*: Margulis, L., Corliss, J.O., Melkonian, M. and Chapman, D. (eds.). Handbook of Protista. Joness and Bartlett. Boston USA.

Dorrance, A.E., Mills, D., Robertson, A.E., Draper, M.A., Giesler, L. and Tenuta, A. 2007. *Phytophthora* root and stem rot of soybean. The Plant Health Instructor. DOI: 10.1094/PHI-I-2007-0830-07.

Dorrance, A.E., Robertson, A.E., Cianzo, S., Giesler, L.J., Grau, C.R., Draper, M.A., Tenuta, A.U. and Anderson, T.R. 2009. Integrated management strategies for *Phytophthora sojae* combining host resistance and seed treatments. Plant Dis. 93: 875-882.

Dos Santos, C.E.P., Santurio, J.M. and Marques, L.C. 2011. Pitiose em animais de produção no Pantanal Matogrossense. Pesq Vet Bras 31: 1083-1089.

Edson, H.A. 1915. *Rheosporangium aphanidermatum*: A new genus and species of fungus parasitic on sugar beets and radishes. J. Agric. Res. 4: 279-292.

El-Tarabily, K.A., Hardy, G.E.S.J., Sivasithamparam, K., Hussein, A.M. and Kurtböke, D.I. 1997. The potential for the biological control of cavity-spot disease of carrots, caused by *Pythium coloratum*, by streptomycete and non-streptomycete actinomycetes. New Phytol. 137: 495-507.

Esker, P.D. and Conley, S.P. 2011. Probability of yield response and breaking even for soybean seed treatments. Crop Sci. 52: 351-359.

Farr, D.F. and Rossman, A.Y. 2012. Fungal nomenclature database. Systematic mycology and microbiology laboratory, ARS, USDA; http: //nt.ars-grin.gov/fungaldatabases/fungushost/FungusHost.cfm.

Fischer, A. 1892. *Phycomyctes*. Rabenhorst. Kryptogamenflora 1: 505.

Fischer, J.R., Pace, L.W., Turk, J.R., Kreeger, J.M., Miller, M.A. and Gosser, H.S. 1994. Gastrointestinal pythiosis in Missouri dogs: Eleven cases. J. Vet. Diagn. Invest. 6: 380-382.

Fravel, D.R. 2005. Commercialization and implementation of biocontrol. Annu Rev Phytopathol 43: 337-359.

Gaastra, W., Lipman, L.J., de Cock, A.W., Exel, T.K., Pegge, R.B., Scheurwater, J., Vilela, R. and Mendoza, L. 2010. *Pythium insidiosum*: An overview. Vet. Microbiol. 146: 1-16.

Gabriel, A.L., Kommers, G.D., Trost, M.E., Barros, C.S.L., Pereira, D.B., Schwendler, S.E. and Santurio, J.M. 2008. Surto de pitiose cutânea em bovinos. Pesq Vet Bras 28: 583-587.

Glassgen, W.E., Rose, A., Madlung, J., Koch, W., Gleitz, J. and Seitz, H.U. 1998. Regulation of enzymes involved in anthocyanin biosynthesis in carrot cell cultures in response to treatment with ultraviolet light and fungal elicitors. Planta 204: 490-498.

Green, H. and Jansen, D.F. 2000. Disease progression by active mycelial growth and biocontrol of *Pythium* var. *ultimum* studied using a rhizobox system. Phytopathology 90: 1049-1055.

Handelsman, J. and Stabb, E.V. 1996. Biocontrol of soilborne plant pathogens. Plant Cell 8: 1855-1869.

Ho, H.H. 2018. The taxonomy and biology of *Phytophthora* and *Pythium*. J. Bacteriol. Mycol. Open Access 6: 40-45.

Hoitink, H.A. and Boehm, M.A. 1999. Biocontrol within the context of soil microbioal communities: A substrate-dependent phenomenon. Annu. Rev. Phytopathol. 37: 427-446.

Hong, C.X. and Moorman, G.W. 2005. Plant pathogens in irrigation water: Challenges and opportunities. Crit. Rev. Plant Sci. 24: 189-208.

Howell, C.R., Stipanovic, R.D. and Lumsden, R.D. 1993. Antibiotic production by strains of *Gliocladium virens* and its relation to the biocontrol of cotton seedlings diseases. Biocontrol Sci. Technol. 3: 435-441.

Intaramat, A., Sornprachum, T., Chantrathonkul, B., Chaisuriya, P., Lohnoo, T., Yingyong, W., Jongruja, N., Kumsang, Y., Sandee, A., Chaiprasert, A., Banyong, R., Santurio, J.M., Grooters, A.M., Ratanabanangkoon, K. and Krajaejun, T. 2016. Protein A/G-based immunochromatographic test for serodiagnosis of pythiosis in human and animal subjects from Asia and Americas. Med. Mycol. 54: 641-647.

Jiang, Y.N., Haudenshield, J.S. and Hartman, G.L. 2012. Characterization of *Pythium* spp. from soil samples in Illinois. Can. J. Plant Pathol. 34: 448-454.

Jindayok, T., Piromsontikorn, S., Srimuang, S., Khupulsup, K. and Krajaejun, T. 2009. Hemagglutination test for rapid serodiagnosis of human pythiosis. Clin. Vaccine. Immunol. 16: 1047-1051.

Kabaluk, J.T., Svircev, A.M., Goettel, M.S. and Woo, S.G. 2010. The use and regulation of microbial pesticides in representative jurisdiction worldwide. IOBC Global: Hong Kong, China, 2010, p. 99.

Kageyama, K. 2014. Molecular taxonomy and its application to ecological studies of *Pythium* species. J. Gen. Plant Pathol. 80: 314-326.

Katsumoto, K. 2010. List of fungi recorded in Japan. Kyoto: The Kanto Branch of the Mycological Society of Japan 840-844.

Kawamura, Y., Yokoo, K., Tojo, M. and Hishiike, M. 2005. Distribution of *Pythium porphyrae*, the causal agent of red rot disease of *Porphyrae* spp. in the Ariake Sea, Japan. Plant Dis. 89: 1041-1047.

Kirk, P.M., Cannon, P.F., Minter, D.W. and Stalpers, J.A. 2008. Ainsworth & Bisby's Dictionary of the Fungi, 10th ed. CAB International, Wallingford, UK.

Kirzhner, M., Arnold, S.R., Lyle, C., Mendoza, L.L. and Fleming, J.C. 2015. *Pythium insidiosum*: A rare necrotizing orbital and facial infection. J. Pediatric. Infect. Dis. Soc. 4: 10-13.

Kobayashi, A., Shirao, Y., Yoshita, T., Yagami, K., Segawa, Y., Kawasaki, K., Shozu, M. and Tseng, S.C. 2003. Temporary amniotic membrane patching for acute chemical burns. Eye (Lond.) 17: 149-158.

Koch, W., Wagner, C. and Seitz, H.U. 1998. Elicitor-induced cell death and phytoalexin synthesis in *Daucus carota* L. Planta 206: 523-532.

Konradt, G., Bassuino, D.M., Bianchi, M.V., Castro, L., Caprioli, R.A., Pavarini, S.A., Santurio, J.M., Azevedo, M.I., Jesus, F.P. and Driemeier, D. 2016. Cutaneous pythiosis in calves: An epidemiologic, pathologic, serologic and molecular characterization. Med. Mycol. Case Rep. 14: 24-26.

Krajaejun, T., Imkhieo, S., Intaramat, A. and Ratanabanangkoon, K. 2009. Development of an immunochromatographic test for rapid serodiagnosis of human pythiosis. Clin. Vaccine. Immunol. 16: 506-509.

Krajaejun, T., Kunakorn, M., Niemhom, S., Chongtrakool, P. and Pracharktam, R. 2002. Development and evaluation of an in-house enzyme-linked immunosorbent assay for early diagnosis and monitoring of human pythiosis. Clin. Diagn. Lab. Immunol. 9: 378-382.

Krajaejun, T., Kunakorn, M., Pracharktam, R., Chongtrakool, P., Sathapatayavongs, B., Chaiprasert, A., Vanittanakom, N., Chindamporn, A. and Mootsikapun, P. 2006a. Identification of a novel 74-kilo Dalton immunodominant antigen of *Pythium insidiosum* recognized by sera from human patients with pythiosis. J. Clin. Microbiol. 44: 1674-1680.

Krajaejun, T., Sathapatayavongs, B., Pracharktam, R., Nitiyanant, P., Leelachaikul, P., Wanachiwanawin, W., Chaiprasert, A., Assanasen, P., Saipetch, M., Mootsikapun, P., Chetchotisakd, P., Lekhakula, A., Mitarnun, W., Kalnauwakul, S., Supparatpinyo, K., Chaiwarith, R., Chiewchanvit, S., Tananuvat, N., Srisiri, S., Suankratay, C., Kulwichit, W., Wongsaisuwan, M. and Somkaew, S. 2006b. Clinical and epidemiological analyses of human pythiosis in Thailand. Clin. Infect. Dis. 43: 569-576.

Kurokawa, K. and Tojo, M. 2010. First record of *Pythium grandisporangium* in Japan. Mycoscience 51: 321-324.

Lacaz, C.S. 2002. Pitiose. pp. 770-744. *In*: Lacaz, C.S., Porto, E., Martins, J.E.C., Heins-Vaccari, E.M. and Melo, N.T. (eds.). Tratado de Micologia Médica Lacaz, 9th ed. Sarvier, São Paulo.

Lerksuthirat, T., Sangcakul, A., Lohnoo, T., Yingyong, W., Rujirawat, T. and Krajaejun, T. 2017. Evolution of the sterol biosynthetic pathway of *Pythium insidiosum* and related oomycetes contributes to antifungal drug resistance. Antimicrob. Agents Chemother. 61(4), doi: 10.1128/AAC.02352-16.

Lévesque, C.A. and de Cock, A.W.A.M. 2004. Molecular phylogeny and taxonomy of the genus *Pythium*. Mycol. Res. 108: 1363-1383.

Liang, X., Liu, Z., Lin, Y., Li, N., Huang, M. and Wang, Z. 2012. A modified symblepharon ring for sutureless amniotic membrane patch to treat acute ocular surface burns. J. Burn Care Res. 33: e32-38.

Martin, F.N. 1995. *Pythium*. pp. 17-36. *In*: Singh, U.S., Kohmoto, K. and Singh, R.P. (eds.). Pathogenesis and Host Specificity in Plant Diseases. Histopathological, Biochemical, Genetic, and Molecular Basis. Elsevier, Tarrytown, NY.

Martin, F.N. and Loper, J.E. 1999. Soilborne plant diseases caused by Pythium spp.: Ecology, epidemiology, and prospects for biological control. Crit. Rev. Plant Sci. 18: 111-181.

Matthiesen, R.L. and Robertson, A.E. 2013. Pathogenicity of *Pythium* species affecting corn and soybean in Iowa at three temperatures using two assay methods. Phytopath. 103: 583-591.

Mendoza, L., Ajello, L. and McGinnis, M.R. 1996. Infections caused by the oomycetous pathogen *Pythium insidiosum*. J. Mycol. Med. 6: 151-164.

Mendoza, L., Mandy, W. and Glass, R. 2003. An improved *Pythium insidiosum*-vaccine formulation with enhanced immunotherapeutic properties in horses and dogs with pythiosis. Vaccine 21: 2797-2804.

Mendoza, L. and Newton, J.C. 2005. Immunology and immunotherapy of the infections caused by *Pythium insidiosum*. Med. Mycol. 43: 477-486.

Mihail, J.D., Hung, L.F. and Buhn, J.N. 2002. Diversity of the *Pythium* community infecting roots of the annual legume *Kummerowia stipulacea*. Soil Biol. Biochem. 34: 585-592.

Miller, R.I. 1985. Gastrointestinal phycomycosis in 63 dogs. J. Am. Vet. Med. Assoc. 186: 473-478.

Miura, M., Hatai, K., Tojo, M., Wada, S., Kobayashi, S. and Okazaki, T. 2010. Visceral mycosis in Ayu *Plecoglossus altivelis* larvae caused by *Pythium flevoense*. Fish Pathol. 45: 24-30.

Moorman, G. 2004. Plant Disease Facts. Penn-State University Cooperative Extension. Pennsylvania, United States of America, pp. 1-5.

Moorman, G.W., Kang, S. and Geiser, D.M. 2002. Identification and characterization of *Pythium* species associated with greenhouse floral crops in Pennsylvania. Plan Dis. 86: 1227-1231.

Moorman, G.W. and Kim, S.H. 2004. Species of *Pythium* from greenhouses in Pennsylvania exhibit resistance to propamocarb and mefenoxam. Plant Dis. 88: 630-632.

Morris, B.M. and Gow, N.A.R. 1993. Mechanism of electrotaxis of zoospores of phytopathogenic fungi. Phytopathol. 83: 877-882.

Narkwiboonwong, T., Trakulhun, K., Watanakijthavonkul, K., Paocharern, P., Singsakul, A., Wongsa, A., Woracharoensri, N., Worasilchai, N., Chindamporn, A., Panoi, A., Methipisit, T. and Sithinamsuwan, P. 2011. Cerebral pythiosis: A case report of *Pythium insidiosum* infection presented with brain abscess. J. Infect. Dis. Antimicrob. Agents 28: 129-132.

Nechwatal, J., Wielgoss, A. and Mendgen, K. 2008. Diversity, host, and habitat specificity of oomycete communities in declining reed stands (*Phragmites australis*) of a large freshwater lake. Mycol. Res. 112: 689-696.

Nelson, E.B. 1990. Exudate molecules initiating fungal responses to seeds and roots. Plant and Soil 129: 61-73.

Neufeld, A., Seamone, C., Maleki, B. and Heathcote, J.G. 2018. *Pythium insidiosum* keratitis: A pictorial essay of natural history. Can. J. Ophthalmol. 53: e48-50.

Nzungize, J.R., Lyumugabe, F., Busogoro, J.-P. and Baudoin, J.-P. 2012. *Pythium* root rot of common bean: Biology and control methods: A review. Biotechnol. Agron. Soc. Environ. 16: 405-413.

Olsen, M.W. and Young, D.J. 1998. Damping-off. Cooperative Extension. College of Agriculture, The University of Arizona, Tucson, Arizona, pp. 1-3.

Otsyula, R.M., Buruchara, R.A., Mahuku, G. and Rubaihayo, P. 2003. Inheritance and transfer of root rots (*Pythium*) resistance to bean genotypes. African Crop Sci. Soc. 6: 295-298.

Pan, J.H., Kerkar, S.P., Siegenthaler, M.P., Hughes, M. and Pandalai, P.K. 2014. A complicated case of vascular *Pythium insidiosum* infection treated with limb-sparing surgery. Int. J. Surg. Case Rep. 5: 677-680.

Paulitz, T.C. and Adams, K. 2003. Composition and distribution of *Pythium* communities in wheat fields in eastern Washington State. Phytopathology 93: 867-873.

Permpalung, N., Worasilchai, N., Plongla, R., Upala, S., Sanguankeo, A., Paitoonpong, L., Mendoza, L. and Chindamporn, A. 2015. Treatment outcomes of surgery, antifungal therapy and immunotherapy in ocular and vascular human pythiosis: A retrospective study of 18 patients. J. Antimicrob. Chemother. 70: 1885-1892.

Phillips, A.J., Anderson, V.L., Robertson, E.J., Secombes, C.J. and van West, P. 2008. New insights into animal pathogenic oomycetes. Trends Microbiol. 16: 13-19.

Pracharktam, R., Changtrakool, P., Sathapatayavongs, B., Jayanetra, P. and Ajello, L. 1991. Immunodiffusion test for diagnosis and monitoring of human pythiosis insidiosi. J. Clin. Microbiol. 29: 2661–2662.

Pringsheim, N. 1858. Beiträgezur Morphology and Systematik der Algen. 2. Die Saprolegníeen. Jb. Wíss. Bot. 1: 284-306.

Radmer, L., Anderson, G., Malvick, D.M. and Kurle, J.E. 2017. *Pythium*, *Phytophthora* and *Pytopythium* spp. associated with soybean in Minessota, their relative aggressiveness on soybean and corn, and their sensitivity to seed treatment fungicides. Plant Dis. 101: 62-72.

Reanpang, T., Orrapin, S., Orrapin, S., Arworn, S., Kattipatanapong, T., Srisuwan, T., Vanittanakom, N., Lekawanvijit, S.P. and Rerkasem, K. 2015. Vascular pythiosis of the lower extremity in Northern Thailand: Ten years' experience. Int. J. Low Extrem. Wounds 14: 245-250.

Rey, P., Floch, Gl., Benhamou, N., Salerno, M.I., Thuillier, E. and Tirilly, Y. 2005. Interactions between the mycoparasite *Pythium oligandrum* and two types of sclerotia of plant pathogenic fungi. Mycol. Res. 109: 779-788.

Ribeiro, T.C., Weiblen, C., de Azevedo, M.I., de Avila, Botton S., Robe, L.J., Brayer Pereira, D.I., Monteiro, D.U., Lorensetti, D.M. and Santurio, J.M. 2017. Microevolutionary analyses of *Pythium insidiosum* isolates of Brazil and Thailand based on exo-1,3-β-glucanase gene. Infection, Gen. Evol. 48: 58-63.

Robideau, G.P., de Cock, A.W.A.M., Coffey, M.D., Voglmayr, H., Brouwer, H., Bala, K., Chitty, D.W., Désaulniers, N., Eggertson, Q.A., Gachon, C.M., Hu, C.H., Küpper, F.C., Rintoul, T.L., Sarhan, E., Verstappen, E.C., Zhang, Y., Bonants, P.J., Ristaino, J.B. and Lévesque, C.A. 2011. DNA barcoding of Oomycetes with cytochrome c oxidase subunit I and internal transcribed spacer. Mol. Ecol. Resour. 11: 1002-1011.

Rodrigues, A., Graça, D.L., Fontoura, C., Cavalheiro, A.S., Hensley, A., Schwendler, S.E., Alves, S.H. and Santurio, J.M. 2006. Intestinal dog pythiosis in Brazil. J. Med. Vet. Mycol. 16: 37-41.

Rojas, J.A., Jacobs, J.L., Napieralski, S., Karaj, B., Bradley, C.A., Chase, T., Esker, P.D., Giesler, L.J., Jardine, D.J., Malvick, D.K., Markell, S.G., Nelson, B.D., Robertson, A.E., Rupe, J.C., Smith, D.L., Sweets, L.E., Tenuta, A.U., Wise, K.A. and Chilvers, M.I. 2017. Oomycete species associated with soybean seedlings in North America – Part II: Diversity and Ecology in Relation to Environmental and Edaphic Factors. Phytopathol. 107: 293-304.

Santurio, J.M., Alves, S.H., Pereira, D.B. and Argenta, J.S. 2006a. Pitiose: Uma micose emergente. Acta Sci.Vet. 34: 1-14.

Santurio, J.M., Leal, A.T., Leal, A.B.M., Alves, S.H., Lübeck, I., Griebeler, J. and Copetti, M.V. 2006b. Teste de ELISA indireto para o diagnóstico sorológico de pitiose. Pesq Vet Bras 26: 47-50.

Santurio, J.M., Argenta, J.S., Schwendler, S.E., Cavalheiro, A.S., Pereira, D.I., Zanette, R.A., Alves, S.H., Dutra, V., Silva, M.C., Arruda, L.P., Nakazato, L. and Colodel, E.M. 2008. Granulomatous rhinitis associated with *Pythium insidiosum* infection in sheep. Vet. Rec. 163: 276-277.

Schroeder, K.L., Martin, F.N., de Cock, A.W.A.M., Lévesque, C.A., Spies, C.F.J., Okubara, P.A. and Paulitz, T.C. 2013. Molecular detection and quantification of *Pyhtium* species: Evolving taxonomy, new tools, and challenges. Plant Dis. 97: 4-20.

Schröter, J. 1893. *Pythiaceae*. Engler & Prantl Nat Pfl Fam 1: 104-105.

Senda, M., Kageyama, K., Suga, H. and Lévesque, C.A. 2009. Two new species of *Pythium*, *P. senticosum* and *P. takayamanum*, isolated from cool-temperature forest soil in Japan. Mycologia 101: 439-448.

Sermsathanasawadi, N., Praditsuktavorn, B., Hongku, K., Wongwanit, C., Chinsakchai, K., Ruangsetakit, Hahtapornsawan, S. and Mutirangura, P. 2016. Outcomes and factors influencing prognosis in patients with vascular pythiosis. J. Vasc. Surg. 64: 411–417.

Shipton, W.A. 1987. *Pythium destruens* sp. nov.: An agent of equine pythiosis. J. Med. Vet. Mycol. 25: 137-151.

Simon, A. and Sivasithampapram, K. 1988a. Microbiological differences between soils suppressive and conductive of the saprophytic growth of *Gaeumannomyces graminis* var. *tritici*. Can. J. Microbiol. 34: 860-864.

Simon, A. and Sivasithampapram, K. 1988b. The soil environment and the suppression of saprophytic growth of *Gaeumannomyces graminis* var. *tritici*. Can. J. Microbiol. 34: 865-870.

Souto, E.P.F., Maia, L.A., Olinda, R.G., Galiza, G.J.N., Kommers, G.D., Miranda-Neto, E.G., Dantas, A.F.M. and Riet-Correa, F. 2016. Pythiosis in the nasal cavity of horses. J. Comp. Pathol. 155: 126-129.

Stewart, S., Abeysekara, N. and Robertson, A.E. 2014. Pathotype and genetic shifts in a population of *Phytophthora sojae* under soybean cultivar rotation. Plant Dis. 98: 614-624.

Supabandhu, J., Fisher, M.C., Mendoza, L. and Vanittanakom, N. 2008. Isolation and identification of the human pathogen *Pythium insidiosum* from environmental samples collected in Thai agricultural areas. Med. Mycol. 46: 41-52.

Teymoori, S., Shahri, M.H., Rahnama, K. and Afzali, H. 2012. Identification and pathogenicity of *Pythium* species on cantaloupe in Khorasan Razavi Province of Iran. J. Crop Prot. 1: 239-247.

Thitithanyanont, A., Mendoza, L., Chuansumrit, A., Pracharktam, R., Laothamatas, J., Sathapatayavongs, B., Lolekha, S. and Ajello, L. 1998. Use of an immunotherapeutic vaccine to treat a life-threatening human arteritic infection caused by *Pythium insidiosum*. Clin. Infect. Dis. 27: 1394-1400.

Tonpitak, W., Pathomsakulwong, W., Sornklien, C., Krajaejun, T. and Wutthiwithayaphong, S. 2018. First confirmed case of nasal pythiosis in a horse in Thailand. JMM Case Reports 2018 5(1): e005136; DOI 10.1099/jmmcr.0.005136

Tyler, B.M., Tripathy, S., Zhang, X., Dehal, P., Jiang, R.H., Aerts, A. et al. 2006. *Phytophthora* genome sequences uncover evolutionary origins and mechanisms of pathogenesis. Science 313: 1261-1266.

Ubiali, D.G., Cruz, R.A., de Paula, D.A., Silva, M.C., Mendonça, F.S., Dutra, V., Nakazato, L., Colodel, E.M. and Pescador, C.A. 2013. Pathology of nasal infection caused by *Conidiobolus lamprauges* and *Pythium insidiosum* in sheep. J. Comp. Pathol. 149: 137-145.

Uzuhashi, S., Okada, G. and Ohkuma, M. 2015. Four new *Pythium* species from aquatic environments in Japan. Antonie van Leeuwenhoek 107: 375-391.

Uzuhashi, S., Tojo, M. and Kakishima, M. 2010. Phylogeny of the genus *Pythium* and description of new genera. Mycosci. 51: 337-365.

Uzuhashi, S., Tojo, M., Kobayashi, S., Tokura, K. and Kakishima, M. 2008. First records *of Pythium aquatile* and *P. macrosporum* isolated from soils in Japan. Mycosci. 49: 276-279.

van West, P., Morris, B.M., Reid, B., Appiah, A.A., Osborne, M.C., Campbell, T.A., Shepherd, S.J., Gow, N.-A.R. 2002. Oomycete plant pathogens use electric fields to target roots. Mol. Plant Microbe. Interact. 15: 790-798.

Vossenkemper, J.P., Nafziger, E.D., Wessel, J.R., Maughan, M.W., Rupert, M.E. and Schmidt, J.P. 2015. Early planting, full-season cultivars, and seed treatments maximize soybean yield potential. Crop Forage Turfgrass Manage. 1: 1-9.

Vurukonda, S.S.K.P., Giovanardi, D. and Stefani, E. 2018. Plant growth promoting and biocontrol activity of *Streptomyces* spp. as endophytes. Int. J. Molec. Sci. 19: 952-977.

Wagner, A., Norris, S., Chatterjee, P., Morris, P.F. and Wildschutte, H. 2018. Aquatic pseudomonads inhibit oomycete plant pathogens of *Glycine max*. Frontiers Microbiol. 9: 1007.

Wanachiwanawin, W., Mendoza, L., Visuthisakchai, S., Mutsikapan, P., Sathapatayavongs, B., Chaiprasert, Suwanagool, P., Manuskiatti, W., Ruangsetakit, C. and Ajello. 2004. Efficacy of immunotherapy using antigens of *Pythium insidiosum* in the treatment of vascular pythiosis in humans. Vaccine 22: 3613-3621.

Wei, L., Xue, A.G., Cober, E.R., Babcock, C., Zhang, J.X., Zhang, S.Z., Li, W.B., Wu, J.J. and Liu, L.J. 2011. Pathogenicity of *Pythium* species causing seed rot and damping-off on soybean under controlled conditions Phytoprot. 91: 3-10.

Weiland, J.E., Santamaria, L. and Grunwald, N.J. 2014. Sensitivity of *Pythium irregulare*, *P. sylvaticum*, and *P. ultimum* from forest nurseries to mefenoxam and fosetyl-Al, and control of *Pythium* damping-off. Plant Dis. 98: 937-942.

Weiland, J.E., Beck, B.R. and Davis, A. 2013. Pathogenicity and virulence of *Pythium* species obtained from forest nursery soils on Douglas-fir seedlings. Plant Dis. 97: 744-748.

Weller, D.M. 1988. Biological control of soil-borne plant pathogens in the rhizosphere with bacteria. Annu. Rev. Phytopathol. 26: 379-407.

Whipps, J.M. 2001. Microbial interactions and biocontrol in the rhizosphere. J. Exp. Bot. 52: 487-511.

Whipps, J.M. and Lumsden, R.D. 1991. Biological control of *Pythium* species. Biocontrol Sci. Technol. 1: 75-90.

Wrather, J.A. and Koenning, S.R. 2006. Estimates of disease effects on soybean yields in the United States 2003 to 2005. J. Nematol. 38: 173-180.

Yellareddygari, S.K.R., Vijay Krishna Kumar, K., Sudini, H., Surendranatha Reddy, E.C., Naga Madhuri, K.V., Suresh, K., Ravindra Reddy, B., Hemalatha, S., Al-Turki, A., El-Meleigi, M.A., Sayyed, R.Z., Han, Y., Sai Sruthi, V. and Reddy, M.S. 2015. Efficacy of various biopesticides in managing *Pythium* root rot of *Petunia*. pp. 450-455. *In*: Reddy, M.S., Ilao, R.I., Faylon, P.S., Dar, W.D., Batchelor, W.D., Sayyed, R., Sudini, H., Kumar, K.V.K., Armanda, A. and Gopalkrishnan, S. (eds.). Recent Advances in Biofertilizers and Biofungicides (PGPR) for Sustainable Agriculture. Cambridge Scholars Publishing, Newcastle.

Zagade, S.N., Deshpande, G.D., Gawade, D.B. and Atnoorkar, A.A. 2013. Studies on pre- and post emergence damping off of chilli caused by *Pythium ultimum*. Indian J. Plant Prot. 41: 332-337.

Zitnick-Anderson, K. and Nelson, B.D. 2015. Identification and pathogenicity of *Pythium* on soybean in North Dakota. Plant Dis. 99: 1.

4

The Genus *Phytopythium*

Abdul Mubeen Lodhi[1]*, Saleem Shahzad[2] and Rehana Naz Syed[3]

[1] Department of Plant Protection, Sindh Agriculture University, Tandojam, Pakistan
[2] Department of Agriculture and Agribusiness Management, University of Karachi, Karachi, Pakistan
[3] Department of Plant Pathology, Sindh Agriculture University, Tandojam, Pakistan

Introduction

Genus *Phytopythium* was described by Bala et al. (2010) with *P. sindhum* as the type species. They suggested that the *Pythium* species from clade-K of Lévesque and de Cock (2004) belongs to this genus, since the placement of species of this clade in the genus *Pythium* has been debated by many scientists (Briard et al. 1995, Panabieres et al. 1997, Dick 2001, Villa et al. 2006, Belbahri et al. 2008). Creation of genus *Phytopythium* in 2010 resolved this issue. Uzuhashi et al. (2010) also proposed a new genus *Ovatisporangium* to settle this issue, but it appeared to be a later synonym of *Phytopythium*. The phylogenetic tree constructed from the ITS sequences of the *Phytopythium* species, two *Phytophthora* species and representatives of different clades of *Pythium,* available in GenBank, clearly shows that *Pythium, Phytopythium* and *Phytophthora* are quite distinct groups (Fig. 4.1).

Based on molecular analysis, Baten et al. (2014) renamed *Phytophthora fagopyri* as *Phytopythium fagopyri,* although Plaats-Niterink (1981) considered it as a synonym of *Pythium helicoides.* Thines (2014) also renamed *Halophytophthora kandeliae* as *Phytopythium kandeliae* on molecular basis. de Cock et al. (2015) transferred 14 species from *Pythium* clade-K to *Phytopythium* and also described a new species *Phytopythium mirpurense.* Two new species *viz., Phytopythium aichiense* and *Phytopythium iriomotense* were added to the genus by Baten et al. (2015), whereas Bennett et al. (2017) described two new species *viz., Phytopythium dogmae* and *Phytopythium leanoi.*

Kirk (2015) published *Phytopythium indigoferae, Phytopythium cucurbitacearum, Phytopythium megacarpum* and *Phytopythium sterile* as new combinations for species previously classified in the genus *Pythium.* Of these, *Phytopythium cucurbitacearum* was considered as an invalid species by Plaats-Niterink (1981). The morphological characters of *Phytopythium indigoferae* described by Butler (1907) are quite different from other species within this genus (de Cock et al. 2015). In the absence of a viable strain, de Cock et al. (2015) concluded that the identity of the species is not confirmed. They also mentioned that some strains with DNA sequences, very close to the strain used by Levesque and de Cock (2004), have been found, but these isolates produce subglobose and proliferating sporangia.

Phylogenetic trees published by various researchers (Baten et al. 2014, Goncalves et al. 2016, de Jesus et al. 2016, Bennett et al. 2017) show that the species of clade K or *Phytopythium* form 3 distinct clades. A phylogenetic tree based on the ITS sequences of 25 species available in the GeneBank depicts six clades in the genus *Phytopythium* (Fig. 4.2). Clad-A includes

*Corresponding author: mubeenlodhi@gmail.com; amlodhi@sau.edu.pk

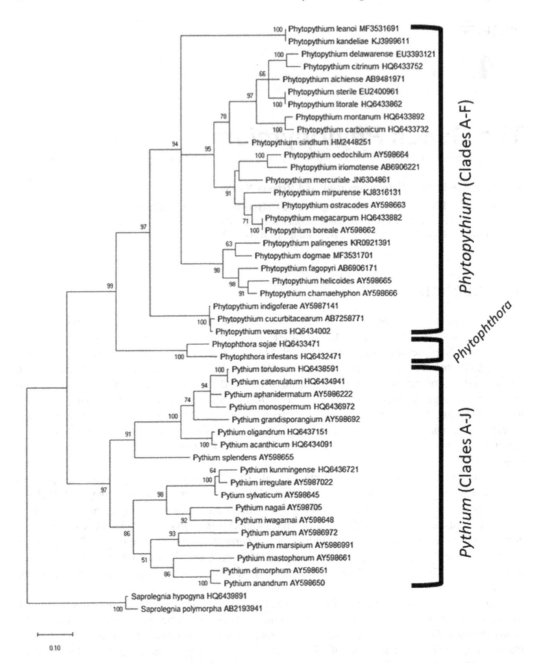

Figure 4.1. Maximum likelihood phylogenetic tree of ITS region of selected oomycetous species showing relationship between *Phytopythium*, *Pythium* and *Phytophthora* spp. The bootstrap consensus tree inferred from 1000 replicates. Maximum likelihood bootstraps supporting values larger than 50 are indicated only. Two distantly related *Saprolegnia* spp. (Oomycetes) are used as outgroup

Phytopythium leanoi and *Phytopythium kandeliae* whereas clade-B consist of seven species *viz.*, *Phytopythium boreale*, *Phytopythium megacarpum*, *Phytopythium ostracodes*, *Phytopythium mirpurense*, *Phytopythium mercuriale*, *Phytopythium iriomotense* and *Phytopythium oedochilum*. Clade-C comprises of seven species *viz.*, *Phytopythium carbonicum*, *Phytopythium montanum*, *Phytopythium citrinum*, *Phytopythium delawarense*, *Phytopythium aichiense*, *Phytopythium litorale*

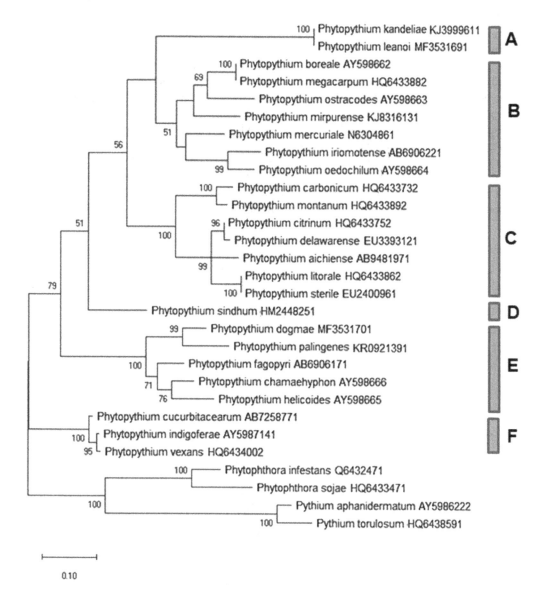

0.10

Figure 4.2. Maximum likelihood phylogenetic tree of ITS region of all 25 *Phytopythium* species (except *Phytopythium polytylum*) distributed in 6 clades, as well as representative *Pythium* and *Phytophthora* spp. The tree with the highest log likelihood is shown. The bootstrap consensus tree inferred from 1000 replicates. Maximum likelihood bootstraps supporting values larger than 50% are indicated only

and *Phytopythium sterile*. *Phytopythium sindhum* was the only member in clade-D, whereas clade-E included five species *viz.*, *Phytopythium dogmae*, *Phytopythium palingenes*, *Phytopythium fagopyri*, *Phytopythium chamaehyphon* and *Phytopythium helicoides*. The last clade F has three species *viz.*, *Phytopythium* cucurbitacearum, *Phytopythium indigoferae* and *Phytopythium vexans*. *Phytopythium polytylum* cannot be placed in any of the six clades, since no viable culture or DNA sequence of the species is available.

Species belonging to *Phytopythium* are morphologically close to the species of *Pythium*, but genetically they are closed to the species of *Phytophthora* (de Cock et al. 2015). Globose to ovoid sporangia are papillate and resemble the sporangia of *Phytophthora*, but the papilla develops at maturity (Fig. 4.3a-c) in contrast to the sporangia of *Phytophthora* that have papilla from the beginning (Fig. 4.3g). The hyaline 'apical' thickening' found in papillae of *Phytophthora* species is absent in species of *Phytopythium* (Blackwell 1949, de Cock et al. 2015). Unlike *Phytophthora* species, the sporangia in *Phytopythium* also show frequent internal proliferation (Fig. 4.3k-l) (Bala et al. 2010). The zoospores are formed and discharged in a *Pythium*-like manner, i.e. a short discharge tube is formed at the apex of the sporangium. The contents of the sporangium are discharged into a membranous vesicle formed at the tip of the discharge tube where differentiation and liberation of zoospores take place (Fig. 4.3d-e). However, instead of the papillae, the sporangia may form discharge tubes from any place, whereas, in some species, the papillae grow into a branched structure (de Cock et al. 2015). *Phytopythium leanoi* and *Phytopythium dogmae* exhibit a mixed *Pythium*- and *Phytophthora*-like zoospore formation where part of the protoplasm moves out into a vesicle through the exit pore and forms zoospores like other *Pythium* and *Phytopythium* species; however, zoospore development also takes place inside the sporangium just like *Phytophthora* species (Bennett et al.

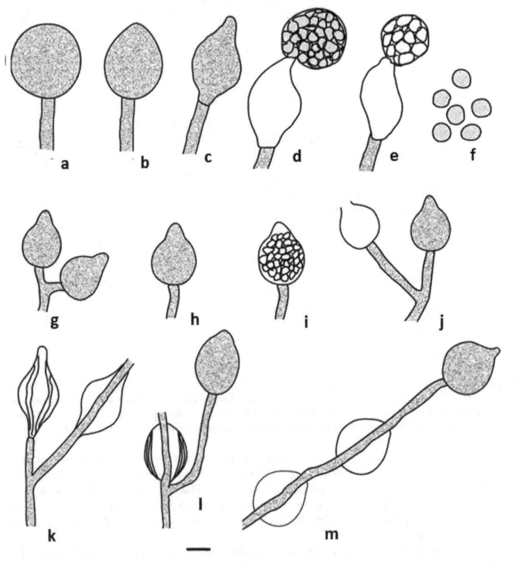

Figure 4.3. Sporangia: (a-f) *Phytopythium* sporangia with *Pythium*-like zoospore release through vesicle, (g-j) *Phytophthora*-like zoospore release, (k-m) proliferating sporangia in *Phytopythium*

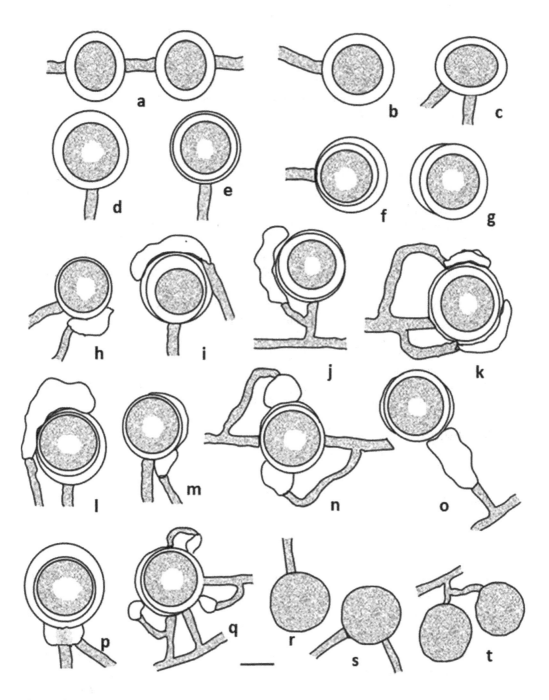

Figure 4.4. Oogonia and antheridia: (a) smooth and intercalary oogonia, (b) terminal oogonium with undifferentiated ooplast, (c) unilaterally intercalary oogonium (d) plerotic oospore, (e) nearly plerotic oospore, (f-g) aplerotic oospore with differentiable ooplast, (h-l) monoclinous or diclinous elongated antheridia with lengthwise application, (m) hypogynous antheridia, (n-o) bell-shaped antheridia, (p) amphigynous antheridia, (q) crook-necked antheridia, (r-t) hyphal swellings

2017). The same happens in *Phytopythium kandeliae* (Marano et al. 2014). *Phytopythium litorale* and *P. mercuriale* also produce hyphal swellings in addition to the zoosporangia (Nechwatal and Mendgen 2006, Belbahri et al. 2008), whereas *P. boreale* and *P. megacarpum* produce hyphal swellings but no zoospore formation was observed (Duan 1985, Paul 2000).

Oogonia of the *Phytopythium* species are smooth (Fig. 4.4a-q). However, *P. carbonicum* forms smooth as well as oogonia with small projections (Paul 2003). The oogonia in the type species *Phytopythium sindhum* are produced usually on short lateral branches, but occasionally the oogonia are terminal or unilaterally intercalary (Fig. 4.4a-c). Oogonia production is not known in *Phytopythium kandeliae, P. litorale, P. sterile* and *P. dogmae* (Fig. 4.4r-t). *Phytopythium ostracodes* produces mostly intercalary oogonia, whereas other species produce terminal, intercalary or unilaterally intercalary oogonia. Sessile oogonia are also known in *P. palingenes* and *P. polytylum.*

P. sindhum produces diclinous as well as monoclinous antheridia that are either elongated with lengthwise application to the oogonium (Fig. 4.4h-l) or crook-necked (Fig. 4.4q), but in either case the contact with the oogonium is apical (Bala et al. 2010). *Phytopythium citrinum* produces bell-shaped hypogynous antheridia (Fig. 4.4m) whereas *P. iriomotense* produces diclinous antheridia which are sometimes amphigynous (Fig. 4.4p). Some species produce either mostly diclinous or monoclinous antheridia, whereas some species produce both types of antheridia. Most of the species usually produce large, elongated antheridia with smooth or irregular contour (Fig. 4.4j and 4.4l). However, crook-necked antheridia are also produced in some species. The crook-necked antheridia produce fertilization tube apically, whereas elongated antheridia produce fertilization tube either apically or laterally. *Phytopythium vexans* produces large and typically bell-shaped or irregular antheridia making apical contact with the oogonia (Fig. 4.4n-o).

The oospores in some species are plerotic (Fig. 4.4d) or nearly plerotic (Fig. 4.4e), whereas some species have aplerotic spore (Fig. 4.4f-g). The oospore wall thickness also shows a great variation in different species, i.e. from up to 1.5 µm in *P. vexans* to 7 µm in *P. megacarpum*. The ooplast in some species is present as a single entity (Fig. 4.4g), whereas in others the ooplast is found as several small globules dispersed in the cytoplasm (Fig. 4.4b).

Keys to *Phytopythium* species

1	Sexual structures not known	2
1′	Sexual structures produced in single cultures	5
2	Sporangia large, 27-55×20-42 µm (av. diameter >35 µm)	*P. kandeliae*
2′	Sporangia relatively small, (av. diameter is ≤30 µm)	3
3	Sporangia not globose, but limoniform, obovoid, obpyriform, av. diameter >29 µm	*P. dogmae*
3′	Sporangia are mostly globose or subglobose, av. diameter <29 µm	4
4	Sporangia are subglobose, terminal, zoospores up to 10 µm	*P. litorale*
4′	Sporangia are globose, terminal and intercalary, zoospores up to 12 µm	*P. sterile*
5	Zoospores not produced	6
5′	Zoospores produced	8
6	Oospores wall very thick, up to 7 µm	*P. megacarpum*
6′	Oospores wall moderately thick, up to 5 µm	7
7	Oogonia with small papillae like outgrowth	*P. carbonicum*
7′	Oogonia without any outgrowth	*P. boreale*
8	Oospores are plerotic	9
8′	Oospores are aplerotic	14
9	Mean oogonia diameter ≤ 20 µm	*P. indigoferae*

9′	Mean oogonia diameter >20 μm	10
10	Mean oogonia diameter >40 μm	*P. leanoi*
10′	Mean oogonia diameter <40 μm	11
11	Mean oogonia diameter <30 μm	12
11′	Mean oogonia diameter >30 μm	13
12	Antheridia hypogynous	*P. citrinum*
12′	Antheridia consists of multiple knots	*P. mercuriale*
13	Oogonia intercalary or terminal, antheridia mostly monoclinous, attachment lengthwise	*P. ostracodes*
13′	Oogonia mostly lateral, occ. terminal, antheridia diclinous/monoclinous, attachment lengthwise/apical	*P. sindhum*
14	Oogonia terminal	15
14′	Oogonia terminal and intercalary	18
15	Mean oogonia diameter >30 μm	16
15′	Mean oogonia diameter <30 μm	17
16	Mean oospores diameter 24.2 μm, antheridia 1-2, crook naked	*P. aichiense*
16′	Mean oospores diameter 28.2 μm, antheridia 1-4, elongate, lengthwise attachment	*P. polytylum*
17	Mean oospores diameter >27 μm	*P. fagopyri*
17′	Mean oospores diameter <26 μm	18
18	Mean oospores diameter <20 μm, antheridia mostly monoclinous, if elongate, lengthwise attachment, if crook naked apical attachment	*P. montanum*
18′	Mean oospores diameter >20 μm, antheridia diclinous, broad lengthwise attachment	*P. delawarense*
19	Antheridial cells hypogynous or amphigynous or crook naked	20
19′	Antheridial cells large, irregular, attached lengthwise or laterally	21
20	Antheridial cells hypogynous	*P. cucurbitacearum*
20′	Antheridial cells amphigynous or crook naked	*P. iriomotense*
21	Mean oogonia diameter <30 μm	22
21′	Mean oogonia diameter >30 μm	23
22	Mean oospores diameter 24.3 μm	*P. chamaehyphon*
22′	Mean oospores diameter 17.3 μm	*P. vexans*
23	Oospores wall very thick (4-6 μm)	*P. helicoides*
23′	Oospores wall moderately thick (less than 4-6 μm)	24
24	Antheridia mostly monoclinous	*P. mirpurense*
24′	Antheridia mostly diclinous	25
25	Oospores wall up to 5 μm thick	*P. oedochilum*
25′	Oospores wall up to 3.5 μm thick	*P. palingenes*

Description of species

1. *Phytopythium aichiense* (Baten and Kageyama 2015)

Description: Sporangia terminal, proliferating, sub-globose, ovoid or limoniform, measuring 25-39×19-27 μm. Oogonia terminal, smooth, measuring 29-40 (av. 34.3) μm. Oospores aplerotic, average diameter 24.2 μm. Antheridia 1-2 per oogonium, diclinous, occasionally monoclinous, crook-necked or inflated.

Distribution and hosts: *Phytopythium aichiense* was isolated from water purification sludge in Aichi prefecture, Japan (Baten et al. 2015).

2. ***Phytopythium boreale* (R.L. Duan) Abad, de Cock, Bala, Robideau, Lodhi and Lévesque 2014**

Synonymy:

Pythium boreale R.L. Duan, Acta Mycol. Sin.: 1 (1985)

Pythium borealis R.L. Duan (1985)

Ovatisporangium boreale (R.L. Duan) Uzuhashi, Tojo and Kakish., Mycoscience 51 (5): 360 (2010)

Description: Sporangia or hyphal swellings globose, subglobose or pyriform, terminal or intercalary. Zoospores not observed. Oogonia smooth, intercalary or terminal, measuring 24-28 (av. 26) μm. Oospores plerotic, measuring 25-26 (av. 25) μm. Antheridia 1-2 per oogonium, monoclinous, occasionally diclinous, sessile, clavate or elongate apply lengthwise but make apical contact with the oogonium.

Distribution and hosts: *P. boreale* was isolated from soil under *Brassica caulorapa* in Beijing, China (Duan 1985). It was also reported from mainland China (Ho 2013). In Pakistan, it was isolated from rose rhizosphere and weeds collected from the lake (Lodhi 2007). In USA, it was associated with soybean (Beckman et al. 2018, Radmer et al. 2017), alder (*Alnus* sp.) (Adams et al. 2009) and leaf litter of streams (Thrailkill 2013).

3. ***Phytopythium carbonicum* (B. Paul) Abad, de Cock, Bala, Robideau, Lodhi and Lévesque 2014**

Synonymy:

Pythium carbonicum B. Paul, FEMS Microbiology Letters 219 (2): 270 (2003)

Ovatisporangium carbonicum (B. Paul) Uzuhashi, Tojo and Kakish., Mycoscience 51 (5): 360 (2010)

Description: Sporangia globose, intercalary occasionally terminal, measuring 18-30 (av. 25.6) μm, germinate directly without zoospores production. Oogonia smooth, intercalary occasionally terminal, measuring 15-42 (av. 26) μm. Oospores plerotic; one occasionally two per oogonium, 14-32 (av. 25.7) μm. Ooplast not differentiated with rest of oospore contents. Antheridia monoclinous as well as diclinous, one to many per oogonium, small knot type wrapping the oogonium from all sides.

Distribution and hosts: Isolated from a spoil heap in northern France (Paul 2003).

4. ***Phytopythium chamaehyphon* (Sideris) Abad, de Cock, Bala, Robideau, Lodhi and Lévesque 2014**

Synonymy:

Pythium chamaehyphon Sideris, Mycologia 24 (1): 33 (1932)

Pythium chamaihyphon Sideris, Mycologia 24 (1): 33 (1932)

Ovatisporangium chamaehyphon (Sideris) Uzuhashi, Tojo and Kakish., Mycoscience 51 (5): 360 (2010)

Description: Sporangia subglobose or oblong, measuring 18-28 μm. Oogonia smooth, terminal as well as intercalary, measuring 24-28 (av. 26.5) μm. Oospores aplerotic, measuring 23-26 (av. 24.3) μm. Antheridia large, irregular, attached lengthwise or laterally applied to oogonium near the oogonial stalk.

Distribution and hosts: *P. chamaehyphon* was originally isolated from *Carica papaya* in Hawaii (Sideris 1932). In Algeria, it was isolated from different substrates collected from the Oubeira Lake (Kachour et al. 2016). Afterwards, it was isolated from soybean (Grijalba et al. 2018) as well as other 27 plant species in Argentina (Palmucci et al. 2011) and from mainland China (Ho 2013). In the Czech Republic, it was associated with either root rot or rhizospheric soil of *Nerium oleander, Quercus* sp., *Rhododendron* sp., *Alnus glutinosa* and *Eunonymus* sp. (Mrázková et al. 2018). Jung et al. (1996) isolated it from declining oak trees (*Quercus robur, Quercus petraea*) and nearby habitats in Germany, Switzerland, Hungary, Italy and Slovenia. It was recovered from affected soybean seedlings in North America (Rojas et al. 2017). It was also isolated from rhizosphere of sugarcane, mango and guava in Pakistan (Lodhi 2007), water samples in Japan (Baten et al. 2013), *Anacardium excelsum* in Panama (Davidson et al. 2000) and mango in Peru (Quiros and Jose 2018). In Africa, it has been reported as a cause of root rot of common bean *Phaseolus vulgaris* in Kenya, Rwanda and Uganda (Nzungize et al. 2011, 2012, Gichuru et al. 2016) as well as recovered from grapevine nurseries and yards in South Africa (Spies et al. 2011, McLeod et al. 2009). In USA, it was isolated from *Camellia japonica* and *Gardenia jasminoides* (DeMott 2015), Pennsylvania greenhouse irrigation water tanks (Choudhary et al. 2016), common bean (USDA-ARS 2016) and wheat seedlings (Broders et al. 2007).

5. *Phytopythium citrinum* **(B. Paul) Abad, de Cock, Bala, Robideau, Lodhi and Lévesque 2014**

Synonymy:
Pythium citrinum B. Paul, FEMS Microbiology Letters 234 (2): 273 (2004)
Ovatisporangium citrinum (B. Paul) Uzuhashi, Tojo and Kakish., Mycoscience 51 (5): 360 (2010)

Description: Sporangia globose, sub-globose, pyriform or lemoniform, terminal or intercalary, measuring 15-35 (av. 24.2) μm, mostly sporulate by zoospores formation, occasionally germinate directly by germ tube. Oogonia smooth, terminal, measuring 20-36 (av. 27.6) μm. Oospores plerotic or nearly plerotic, measuring 18-34 (av. 24.9) μm. Ooplast not differentiated with rest of oospore contents. Antheridia bell shaped, hypogynous, persistent.

Distribution and hosts: It was first isolated from burgundian vineyards in France (Paul 2004). Afterwards, it was isolated from waterways in Australia (Huberli et al. 2013) and narrow-leaved ash (*Fraxinus angustifolia* Vahl) in Croatia (Kranjec et al. 2017). It was also found to be associated with root rot or rhizosphere of *Populus tremula, Fragaria* sp., *Tilia cordata, Abies grandis, Populus* sp., *Quercus robur, Alnus glutinosa, Acer pseudoplatanus, Malus domestica, Pyrus* sp. and *Malus* sp. (Mrázková et al. 2018) as well as different fruit orchards in Czech Republic (Hrabětová et al. 2017), and from declining reed stands (*Phragmites australis*) in Germany (Nechwatal et al. 2008). In Poland, it was isolated from oak forests in southern Poland (Jankowiak et al. 2015), rivers of the Polish-Ukrainian border areas (Matsiakh et al. 2016), water courses use for forest (Oszako et al. 2016) and forest soils (Lefort et al. 2013). This species was isolated from sediment sample collected from Southwest Indian Ocean (Wen et al. 2014) and from soybean in USA (Coffua et al. 2016).

6. *Phytopythium cucurbitacearum* **S. Takim. ex P.M. Kirk 2015**
Synonymy:
Pythium cucurbitacearum S. Takim., Annals of the Phytopathological Society of Japan 11: 91 (1941)

Description: Sporangia spherical to oval, papillate, measuring 12-15×10-13 μm. Oogonia smooth, 22-25 μm in diam. Antheridia nearly spherical, applied near the base of the oogonium.

Distribution and hosts: It was isolated as the causal agent of damping-off of seedling and fruit rot of cucumber, watermelon, melon and rag gourd in Japan (Takimoto 1941), stem rot and root rot of papaya in Alagoas, Brazil (Santos 2015) and avocado wilt complex disease in Colombia (Ramírez-Gil 2018). It was also isolated from African oil palm (*Elaeis guineensis*) (Azni et al. 2017), leaves and sea water collected from Hainan Island, China (Ho et al. 2012), broccoli, carnation, onion, papaya, pepper, Persian lemon and ponytail grass in Guatemala (Abad-Campos et al. 2008), fruit rot disease of pointed gourd *Trichosanthes dioica* in India (Chaudhuri 1975, Misra and Hall 1996), and durian (*Durio* sp.) in Indonesia (Santoso et al. 2015). In Japan, *P. cucurbitacearum* has also been found to be pathogenic to tomato, *Hibiscus, Ipomoea* and potato (Takahashi et al. 1970, Reen 1971a, b, Takahashi 1973, Takimoto 1941) as well as found associated with *Citrullus vulgaris, Cucumis melo* and *Cucumis sativus* in Japan (Watson 1971). It was also recovered from cocoa farm soils in

Nigeria (Aigbe and Woodward 2018), from citrus and mango in Peru (Quiros and Jose 2018), from common bean (*Phaseolus vulgaris*) in Rwanda (Nzungize et al. 2011), from durian orchard (Suksiri et al. 2018) and chilli in Thailand (Sawangphiphop et al. 2016), and common bean (USDA-ARS 2016) and ornamental plants in USA (Woodward and DeMott 2014). It also causes root rot diseases of *Acacia mangium* and *Acacia hybrid* in North Vietnam (Thu 2016).

7. *Phytopythium delawarense* (Broders, P.E. Lipps, M.L. Ellis and Dorrance) Abad, de Cock, Bala, Robideau, Lodhi and Lévesque 2014
   Synonymy:
   Pythium delawarii Broders, Lipps and Dorrance, Mycologia 101 (2): 233 (2009)
   Pythium delawarei Broders, Lipps and Dorrance, Mycologia 101 (2): 233 (2009)
   Pythium delawarense Broders, P.E. Lipps, M.A. Ellis and Dorrance, Mycologia 104 (3): 789 (2012)
   Description: Sporangia globose, lemoniform or ovoid, mostly terminal occasionally intercalary, measuring 25-40 (av. 32.5) μm in diameter. Oogonia smooth, terminal, measuring 26-29 (av. 27.2) μm. Oospores aplerotic, measuring 21-26 μm. Antheridia mostly one occasionally two per oogonium, diclinous, occasionally monoclinous, making broad lengthwise contact with the oogonia.
   Distribution and hosts: It was isolated from soybean (*Glycine max*) in Ohio, USA (Broders et al. 2009). Later on, it was also reported from alder (*Alnus* sp.) in USA (Adams et al. 2009).

8. *Phytopythium dogmae* R. Bennett and Thines 2017
   Description: Sporangia terminal, ovoid to obovoid, limoniform, obpyriform or pyriform, measuring 25-27×29-37 (av. 26×33) μm. Sexual structures such as oogonia, oospores and antheridia not observed.
   Distribution and hosts: It was isolated from decaying mangrove leaf-litter in Philippines (Bennett et al. 2017).

9. *Phytopythium fagopyri* (S. Takim. ex S. Ito and Tokun.) Kageyama and Baten 2014
   Synonymy:
   Phytopythium fagopyri (S. Takim. ex S. Ito and Tokun.) Kageyama and M.A. Baten, Mycological Progress 13 (4/1003): 1153 (2014)
   Pythium helicoides Drechsler, Journal of the Washington Academy of Sciences 20: 413 (1930)
   Description: Sporangia sub-globose or pyriform, measuring 27-37×17-39 (av. 32×28.5) μm. Oogonia smooth, terminal, 28-31 (av. 29) μm. Antheridia mostly diclinous, usually one, occasionally 2-3 per oogonium, antheridial cells large, wavy or smooth, laterally applied. Oospores aplerotic, 23-25 (av. 24.3) μm.
   Distribution and hosts: It was originally isolated from damped-off buckwheat seedlings (*Fagopyrum esculentum*) in Japan (Takimoto 1935). Afterwards, it was repeatedly isolated from the same host in Japan (Watson 1971, Baten et al. 2014) as well as in India (Joshi and Paroda 1991) and Korea (Cho and Shin 2004).

10. *Phytopythium helicoides* (Drechsler) Abad, de Cock, Bala, Robideau, Lodhi and Lévesque 2014
    Synonymy:
    Pythium helicoides Drechsler, Journal of the Washington Academy of Sciences 20: 413 (1930)
    Ovatisporangium helicoides (Drechsler) Uzuhashi, Tojo and Kakish., Mycoscience 51 (5): 360 (2010)
    Phytophthora fagopyri S. Takim. ex S. Ito and Tokun., Transactions of the Sapporo Natural History Society 14 (1): 15 (1935)
    Description: Sporangia subglobose or obovoid, terminal, measuring 27-38×16-27 μm (av. 31×21) μm. Oospores aplerotic measuring 27-32 (av. 30.5) μm. Antheridia 1-4 per oogonium, elongate, applied length wise or laterally to the oogonium.

Distribution and hosts: Initially, it was isolated from *Dahlia* sp. in USA (Dreshsler 1930). It was also isolated from different substrates collected from Oubeira Lake, Algeria (Kachour et al. 2016) and soybean in Buenos Aires, Argentina (Grijalba et al. 2018). In Australia, it was also associated with *Gossypium hirsutum*, damping-off of *Medicago sativa* (Simmonds 1966), seedlings of *Gossypium* sp., *Medicago sativa* and *Pinus elliottii* (Teakle 1960) and Ord River irrigation area in Kununurra, Western Australia (Zappia et al. 2012). It was also frequently recovered from hydroponic substrates and water in Brazil (de Jesus et al. 2016, Goncalves et al. 2016). In China, it causes rhizome rot of Asian lotus *Nelumbo nucifera* (Yin et al. 2016), root and collar rot on kiwifruit (*Actinidia chinensis*) (Wang et al. 2015), stem rot of citrus mandarin (Chen et al. 2016), stem rot of Shatangju mandarin (*Citrus reticulate*) seedlings (Chen et al. 2016). It was also isolated from soils (Ho et al. 2012) as well as from leaves of lotus (*Nelumbo nucifera*) (Chen and Kirschner 2018) and Victoria (Longwood hybrid) (Ho 2013), soil, leaves and sea water from Hainan Island (Ho et al. 2012). It was also found associated with oil palm *Elaeis guineensis* in Colombia (Azni et al. 2017), *Sarracenia flava* in Czech Republic (Mrázková et al. 2018), with declining reed stands (*Phragmites australis*) in Germany (Nechwatal et al. 2008), and root rot and leaf blight of elephant foot yam (*Amorphophallus paeonifolius*) in India (Guha and Hong 2008). In Japan, it was isolated from water samples (Baten et al. 2013), whereas it also caused root rot in *Rosa* sp. (Kobayashi 2007), soybean seedlings (Kato et al. 2013), kalanchoe (Watanabe et al. 2007), miniature roses (Kageyama et al. 2002), roses (Kageyama et al. 2003), and strawberry (Watanabe et al. 2005, Suzuki et al. 2005), root and stem rot of chrysanthemum (Tsukiboshi et al. 2007) and wilt and root rot of poinsetta (Miyake et al. 2014). It was also isolated from indigenous forests in Kenya (Mukundi 2011), in Malaysia from *Glycine* max, *Hibiscus* sp. and *Piper nigrum* (Liu 1977), from *Hibiscus rosa-sinensis* and *Piper nigrum* (Liu 1977). It was also recovered from soybean seedlings in North America (Rojas et al. 2017), submerged plants of *Myriophyllum verticillatum* and *Ceratophyllum submersum* (Czeczuga et al. 2005), forest water courses in Poland (Oszako et al. 2016), from *Fragaria ananatis* and *Vitis vinifera* in South Africa (McLeod et al. 2009) as well as South African grape vine nurseries and yards (McLeod et al. 2009, Spies et al. 2011). It was also reported from Taiwan (Ho 2013).

In USA, *P. helicoides* was repeatedly isolated from a large number of hosts such as *Citrullus vulgaris, Pisum sativum, Spinacia oleracea* (Middleton 1943), soil and roots of *Ilex* sp., *Pyracantha, Photinia* and *Azalea* sp. (Hendix and Campbell 1966, 1970), pecan trees (Hendix and Powell 1968), sugar-cane rootlets (Rands and Dopp 1938), *Phaseolus vulgaris,* kidney bean, carnation, nemesia, soybean, kalanchoe, dianthus, curcuma, peanut, cianthus, rose, chrysanthemum, strawberry, *Erica Formosa,* begonia and poinsettia (Li et al. 2014), root rot of *Buxus sempervirens* cv. Suffruticosa (Grand 1985), *Capsicum annuum* (Chellemi et al. 2000), fruit rot of *Citrullus vulgaris, Spinacia oleracea, Saccharum officinarum,* leaf rot of *Typha latifolia* (1960), damping-off of *Phaseolus vulgaris* (Levesque et al. 1998, USDA-ARS 2016, Jung et al. 2017), root rot of rau ram (*Polygonum odoratum*) (Rosskopf et al. 2005), from *Hydrangea arborescens, Leucanthemum* × *superbum* (Shasta daisy), *Thymus praecox* (creeping thyme), *Tagetes patula* (DeMott 2015), seedlings of *Glycine max* (Broders et al. 2007), *Spinacia oleracea* (French 1989), root rot of pistachio rootstock *Pistacia atlantica-integerrima* (Fichtner et al. 2016), from floricultural greenhouses (Haro 2014), Pennsylvania greenhouse irrigation water tanks (Choudhary et al. 2016), irrigation water used for fruit and vegetable crops in New York State (Jones et al. 2014), and ornamental plants (Woodward and DeMott 2014). In Vietnam, it was associated with root rot diseases of *Acacia mangium* and *Acacia* hybrid (Thu 2016).

11. *Phytopythium indigoferae* (E.J. Butler) P.M. Kirk, Index Fungorum 280: 1 (2015)
Synonymy:
Pythium indigoferae E.J. Butler, Memoirs of the Department of Agriculture India 1(5): 73 (1907)
Ovatisporangium indigoferae (E.J. Butler) Uzuhashi, Tojo & Kakish., Mycoscience 51(5): 360 (2010)

Nematosporangium indigoferae (E.J. Butler) Sideris, Mycologia 23(4): 290 (1931)

Description: Sporangia originally reported as inflated filamentous, but it has mostly been mistaken or due to mix culture. Oogonia smooth, terminal, measuring 10-20 µm. Oospores plerotic, up to 18 µm. Antheridia monoclinous, occasionally diclinous, antheridial cell very broad and irregular but making narrow apical contact with oogonium (Butler 1907).

Distribution and hosts: *P. indigoferae* was originally isolated from *Indigofera arrecta* in Calcutta, India (Butler 1907). In Brazil, it was also isolated from *Luffa acutangula* (Chou and Schmitthenner 1974) and *Stapelia grandiflora* (Mendes et al. 1998) as well as root rot of pineapple (*Ananas comosus* L.) and agricultural fields (Milanez et al. 2007, Pereira 2008, Pereira and Roch 2008, Forzza et al. 2010). In China, it was reported from *Brassica* sp. (Yu 1998), soil, leaves and fresh water from Hainan Island of South China (Ho et al. 2012) and from Hainan Islands (Ho 2013). It was repeatedly reported from India (Sydow and Butler 1907, Butler and Bisby 1931), from *Cucumis sativus* (Misra and Hall 1996) and *Indigofera arrecta* (Misra and Hall 1996, Pande and Rao 1998). In Poland, it was isolated from *Menyanthes trifoliata* (Czeczuga et al. 2007) as well as water submerged plants of *Fontinalis antipyretica*, *Potamogeton filiformis* and *Ceratophyllum submersum* (Czeczuga et al. 2005). It was also isolated from common bean *Phaseolus vulgaris* causing root rot in Rwanda (Nzungize et al. 2011) in Kenya, Rwanda and Uganda (Nzungize et al. 2012) and root rot of *Ananas comosus* in USA (Anonymous 1960, Raabe et al. 1981, USDA-ARS 2016).

12. *Phytopythium iriomotense* Baten and Kageyama 2015

Description: Sporangia terminal, globose to sub-globose, 27-42×18-31 (av. 32.7×25.8) µm. Oogonia smooth, terminal, rarely intercalary, measuring 18-32 (av. 27.3) µm. Antheridia 1-2 per oogonium, diclinous occasionally amphigynous, crook-necked making apical contact with oogonium. Oospores aplerotic as well as plerotic, measuring 17-29 (av. 24.4) µm.

Distribution and hosts: It was isolated from river water of Iriomote Island, Okinawa Prefecture, Japan (Baten et al. 2015).

13. *Phytopythium kandeliae* (H.H. Ho, H.S. Chang and S.Y. Hsieh) A.V. Marano, A.L. Jesus and C.L.A, Pires-Zottarelli, Mycosphere 5(4): 518 (2014)

Synonymy:

Halophytophthora kandeliae H.H. Ho, S.Y. Hsieh and H.S. Chang, Abstracts IMC-4, Regensburg (1990)

Halophytophthora kandelii H.H. Ho, S.Y. Hsieh and H.S. Chang (1990)

Halophytophthora kandeliae H.H. Ho, H.S. Chang and S.Y. Hsieh, Mycologia 83: 419 (1991)

Phytopythium kandeliae (H.H. Ho, H.S. Chang and S.Y. Hsieh) Thines, European Journal of Plant Pathology 138: 435 (2014)

Description: Sporangia obovate to turbinate, measuring (25-)38(-49)×(20-)29(-36) µm. Sexual structures such as oogonium, oospores and antheridium not observed.

Distribution and hosts: Originally, this species was isolated from fallen leaves of mangrove plants *Kandelia candel* in Taiwan (Ho et al. 1991). Afterwards, it was also isolated from different mangrove species from various places. In Brazil, it was isolated from leaves of *Laguncularia racemosa* collected from Perequê river (Marano et al. 2014). In Japan, it was isolated from submerged leaves of another mangrove species *Rhizophora stylosa* (Nakagiri 2000). In Philippines, it was found associated with leaves of *Avicennia lanata, Rhizophora apiculata* and *Sonneratia* sp (Leafio 2001). It was isolated from tropical mangroves in Singapore (Tan and Pek 1997) as well as in USA (Newell and Fell 1995).

14. *Phytopythium leanoi* R. Bennett and Thines 2017

Description: Sporangia obovoid to pyriform, measuring 22-34×27-43 (av. 28×35) µm. Oogonia smooth, lateral, terminal or intercalary, measuring 36-46 (av. 41) µm. Oospores plerotic, occasionally aplerotic, measuring 31-41 (av. 36) µm. Antheridia monoclinous or diclinous, one occasionally two per oogonium, elongate or cylindrical, constricted, antheridial cells lobate, laterally applied to the oogonium.

Distribution and hosts: This recently described species was isolated from mangrove leaf litter, collected from Pagbilao Mangrove Forest, Philippines (Bennett et al. 2017).

15. *Phytopythium litorale* **(Nechw.) Abad, de Cock, Bala, Robideau, Lodhi and Lévesque 2014**
Synonymy:
Pythium litorale Nechw., FEMS Microbiology Letters 255: 99 (2006)
Ovatisporangium litorale (Nechw.) Uzuhashi, Tojo and Kakish., Mycoscience 51 (5): 360 (2010)
Description: Sporangia terminal, subglobose, broad ovoid or ob-pyriform, measuring 22.8-32.2×20.8-29.1 (av. 28.5×26.7) μm. Hyphal swellings globose, terminal or intercalary, on average 28.8 μm in diam. Oogonia and antheridia production not observed.

Distribution and hosts: *P. litorale* was originally described by Nechwatal and Mendgen (2006) from reed stands *Phragmites australis* in Germany. In Czech Republic, it was isolated from *Fagus sylvatica, Alnus glutinosa, Tilia europaea, Quercus robur, Juniperus* sp.*, Malus domestica, Prunus domestica, Pyrus* sp.*, Prunus domestica* 'Althanova renklóda' and water course and water tanks (Mrazkova et al. 2018), and from fruit orchards (Hrabetova et al. 2017). It was isolated from declining reed stands (*Phragmites australis*) in Germany (Nechwatal et al. 2008), from *Juncus* sp. (Bouket et al. 2016), rice in Iran (Salmaninezhad and Ghalamfarsa 2017), water samples in Japan (Baten et al. 2013), and narrow-leaved ash (*Fraxinus angustifolia*) in Croatia (Kranjec et al. 2017). It was recovered from a wide variety of habitats in Pakistan such as rhizosphere of spinach, date palm, sea weeds, irrigation ponds and canals, water courses as well as fish ponds (Lodhi 2007, Abdul Haq 2015). It was also isolated from cherimoya *Annona cherimola* in Peru (Quiros and Jose 2018), grapevines nurseries and yards in South Africa (McLeod et al. 2009, Spies et al. 2011), laurel forest of Canary Islands in Spain (Beltrán et al. 2012), and Beskid's rivers in Poland (Matsiakh et al. 2012). It was also found associated with conifer (*Cupresso-cyparis*) root rot and dieback in UK (Wedgwood 2011). In USA, it was isolated from *Cucurbita* sp. (Parkunan and Pingsheng 2013), soybean (Coffua et al. 2016, Radmer et al. 2017, Rojas et al. 2017), *Rosmarinus officinalis* (DeMott 2015), ornamental plants (Woodward and DeMott 2014), floricultural greenhouses (Haro 2014), greenhouse irrigation water tanks (Choudhary et al. 2016) and irrigation ponds used for vegetable production (Parkunan and Pingsheng 2013).

16. *Phytopythium megacarpum* **B. Paul ex Uzuhashi, Tojo and Kakish. ex P.M. Kirk 2015**
Synonymy:
Ovatisporangium megacarpum (B. Paul) Uzuhashi, Tojo and Kakish., Mycoscience 51 (5): 360 (2010)
Pythium megacarpum B. Paul, FEMS Microbiology Letters 186 (2): 231 (2000)
Description: Sporangia spherical to ovoid, terminal to sub-terminal, germinating directly by germ tubes, zoospores not observed. Oogonia smooth, terminal or intercalary, measuring 20-45 (av. 28) μm. Oospores plerotic, measuring 19-36 (av. 26.8) μm. Antheridia monoclinous, branched, elongated, mostly applied lengthwise to oogonium making apical contact. Oospore wall up to 7 (av. 5) μm thick.

Distribution and hosts: It was originally isolated from a soil sample taken from a wheat field in Lille in northern France (Paul 2000). It was frequently isolated from soybean in USA (Radmer et al. 2017, Rojas et al. 2017, Brenk et al. 2016) and in North America (Rojas et al. 2017) as well as from strawberry in USA (Shennan et al. 2018).

17. *Phytopythium mercuriale* **(Belbahri, B. Paul and Lefort) Abad, de Cock, Bala, Robideau, Lodhi and Lévesque 2014**
Synonymy:
Pythium mercuriale Belbahri, B. Paul and Lefort, FEMS Microbiology Letters 284 (1): 17-27 (2008)

Ovatisporangium mercuriale (Belbahri, B. Paul and Lefort) Uzuhashi, Tojo and Kakish., Mycoscience 51 (5): 360 (2010)

Description: Sporangia subglobose to obovoid, 18-23×23-32 (av. 22×26) μm. Oogonia smooth, 22-37 (av. 29.8) μm, terminal or lateral on short branches. Antheridia multiple, mostly diclinous, knotted around the oogonia. Oospores plerotic.

Distribution and hosts: Originally, it was isolated from multiple locations and habitats such as grapevine (*Vitis vinifera*) and macadamia nut (*Macadamia integrifolia*) in South Africa, *Quercus ilex* in Spain, soil in Poland and France (Belbahri et al. 2008). Afterwards, it was isolated from diseased soybean seedlings in Japan (Kato et al. 2013), declining forests in Poland (Cordier et al. 2009), grapevine nurseries and yards in South Africa (McLeod et al. 2009, Spies et al. 2011), forest in Spain (Beltrán et al. 2012), citrus trees infected with gummosis disease in Tunisia (Benfradj et al. 2017), soybean, floricultural greenhouses and irrigation water in USA (Haro 2014, Jones et al. 2014, Rojas et al. 2017).

18. *Phytopythium mirpurense* Lodhi, De Cock, Lévesque and Shahzad 2014

Description: Sporangia subglobose, limoniform, obovoid or ovoid, measuring 20-25 μm. Oogonia smooth, terminal or intercalary, measuring (27-)34-37(-40) (av. 34) μm. Oospores aplerotic or nearly plerotic, measuring (22-)29-32(-34) (av. 29.45) μm. Antheridia 1-3 per oogonium, mostly monoclinous occasionally diclinous. Antheridia apply lengthwise to the oogonium producing lateral or occasionally apical fertilization tubes.

Distribution and hosts: It was isolated from water samples in Pakistan (de Cock et al. 2015).

19. *Phytopythium montanum* (Nechw.) Abad, de Cock, Bala, Robideau, Lodhi and Lévesque 2014

Synonymy:

Pythium montanum Nechwatal, Mycological Progress 2 (1): 79 (2003)

Ovatisporangium montanum (Nechw.) Uzuhashi, Tojo and Kakish., Mycoscience 51 (5): 360 (2010)

Description: Sporangia subglobose, broad ovoid or ob-pyriform, measuring 19-29×17-25 (av. 24.0×21.1) μm. Oogonia smooth, mostly terminal, measuring 21.2-27.2 (av. 24.3) μm. Oospores strictly aplerotic, measuring 16.1-21.3 (av. 18.7) μm. Antheridia monoclinous, occasionally diclinous, mostly one rarely two per oogonium, variable in size and attachment, either elongated, lobate, curved and wavy in contour and applied lengthwise to the oogonium, or globose to clavate or crook-necked and making blunt apical contact to the oogonium.

Distribution and hosts: This species was originally isolated from rhizosphere soil of declining spruce stand (*Picea abies*) in Bavarian Alps, Germany (Nechwatal and Oßwald 2003). It was also reported from conifer species (*Taxus baccata*) root rot and dieback in UK (Wedgwood 2011).

20. *Phytopythium oedochilum* (Drechsler) Abad, de Cock, Bala, Robideau, Lodhi and Lévesque 2014

Synonymy:

Pythium oedochilum Drechsler, J. Wash. Acad. Sci. 20: 414. (1931).

Ovatisporangium oedochilum (Drechsler) Uzuhashi, Tojo and Kakish., Mycoscience 51: 360. (2010) (as 'oedichilum').

Description: Sporangia subglobose, limoniform, obovoid or ovoid. Oogonia smooth, terminal, occasionally intercalary, measuring (22-)31-36(-39) (av. 32.8) μm. Oospores aplerotic, (20-)28-34 (-38) (av. 30.3) μm. Antheridia diclinous, occasionally monoclinous, 1-2(-4) per oogonium, curved, elongate, often wavy in contour, applied lengthwise to the oogonium.

Distribution and hosts: *P. oedochilum* was first described by Drechsler (1930) from USA, he isolated it from *Dahlia* sp. Afterwards it was isolated from different countries where it has been associated with rhizosphereic soil of *Eucalyptus* spp. (Pratt and Heather 1973), rice seedlings (Cother and Gilbert 1993), *Melaleuca ericifolia* (Sampson and Walker 1982) in Australia, soil in

China (Ho et al. 2012), mango, peppers and gombo in the Congo Republic (Ravisé et al. 1969), *Brassica oleracea* in India (Ragunathan 1968, Sarbhoy et al. 1971, Misra and Hall 1996), and *Cirsium* sp. in Iran (Bouket et al. 2016). In Japan, it was frequently associated with stunt disease of strawberry (*Fragaria* sp.) (Takahashi and Kawase 1965, Watanabe et al. 1977), root rot of yakon (*Polymnia sonchifolia*) (Ohgami and Kageyama 2005), and root and stem rot of *Chrysanthemum* (Tsukiboshi et al. 2007). It was also isolated from soil in Netherlands (Plaats-Niterink 1975), mud and algae collected from the pond, rhizosphere of mustard, rhizosphere of gram and from weeds in Pakistan (Lodhi 2007), and from rice seedlings in Philippines (Kessank 1987). In Poland, it was isolated from different sources such as from water of river, pond, springs and lakes (Czeczuga and Snarska 2001), water submerged feathers of white-tailed eagle *Haliaeetus albicilla*, water bodies (Czeczuga et al. 2004), and *Nitella mucronata* submerged in water (Czeczuga et al. 2005). It was also reported from *Cynara cardunculus* in South Africa (McLeod et al. 2009), and from Taiwan (Ho 2013). It is widely distributed in USA, isolated from soil (Hendrix and Campbell 1970), root rot of *Bidens* sp., *Dahlia variabilis* and damping-off of *Matthiola incana* (Anonymous 1960), *Citrullus vulgaris* (Drechsler 1939), *Bidens aristosa* (Drechsler 1941), *Citrullus vulgaris*, *Daphne odora* and *Matthiola incana* (Middleton 1943), floricultural greenhouses (Haro 2014) and from surface water used for irrigation of fruit and vegetable crops (Jones et al. 2014).

21. *Phytopythium ostracodes* (Drechsler) Abad, de Cock, Bala, Robideau, Lodhi and Lévesque 2014

Synonymy:

Pythium ostracodes Drechsler, Phytopathology 33: 286 (1943)

Ovatisporangium ostracodes (Drechsler) Uzuhashi, Tojo and Kakish., Mycoscience 51 (5): 360 (2010)

Description: Sporangia terminal, subglobose, measuring 25-55×16-38 μm. Oogonia smooth, intercalary or terminal, measuring (14-)35-38(-43) (av. 35) μm. Oospores plerotic or nearly plerotic, measuring (13-)30-34(-41) (av. 32.5) μm. Antheridia monoclinous, occasionally diclinous, 1-2 per oogonium, antheridial cells laterally applied or lengthwise to the oogonium.

Distribution and hosts: *P. ostracodes* was first isolated from wheat in Texas (Drechsler 1943). It was also found to be associated with various soil samples in China (Ho et al. 2012), *Citrus nobilis* in Egypt (Anonymous 2019), cultivated soils in France (Paul 1994) and Great Britain (Anonymous 2019). In Iran, it was isolated from diseased sugarbeet seedlings (Babai-Ahary et al. 2004), rice (Salmaninezhad and Ghalamfarsal 2017) and cultivated soil of Fars province (Ghalamfarsa and Banihashemi 2005). It was also isolated from rhizomes of lotus (Takahashi and Kawase 1965), *Nelumbo nucifera* in Japan (Chen and Kirschner 2018), soil of indigenous forests and *Zea mays* soil in Kenya (Mukundi et al. 2009, Mukundi 2011), from soil of guava, chillies, mustard, lake and irrigation water in Pakistan (Lodhi et al. 2005, Lodhi 2007). In Poland, it was associated with *Rorippa amphibian*, dead benthos crustaceans, muscles of fishes, *Juncus effusus* (Czeczuga et al. 1999, 2002, 2007). It was also reported from soil in Spain (Asano et al. 2010) and Switzerland (Ko et al. 2010), from *Rosmarinus officinalis* in Taiwan (Ting et al. 2015), soybean in Thailand (Vannarug 1993), *Gossypium hirsutum* and *Triticum aestivum* in USA (Sprague 1950, Anonymous 1960, Mitchell 1978, Alfieri et al. 1984).

22. *Phytopythium palingenes* (Drechsler) Abad, de Cock, Bala, Robideau, Lodhi and Lévesque 2014

Synonymy:

Pythium palingenes Drechsler, Journal of the Washington Academy of Sciences 20: 416 (1930)

Description: Sporangia terminal, subglobose or ovoid, measuring 24-42×18-33 (av. 33×29) μm. Oogonia smooth, terminal or intercalary, unilaterally intercalary, measuring (19-)28-40(-41) (av. 34) μm. Oospores aplerotic, measuring (18-)26-36(-37) (av. 31.3) μm. Antheridia diclinous, (1-)2(-4) per oogonium, antheridial cells cylindrical, often wavy or irregular in contour, applied lengthwise to the oogonium over its entire length.

Distribution and hosts: *P. palingenes* was originally isolated in USA from discoloured roots of *Ambrosia trifida* and later from discoloured roots of *Senecio jacobaea* (Drechsler 1941). It was very frequently isolated from different water samples in Brazil (Rocha et al. 2001, Rocha 2002, Milanez et al. 2007, Negreiros 2008, Forzza et al. 2010, Trindade-Junior 2013, de Jesus et al. 2016, Rios and Rocha 2018, Trindade-Junior and Rocha 2018) as well as root rot of cassava (Matias 2012). In Poland, it was found to be associated with *Butomus umbellatus* and *Lathyrus palustris* (Czeczuga et al. 2007) as well as water submerged plant species such as *Hippuris vulgaris* f. *submersa*, *Myriophyllum verticillatum*, *Potamogeton crispus*, *Potamogeton filiformis* and *Potamogeton rutilus* (Czeczuga et al. 2005) and water of river, pond, springs and lakes (Czeczuga and Snarska 2001). It was also reported from Taiwan causing root rot of *Tibouchina semidecandra* (Huang 2008) and isolated from other sources (Ho 2009, 2013). In USA, it was isolated from *Ambrosia trifida*, root necrosis of *Prunella vulgaris* (Anonymous 1960) and infected buckwheat (Mihail 1993).

23. *Phytopythium polytylum* (Drechsler) Abad, de Cock, Bala, Robideau, Lodhi and Lévesque 2014

Synonymy:

Pythium polytylum Drechsler, Journal of the Washington Academy of Sciences 20: 415 (1930)

Description: Sporangia terminal, rarely intercalary, subglobose, measuring 28-33 µm. Oogonia smooth, terminal or lateral, measuring (26-)29-37(-40) (av. 32.5) µm. Oospores aplerotic, (23-) 25-33(-35) (av. 28.8) µm. Antheridia diclinous, occasionally monoclinous, 1(-4) per oogonium, curved, elongate or cylindrical, mostly wavy in contour, applied lengthwise to the oogonium.

Distribution and hosts: *P. polytylum* was originally isolated from decaying roots of *Prunella vulgaris* in the USA (Drechsler 1930). In India, it was found associated with *Malus* sp. (Misra and Hall 1996) and collar rot of apple (Rana and Gupta 1984). It was also isolated from different ornamentals in Mexico (Trigos et al. 2006), cultivated fields in Nepal (Verma and Khulble 1985), *Spinacia oleracea* in USA (French 1989, Middleton 1943), root necrosis of *Prunella vulgaris* (Anonymous 1960) and stem and root rot of *Chrysanthemum morifolium* (Lumsden and Haasis 1964).

24. *Phytopythium sindhum* Lodhi, Shahzad and Lévesque 2010

Description: Sporangia subglobose, terminal, occasionally intercalary, or unilaterally intercalary, measuring 15×20-35×40 µm. Oogonia smooth, laterally on a short stalk, occasionally terminal and unilaterally intercalary, 30-39 (av. 34.5) µm diam. Oospores smooth, mostly plerotic or nearly plerotic, occasionally aplerotic, 30-38 (av. 34) µm, mostly one occasionally two per oogonium. Antheridia diclinous or monoclinous, variable in shape and attachment, if elongate, applied lengthwise, if crook necked, making narrow apical contact with the oogonium.

Distribution: It was isolated from Pakistan in banana rhizosphere (Bala et al. 2010, Lodhi 2007) and rice rhizosphere (Abdul Haq 2015).

25. *Phytopythium sterile* Belbahri and Lefort ex Uzuhashi, Tojo and Kakish. ex P.M. Kirk 2015

Synonymy:

Pythium sterile Belbahri and Lefort: 210 (2006)

Pythium sterilum Belbahri and Lefort, FEMS Microbiology Letters 255 (2): 210 (2006)

Description: Sporangia terminal or intercalary, globose or pyriform, measuring 16-32 (av. 25.7) µm. Sexual structures not observed.

Distribution and hosts: It was originally isolated from *Quercus ilex* stand in Spain, from soil samples of an alder stand in Poland, from soil samples of vineyard in France (Belbahri et al. 2006), alder (*Alnus* sp.) in USA (Adams et al. 2009) and from rice in Iran (Salmaninezhad and Ghalamfarsal 2017).

26. *Phytopythium vexans* (de Bary) Abad, de Cock, Bala, Robideau, Lodhi and Lévesque 2014
Synonymy:
Pythium vexans de Bary, Journal of the Royal Agricultural Society of England 12: 255 (1876)
Ovatisporangium vexans (de Bary) Uzuhashi, Tojo and Kakish., Mycoscience 51 (5): 360 (2010)
Pythium complectens M. Braun, Journal of Agricultural Research 29: 415 (1924)
Pythium allantocladon Sideris, Mycologia 24 (1): 27 (1932)
Pythium ascophallon Sideris, Mycologia 24 (1): 29 (1932)
Pythium polycladon Sideris, Mycologia 24 (1): 32 (1932)
Pythium euthyhyphon Sideris, Mycologia 24 (1): 34 (1932)
Pythium piperinum Dastur, Proceedings of the Indian Academy of Sciences Section B 1 (11): 803 (1935)

Description: Sporangia intercalary or terminal, subglobose, ovoid or pyriform, measuring 18-23×15-21 (av. 20.2×17.9) μm. Oogonia smooth, terminal occasionally lateral or intercalary, measuring (16-)18-23(-24) (av. 20.0) μm. Antheridia monoclinous, occasionally diclinous, 1 occasionally 2 per oogonium, antheridial cells large, bell-shaped. Oospores aplerotic, (14-)16-19 (-20) (av. 17.3) μm.

Distribution and hosts: *Pythium vexans* was first described by de Bary in 1876 from Paris. *P. vexans* is worldwide in distribution and it has been isolated from different regions of the world. It was isolated from different substrates collected from Oubeira Lake in Algeria (Kachour et al. 2016) and from ornamentals and *Citrullus vulgaris* in Argentina (Frezzi 1956). In Australia, it was associated with *Casuarina* sp. (Boa and Lenné 1994), durian decline (Vawdrey et al. 2005), tomato seedlings (Kerling 1947), pawpaw, *Annona*, *Brassica* and *Persea* (Teakle 1957, 1960), dieback of *Parkinsonia aculeata* (Steinrucken et al. 2017), with *Actinidia chinensis*, *Trifolium subterraneum* and *Passiflora edulis* (Shivas 1989), *Carica papaya*, *Hakea purpurea*, *Leptospermum flavescens* and *Callistemon pinifolius* (Simmonds 1966), *Citrus sinensis*, *Malus sylvestris*, *Pinus radiata* and *Triticum aestivum* (Cook and Dube 1989). In Brazil, it was isolated from *Strelitzia* sp. (Carvalho 1965), water samples collected from Parque Nacional de Sete Cidades (Rocha et al. 2001), *Solanum tuberosum*, *Solanum melongena*, *Solanum gilo*, *Saccharum officinarum*, *Capsicum* sp. (Mendes et al. 1998, Rocha 2002, Milanez et al. 2007, Pereira 2008, Trajano 2009, Forzza et al. 2010), from hydroponic system and water samples (Goncalves et al. 2016, Baptista et al. 2004). It was also associated with seedling blight and root rot *Piper nigrum* and *Zea mays* in Brunei Darussalam (Peregrine and Ahmad 1982), and also isolated from *Cucumis vexans*, *Lupinus* sp., *Lycopersicon esculentum* and *Dianthus caryophyllus* causing root rot in Bulgaria (Vanev et al. 1993, Ilieva 1995), in Cameroon (Charlemagne and Xu-Tong 1997), *Pinus silvestris*, *Quercus robur* and *Pisum sativum* in Canada (Vaartaja 1968, Ginns 1986), avocado orchards in Canary Islands (Padron et al. 2018), soil, leaves, fresh water (Ho et al. 2012), *Dendrobium aurantiacum*, *D. chrysanthum*, *D. chrysotoxum*, *D. thyrsiflorum*, *Dioscorea opposita*, *Setaria italica*, *Solanum tubersoum*, *Sorghum vulgaris*, *Zea may* (Ho 2013), *Solanum tuberosum* (Yu 1998), from brown root rot on ramie *Boehmeria nivea* (Yu et al. 2016), patch canker of rubber trees *Hevea brasiliensis* (Zeng et al. 2005), stem rot of *Dendrobium* sp. in China (Tao et al. 2011), avocado wilt complex disease in Colombia (Ramírez-Gil 2018) and from various cultivated plants (Ravisé et al. 1969).

In Czech Republic, it was associated with *Cypripedium calceolus*, *Alnus glutinosa*, *Juniperus horizontalis*, *Fagus sylvatica*, *Fagus* sp., *Picea omorika*, *Rhododendron hybridum*, *Tilia cordata*, *Rhododendron* sp., *Tilia europaea*, *Aesculus hippocastanum*, *Prunus cerasifera nigra*, *Prunus armeniaca*, *Acer pseudoplatanus*, *Prunus* sp., *Malus* spp. (Mrázková et al. 2018), fruit orchards (Hrabětová et al. 2017) and soil (Cejp 1962).

P. vexans was also isolated from cereals and soil in England (Salt 1977, Butler 1907), *Cocos nucifera* in Fiji (Firman 1972, Dingley et al. 1981), *Solanum tuberosum* (de Bary 1876, 1881)

Lupinus sp. and *Medicago* sp. (Schultz 1939) and soil in Germany (Remy 1950), *Humulus lupulus* in Greece (Pantidou 1973), soil in Guatemala (Tejada 1983), *Chrysanthemum × morifolium* in Hong Kong (Lu et al. 2000). In India, it is reported from *Zingiber, Eletteria, Cinchona*, apple grafts, *Pelargonium* and *Piper* spp. (Park 1936, Ramakrishna 1949, Butler and Bisby 1960, Rao 1963), potato (Cunningham 1897) and *Aframomum* sp. (Wilson and Rahim 1978), *Cinchona officinalis* and *Cinchona pubescens* (Spaulding 1961), *Atramomum melegueta, Cinchona ledgeriana, Cinchona officinalis, Cinchona* sp., *Cinchona succirubra, Elettaria cardamomum, Eucalyptus tereticornis, Malus* sp., *Pelargonium* sp., *Piper betle, Piper longum, Pyrus malus* and *Zingiber officinale* (Misra and Hall 1996).

It was isolated from durian (*Durio* sp.) in Indonesia (Santoso et al. 2015), soil (Ghalamfarsa and Banihashemi 2005, Levesque et al. 1998), and *Juglans regia, Lolium perenne, Pinus nigra, Vitis vinifera* in Iran (Ershad 1977, Ghaderi and Banihashemi 2008, 2011, Marvasti and Banihashemi 2010), from soil in Iraq (Al-Doory et al. 1959), Iceland (Johnson 1971), from cultivated and uncultivated soil (Uzuhashi et al. 2010, Watanabe 1989, Bandoni and Barr 1976, Takahashi 1973) and water samples in Japan (Baten et al. 2013), from *Nicotiana* and soil in Java (Raciborski 1900), root rot of common bean in Kenya, Rwanda and Uganda (Nzungize et al. 2012), soil in Lebanon (Ahrens 1971) and Madeira (Hohnk 1962). In Malaysia, it was isolated from *Theobroma, Hevea, Thea, Camellia, Dianthus, Carica, Citrus, Piper, Saccharum, Vicia, Vigna, Vitis, Zingiber* and a number of ornamentals (Thompson 1929, 1936, 1939, Sharples 1930, Thompson and Johnston 1953, Johnston 1960, Liu 1977), damping-off and root rot of *Hevea brasiliensis, Anacardium occidentale, Piper nigrum* and *Vigna sinensis* (Williams and Liu 1976), *Carica papaya, Citrus nobilis, Begonia* sp., *Eugenia* sp., *Fortunella* sp., *Elaeis guineensis, Manihot esculenta, Sauropus androgynus, Syzygium aromaticum, Theobroma cacao, Vicia faba, Vitis vinifera, Hevea brasiliensis, Pinus taiwanensis, Saccharum officinarum* and *Zingiber officinale* (Liu 1977), and *Durio zibethinus* (Spaulding 1961).

It was also found to be associated with *Linum* (Diddens 1932) and *Hydrangea* in Netherlands (Plaats-Niterink 1975), *Malus domestica* (Pennycook 1989) and *Malus domestica* (Gadgil et al. 2005) and soil in New Zealand (Robertson 1980), cocoa farm soils in Nigeria (Aigbe and Woodward 2018), soybean seedlings in North America (Rojas et al. 2017), from citrus and betelvine field (Lodhi 2007) and from *Dianthus* in Pakistan (Azeem 1968), soil in Palestine (Ali-Shtayeh et al. 2003), Panama (Davidson et al. 2000), *Cocos* and *Theobroma* (Stamps et al. 1972), *Ananas comosus, Cocos nucifera, Musa* sp., *Theobroma cacao, Xanthosoma* sp. in Papua New Guinea (Shaw 1984), citrus, avocado and mango in Peru (Quiros and Jose 2018), tomato and potato (Savulescu 1940), root rot of common bean *Phaseolus vulgaris* in Rwanda (Nzungize et al. 2011), citrus, pawpaw and avocado (Wager 1932, 1941, 1942), grapevines nurseries and yards (McLeod et al. 2009, Spies et al. 2011), *Carica papaya, Persea Americana, Limonium vulgare, Prunus persica* (Crous et al. 2000), *Encephalartos middelburgensis, Acacia xanthophloea, Persea americana* in South Africa (McLeod et al. 2009), laurel forest (Beltrán et al. 2012), root rot of avocado in Spain (Rodriguez-Padron et al. 2018), soil in Sri Lanka (Rajapakse and Kulasekera 1981), *Pinus luchuensis* and soil in Taiwan (Anonymous 1979, Chang 1993, Ho 2013), *Carica papaya* in Tanzania (Riley 1960), Thailand (Vannarug 1993), citrus trees infected with gummosis disease in Tunisia (Benfradj et al. 2017), root and collar rot of kiwi fruit *Actinidia deliciosa* (Polat et al. 2017), *Abies nordmanniana* sub. sp. *bornmuelleriana, Cupressus sempervirens, Pinus brutia, Pinus nigra, Platanus orientalis, Thuja occidentalis* and *Thuja orientalis* in Turkey (Lehtijarvi et al. 2017), *Pelargonium* in Uganda (Hansford 1938), conifer (Cupressocyparis and Chamaecyparis) root rot and dieback in UK (Wedgwood 2011).

In USA, it was isolated from *Citrus* (Harvey 1944a, Middleton 1943), *Hevea* (Chee 1968), *Metrosideros* (Kliejunas and Ko 1975), *Persea* (Harvey 1944a, b, 1945), *Saccharum officinarum* (Rands 1930, Rands and Dopp 1938, Stevenson and Rands 1938), grasses (Hendrix et al. 1970, Sprague 1950), conifers (Calderone and True 1967, Sutherland et al. 1966), ornamentals (Braun 1924, Middleton 1938, 1943, Hendrix and Campbell 1966, Biesbrock and Hendrix 1970a), pear trees (Royle and Hickman 1964), peach (Lorio 1966, Mircetich and Fogle 1969, Mircetich 1971, Biesbrock and Hendrix 1970b), pecan trees (Hendrix et al. 1966, Hendrix and Powell 1968), soil (Meredith 1940, Middleton 1943, Scott 1960, Nichols et al. 1964, Powell et al. 1965, Lorio 1966, Campbell and Hendrix 1967, Vaartaja 1968, Benjamin and Slot 1970, Hendrix and Campbell 1970, Hendrix et al. 1971, Lumsden and Ayers 1975), *Ananas* (Sideris 1931, 1932, Klemmer and Nakano 1964), ornamentals, *Spinacia, Ricinus, Carica* (Sideris 1931, 1932), common bean USA (USDA-ARS 2016), *Citrus nobilis, Citrus* sp., *Dianthus caryophyllus, Pelargonium* × *domesticum, Capsicum frutescens, Coleus* sp., *Antirrhinum majus, Ananas comosus, Delphinium ajacis, Lycopersicon esculentum, Persea borbonia, Pisum sativum, Pelargonium hortorum, Pelargonium graveolens, Phaseolus vulgaris* and *Saccharum officinarum* (Anonymous 1960), floricultural greenhouses (Haro 2014), *Ananas comosus, Allium cepa, Ricinus communis* (Raabe et al. 1981), oil palm *Elaeis guineensis* (Turner 1971), *Citrus* sp., *Coleus* sp., *Clarkia grandiflora, Consolida ambigua, Dianthus caryophyllus, Antirrhinum majus, Antirrhinum* sp., *Delphinium ajacis, Dianthus* sp., *Matthiola incana, Matthiola* sp., *Persea americana* (French 1989), alfalfa *Medicago sativa* (Hancock 1985), *Zea mays* (Hanlin et al. 1978), *Metrosideros collina* sub. sp. *polymorpha, Carica papaya, Spinacia oleracea, Rhododendron 'Jennifer' (Kurume azalea)* and *Sedum* sp., (DeMott 2015), *Anthurium andraeanum* (Guo and Ko 1994), irrigation water (Jones et al. 2014), peanuts *Arachis hypogaea* (Wheeler et al. 2005), *Protea eximia* and *Protea neriifolia* (Crous et al. 2013), ornamental plants (Woodward and DeMott 2014), pear trees *Pyrus communis* (Sprague 1957). It was also isolated from root rot diseases of *Acacia mangium* and *Acacia* hybrid in Vietnam (Thu 2016).

References

Abad-Campos, P., Perez-Sierra, A., Alvarez, L.A., Armengol, J., Lopez-Pineda, R., Sanchez-Perez, A., Rodriguez-Quezada, E. and Alvarez-Valenzuela, G. 2008. A survey of *Phytophthora* and *Pythium* associated to export crops in Guatemala. *Phytophthora/Pythium* and related genera. Third International Workshop, August 23-24, 2008. Turin, Italy. pp. 24-25. http://www.phytophthoradb.org/pdf/015Abad.pdf, modified 24/6/14.

Abdul Haq, M. 2015. Studies on Biodiversity of Oomycetes of Bajaur Agency, Fata, Pakistan. PhD thesis, submitted to Department of Botany, University of Karachi, Karachi, Pakistan.

Adams, G.C., Catal, M. and Trummer, L. 2009. Distribution and severity of alder *Phytophthora* in Alaska. *In*: Proceedings of the sudden oak death fourth science symposium, June 15-18, 2009. Santa Cruz, USA. https://www.fs.fed.us/psw/publications/documents/psw_gtr229/psw_gtr229_029.pdf

Ahrens, C. 1971. Untersuchungen zur taxonomie and nur geographischen verbreitung der gattung *Pythium* Pringsheim. Diss. Univ. Bonn.

Aigbe, S.O. and Woodward, S. 2018. Isolation and identification of Oomycetes species from cocoa farm soils in Nigeria based on PCR analysis. Phytopathol. 108(10): 48-49.

Al-Doory, Y., Tolba, M.K. and Al-Ani, H. 1959. On the fungal flora of Iraqi soils. Mycologia 51: 429-439.

Alfieri Jr., S.A., Langdon, K.R., Wehlburg, C. and Kimbrough, J.W. 1984. Index of plant diseases in Florida (Revised). Florida Dept. Agric. & Consumer Serv. Div. Plant Ind. Bull. 11: 1-389.

Ali-Shtayeh, M.S., Salahl, A.M.A. and Jamous, R.M. 2003. Ecology of hymexazol-insensitive *Pythium* species in field soils. Mycopathologia 156: 333-342.

Anonymous. 2019. Herb. IMI records for fungus *Phytopythium ostracodes* http://www.herbimi.info/herbimi/results.htm?name=Phytopythium%20ostracodes

Anonymous. 1960. Index of plant diseases in the United States. U.S.D.A. Agric. Handb. 165: 1-531.

Anonymous. 1979. List of plant diseases in Taiwan. Plant Protect. Soc. Republ. of China. 404 pages.

Asano, T., Senda, M., Suga, H. and Ageyama K.K. 2010. Development of multiplex PCR to detect five *Pythium* species related to turfgrass diseases. J. Phytopathol. 158: 609-615.

Azeem, S. 1968. Study on the pathogenicity of *Pythium vexans* de Bary, *Phytophthora* sp. and *Phytophthora cactorum*. Pak. J. Sci. Ind. Res. 11: 446-448.

Azni, I.N.A.M., Sundram, S., Ramachandran, V. and Seman, I.A. 2017. An *in vitro* investigation of Malaysian *Phytophthora palmivora* isolates and pathogenicity study on oil palm. J. Phytopathol. 165(11-12): 800-812.

Babai-Ahary, A., Abrinnia, M. and Heravan, I.M. 2004. Identification and pathogenicity of *Pythium* species causing damping-off in sugarbeet in northwest Iran. Aus. Plant Pathol. 33(3): 343-347.

Bala, K., Robideau, G.P., Lévesque, A., de Cock, A.W.A.M., Adad, Z.G., Lodhi, A.M., Shahzad, S., Ghaffar, A. and Coffey, M.D. 2010. *Phytopythium* Abad, de Cock, Bala, Robideau, Lodhi and Lévesque gen. nov. and *Phytopythium sindhum* Lodhi, Shahzad and Lévesque sp. nov. Persoonia 24: 136-137.

Bandoni, R.J. and Barr, D.J.S. 1976. On some zoosporic fungi from washed terrestrial litter. Trans. Mycol. Soc. Japan 17: 220-225.

Baptista, F.R., Pires-Zottarelli, C.L.A., Rocha, M. and Milanez, A.I. 2004. The genus *Pythium* Pringsheim from Brazilian cerrado areas, in the state of São Paulo, Brazil. Braz. J. Bot. 27(2): 281-290.

Baten, M.A., Asano, T., Motohashi, K., Ishiguro, Y., Rahman, M.Z., Inaba, S., Suga, H., Kageyama, K. 2014. Phylogenetic relationship among *Phytopythium* species, and re-evaluation of *Phytopythium fagopyri* comb. nov., recovered from damped-off buckwheat seedlings in Japan. Mycol. Progr. 13: 1145-1156.

Baten, M.A., Suga, H. and Kageyama, K. 2013. Abundance and distribution of *Phytopythium* species in Japan, (Abstract) presented at the 57th annual meeting of the Mycological Society of Japan. https://doi.org/10.11556/msj7abst.57.0_40

Baten, M.D.A., Mingzhu, L., Motohashi, K., Ishiguro, Y., Rahman, S., Suga, H. and Kageyama, K. 2015. Two new species, *Phytopythium iriomotense* sp. nov. and *P. aichiense* sp. nov., isolated from river water and water purification sludge in Japan. Mycol. Progr. 14: 1-12.

Beckman, N., Anderson, G.M. and Kurle, J.E. 2018. Is *Phytopythium boreale* a pathogen of soybeans? (Abstr.). Phytopathol. 108: S2.29. https://doi.org/10.1094/PHYTO-108-12-S2.20

Belbahri, L., Calmin, G., Hernandez, E.S., Oszako, T. and Lefort, F. 2006. *Pythium sterilum* sp. nov. isolated from Poland, Spain and France: Its morphology and molecular phylogenetic position. FEMS Microbiol. Lett. 255: 209-214.

Belbahri, L., McLeod, A., Paul, B., Calmin, G.R., Moralejo, E., Spies, C.F.J., Botha, W.J., Clemente, A., Descals, E., Sánchez-Hernández, E. and Lefort, F. 2008. Intraspecific and within-isolate sequence variation in the ITS rRNA gene region of *Pythium mercuriale* sp. nov. (Pythiaceae). FEMS Microbiol. Lett. 284(1): 17-27.

Beltran, A., Leon, M., Abad-Campos, P., Siverio, F., Gallo, L. and Perez-Sierra, A. 2012. Diversity of oomycetes detected in the laurel forest in Tenerife (Canary Islands). *In*: Phytophthoras in Forests and Natural Ecosystems, Proceedings of the 6th meeting of the International Union of Forest Research Organizations (IUFRO), September 9-14, 2012, Cordoba, Spain.

Benfradj, N., Migliorini, D., Luchi, N., Santini, A. and Boughalleb, N. 2017. Occurrence of *Pythium* and *Phytopythium* species isolated from citrus trees infected with gummosis disease in Tunisia. Arch. Phytopathol. Plant Prot. 50(5-6): 286-302.

Benjamin, C.R. and Slot, A. 1970. Fungi of Haiti. Sydowia 23: 125-163.

Bennett, R.M., Nam, B., Dedeles, G.R. and Thines, M. 2017. *Phytopythium leanoi* sp. nov. and *Phytopythium dogmae* sp. nov., *Phytopythium* species associated with mangrove leaf litter from the Philippines. Acta Mycol. 52(2): 1103. https://doi. org/10.5586/am.

Biesbrock, J.A. and Hendrix, F.F. Jr. 1970a. Influence of soil water and temperature on root necrosis of peach caused by *Pythium* spp. Phytopathol. 60: 880-882.

Biesbrock, J.A. and Hendrix, F.F. Jr. 1970b. Influence of continuous and periodic soil water conditions on root necrosis of holly caused by *Pythium* sp. Can. J. Bot. 48: 1641-1645.

Blackwell, E. 1949. Terminology in *Phytophthora*. Mycol. Papers 30: 1-24.

Boa, E. and Lenne, J. 1994. Diseases of Nitrogen Fixing Trees in Developing Countries. An annotated list. Natural Resources Inst., Kent, United Kingdom. 82 pages.

Bouket, C.A., Babai-Ahari, A., Mahdi, A. and Motoaki, T. 2016. Morphological and molecular characterization of *Phytopythium litorale* and *P. oedochilum* from Iran. Nova Hedwigia 102. 10.1127/nova_hedwigia/2015/0307.

Braun, H. 1924. Geranium stem rot caused by *Pythium complectens* n. sp. Host resistance reactions, significance of *Pythium*-type of sporangial germination. J. Agric. Res. 29: 399-419.

Brenk, V., Anderson, G.A. and Kurle, J.E. 2016. *Pythium* and *Phytopythium* spp. isolated from soybean in Minnesota exhibit symbiotic behaviors ranging from parasitism to mutualism (Abstr.). Phytopathol. 106: S4.185. http://dx.doi.org/10.1094/PHYTO-106-12-S4.184

Briard, M., Dutertre, M., Rouxel, F. and Brygoo, Y. 1995. Ribosomal RNA sequence divergence within the Pythiaceae. Mycol. Res. 99: 1119-1127.

Broders, K.D., Lipps, P.E., Ellis, M.L. and Dorrance, A.E. 2009. *Pythium delawarii* a new species isolated from soybean in Ohio. Mycologia 101(2): 232-238.

Broders, K.D., Lipps, P.E., Paul, P.A. and Dorrance, A.E. 2007. Characterization of *Pythium* spp. associated with corn and soybean seed and seedling disease in Ohio. Plant Dis. 91: 727-735.

Butler, E.J. and Bisby, G.R. 1931. The Fungi of India. Scient. Monogr. Conn. Agric. Res. India 1: 1-237.

Butler, E.J. 1907. An account of the genus *Pythium* and some Chytridiaceae. Mem. Dep. Agric. India Bot. Ser. 1(5): 1-162.

Butler, E.J. and Bisby, G.R. 1960. The Fungi of India, revised by R.S. Vasudeva. The Indian Council of Agric. Res., New Delhi. 552 pages.

Calderone, R.A. and True, R.P. 1967. Pathogenicity and physiology of *Pythium vexans*. Phytopathol. 57: 806.

Campbell, W.A. and Hendrix, F.F. Jr. 1967. *Pythium* and *Phytophthora* species in forest soils in the southeastern United States. Plant Dis. Reptr. 51: 929-932.

Carvalho, P.C.T. 1965. Ocorrência no Brasil de algumas espécies de *Pythium* Pringsheim de interesse à olericultura. Rickia 2: 89-106.

Cejp, K. 1962. *Pythium megalacanthum*. Ceska Mykol. 16: 23-26.

Chang, T.T. 1993. Investigation and pathogenicity tests of *Pythium* species from rhizosphere of *Cinnamomum osmophloeum*. Plant Pathol. Bull. 2(2): 66-70.

Charlemagne, T.T.J. and Xu-Tong. 1997. *Pythium* species from cocoyam farm soils in Cameroon. Fungal Sci. 12(1/2): 9-15.

Chaudhuri, S. 1975. Fruit rot of *Trichosanthes dioica* caused by *Pythium cucurbitacearum* Takimoto in West Bengal. Curr. Sci. 44: 68.

Chee, K.H. 1968. Patch canker of *Hevea brasiliensis* caused by *Phytophthora palmivora*. Plant Dis. Reptr. 52: 132-133.

Chellemi, D.O., Mitchell, D.J., Kannwischer-Mitchell, M.E., Rayside, P.A. and Rosskopf, E.N. 2000. *Pythium* spp. associated with bell pepper production in Florida. Plant Dis. 84(12): 1271-1274.

Chen, K.L. and Kirschner, R. 2018. Fungi from leaves of lotus (*Nelumbo nucifera*). Mycol. Progr. 17(1-2): 275-293.

Chen, X.R., Liu, B.B., Xing, Y.P., Cheng, B.P., Liu, M.L., Tong, Y.H. and Xu, J.Y. 2016. Identification and characterization of *Phytopythium helicoides* causing stem rot of Shatangju mandarin seedlings in China. Eur. J. Plant Pathol. 146: 715-727.

Cho, W.D. and Shin, H.D. (ed.). 2004. List of Plant Diseases in Korea. 4th edition. Korean Soc. Plant Pathol. 779 pages.

Chou, L.G. and Schmitthenner, A.F. 1974. Effect of *Rhizobium japonicum* and *Endogone mosseae* on soybean root rot caused by *Pythium ultimum* and *Phytophthora megasperma* var. *sojae*. Plant Dis. Reptr. 58: 221-225.

Choudhary, C.E., Burgos Garay, M.L., Moorman, G.W. and Hong, C. 2016. *Pythium* and *Phytopythium* species in two Pennsylvania greenhouse irrigation water tanks. Plant Dis. 100(5): 926-932.

Coffua, L.S., Veterano, S.T., Clipman, S.J., Mena-Ali, J.I. and Blair J.E. 2016. Characterization of *Pythium* spp. associated with asymptomatic soybean in South Eastern Pennsylvania. Plant Dis. 100(9): 1870-1879.

Cook, R.P. and Dube, A.J. 1989. Host-pathogen index of plant diseases in South Australia. South Aus. Dep. Agr. 1-142.

Cordier, T., Belbahri, L., Calmin, G., Oszako, T., Nowakowska, J. and Lefort, F. 2009. Emerging *Phytophthora* and *Pythium* species in Polish declining forests. Established and emerging

Phytophthora: Increasing threats to woodland and forest ecosystems in Europe. First working groups meeting, 16–19, 2009. April, Novy Smokovec, Slovakia.

Cother, E.J. and Gilbert, R.L. 1993. Comparative pathogenicity of *Pythium* species associated with poor seedling establishment of rice in Southern Australia. Plant Pathol. 42: 151-157.

Crous, P.W., Phillips, A.J.L. and Baxter, A.P. 2000. Phytopathogenic Fungi from South Africa. University of Stellenbosch, Dep. Plant Pathol. Press. 358 pages.

Crous, P.W., Denman, S., Taylor, J.E., Swart, L., Bezuidenhout, C.M., Hoffman, L., Palm, M.E. and Groenewald, J.Z. 2013. Cultivation and Diseases of Proteaceae: *Leucadendron, Leucospermum,* and *Protea.* Second edition. CBS Biodiversity Series. 360 pages.

Cunningham, D.D. 1897. On certain diseases of fungal and algal origin affecting economic plants in India. Sci. Mem. Med. Off. Army India 10: 95.

Czeczuga, B., Kozlowska, M. and Godlewska, A. 1999. Zoosporic fungus species growing on dead benthos crustaceans. Pol. J. Environ. Stud. 8(6): 377-382.

Czeczuga, B. and Snarska, A. 2001. *Pythium* species in 13 various types of water bodies of N-E Poland. Acta. Soc. Bot. Pol. 70(1): 61-69.

Czeczuga, B., Kiziewicz, B. and Danilkiewicz, Z. 2002. Zoosporic fungi growing on the specimens of certain fish species recently introduced to Polish waters. Acta. Ichthyol. Piscat. 32(2): 117-125.

Czeczuga, B., Godlewska, A. and Kiziewicz, B. 2004. Aquatic fungi growing on feathers of wild and domestic bird species in limnologically different water bodies. Pol. J. Environ. Stud. 13(1): 21-31.

Czeczuga, B., Mazalska, B., Godlewska, A. and Muszynska, E. 2005. Aquatic fungi growing on dead fragments of submerged plants. Limnologica 35: 283-297.

Czeczuga, B., Muszynska, E., Godlewska, A. and Mazalska, B. 2007. Aquatic fungi and straminipilous organisms on decomposing fragments of wetland plants. Mycol. Balc. 4: 31-44.

Davidson, J.M., Rehner, S.A., Santana, M., Lasso, E., Urena de Chapet, O. and Herre, E.A. 2000. First report of *Phytophthora heveae* and *Pythium* spp., on tropical tree seedlings in Panama. Plant Dis. 84(6): 706.

de Bary, A. 1876. Researches into the nature of the potato fungus, *Phytophthora infestans.* J. Bot. Paris 14: 105-126.

de Bary, A., 1881. Zur Kenntniss der Peronosporeen. Bot. Ztg. 39: 521-625.

de Cock, A.W.A.M, Lodhi, A.M., Rintoul, T.L., Bala, K., Robideau, G.P., Abad, Z.G., Coffey, M.D., Shahzad, S. and Levesque, C.A. 2015. *Phytopythium*: Molecular phylogeny and systematics. Persoonia 34: 25-39.

de Jesus, A.L., Gonçalves, D.R., Rocha, S.C.O., Marano, A.V., Jerônimo, G.H., De Souza, J.I., Boro, M.C. and Pires-Zottarelli, C.L.A. 2016. Morphological and phylogenetic analyses of three *Phytopythium* species (Peronosporales, Oomycota) from Brazil. Cryptogam Mycol. 37: 117-128.

DeMott, M.E. 2015. Identification and mefenoxam sensitivity of Oomycete root pathogens recovered from ornamental plants in Georgia. Master Dissertation, University of Georgia. https://getd.libs.uga.edu/pdfs/demott_max_e_201505_ms.pdf

Dick, M.W. 2001. Straminipilous Fungi: Systematics of the Peronosporomycetes including Accounts of the Marine Straminipilous Protists, the Plasmodiophorids and Similar Organisms. Kluwer Academic Publishers, Dordrecht.

Diddens, H.A. 1932. Untersuchungen über den Flachsbrand, verursacht durch *Pythium megalacanthum.* Phytopath. Z. 4: 291-313.

Dingley, J.M., Fullerton, R.A. and McKenzie, E.H.C. 1981. Records of fungi, bacteria, algae, and angiosperms pathogenic on plants in Cook Islands, Fiji, Kiribati, Niue, Tonga, Tuvalu, and Western Samoa. Survey of Agricultural Pests and Diseases. Technical Report, Vol. 2. FAO, Rome, Italy. 485 pages.

Drechsler, C. 1930. Some new species of *Pythium.* J. Wash. Acad. Sci. 20: 398-418.

Drechsler, C. 1939. Several species of *Pythium* causing blossom-end rot of watermelon. Phytopathol. 29: 391-422.

Drechsler, C. 1941. Three species of *Pythium* with proliferous sporangia. Phytopathol. 31: 478-507.

Drechsler, C. 1943. Two species of *Pythium* occurring in southern states. Phytopathol. 33: 261-299.

Duan, R.L. 1985. A new species and two new records of *Pythium* in China. Acta Mycol. Sin. 4(1): 1-4.

Ershad, D. 1977. Contribution to the knowledge of *Pythium* species of Iran. Iran J. Plant Pathol. 13(3-4): 55-74.

Fichtner, E.J., Browne, G.T., Mortaz, M., Ferguson, L. and Blomquist, C.L. 2016. First report of root rot caused by *Phytopythium helicoides* on pistachio rootstock in California. Plant Dis. 100(11): 23-37.

Firman, I.D. 1972. A list of fungi and plant parasitic bacteria, viruses and nematodes in Fiji. Phytopathological Papers No. 15. Commonwealth Mycological Institute. 36 pages.

Forzza, R.C., Baumgratz, J.F.A., Bicudo, C.E.M., Carvalho-Jr., A.A., Costa, A., Costa, D.P., Hopkins, M., Leitman, P.M., Lohmann, L., Maia, L.C., Martinelli, G., Menezes, M., Morim, M.P., Coelho, M.A.N., Peixoto, A.L., Pirani, J.R., Prado, J., Queiroz, L.P., Souza, V.C., Stehmann, J.R., Sylvestre, L.S., Walter, B.N.T. and Zappi, D. 2010. Catálogo de plantas e fungos do Brasil. Vol. 1. Rio de Janeiro, Instituto de Pesquisas Jardim Botânico do Rio de Janeiro.

French, A.M. 1989. California Plant Disease Host Index. Calif. Dept. Food Agric., Sacramento, CA, USA. 394 pages.

Frezzi, M.J. 1956. Especies de *Pythium* fitopatogenas identificadas en la Republica Argentina. R. Inv. Agr. 10: 113-241.

Gadgil, P.D., Dick, M.A., Hood, I.A. and Pennycook, S.R. 2005. Fungi on Trees and Shrubs in New Zealand. Fungal Diversity Research Series 16. Fungal Diversity Press, Hong Kong. 437 pages.

Ghaderi, F. and Banihashemi, Z. 2011. Identification and pathogenicity of *Pythium* spp. isolated from walnut seedlings in Fars province nurseries. Appl. Entomol. Phytopathol. 78: 237-256.

Ghaderi, F. and Banihashemi, Z. 2008. Identification and pathogenicity of *Pythium* species isolated from walnut seedlings in Fars province. Proceedings of 18th Iran Plant Protection Congress Vol. II, Aug 24–27, Hamedan, Iran. 227 pages.

Ghalamfarsa, M.R. and Banihashemi, Z. 2005. Identification of soil *Pythium* species in Fars province of Iran. Iran J. Sci. Technol. A. 29(A1): 79-87.

Gichuru, V., Buruchara, R. and Okori, P. 2016. Pathogenic and molecular characterisation of *Pythium* spp. inducing root rot symptoms in other crops intercropped with beans in Southwestern Uganda. J. Appl. Biosci. 104: 9955- 9964.

Ginns, J.H. 1986. Compendium of plant disease and decay fungi in Canada 1960-1980. Publication 1813, Research Branch, Canada Department of Agriculture. 416 pages.

Gonçalves, D., Ana, J. and Carmen, Z. 2016. *Pythium* and *Phytopythium* species associated with hydroponically grown crops around the city of São Paulo, Brazil. Trop. Plant Pathol. 41(6): 397-405.

Grand, L.F. (ed.). 1985. North Carolina plant disease index. North Carolina Agric. Res. Serv. Techn Bull. 240: 1-157.

Grijalba, P.E., del Carmen Ridao, A. and Steciow, M. 2018. *Pythium and Phytopythium* associated with soybean in Buenos Aires (Argentina). (Abst.). Phytopathol. 108(10): 48.

Guha, R.S. and Hong, C.X. 2008. The first finding of Pythium root rot and leaf blight of elephant foot yam (*Amorphophallus paeonifolius*) in India. Plant Pathol. 57: 369.

Guo, L.Y. and Ko, W.H. 1994. Survey of root rot of *Anthurium*. Plant Pathol. Bull. 3: 18-23.

Hancock, J.G. 1985. Fungal infection of feeder rootlets of alfalfa. Phytopathol. 75: 1112-1120.

Hanlin, R.T., Foudin, L.L., Berisford, Y., Glover, S.U., Jones, J.P. and Huang, L.H. 1978. Plant disease index for maize in the United States, Part I: Host index. Agric. Exp. Sta. Univ. Georgia. Res. Rep. 277: 1-62.

Hansford, C.G. 1938. Annotated host list of Uganda parasitic fungi and plant diseases (parts 1-5). E. Afr, Agric, J. 4: 235-240.

Haro, P.G. 2014. Molecular characterization of *Pythium* populations in ornamental greenhouse and nursery crops. Master Dissertation, Oklahoma State University.

Harvey, J.V. 1944a. Fungi associated with the decline of citrus and avocado in California. Plant Dis. Reptr. 28: 565-568.

Harvey, J.V. 1944b. Fungi associated with the decline of avocado and citrus in California. Plant Dis. Reptr. 28: 1028-1031.

Harvey, J.V. 1945. Fungi associated with decline of avocado and citrus in California. Plant Dis. Reptr. 29: 110-113.

Hendrix, F.F. Jr. and Campbell, W.A. 1966. Root rot organisms isolated from ornamental plants in Georgia. Plant Dis. Reptr. 50: 393-395.

Hendrix, F.F. Jr., Powell, W.M. and Owen, J.H. 1966. Relation of root necrosis caused by *Pythium* species to peach tree decline. Phytopathol. 56: 1229-1232.

Hendrix, F.F. Jr. and Powell, W.M. 1968. Nematode and *Pythium* species associated with feeder root necrosis of pecan trees in Georgia. Plant Dis. Reptr. 52: 334-335.

Hendrix, F.F. Jr. and Campbell, W.A. 1970. Distribution of *Phytophthora* and *Pythium* species in soils in the continental United States. Can. J. Bot. 48: 377-384.

Hendrix, F.F. Jr., Campbell, W.A. and Moncrief, J.B. 1970. *Pythium* species associated with golf turfgrasses in the South and Southeast. Plant Dis. Reptr. 54: 419-421.

Hendrix, F.F. Jr., Campbell, W.A. and Chien, C.Y. 1971. Some Phycomycetes indigenous to soils of old growth forests. Mycologia 63: 283-289.

Ho, H.H. 2009. The genus *Pythium* in Taiwan, China (1) – A synoptic review. Front. Biol. 4: 15-28.

Ho, H. 2013. The genus *Pythium* in mainland China. Mycosystema 32(Suppl): 20-44.

Ho, H.H., Chang, H.S. and Hsieh, S.Y. 1991. *Halophytophthora kandeliae*, a new marine fungus from Taiwan. Mycologia 83: 419-424.

Ho, H.H., Chen, X.X., Zeng, H.C. and Zh, F.C. 2012. The occurrence and distribution of *Pythium* species on Hainan Island of South China. Bot. Stud. 53: 525-534.

Hohnk, W. 1962. Über die Phycomyceten der Insel Madeira. Veröff. Inst. Meeresforsch. Bremerh. 8: 99-108.

Hrabětová, M., Mrazkova, M., Havrdova, L. and Cerny, K. 2017. Patogeny rodu *Phytophthora* v ovocnych vysadbach CR. 5. Cesko-Slovenska Mykologicka Konference, 28–30 August. Brno.

Huang, J.H. 2008. Root rot of *Tibouchina semidecandra* caused by *Pythium palingenes*. Plant Pathol. Bull. 17: 69.

Huberli, D. Hardy, G.E.St.J., White, D., Williams, N. and Burgess, T.I. 2013. Fishing for *Phytophthora* from Western Australia's waterways: A distribution and diversity survey. Aus. Plant. Pathol. 42: 251-260.

Ilieva, E.I. 1995. Two species of *Pythium* isolated from tomatoes, cucumbers and carnation. Bulgarian J. Agri. Sci. 1(1): 1-6.

Jankowiak, R., Stępniewska, H. and Bilanski, P. 2015. Notes on some *Phytopythium* and *Pythium* species occurring in oak forests in southern Poland. Acta Mycol. 50(1): 1052. http://dx.doi.org/10.5586/am.1052

Johnson, T.W. Jr. 1971. Aquatic fungi of Iceland: *Pythium*. Mycologia 63: 517-536.

Johnston, A. 1960. A supplement to a host list of plant diseases in Malaya. Mycol. Pap. 77: 1-30.

Jones, L.A., Worobo, R.W. and Smart, C.D. 2014. Plant-pathogenic oomycetes, *Escherichia coli* strains, and *Salmonella* spp. frequently found in surface water used for irrigation of fruit and vegetable crops in New York State. Appl. Environ. Microbiol. 80: 4814-4820.

Joshi, B.D. and Paroda, R.S. 1991. Buckwheat in India. National Bureau of Plant Genetic Resources, New Delhi, India. 160 pages.

Jung, T., Blaschke, H. and Neumann, P. 1996. Isolation, identification and pathogenicity of *Phytophthora* species from declining oak stands. Eur. J. Forest. Pathol. 26(5): 253-272.

Jung, T., Scanu, B., Bakonyi, J., Seress, D., Kovacs, G.M., Duran, A., Sanfuentes von Stowasser, E., Schena, L., Mosca, S., Thu, P.Q., Nguyen, C.M., Fajardo, S., Gonzalez, M., Perez-Sierra, A., Rees, H., Cravador, A., Maia, C. and Horta Jung, M. 2017. *Nothophytophthora* gen. nov., a new sister genus of *Phytophthora* from natural and semi-natural ecosystems. Persoonia 39: 143-174.

Kachour, L., Gacemi-Kirane, D., Loucif, L. and Alayat, H. 2016. First survey of aquatic microbial fungi-like Pythiaceae predominantly colonizing the South-Mediterranean freshwater wetlands. Res. J. Pharm. Biol. Chem. Sci. 7(6): 3067-3078.

Kageyama, K., Aoyagi, T., Sunouchi, R. and Fukui, H. 2002. Root rot of miniature roses caused by *Pythium helicoides*. J. Gen. Plant Pathol. 68: 15-20.

Kageyama, K., Suzuki, M., Priyatmojo, A., Oto, Y., Ishiguro, K., Suga, H., Aoyagi, T. and Fukui, H. 2003. Characterization and identification of asexual strains of *Pythium* associated with root rot of rose in Japan. J. Phytopathol. 151: 485-491.

Kato, M., Minamida, K., Tojo, M., Kokuryu, T., Hamaguchi, H. and Shimada, S. 2013. Association of *Pythium* and *Phytophthora* with pre-emergence seedling damping-off of soybean grown in a field converted from a paddy field in Japan. Plant Prod. Sci. 16(1): 95-104.

Kerling, L.C.P. 1947. *Pythium debaryanum* and related species in South Australia. Trans. R. Soc. Aust. 71: 253-258.

Kessank, S. 1987. Taxonomy, distribution and pathogenicity of *Pythium* spp., associated with lowland rice field soils. University of the Philippines Los Baños, Philippines. http://agris.fao.org/agris-search/search.do?recordID=PH8810078

Kirk, P.M. 2015. Nomenclatural novelties. Index Fungorum. 280: 1-1. http://www.indexfungorum.org/Publications/Index%20Fungorum%20no.280.pdf.

Klemmer, H.W. and Nakano, R.Y. 1964. Distribution and pathogenicity of *Phytophthora* and *Pythium* in pineapple soils of Hawaii. Plant Dis. Reptr. 48: 848-852.

Kliejunas, J.T. and Ko, W.H. 1975. The occurrence of *Pythium vexans* in Hawaii and its relation to Ohia decline. Plant Dis. Reptr. 59: 392-395.

Ko, W.H., Lin, M., Hu, C.Y. and Pao-Jen Ann. 2010. *Aquaperonospora taiwanensis* gen. et sp. nov. in Peronophythoraceae of Peronosporales. Bot. Stud. 51: 343-350.

Kobayashi, T. 2007. Index of Fungi Inhabiting Woody Plants in Japan. Host, Distribution and Literature. Zenkoku-Noson-Kyoiku Kyokai Publishing Co., Ltd. 1227 pages.

Kranjec, J., Milotic, M., Hegol, M. and Diminic, D.O. 2017. Fungus-like organisms in the soil of declining narrow-leaved ash stands (*Fraxinus angustifolia* Vahl). Sumar. List 141(3-4): 115-122.

Leafio, E.M. 2001. Straminipilous organisms from fallen mangrove leaves from Panay Island, Philippines. Fungal Divers. 6: 75-81.

Lefort, F., Pralon, T., Nowakowska, J. and Oszako, T. 2013. Screening of bacteria and fungi antagonist to *Phytophthora* and *Pythium* species pathogenic of forest trees. Biological Control of Fungal and Bacterial Plant Pathogens. IOBC-WPRS Bull. 86: 185-186.

Lehtijarvi, A., Aday, G., Woodward Jung, S. and Dogmus-Lehtijarvi, T. 2017. Oomycota species associated with deciduous and coniferous seedlings in forest trees nurseries of Western Turkey. Forest Pathol. 47(3): e12363.

Lévesque, C.A. and de Cock, A.W.A.M. 2004. Molecular phylogeny and taxonomy of the genus *Pythium*. Mycol. Res. 108: 1363-1383.

Lévesque, C.A., Harlton, C.E., de Cock, A.W.A.M. 1998. Identification of some oomycetes by reverse dot blot hybridization. Phytopathol. 88: 213-222.

Li, Mingzhu, Yasushi, I., Kayoko, O., Hirofumi, S., Tomoko, T., Noriyuki, M., Hirofumi, N., Haruhisa, S. and Koji, K. 2014. Monitoring by real-time PCR of three water-borne zoosporic *Pythium* species in potted flower and tomato greenhouses under hydroponic culture systems. Eur. J. Plant Pathol. 140. 10.1007/s10658-014-0456-z.

Liu, P.S.W. 1977. A supplement to a host list of plant diseases in Sabah, Malaysia. Phytopathol. Pap. 21: 1-49.

Lodhi, A.M. 2007. Taxonomic Studies on Oomycetous Fungi from Sindh. PhD Dissertation, University of Karachi, Karachi, Pakistan. http://prr.hec.gov.pk/jspui/handle/123456789/5097

Lodhi, A.M., Shahzad, S. and Ghaffar, A. 2005. First report of *Pythium ostracodes* Drechsler, a new record from Pakistan. Pak. J. Bot. 37(1): 171-175.

Lorio, P.L. Jr. 1966. *Phytophthora cinnamomi* and *Pythium* species associated with loblolly pine decline in Louisiana. Plant Dis. Reptr. 50: 596-597.

Lu, B., Hyde, K.D., Ho, W.H., Tsui, K.M., Taylor, J.E., Wong, K.M. and Yanna Zhou, D. 2000. Checklist of Hong Kong fungi. Fungal Diversity Press, Hong Kong. 207 pages.

Lumsden, R.D. and Ayers, W.A. 1975. Influence of soil environment on the germinability of constitutively dormant oospores of *Pythium ultimum*. Phytopathol. 65: 1101-1107.

Lumsden, R.D. and Haasis, F.A. 1964. Pythium root and stem diseases of chrysanthemum in North Carolina. North Carolina Agric. Exp. Sta. Tech. Bull. 158: 1-27.

Marano, A.V., Jesus, A.L., de Souza, J.I., Leano, E.M., James, T.Y., Jeronimo, G.H., de Cock, A.W.A.M. and PiresZottarelli, C.L.A. 2014. A new combination in *Phytopythium*: *P. kandeliae* (Oomycetes, Straminipila). Mycosphere 5(4): 510-522.

Marvasti, F.B. and Banihashemi, Z. 2010. Study of pathogenicity of turf grass-infecting fungi in Shiraz landscape. Proceedings of 19th Iran Plant Protection Congress. Vol. II, July 31-Aug 3, Tehran, Iran. Tehran. 226 pages.

Matias, A.O. 2012. Caracterização de podridão radicular e identificação de genótipos resistentes de mandioca (Manihot Crantz) na produção de agricultores familiares no município de Brejo, Maranhão. (Dissertação de Mestrado). Universidade Federal do Piauí. Piauí. https://sigaa.ufpi.br/sigaa/public/programa/noticias_desc.jsf?lc=lc=pt_BRandid=340andnoticia=3564970

Matsiakh, I., Oszako, T., Kramarets, V. and Nowakowska, J.A. 2016. *Phytophthora* and *Pythium* species detected in rivers of the Polish-Ukrainian border areas. Balt. For. 22(2): 230-238.

Matsiakh, I., Kramarets, V., Orlikowski, L.B. and Trzewik, A. 2012. First report on *Phytophthora* spp. and *Pythium litorale* occurrence in Beskid's rivers (of the Ukrainian Carpathians)]. Prog. Plant Prot. 52(4): 1200-1203.

McLeod, A., Botha, W.J., Meitz, J.C., Spies, C.F.J., Tewoldemedhin, Y.T. and Mostert, L. 2009. Morphological and phylogenetic analyses of *Pythium* species in South Africa. Mycol. Res. 113: 933-951.

Mendes, M.A.S., da Silva, V.L., Dianese, J.C., Ferreira, M.A.S.V., Santos, C.E.N., Gomes Neto, E., Urben, A.F. and Castro, C. 1998. Fungos em Plants no Brasil. Embrapa-SPI/Embrapa-Cenargen, Brasilia. 555 pages.

Meredith, C.H. 1940. A quick method of isolating certain phycomycetous fungi from the soil. Phytopathol. 30: 1055-1056.

Middleton, J.T. 1938. Plant diseases caused by *Pythium* spp. observed in California in 1938. Plant. Dis. Reptr. 22: 354-356.

Middleton, J.T. 1943. The taxonomy, host range and geographic distribution of the genus *Pythium*. Mem. Torrey Bot. Club. 20: 1-171.

Mihail, J.D. 1993. Diseases of alternative crops in Missouri. Can. J. Plant Pathol. 15(2): 119-122.

Milanez, A.I., Pires-Zotarelli, C.L.A. and Gomes, A.L. 2007. Brazilian Zoosporic Fungi. São Paulo. WinnerGraph.

Mircetich, S.M. 1971. The role of *Pythium* in feeder roots of diseased and symptomless peach trees and in orchard soils in peach tree decline. Phytopathol. 61: 357-360.

Mircetich, S.M. and Fogle, H.W. 1969. Role of *Pythium* in damping-off of peach. Phytopathol. 59: 356-358.

Misra, J.K. and Hall, G.S. 1996. Occurrence and distribution of the genus *Pythium* in India: A review. Kavaka 24: 57-119.

Mitchell, D.J. 1978. Relationships of inoculum levels of several soilborne species of *Phytophthora* and *Pythium* to infection of several hosts. Phytopathol. 68: 1754-1759.

Miyake, N., Nagai, H. and Kageyama, K. 2014. Wilt and root rot of poinsettia caused by three high-temperature-tolerant *Pythium* species in ebb-and-flow irrigation systems. J. Gen. Plant Pathol. 80: 479-489.

Mrazkova, M., Hrabetova, M. and Fedusiv, L. 2018. Czech collection of phytopathogenic Oomycetes, RILOG Pruhonice. http://www.vukoz.cz/dokumenty/056/Sbirka_Katalog_Jan_2018.pdf

Mukundi, D.N. 2011. Distribution and diversity of *Pythium* spp. in indigenous forests and adjacent farm lands in Taita and Embu districts in Kenya. Masters Dissertation, School of Biological Sciences, University of Nairobi. http://erepository.uonbi.ac.ke:8080/xmlui/handle/123456789/12276

Mukundi, N.D., Okoth, S.A. and Mibey, R.K. 2009. Influence of land use on the distribution and diversity of *Pythium* spp. Trop. Subtrop. Agroecosyst. 11(2): 347-352.

Nakagiri, A. 2000. Ecology and diversity of *Halophytophthora* species. Fungal Divers. 5: 153-164.

Nechwatal, J. and Oßwald, W.F. 2003. *Pythium montanum* sp. nov., a new species from a spruce stand in the Bavarian Alps. Mycol. Prog. 2(1): 73-80.

Nechwatal, J. and Mendgen, K. 2006. *Pythium litorale* sp. nov., a new species from the littoral of Lake Constance, Germany. FEMS Microbiol. Lett. 255: 96-101.

Nechwatal, J., Wielgoss, A. and Mendgen, K. 2008. Diversity, host, and habitat specificity of oomycete communities in declining reed stands (*Phragmites australis*) of a large freshwater lake. Mycol. Res. 112: 689-696.

Negreiros, N.C. 2008. Uso sustentável de culturas agrícolas suscetíveis a oomicetos (Oomycota) fitopatogênicos às margens do rio Parnaíba no município de Floriano, Piauí. (Dissertação de Mestrado). Universidade Federal do Piauí. Piauí.

Newell, S.Y. and Fell, J.W. 1995. Do halophytophthoras (marine Pythiaceae) rapidly occupy fallen leaves by intraleaf mycelial growth? Can. J. Bot. 73(5): 761-765.

Nichols, C.W., Garnsey, S.M., Rackham, R.L., Gotan, S.M. and Mahannah, C.N. 1964. Pythiaceous fungi and plant-parasitic nematodes in California pear orchards. 1. Occurrence and pathogenicity of pythiaceous fungi in orchard soils. Hilgardia 35: 577-602.

Nzungize, J., Gepts, P., Buruchara, R., Buah, S., Ragama, P., Busogoro, J.P. and Baudoin, J.P. 2011. Pathogenic and molecular characterization of *Pythium* species inducing root rot symptoms of common bean in Rwanda. Afr. J. Microbiol. Res. 5(10): 1169-1181.

Nzungize, J.R., Lyumugabe, F., Busogoro, J.P. and Baudoin, J.P. 2012. Focus on Pythium root rot of common bean: Biology and control methods. A review. Biotechnol. Agron. Soc. Environ. 16(3): 405-413.

Ohgami, D. and Kageyama, K. 2005. Root rot of yakon (*Polymnia sonchifolia*). (Abst.). Meeting of the Hokkaido Division, The Phytopathological Society of Japan, Sapporo, October 20-21, 2005, http://repository.lib.gifu-u.ac.jp/bitstream/20.500.12099/27966/1/c200600711.pdf

Oszako, T., Katarzyna, S., Belbahri, L. and Justyna, N. 2016. Molecular detection of oomycetes species in water courses. Folia For. Po. 58: 246-251.

Padron, R.C., Felipe, S., Perez-Sierra, A. and Rodriguez, A. 2018. Isolation and pathogenicity of *Phytophthora* species and *Phytopythium vexans* recovered from avocado orchards in the Canary Islands, including *Phytophthora niederhauserii* as a new pathogen of avocado. Phytopathol. Mediterr. 57 (1): 10.14601/Phytopathol_Mediterr-22022.

Palmucci, H., Grijalba, P., Wolcan, S., Herrera, C., Elisa, F., Steciow, M. and Abad, G. 2011. Morphological-molecular characterization of *Phytophthora*, *Pythium* and *Phytopythium* on intensive crops in Buenos Aires, Argentina. Phytopathol. 101: S136.

Panabieres, F., Ponchet, M., Allasia, V., Cardin, L. and Ricci, P. 1997. Characterization of border species among Pythiaceae: Several *Pythium* isolates produce elicitins, typical proteins from *Phytophthora* spp. Mycol. Res. 101: 1459-1468.

Pande, A. and Rao, V.G. 1998. A Compendium of Fungi on Legumes from India. Scientific Publishers Jodhpur, India. 188 pages.

Pantidou, M.E. 1973. Fungus-host index for Greece. Benaki Phytopathol. Inst., Kiphissia, Athens. 382 pages.

Park, M. 1936. Report on the work of the mycological division. Administration Report Director Agricultural. D28-D35.

Parkunan, V. and Pingsheng, J. 2013. Isolation of *Pythium litorale* from irrigation ponds used for vegetable production and its pathogenicity on squash. Cana. J. Plant Pathol. 35: 415-423.

Paul, B. 1994. Some species of *Pythium* isolated from cultivated soils in northern France. Cryptogamie Mycol. 15(4): 263-271.

Paul, B. 2000. ITS1 region of the rDNA of *Pythium megacarpum* sp. nov., its taxonomy, and its comparison with related species. FEMS Microbiol. Lett. 186(2): 229-233.

Paul, B. 2003. *Pythium carbonicum*, a new species isolated from a spoil heap in northern France, the ITS region, taxonomy and comparison with related species. FEMS Microbiol. Lett. 219: 269-274.

Paul, B. 2004. A new species of *Pythium* isolated from burgundian vineyards and its antagonism towards *Botrytis cinerea*, the causative agent of the grey mould disease. FEMS Microbiol. Lett. 234(2): 269-274.

Pennycook, S.R. 1989. Plant Diseases Recorded in New Zealand. Plant. Dis. Div., DSIR. Auckland, N.Z.

Peregrine, W.T.H. and Ahmad, K.B. 1982. Brunei: A first annotated list of plant diseases and associated organisms. Phytopathol. Pap. 27: 1-87.

Pereira, A.A. 2008. Oomicetos (Oomycota) no campo agrícola de Nazária, Piauí- Sustentabilidade na prevenção e controle dos fitopatógenos em agricultura familiar. (Dissertação de Mestrado). Universidade Federal do Piauí, Piauí.

Pereira, A.A. and Rocha, J.R.S. 2008. *Pythium* (Pythiaceae): Três novos registros para o nordeste do Brasil. Acta Bot. Malacit. 33: 347-350.

Plaats-Niterink and Van Der, A.J. 1975. Species of *Pythium* in the Netherlands. Neth. J. Plant Path. 81: 22-37.

Plaats-Niterink and Van Der, A.J. 1981. Monograph of the genus *Pythium*. Studies in Mycology 21: 1-242.

Polat, Z., Awan, Q.N., Hussain, M. and Akgul, D.S. 2017. First report of *Phytopythium vexans* causing root and collar rot of kiwifruit in Turkey. Plant Dis. 101(6): 1058-1059.

Powell, W.M., Owen, J.H. and Campbell, W.A. 1965. Association of phycomycetous fungi with peach tree decline in Georgia. Plant Dis. Reptr. 49: 279.

Pratt, B.H. and Heather, W.A. 1973. Recovery of potentially pathogenic *Phytophthora* and *Pythium* spp. from native vegetation in Australia. J. Biol. Sci. 26: 575-582.

Quiros, O. and Jose, J. 2018. Biomolecular identification of root Peronosporals in avocado, mango, custard apple and mandarin on the Peruvian coast. Master's Dissertation, National Agrarian University La Molina, Lima, Peru.

Raabe, R.D., Conners, I.L. and Martinez, A.P. 1981. Checklist of Plant Diseases in Hawaii. College of Tropical Agriculture and Human Resources, University of Hawaii. Information Text Series No. 22. Hawaii Inst. Trop. Agric. Human Resources. 313 pages.

Raciborski, M. 1900. Parasitische Algen and Pilze Java's. I Theil. Bot. Inst. Buitenzorg, Batavia. 39 pages.

Radmer, L., Anderson, G., Malvick, D.K., Kurle, J.E., Rendahl, A. and Mallik, A. 2017. *Pythium, Phytophthora,* and *Phytopythium* spp. associated with soybean in Minnesota, their relative aggressiveness on soybean and corn, and their sensitivity to seed treatments fungicides. Plant Dis. 101(1): 62-72.

Ragunathan, V. 1968. Damping-off of green gram, cauliflower, daincha, ragi and clusterbean. Indian Phytopath. 21: 456-457.

Rajapakse, R.H.S. and Kulasekera, V.L. 1981. Rhizome rot of cardamom *Elettaria cardamomum* (Maton). J. Plant. Crops 9(1): 56-58.

Ramakrishna, T.S. 1949. The occurrence of *Pythium vexans* de Bary in South India. Indian Phytopath. 2: 27-30.

Ramirez-Gil, J.G. 2018. Avocado wilt complex disease, implications and management in Colombia. Rev. Fac. Nac. Agron. Medellin. 71(2): 8525-8541.

Rana, K.S. and Gupta, V.K. 1984. Occurrence of pythiaceous fungi in collar rot affected apple soils of Himachal Pradesh. Indian Phytopathol. 37: 39-42.

Rands, R.D. 1930. Fungi associated with root rots of sugar cane in the southern United States. Proc. Int. Soc. Sugar Cane Technol. 3: 119-131.

Rands, R.D. and Dopp, E. 1938. Pythium root rot of sugarcane. Tech. Bull. US Dep. Agric. 666: 1-96.

Rao, V.G. 1963. An account of the genus *Pythium* in India. Mycopath. Mycol. Appl. 21: 45-49.

Ravisé, A., Boccas, B. and Dihoulou, P. 1969. Première liste annotée des Pythiacées parasites des plantes cultivées au Congo. Cahiers de la Maboké 7: 41-69.

Reen, L. 1971a. Studies on the factors influencing virulence and enzyme activity of *Pythium* spp. on potato tubers. Indian Phytopath. 24: 74-87.

Reen, L. 1971b. Virulence of *Pythium* spp. on potato tuber and their capacity to produce pectic enzymes. Indian Phytopath. 24: 88-100.

Remy, E. 1950. Ober niedere Bodenphycomyceten. Arch Mikrobiol. 14: 212-239.

Riley, E.A. 1960. A revised list of plant diseases in Tanganyika Territory. Mycol. Pap. 75: 1-42.

Rios, T.L. and Rocha, J.R.S. 2018. Potencial patogênico de espécies do complexo *Pythium* (Oomycota) para a agricultura familiar no estado do piauí. Pesqui. Bot. 71: 147-158.

Robertson, G.I. 1980. The genus *Pythium* in New Zealand. New Zeal. J. Bot. 18(1): 73-102.

Rocha, J.R.S. 2002. Fungos zoosporicos em area de cerrado no Parque Nacional de Sete Cidades, Piaui, Brasil (Tese de Doutorado). Universidade de São Paulo. São Paulo.

Rocha, J.R.S., Milanez, A.I. and Pires-Zottarelli, C.L.A. 2001. O genero *Pythium* (Oomycota) em areas de cerrado no Parque Nacional de Sete Cidades, Piaui, Brasil. Hoehnea 28: 209-230.

Rodríguez Padrón, C., Siverio, F., Pérez-Sierra, A. and Rodriguez, A. 2018. Isolation and pathogenicity of *Phytophthora* species and *Phytopythium vexans* recovered from avocado orchards in the Canary Islands, including *Phytophthora niederhauserii* as a new pathogen of avocado. Phytopathol. Mediterr. 57. 10.14601/Phytopathol_Mediterr-22022.

Rojas, J.A., Jacobs, J.L., Napieralski, S., Karaj, B., Bradley, C.A., Chase, T., Esker, P.D., Giesler, L.J., Jardine, D.J., Malvick, D.K., Markell, S.G., Nelson, B.D., Robertson, A.E., Rupe, J.C., Smith, D.L., Sweets, L.E., Tenuta, A.U., Wise, K.A. and Chilvers, M.I. 2017. Oomycete species associated with soybean seedlings in North America. Part I: Identification and pathogenicity characterization. Phytopathol. 107: 280-292.

Rosskopf, E.N., Yandoc, C.B., Stange, B., Lamb, E.M. and Mitchell, D.J. 2005. First report of Pythium root rot of rau ram (*Polygonum odoratum*). Plant Dis. 89: 340.

Royle, D.J. and Hickman, C.J. 1964. Analysis of factors governing *in vitro* accumulation of zoospores of *Pythium aphanidermatum* on roots. 1. Behaviour of zoospores. 2. Substances causing response. Can. J. Microbiol. 10: 151-162, 201-219.

Salmaninezhad, F. and Ghalamfarsa, M.R. 2017. Phylogeny of *Phytophthora* and *Phytopythium* species associated with rice in Fars province (Iran). Rostaniha 18(1): 1-15.

Salt, G.A. 1977. A survey of fungi in cereal roots at Rothamsted, Woburn and Saxmundham, 1970-1975. Rep. Rothamsted Exp. Stn. 1976, Part 2, 153-168.

Sampson, P.J. and Walker, J. 1982. An Annotated List of Plant Diseases in Tasmania. Dep. Agri. Tasmania. 121 pages.

Santos, M.Q.C. dos. 2015. Caracterização de *Pythium cucurbitacearum*, e controle alternativo da podridão do colo e das raízes do mamoeiro, grupo solo, cultivar Golden. Dissertation Universidade Federal de Alagoas Centro De Ciências Agrárias Programa De Pós-Graduação Em Proteção de Plantas.

Santoso, P.J., Aryantha, I.N.P., Pancoro, A. and Suhandono, S. 2015. Identification of *Pythium* and *Phytophthora* associated with durian (*Durio* sp.) in Indonesia: Their molecular and morphological characteristics and distribution. Asian J. Plant Pathol. 9: 59-71.

Sarbhoy, A.K., Lal, G. and Varshney, J.L. 1971. Fungi of India (1967-71). Navyug Traders, New Delhi. 148 pages.

Savulescu, T. 1940. Etude systématique du genre *Pythium* en Roumanie. Bull. Acad. Sci. Roumanie, Sect Sci. 23: 1-8.

Sawangphiphop, C., Petcharat, V. and Plodpai, P. 2016. Fungicidal activity of formulations of Chinese desmos leaf crude extract against damping-off pathogens. Songklanakarin J. Plant Sci. 3: M09/132-137.

Schultz, H. 1939. Untersuchungen über die Rolle von *Pythium-Arten* als Erreger der Fußkrankheit der Lupine, 1. Phytopath. Z. 12: 405-420.

Scott, W.W. 1960. A study of some soil-inhabiting Phycomycetes from Haiti. Virginia J. Sci. 2: 19-24.

Sharples, A. 1930. Division of mycology. Annual Report for 1929. Rep. Dep. Agric. Staits Settlements and Fed Malaya States, 1929, Gen. Ser. Bull. 3: 62-72.

Shaw, D.E. 1984. Microorganisms in Papua New Guinea. Dept. Primary Ind., Res. Bull. 33: 1-344.

Shennan, C., Muramoto, J., Koike, S., Baird, G., Fennimore, S., Samtani, J., Bolda, M., Dara, S., Daugovish, O., Lazarovits, G., Butler, D., Rosskopf, E., Kokalis-Burelle, N., Klonsky, K. and Mazzola, M. 2018. Anaerobic soil disinfestation is an alternative to soil fumigation for control of some soilborne pathogens in strawberry production. Plant Pathol. 67: 51-66.

Shivas, R.G. 1989. Fungal and bacterial diseases of plants in Western Australia. J. Roy. Soc. West Aus. 72: 1-62.

Sideris, C.P. 1931. Pathological and histological studies on pythiaceous root rots of various agricultural plants. Phytopathol. Z. 3: 137-161.

Sideris, C.P. 1932. Taxonomic studies in the family Pythiaceae. II. *Pythium*. Mycologia 24: 14-61.

Simmonds, J.H. 1966. Host Index of Plant Diseases in Queensland. Queensland Dep Primary Ind., Brisbane. 111 pages.

Spaulding, P. 1961. Foreign diseases of forest trees of the world. USDA Agric. Handbook 197: 1-361.

Spies, C.F.J., Mazzola, M. and McLeod, A. 2011. Characterisation and detection of *Pythium* and *Phytophthora* species associated with grapevines in South Africa. Eur. J. Plant Pathol. 131: 103-119.

Sprague, R. 1950. Diseases of Cereals and Grasses in North America. Ronald Press Co., New York. 538 pages.

Sprague, R. 1957. Fungi isolated from roots and crowns of pear trees. Plant Dis. Reptr. 41: 74-76.

Stamps, D.J., Shaw, D.E. and Cartledge, E.G. 1972. Species of *Phytophthora* and *Pythium* in Papua-New Guinea. Papua New Guin. Agric. J. 23: 41-45.

Steinrucken, T. V., Aghighi, S., Hardy, G.S.J., Bissett, A., Powell, J.R. and Van Klinken, R.D. 2017. First report of oomycetes associated with the invasive tree *Parkinsonia aculeata* (Family: Fabaceae). Aus. Plant Pathol. 46(4): 313-321.

Stevenson, J.A. and Rands, R.D. 1938. An annotated list of the fungi and bacteria associated with sugarcane and its products. Hawaii Planters' Record 12: 247.

Suksiri, S., Laipasu, P., Soytong, K. and Poeaim, S. 2018. Isolation and identification of *Phytophthora* sp. and *Pythium* sp. from durian orchard in Chumphon province, Thailand. Int. J. Agr. Technol. 14(3): 389-402.

Sutherland, J.R., Adams, R.E. and True, R.P. 1966. *Pythium vexans* and other conifer seedbed fungi isolated by the apple technique following treatments to control nematodes. Plant Dis. Reptr. 50: 545-547.

Suzuki, M., Togawa, M. and Yoneyama, C. 2005. Occurrence of Pythium root rot of strawberry caused by *Pythium helicoides*. (Abst.). Jpn. J. Phytopathol. 71: 209.

Sydow, H. and Butler, E.J. 1907. Fungi Indiae Orientalis. Part II. Annu. Mycol. 5: 485-515.

Takahashi, M. 1973. Ecologic and taxonomic studies on *Pythium* in Japan. Rev. Plant Prot. Res. Tokyo 6: 132-144.

Takahashi, M. and Kawase, Y. 1965. Ecologic and taxonomic studies on *Pythium* as pathogenic soil fungi. 5. Several species of *Pythium* causing root rot of strawberry. Ann. Phytopath. Soc. Japan. 30: 181-185.

Takahashi, M., Ichitani, T., Akaji, K. and Kawase, T. 1970. Ecologic and taxonomic studies on *Pythium* as pathogenic soil fungi. 8. Differences in pathogenicity of several species of *Pythium*. Bull. Univ. Osaka Pref., Ser. B 22: 87-93.

Takimoto, S. 1935. Phytophthorose of buckwheat. Bull. Sci. Fac. Terk. Kyushu Univ. 6: 105-110.

Takimoto, S. 1941. On the *Pythium* causing damping-off of seedling and fruit rot of cucumber. Ann. Phytopath. Soc. Japan 11: 89-91.

Tan, T.K. and Pek, C.L. 1997. Tropical mangrove leaf litter fungi in Singapore with an emphasis on *Halophytophthora.* Mycol. Res. 101(2): 165-168.

Tao, Y., Zeng, F., Ho, H., Wei, J., Wu, Y., Yang, L. and He, Y. 2011. *Pythium vexans* causing stem rot of *Dendrobium* in Yunnan province, China. J. Phytopathol. 159: 255-259.

Teakle, D.S. 1957. Papaw root rot caused by *Phytophthora palmivora*. Qd. J. Agric. Sci. 14: 81-91.

Teakle, D.S. 1960. Species of *Pythium* in Queensland. Qd. J. Agric. Sci. 17: 15-31.

Tejada, C.J.R. 1983. Preliminary note on diseases of rubber observed in Guatemala. Informe preliminar sobre las enfermedades observadas en el hule hevea en Guatemala. Revista Cafetalera 230: 13-15.

Thines, M. 2014. Phylogeny and evolution of plant pathogenic oomycetes—A global overview. Eur. J. Plant Pathol. 138: 431-447.

Thompson, A. 1929. *Phytophthora* species in Malaya. Malaya Agric. J. 14: 1-6.

Thompson, A., 1936. The division of mycology. Rep. Dep. Agric. Malaya 1935: 64-66.

Thompson, A., 1939. Notes on plant diseases in 1937-1938. Malaya Agric. J. 27: 84-98.

Thompson, A. and Johnston, A. 1953. A host list of plant diseases in Malaya. Mycol. Pap. 52: 1-38.

Thrailkill, K.Q. 2013. Wetting and drying cycles and the fungal communities on leaf litter in streams. Masters Theses. Missouri University of Science and Technology, 5443. http://scholarsmine.mst.edu/masters_theses/5443

Thu, P.Q. 2016. Surveys of Pythiaceae causing root rot diseases of *Acacia mangium* and *Acacia hybrid* in some provinces of North Vietnam. Tạp Chí Khln. 1: 4251-4256.

Ting, Q.D., Jinzhi, C., Chia-Hsin, T., Thick, C. and Fang, X. 2015. Taiwan's New Emerging Crop Disease and its Prevention. Taiwan Agri. Res. Inst. 208 pages.

Trajano, H.M.R. 2009. Resistência de espécies de pimenta (*Capsicum* sp.) a *Pythium* sp. Master Dissertation. Federal University of Piauí, Piauí, Brazil.

Trigos, A., Castellanos Onorio, O., Salinas, A., Espinoza, C. and Yáñez Morales, M.J. 2006. Antibiotic activity of several phytopathogenic fungi. Micol. Aplicada Int. 18(1): 3-6.

Trindade-Junior, O.C. 2013. Riscos socioambientais e diversidade de fungos zoospóricos em lagoas de Teresina, Piauí (Dissertação de Mestrado). Universidade Federal do Piauí, Piauí.

Trindade-Junior, O.C. and Rocha, J.R.S. 2018. Diversidade, abundância e frequência de Oomycota, Blastocladiomycota e Chytridiomycota em lagoas de Teresina, Piauí. Gaia Scientia 12: 1-11.

Tsukiboshi, T., Chikuo, Y., Ito, Y. Matsushita, Y. and Kageyama, K. 2007. Root and stem rot of *Chrysanthemum* caused by five *Pythium* species in Japan. J. Gen. Plant Pathol. 73(4): 293-296.

Turner, P.D. 1971. Microorganisms associated with oil palm (*Elaeis guineensis* Jacq.). Phytopathol. Pap. 14: 1-58.

USDA-ARS. 2016. USDA-ARS fungal database. Fungal Databases - Fungus-Host Distrib. Available at: http://nt.ars-grin.gov/fungaldatabases/fungushost/FungusHost.cfm

Uzuhashi, S., Tojo, M. and Kakishima, M. 2010. Phylogeny of the genus *Pythium* and description of new genera. Mycosci. 51: 337-365.

Vaartaja, O. 1968. *Pythium* and *Mortierella* in soils of Ontario forest nurseries. Can. J. Microbiol. 14: 265-269.

Vanev, S.G., Dimitrova, E.G. and Ilieva, E.I. 1993. Fungi Bulgaricae, 2 tomus, ordo Peronosporales. In Aedibus Academiae Scientiarum Bulgaricae, Serdicae. 195 pages.

Vannarug, S. 1993. Vegetable soybean diseases caused by *Pythium* spp., and their biological control. Thai National AGRIS Centre. 161 leaves.

Vawdrey, L.L., Langdon, P. and Martin, T. 2005. Incidence and pathogenicity of *Phytophthora palmivora* and *Pythium vexans* associated with durian decline in far Northern Queensland. Aust. Plant Pathol. 34: 127-128.

Verma, B.L. and Khulble, R.D. 1985. *Pythium polytylum* Deechsler and *Pythium rostratum* Butler in some crop field of Tarai. Nainital Sci. Cul. 51: 156-157.

Villa, N.O., Kageyama, K., Asano, T. and Suga, H. 2006. Phylogenetic relationships of *Pythium* and *Phytophthora* species based on ITS rDNA, cytochrome oxidase II and beta-tubulin gene sequences. Mycologia 98: 410-422.

Wager, V.A. 1932. Foot-rot diseases of papaws. Fmg. S. Afr. Jan. 1932: 6.

Wager, V.A. 1941. Descriptions of the South African Pythiaceae with records of their occurrence. Bothalia 4: 3-35.

Wager, V.A. 1942. Pythiaceous fungi on citrus. Hilgardia 14: 535-548.

Wang, K.X., Xie, Y.L., Yuan, G.Q., Li, Q.Q. and Lin, W. 2015. First report of root and collar rot caused by *Phytopythium helicoides* on kiwifruit (*Actinidia chinensis*). Plant Dis. 99: 725.

Watanabe, H., Horinouchi, H., Tanahashi, I. and Kageyama, K. 2005. Occurrence of Pythium root rot of strawberry caused by *Pythium helicoides*, and pathogenicity to several crops (abstract), Jpn. J. Phytopathol. 71: 209-210.

Watanabe, H., Yoshihiro, T., Mitsuro, H. and Kageyama, K. 2007. *Pythium* and *Phytophthora* species associated with root and stem rots of kalanchoe. J. Gen. Plant Pathol. 73: 81-88.

Watanabe, T., Hashimoto, K. and Sato, M. 1977. *Pythium* species associated with strawberry roots in Japan, and their role in the strawberry stunt disease. Phytopathol. 67: 1324-1332.

Watanabe, T. 1989. Further study of *Pythium* species isolated from soils in the Ryukyu Islands. Ann Phytopathol. Soc. Japan 55(3): 349-352.

Watson, A.J. 1971. Foreign bacterial and fungus diseases of food, forage, and fiber crops. USDA Agri. Res. Service. 111 pages.

Wedgwood, E.F. 2011. AHDB Project HNS 181, Survey, detection and diagnosis of Phytophthora root rot and other causes of die-back in conifers. Agriculture and Horticulture Development Board. https://horticulture.ahdb.org.uk/sites/default/files/research_papers/HNS%20181%20Final%20Report_0.pdf

Wen, C., Guo, W. and Chen, X. 2014. Purification and identification of a novel antifungal protein secreted by *Penicillium citrinum* from the Southwest Indian Ocean. J. Microbiol. Biotechnol. 24(10): 1337-1345.

Wheeler, T.A., Howell, C.R., Cotton, J. and Porter, D. 2005. *Pythium* species associated with pod rot on West Texas peanuts and *in vitro* sensitivity of isolates to Mefenoxam and Azoxystrobin. Peanut Sci. 32: 9-13.

Williams, T.H. and Liu, P.S.W. 1976. A host list of plant diseases in Sabah, Malaysia. Phytopathol. Pap. 19: 1-67.

Wilson, K.I. and Rahim, M.A. 1978. *Pythium vexans* causing fruit and rhizome rot of *Aframomum melegueta*. Indian Phytopath. 31: 238.

Woodward, J.L.W. and DeMott, M.E. 2014. Mefenoxam insensitivity in *Pythium* and *Phytophthora* isolates from ornamental plants in Georgia. 163-167. SNA Research Conference Vol. 59 Pathology and Nematology (Alan Windham Section Editor) https://sna.org/Resources/Documents/14resprosec06. pdf

Yin, X., Li, X.Z., Yin, J.J. and Wu, X. 2016. First report of *Phytopythium helicoides* causing rhizome rot of Asian lotus in China. Plant Dis. 100(2): 532-533.

Yu, Y.N. (Ed.). 1998. Flora Fungorum Sinicorum. Vol. 6. Peronosporales. Sci. Press, Beijing. 530 pages.

Yu, Y.T., Chen, J., Gao, C.S., Zeng, L.B., Li, Z.M., Zhu, T.T., Sun, K., Cheng, Y., Sun, X.P., Yan, L., Yan, Z. and Zhu, A. 2016. First report of brown root rot caused by *Pythium vexans* on ramie in Hunan, China. Can. J. Plant Pathol. 38(3): 405-410.

Zappia, R., Hüberli, D., Hardy, G.E.S.J. and Bayliss, J.K.L. 2012. Root pathogens detected in irrigation water of the Ord river irrigation area. (Abst.). 7th Australasian Soilborne Disease Symposium. Fremantle, Australia.

Zeng, H.C., Ho, H.H. and Zheng, F.C. 2005. *Pythium vexans* causing patch canker of rubber trees on Hainan Island, China. Mycopathol. 159: 601-606.

5

Top Three Plant Pathogenic *Pythium* Species

Amal-Asran[1*] and Kamel A. Abd-Elsalam[1,2]

[1] Plant Pathology Research Institute, Agricultural Research Center (ARC), Egypt
[2] Unit of Excellence in Nano-Molecular Plant Pathology, Plant Pathology Research Institute,
9 Gamaa St., 12619 Giza, Egypt

Introduction

Pythium Pringsh. is a fungal genus in the class Oomycota and kingdom Chromalveolata, subgroup-Stramenopiles/Heterokontophyta (Beakes et al. 2012). Pythium originates from the Greek word "pythein" – the causal organism of root-rot (Merriam-Webster 2014), and it is quite appropriately termed. The genus *Pythium* is a complex fungal genus containing over 355 identified species (www. mycobank.org), designated species that inhabit a spread of terrestrial and aquatic ecological habitats (Dick 2001). Possibly the most economically significant and aggressive species such as *P. aphanidermatum*, and *P. irregulare* of the current pathogen were able to infect different plant species (Hendrix and Compbell 1973). The new species that presently lack type specimens and recognized descriptions have increased the total number species up to 400. Two features have contributed to the quick growth in the number of designated species. The first one is an increase in the diversity of agricultural zones from which new species are being isolated. The number of host ranges is also susceptible to different *Pythium* species. These include diseased plants and infested soils in both cultivated lands and non-cultivated fields and forests (Kageyama 2014). Numerous *Pythium* species are also recognized to cause plant diseases, which mainly include *P. aphanidermatum* and *P. irregulare* (Sutton et al. 2006, Ivors and Moorman 2014, Kageyama 2014). Certainly, *P. aphanidermatum* is one of the most distressing pathogenic species in China, particularly through the summer, with the widest host range, to be followed by *P. ultimum* and *P. irregulare* (Ho et al. 2008).

The causal organism

Pythium is a common term used herein to symbolize some species of fungal-like organisms in the Genus *Pythium*. The fungus *Pythium* belongs to the phylum Oomycota and the genus is believed to incorporate more than 355 recorded species (Ho 2018, Webster and Weber 2007, www.mycobank. org). In the 1930s, *P. anandrum* was the first *Pythium* sp. investigated and detected in the United States. These species showcase an extensive range of interactions, from aggressive plant pathogens, saprophytes, weak root pathogens as well as biocontrol agents (Martin and Loper 1999, Le et al. 2015). When the environment is favourable for these organisms and negative for plant life, infection of plants may also occur. *Pythium* spp. as plant pathogens are able to infect a large number of plants,

---

*Corresponding author: asran.amal@gmail.com

causing seed, root and lower stem-rot, as well as seedling damping-off of many plants (Broders et al. 2007, Martin and Loper 1999). Plant infections commonly arise in roots, lower stems, and fruit close to the soil, and soft plant tissues which exist naturally in seedlings of plants before or quickly after emergence from the soil. The detection and classification key of *Pythium* species are generally hard because the shortage of table morphological character types during the first phase isn't enough to supply essential recognition keys to the genus *Pythium* (Dick 1990). The populations of *Pythium* species are genetically varied (Martin 2009) and reveal a significant difference in terms of virulence, host range, and geographic distribution (Van Der Plaats-Niterink 1981). Despite being members of the fungus lineage that produces filiforms sporangia, *P. aphanidermatum* and *P. arrhenomanes* have distinct temperature optima and levels of virulence (Martin 2000). The infection mechanisms caused by the current pathogens to induce damping-off diseases are still not clear and may severely decrease plant survival (Raaijmakers et al. 2009). Several fungal species seem to be extremely aggressive, as a result of the induction of necrotrophic infections (Jarosz and Davelos 1995, Raaijmakers et al. 2009). This infectious agent contains a broad host spectrum and a good geographical distribution (Hendrix and Campbell 1973, Schüler et al. 1989, Francis and St-Clair 1997, Paulitz and Adams 2003, Levesque and De Cock 2004).

Economic significance

Pythium species are plant pathogens of economic significance in agricultural commodities. The host range comprising 180 plant hosts infected with distinct *Pythium* species has been discussed elsewhere in the book. *Pythium* species are extensively disbursed, and maximum species are saprobic in the soil, water, and remains of dead plants and animals. *Pythium* includes plant pathogens such as *P. aphanidermatum*, animal pathogens like *P. insidiosum*, algal pathogens such as *P. porphyrae*, mycoparasites such as *P. oligandrum*, and saprophytes like *P. intermedium*. Extensively disbursed all over the world, they occur in freshwater, seawater, and terrestrial habitats (Kageyama 2014). *Pythium* species are vital pathogens of plant species and animals, and some can be useful as biocontrol agents – protecting plants from pathogenic fungi (van der Plaats-Niterink 1981, Ali-Shtayeh and Saleh 1999). They are pathogens of monocotyledonous plants, causing severe damage to numerous flora, inclusive of cereal crops, turf grasses, forest, and fruit trees; they lead to rot of fruit, the rot of roots, and stems, and pre- or post-emergence damping-off. *Pythium* spp. are normally considered to be root pathogens, but from time to time, some species of this fungus additionally cause severe foliage blight and root necrosis of higher plants. They are accountable for poor germination damping-off of seedlings in crops and also of ornamentals, vegetable plants, and forest trees and in nurseries. Some species result in decay and decomposition of plant remains within the soil. *Pythium* infects the seed at or prior to germination, attacking the younger seedling before or just after emergence. Fungal pathogens destroy the root system by causing damping-off, which results in deleterious effects on seed germination and stunted plants with shortened or distorted leaves, fewer tillers and smaller heads. Bean root rot caused by *Pythium* species resulted in a decline in soil fertility leading to bean yield losses (Miklas et al. 2006). In Western Africa, *Pythium* spp. are the fungal pathogens frequently isolated from severe root-rot epidemics (Rusuku et al. 1997). Crop losses of up to 70% in marketable bean cultivars have been described in Rwanda and Kenya (Otsyula et al. 2003, Nzungize et al. 2012). In further studies carried out in these republics, the same species have been collected from bean samples affected by root rot symptoms. It is not easy to manage the *Pythium* species once the rot begins, and the rapid death of crops occurs when infected. The economic impact of different *Pythium* types can derive from stand reduction, poor flower vigour, delayed germination, poor plant losses and a decrease in yield crop.

Taxonomy and biological characteristics of *Pythium* spp.

The genus *Pythium* belongs to the family Pythiaceae, order Pythiales, class Oomycetes, Phylum

Oomycota and kingdom Chromista (Kirk et al. 2008). There are almost 355 *Pythium* species described (www. mycobank.org) (Table 5.1). Hence, its taxonomy is difficult, which is classified in the Kingdom Straminopila; some selected species are shown in Table 5.2 (Webster and Weber 2007, Ho 2018). Historically, there has also been great confusion regarding the validity of *Pythium* as a distinct genus (Ho 2018). The genus *Pythium* Pringsheim, recognized by Pringsheim in 1858, belongs to Pythiaceae, Pythiales, Oomycetes, Oomycota, and Straminipila (Hulvey et al. 2010). The observation of genus *Pythium* is still focus on the absence of identity guidelines for the species such as, the absence of extensive check-lists in the species, their very own dispersal and matrices, the insufficiency of or the inability of the types and information (Bala et al. 2010, Uzuhashi et al. 2010, WenHsiung et al. 2010).

How can the Genus *Pythium* survive in the soil?

Diverse *Pythium* spp. are able to live and persist in soil for long periods of time by two different types of the spores. The main survival structures of most *Pythium* species are oospores and sporangia; *Pythium* species can live in the soil for up to 10 years because oospores naturally have thickened walls (Martin et al. 1999). When soils are water saturated, oospores can germinate and produce swimming spores called zoospores or directly infect susceptible plant tissues within four hours of contact with plant exudates (Levesque et al. 1993). Fungal oospores can continue to be found in former plant debris that includes undecomposed root beginnings and origins which were previously contaminated and infected with plant cells or perhaps infested land in the following season. For instance, the culture-produced oospores of *P. ultimum* survived air-drying and were still viable after eight months of storage (Lumsden and Ayers 1975). Hoppe (1966) described that oospores of a *Pythium* sp. were still viable after 12 years in an air-dried muck soil. Constitutive latency of oospores increases the biological fitness of *Pythium* spp. because it prevents the germination of all spores at the same time, which is unfavourable for sustained growth and survival of fungal pathogens. In some species, like *P. aphanidermatum*, the transformation to germinability is not related to the change in wall thickness of oospores (Lumsden et al. 1987). A motile asexual spore can swim for 20 to 30 hours and transfer three or more inches inside the soil. Normally, zoospores swim against earth gravity and after reaching the surface of flooded soil, may be passed to new roots if the water is moving. Zoospores may die or survive for up to a week without free water in the soil. Zoospores are then attracted by chemicals to root tips where infection typically occurs. Transfer irrigation water infested with *Pythium* spp. spores provides an effective method for spread of *Pythium* in different growing areas.

Asexual reproductive structures

Sporangia and hyphal swellings are asexual reproductive structures in *Pythium* spp. Their morphology varies from spherical hyphal swellings created by *P. ultimum*, *P. sylvaticum*, and *P. heterothallic* to inflated filamentous or lobate forms typified by sporangia of *P. aphanidermatum*. A spherical structure can survive long periods of time in dry soil and therefore function as an important inoculum for saprophytic colonization of substrates or attack of a susceptible plant. For example, the spherical hyphal structure of *P. ultimum* survived for one year in an air-dried soil (Stanghellini and Hancock 1971). On the contrary, inflated filamentous or lobulated sporangia are assumed to be short-term in the soil. However, a long persistent filamentous and lobulated sporangia in the soil has been described. For instance, the sporangia of *P. aphanidermatum* and *P. graminicola* stay alive in soil for six and four weeks, respectively (Peethambaran and Singh 1977). Van der Plaats-Niterink (1981) and Dick (1990) used sporangia to identify asexual structures that consist of zoospores and hyphal swellings. However, some reviews used sporangia term to detect the asexual structures and zoosporangia to discriminate *Pythium* species that produce zoospores.

Table 5.1. List of the identified *Pythium* species (www.mycobank.org)

Pythium acanthicum	*Pythium erinaceum*	*Pythium ornacarpum*
Pythium acanthophoron	*Pythium flevoense*	*Pythium orthogonon*
Pythium acrogynum	*Pythium folliculosum*	*Pythium ostracodes*
Pythium adhaerens	*Pythium glomeratum*	*Pythium pachycaule*
Pythium amasculinum	*Pythium graminicola*	*Pythium pachycaule*
Pythium anandrum	*Pythium grandisporangium*	*Pythium paddicum*
Pythium angustatum	*Pythium guiyangense*	*Pythium paroecandrum*
Pythium aphanidermatum	*Pythium helicandrum*	*Pythium polare*
Pythium apleroticum	*Pythium helicoides*	*Pythium polymastum*
Pythium aquatile	*Pythium heterothallicum*	*Pythium porphyrae*
Pythium aristosporum	*Pythium hydnosporum*	*Pythium prolatum*
Pythium arrhenomanes	*Pythium hypogynum*	*Pythium proliferatum*
Pythium attrantheridium	*Pythium indigoferae*	*Pythium pulchrum*
Pythium bifurcatum	*Pythium inflatum*	*Pythium pyrilobum*
Pythium boreale	*Pythium insidiosum*	*Pythium quercum*
Pythium buismaniae	*Pythium intermedium*	*Pythium radiosum*
Pythium butleri	*Pythium irregulare*	*Pythium ramificatum*
Pythium camurandrum	*Pythium iwayamae*	*Pythium regulare*
Pythium campanulatum	*Pythium jasmonium*	*Pythium rhizo-oryzae*
Pythium canariense	*Pythium kunmingense*	*Pythium rhizosaccharum*
Pythium capillosum	*Pythium litorale*	*Pythium rostratifingens*
Pythium carbonicum	*Pythium longandrum*	*Pythium rostratum*
Pythium carolinianum	*Pythium longisporangium*	*Pythium salpingophorum*
Pythium catenulatum	*Pythium lutarium*	*Pythium scleroteichum*
Pythium chamaehyphon	*Pythium macrosporum*	*Pythium segnitium*
Pythium chondricola	*Pythium mamillatum*	*Pythium spiculum*
Pythium citrinum	*Pythium marinum*	*Pythium spinosum*
Pythium coloratum	*Pythium marsipium*	*Pythium splendens*
Pythium conidiophorum	*Pythium mastophorum*	*Pythium sterilum*
Pythium contiguanum	*Pythium megacarpum*	*Pythium stipitatum*
Pythium cryptoirregulare	*Pythium middletonii*	*Pythium sulcatum*
Pythium cucurbitacearum	*Pythium minus*	*Pythium terrestris*
Pythium cylindrosporum	*Pythium monospermum*	*Pythium torulosum*
Pythium cystogenes	*Pythium montanum*	*Pythium tracheiphilum*
Pythium debaryanum	*Pythium multisporum*	*Pythium ultimum*
Pythium deliense	*Pythium myriotylum*	*Pythium ultimum* var. *ultimum*
Pythium destruens	*Pythium nagaii*	*Pythium uncinulatum*
Pythium diclinum	*Pythium nodosum*	*Pythium undulatum*
Pythium dimorphum	*Pythium nunn*	*Pythium vanterpoolii*
Pythium dissimile	*Pythium oedochilum*	*Pythium viniferum*
Pythium dissotocum	*Pythium okanoganense*	*Pythium violae*
Pythium echinulatum	*Pythium oligandrum*	*Pythium volutum*
Pythium emineosum	*Pythium oopapillum*	*Pythium zingiberis*
		Pythium zingiberu

Table 5.2. Taxonomic position of three major plants pathogenic *Pythium* species

Taxonomic Hierarchy
Domain: Eukaryota
Kingdom: Chromista
Phylum: Oomycota
Class: Oomycetes
Order: Pythiales
Family: Pythiaceae
Genus: Pythium
Species: *Pythium aphanidermatum* (Edson) Fitzp.
Species: *Pythium ultimum* Trow
Species: *Pythium debaryanum* R. Hesse

Major three plant pathogenic *Pythium* species

Pythium aphanidermatum

The first major *Pythium* species is *P. aphanidermatum* which can produce two types of spores including filamentous sporangia with reniform biciliate zoospores. Oogonia is terminal, smooth, and spherical; oospores are smooth and spherical. Antheridia are stalked or intercalary; usually, one, or rarely two, antheridia are formed per oogonium (Fig. 5.1). *P. aphanidermatum* has a wide range of plant hosts and also infect plant nematodes (Tzean et al. 1981) besides human wounds (Calvano et al. 2011), although in the USA greenhouses it is most frequently found infecting *poinsettias* (Moorman et al. 2002). Most disease symptoms caused by *P. aphanidermatum* are root rot in plants grown hydroponically (Sutton et al. 2006) or soilless culture of cucumber in Rockwool (Moulin et al. 1994). *P. aphanidermatum* is isolated from a fast-growing garden plant grown in horticultural substrates based on peat containing variable percentages of pine wood chips, and an industry standard potting mix. Disease severity of *P. aphanidermatum* is often less widespread than in an old-style potting mix, which offers an extra indication of the capability of pine wood chips to be used as a replacement to perlite in substrates as effective procedures for improving production in the greenhouses (Laura et al. 2013). Also, *P. aphanidermatum* is the main causal organism for a leak,

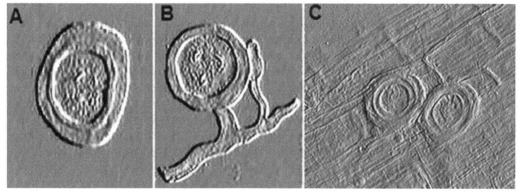

Figure 5.1. Some *Pythium aphanidermatum* structures, A) oospores, B) antheridium oogonium C) oospores inside plant host tissue

root rot, damping-off of cucurbits, also legumes, and different plant hosts (Fig. 5.2). The fungus develops water-soaked soft yellow to light-brown rot on fruit. The fungal pathogen can survive in the soil; primary infection comes from field and secondary infestation from contact in transit or storage.

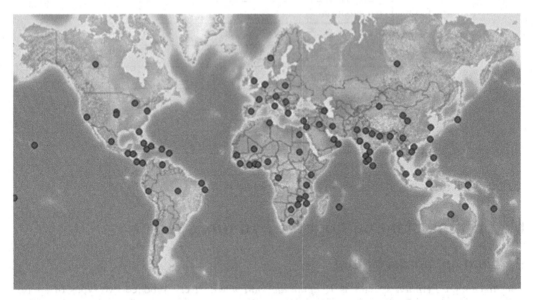

Figure 5.2. The distribution map of *P. aphanidermatum*, the causal organism of damping-off. (Reprinted from https://www.cabi.org/isc/datasheet/46141)

Economic impact of P. aphanidermatum

Until now, there have not been many investigations related to the economic impact of *P. aphanidermatum* on different type of crops. Decreased crop yields are due to the damping-off caused by this pathogen alone, independent of another *Pythium* spp. or general root rotting fungal pathogens are not commonly described. In addition to plant death, crop losses can also be caused by high infections of the root system, which decreases the growth vigour of the plant host and effect on crop yield. However, given its wide spectrum of plants and various geographic distribution (Fig. 5.3), it is estimated that losses attributed to *P. aphanidermatum* could be important, particularly in the warm growing regions of the world.

P. aphanidermatum hosts range

Among the *Pythium* species, *P. aphanidermatum* is diverse in distribution and one of the most important soil-borne fungus pathogenic to different plant hosts in warmer parts of the world. *P. aphanidermatum* is known to cause infection on a wide range of plant species, belonging to more than 65 different plant species such as: *Abelmoschus esculentus* (okra), *Acacia nilotica* (gum arabic tree), *Agrostis stolonifera* var. *palustris* (bent grass), Amaranthus (amaranth), *Amaranthus cruentus* (redshank), *Arachis hypogaea* (groundnut), *Beta vulgaris* (beetroot), *Beta vulgaris* var. *saccharifera* (sugarbeet), *Brassica juncea* var. *juncea* (Indian mustard), *Brassica oleracea* var. *botrytis* (cauliflower), *Brassica oleracea* var. *capitata* (cabbage), *Brassica rapa* subsp. *Pekinensis, Cannabis sativa* (hemp), *Capsicum* (peppers), *Capsicum annuum* (bell pepper), *Carica papaya* (pawpaw), *Carthamus tinctorius* (safflower), *Catharanthus roseus* (Madagascar periwinkle), *Citrullus lanatus* (watermelon), *Crotalaria juncea* (sunn hemp), *Cucumis sativus* (cucumber), *Curcuma longa* (turmeric), *Cymbopogon winterianus* (java citronella grass), *Cynara cardunculus* var. *scolymus* (globe artichoke), *Daucus carota* (carrot), *Dianthus caryophyllus* (carnation), *Elettaria*

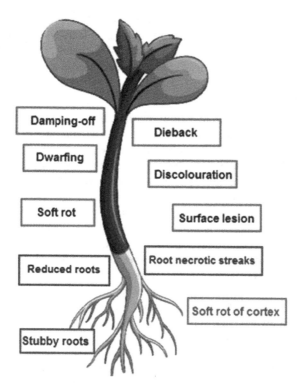

Figure 5.3. Disease's symptoms caused by *P. aphanidermatum* appeared on three main parts in plant species, such as Roots symptoms include: necrotic streaks or lesions, reduced root system, soft rot of cortex. Foliar parts symptoms include: internal rotting or discolouration, soft rot, surface lesions or discolouration. Whole plant symptoms include: damping off, dwarfing and plant dead, dieback and other symptoms.

cardamomum (cardamom), *Euphorbia pulcherrima* (poinsettia), *Ficus benjamina* (weeping fig), *Fragaria ananassa* (strawberry), *Geranium* (cranesbill), *Glycine max* (soybean), *Gossypium* (cotton), *Gypsophila paniculata* (baby's breath), *Hordeum vulgare* (barley), *Lactuca sativa* (lettuce), *Lampranthus spectabilis, Lens culinaris* subsp. *culinaris* (lentil), *Linum usitatissimum* (flax), *Luffa aegyptiaca* (loofah), *Mentha piperita* (Peppermint), *Mesembryanthemum crystallinum* (crystalline iceplant), *Nicotiana tabacum* (tobacco), *Nopalea cochenillifera* (cochineal cactus), *Opuntia ficus-indica* (prickly pear), *Parthenium argentatum* (Guayule), *Phaseolus vulgaris* (common bean), *Pinus* (pines), *Pisum sativum* (pea), *Solanum lycopersicum* (tomato), *Solanum melongena* (aubergine), *Solanum tuberosum* (potato), *Spinacia oleracea* (spinach), *Triticum aestivum* (wheat), *Vigna unguiculata* (cowpea), *Zea mays* (maize), *Zingiber officinale* (ginger), *Ziziphus mauritiana* (jujube). (https://www.plantwise.org/).

Pythium ultimum

P. ultimum (syn. *Globisporangium ultimum*) belongs to the peronosporalean lineage of oomycetes (Uzuhashi et al. 2010). *P. ultimum* is able to propagate saprotrophically in soil and plant residue. This species produces spherical sporangia, which are generally terminal but maybe intercalary and zoospores are never formed. Oogonia is smooth and usually terminal, rarely intercalary and spherical. Antheridia arises from a cluster of hyphae under the oogonium; normally, one antheridium is formed per oogonium. *P. ultimum* is also homothallic, has male and female reproductive structures on the same thallus and is able to produce oospores capable of long-term persistence in low water content (Martin and Loper 1999). The first record of *P. ultimum* was made by Wager in the Union of South Africa during 1931 (Wager 1931). But the occurrence of *P. ultimum* was not described as a disease in the U.S. until the 1940s when some plant pathologists reported that *P. ultimum* can infect rhubarb grown in California (Hendrix and Campbell 1973). *P. ultimum* is one of the most common plant pathogenic fungus recorded from soil and various plant hosts (Plaats-Niterink 1981, Levesque et al. 2010). *P. ultimum* is one of the most damaging members of the genus, like *P. aphanidermatum* and *P. irregulare,* also topping lists of important *Pythium* species (Martin and Loper 1999).

Molecular characterization

Genome sequence analysis of the *P. ultimum* proposes that not all oomycete plant pathogens contain a comparable 'toolkit' for persistence and pathogenesis. *P. ultimum* has a distinct effector range compared to *Phytophthora* spp. The collection of metabolic genes within the *P. ultimum* genome displays its pathogenic lifestyle. Widely distributed pathogens are able to infect young seedlings and plant roots with little or no cuticle or heavily suberized tissues, reliable for lack of cutinase encoding genes. It is a poor competitor against secondary pathogens of damaged plant tissues and soil causal organisms with improved saprobic capacity. *P. ultimum* is really a composite species that has a varied genetic diversity that includes *P. ultimum* var. *ultimum* and var. *sporangiiferum*, which has not differentiated the use of internal transcribed spacer (ITS) sequences (Schroeder et al. 2013). Fungus is able to produce oogonia but sporangia and zoospores are formed hard. *P. ultimum* var. *ultimum* forms sporangia-like hyphal swellings which germinate to yield infective hyphae (Stanghellini and Hancock 1971). Var. *sporangiiferum* is not a common species able to produce zoospores at room temperature (Drechsler 1960, Plaats-Niterink 1981). Correct taxonomic classification is needed for *Pythium* species when the pathogenicity varies inside a species complex, primarily after specific disease administration methods are used. The existence or lack of sporangia and zoospore development can perform an important role intended for dispersal probable, infestation approaches and pathogen persistence (Jeger et al. 2008). *P. irregulare* forms structures similar to *P. ultimum* var. *ultimum*. They form a globose sporangia, which is extremely virulent at low-temperatures (Martin and Loper 1999). They are widely distributed, and show high genetic and morphological variations (Harvey et al. 2000, 2001, Spies et al. 2011).

P. ultimum host range

P. ultimum is a soil-borne pathogen which develops symptoms such as damping off, root rot and lower stem parts of various plants hosts inclusive of corn, cabbage, carrot, cucumber, melon, soybean, potato, wheat, fir, forests, ornamentals, and lots of other crops (Hendrix and Campbell 1973, Plaats-Niterink 1981, Palmucci et al. 2011). For example, *P. ultimum* is a causal agent of the Pythium blight of turfgrass, which causes serious harm to golf courses (Allen et al. 2004). *P. irregulare, P. ultimum* var. *ultimum* and *P. ultimum* var. *sporangiiferum* are the most aggressive *Pythium* species affecting soybean in the USA (Broders et al. 2007). *P. ultimum* has been found in Australia, Brazil, Canada, China, Japan, Korea, South Africa and many other countries all over the world. Similar to the *Pythium ultimum* var. *ultimum, P. aphanidermatum* has a wide range of plant hosts and is isolated from greenhouses and high-temperature environments (Gold and Stanghellini 1985, Koike et al. 2006). In general, abundant soil moisture and high soil temperature are the two most important environmental factors that regulate the distribution of *P. ultimum*. A warm greenhouse could be attributed to the introduction of *P. ultimum* into the cold regions, where the incidence of *P. ultimum* has been rarely seen due to very low temperature (Tojo et al. 2001). On the other hand, the host range of *P. arrhenomanes* is specific to monocot plants (Martin and Loper 1999).

Symptoms

P. ultimum induces diverse signs and symptoms which include root rot, damping off, often stunted or chlorotic, which might be similar to nitrogen deficiency. Moderate infections can induce plant wilt, reduce plant populations, and retard maturation. Excessive infections frequently result in plant downfall and death. For instance, early symptoms of turfgrass infected through *P. ultimum* are dark green, and then excessive harm results in rounds or sporadically formed patches of turfgrass swards (Allen et al. 2004). Symptoms are much more apparent below optimum environmental conditions like warm weather and excessive soil moisture. The roots of the diseased plants appear to be brown and round, with thick-walled spores that might be located in root cells.

Pythium debaryanum

Pythium debaryanum host range

Arachis hypogaea (groundnut), *Banksia marginata* (silver banksia), *Beta vulgaris* (beetroot), *B. vulgaris* var. *saccharifera* (sugarbeet), *Brassica nigra* (black mustard), *Brassica oleracea* var. *capitata* (cabbage), Cactaceae (cacti), *Capsicum annuum* (bell pepper), *Carica papaya* (pawpaw), *Citrullus lanatus* (watermelon), Citrus, Clarkia (satin flowers), *Colocasia esculenta* (taro), *Cucumis sativus* (cucumber), Cyclamen, *Eucalyptus globulus* (Eucalyptus*), Euphorbia pulcherrima* (poinsettia), *Fragaria vesca* (wild strawberry), *Hordeum vulgare* (barley), Malus (ornamental species apple), *Musa x paradisiaca* (plantain), *Nicotiana tabacum* (tobacco), Nothofagus, *Petroselinum crispum* (parsley), Phaseolus (beans), Pinus (pines), *Pinus caribaea* (Caribbean pine), *Pinus elliottii* (slash pine), *Pinus taeda* (loblolly pine), *Prunus persica* (peach), *Saccharum officinarum* (sugarcane), *Schefflera actinophylla* (umbrella tree), *Solanum lycopersicum* (tomato), *Solanum melongena* (aubergine), *Syzygium aromaticum* (clove), *Theobroma cacao* (cocoa), *Trifolium subterraneum* (subterranean clover), Triticum (wheat), Tulipa (tulip), *Ulmus americana* (American elm), *Vigna mungo* (black gram), *Vigna radiata* (mung bean), Viola wittrockiana hybrids, *Zea mays* (maize) (https://www.plantwise.org/).

Disease symptoms

P. debaryanum is one of the most important causal organisms of seedling diseases known as damping-off. Damping-off symptoms can be detected from seeding until the 4th to the 6th week of cultivation (Horst 2013). The disease symptoms may be divided into two phases primarily based on the time of its presence (Table 5.3). They arise while seeds decay prior to emergence. This can occur (i) before seed germination, or when (ii) the germinating seeds are killed by means of biotic stresses while shot tissues are still below ground (Crous 2002, Horst 2013). In the first stage, seeds turn out to be smooth, rotten and fail to germinate. In the second stage, stems of germinating seeds are affected by function water-soaked lesions formed at or under the soil line (Cram 2003, Landis 2013). With the development of the disorder, those lesions may additionally darken to reddish-brown, brown, or black. Growing lesions quickly girdle younger and smooth stems. Seedlings may possibly wilt and die soon before emergence. The preferred random wallet of negative seedling emergence is an illustration of the pre-emergence damping-off. Post-emergence damping-off symptoms appear while seedlings decline, wilt, and die after emergence (Horst 2013). In most instances, all symptoms bring about the crumble and loss of life of at least some seedlings in any given seedling population. In the case of the soil-borne pathogen, there may be the death of seedlings in groups in kind of round patches and the seedlings might also have stem lesions at a ground degree. Overall, the signs and symptoms on the stem of the seedlings consist of water-soaked, a sunken lesion at or barely below the floor degree and sometimes also below the ground line (i.e. on the roots), causing the plant to fall over (Filer and Peterson 1975). Living plants are stunted, and affected areas often show irregular growth, moreover causing a watery leak of potatoes. Leak starts as a brown discolouration around a wound and briefly spreads to cover all potato plants, which are soft, simply crumpled, and drips a brown liquid with the slightest pressure. Entrance to the tuber is normally through harvest wounds. *Pythium* hyphae grow through the soil in incredible profusion and may input seedlings through both stomata and undestroyed epidermis. Furthermore, seedlings root browning results in the development of the necrotrophic segment of the disorder, which is related to the sudden decline in the growth rate of the roots and foliar part of affected pepper plants (Sutton et al. 2006, Aliyu et al. 2012). The disease's symptoms of *P. aphanidermatum* appeared on three main parts of the plant species, also infected several hosts including African-violet rot, aloe root rot, black rot of orchids, begonia root rot, coleus blackleg, geranium reducing rot, bean and basil root rot, rhubarb crown rot, mottle necrosis of sweet potato, and other forms of rot.

Table 5.3. List of the most important plant diseases caused by *Pythium* spp.

No.	Name of disease	Pathogen
1	Damping off	*Pythium debaryanum*
2	Soybean seedling blight	*Pythium* sp.
3	Damping-off on *Cucumis melo*	*Pythium spinosum*
4	Pythium root dysfunction in turfgrass	*P. volutum, P. aristosporum, P. arrehenomanes*
5	Pythium blight in turfgrass	*P. aphanidermatum*
6	Pythium root rot in turfgrass	*Pythium* spp.
7	Wilt and vascular root rot	*Pythium tracheiphilum*
8	Wilt on peanut	*Pythium myriotylum*
9	Wilt on lettuce and also leaf blight	*Pythium tracheiphilum*
10	Wilt of Nicotiana	*Pythium aphanidermatum*
11	Pythium root rot on Hydroponic Lettuce	*Pythium dissotocum*
12	Root seedling in rice	*P. arrhenomanes, P. graminicola, P. inflatum*
13	Dying soybean seedlings	*Pythium debaryanum, Pythium aphanidermatum P. irregular, P. myriotylum, P. paroecandrum, P. spinosum, P. sylvaticum, P. torulosum*
14	Dying corn seedlings	*P. paroecandrum, P. catenulatum*
15	Root and hypocotyl disease in common bean	*P. conidiophorum, P. diclinum, P. intermedium, P. irregulare, P. lutarium, P. mamillatum, P. pachycaule*
16	Root rot of cucurbits fruits	*Pythium acanthicum, P. myriotylum, P. periplocum*
17	Root rot of Stevia	*Pythium myriotylum.* In field plantings, *P. irregulare, P. aphanidermatum*
18	Root rot of bean.	*Pythium aristosporum*
19	Root rot of tomato, broadleaf signalgrass, large crabgrass, barnyardgrass, nutsedge, goosegrass, itchgrass and johnsongrass	*Pythium arrhenomanes*
20	Root rot of bean	*Pythium carolinianum, Pythium catenulatum*
21	Root rot of bean, and spinach	*Pythium dissotocum*
22	Melon Root rot and Fruit rots of other cucurbits in cool weather and Seed Decay of corn. Root rot and Crown rots of clovers and basil	*Pythium irregulare*
23	Damping-off on celery	*Pythium mastophorum*
24	Root rot of tomato	*Pythium myriotylum*
25	Stem, Crown and Root rot on lupine	*Pythium paroecandrum*
26	Bottom rot and Damping-off on cabbage	*Pythium polymastum*
27	Root rot of Chinese evergreen, peperomia, and philodendron	*Pythium splendens*
28	Fruit rot of muskmelon, often with luxuriant white fungus growth; Damping-off, Root rot of many seedlings in greenhouse and field	*Pythium ultimum*
29	Root rot and Crown rot of clovers. Crown rot of impatiens. Root rot of kiwi. Stunt and Leaf Yellowing on lettuce.	*Pythium uncinulatum*

Loop-mediated isothermal amplification (LAMP)

On-site detection methods such as Loop-mediated isothermal amplification (LAMP) have been observed as an innovative gene amplification tool and appear as unconventional to PCR-based procedures in both clinical laboratory and food safety diagnostics assay. Currently, LAMP has been used for detection and identification of various food pathogens collected from agricultural commodities, as it showed important advantages such as high sensitivity, specificity, and quickness (Li et al. 2017). To offer practical support for the screening and management of plant diseases caused by *P. aphanidermatum*, a fast and precise loop-LAMP technique is applied for detection of *P. aphanidermatum* in different samples. A specific primer set from the rDNA ITS sequence of *P. aphanidermatum* was designed. Primer specificity was checked using 57 *Pythium* species, the LAMP method provided no cross-reactions in the other 39 *Pythium* species, and 11 strains of *Phytophthora* spp. as well as eight other soil-borne fungi. The sensitivity limit was 10 fg of genomic DNA (Fukuta et al. 2013). Li and his co-workers also used a gene encoding Trypsin protease to develop a new specific primer for *P. aphanidermatum*. The specificity investigation indicated that positive results were only obtained from *P. aphanidermatum* isolates. The detection limit was 100 pg.μL^{-1}, which was similar to real-time PCR assay, and was nearly more than 100 times PCR-based technique (Li et al. 2018). In another *Pythium* species, a specific primer set was designed from the ITS region to detect *P. myriotylum* fungal pathogen in a field sample. Analysis of the annealing curve of the LAMP reaction products increases the reliability of the LAMP diagnosis. The detection limit was almost 100 fg of genomic DNA, which was as sensitive as PCR. The on-site technique can detect *P. myriotylum* in hydroponic solution samples and in almost the same results agreed with the conventional plating procedures. The LAMP test was used as a diagnostic method for observing *P. myriotylum* in open field (Fukuta et al. 2014). Species-specific, LAMP test was used for the fast detection of *P. helicoides*. The primers were designed using the ribosomal DNA ITS sequence. 40

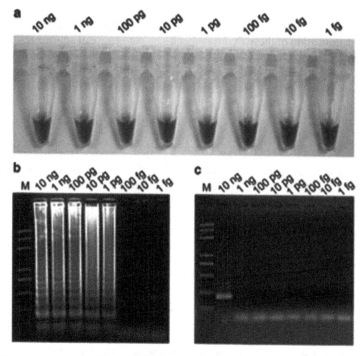

Figure 5.4. The sensitivity of the LAMP and conventional PCR assays for detecting *P. ultimum*. a) Detection of LAMP products using HNB as a visual indicator. b) Detection of LAMP products by agarose gel electrophoresis. c) Detection of conventional PCR products by agarose gel (Reprinted from Shen 2017). Seek permission from the publisher

Pythium species including *P. helicoides*, 11 *Phytophthora* species and eight other fungal pathogens were screened for specificity. A sensitivity test was approved using fungal DNA extracted from *P. helicoides*, and the detection limit was 100 fg which is compared with traditional PCR (Takahashi et al. 2014). LAMP PCR primers designed from the sequences of the *P. irregulare* ribosomal DNA ITS region were used for fast detection of the fungal pathogen. The specificity of the primers for *P. irregulare* was screened using 50 isolates of 40 *Pythium* species, 11 *Phytophthora* isolates and eight isolates of seven other soil-borne fungi. The developed test indicated that the limit of sensitivity of the LAMP technique was 100 fg of pure DNA similar to PCR-based DNA (Feng et al. 2015). The LAMP-PCR method was applied for precise detection of *P. ultimum*, and positive results were achieved only with *P. ultimum* isolates (Fig. 5.4). The detection limit of the LAMP system was 1 pg μL^{-1} DNA, which is 1000 times more sensitive than PCR-based technique (Shen et al. 2017).

Conclusion and future perspectives

Several *Pythium* species are plant pathogenic fungi on several economic crops and in agricultural commodities. The current pathogen is responsible for poor germination and has lost a high number of seedlings in field crops as well as ornamentals, vegetable plants, and forest trees in nurseries and hydroponics. Members of the genus *Pythium* are soil-borne fungi that appear all over the world. This chapter discusses the top three pathogenic *Pythium* spp. with the intent of initiating discussion and debate amongst the plant mycology community. Loop-mediated isothermal amplification is a rather novel method of enzymatic deoxyribonucleic acid amplification which can be applied for the examination of soil-borne fungi. Koji Kageyama's Lab. in Japan developed on-site detection techniques for fast diagnosis of the most important pathogenic species of *Pythium* such as *P. aphanidermatum, P. myriotylum, P. helicoides, P. irregulare,* and *P. ultimum*. The specific primers were designed using the ribosomal DNA ITS sequence and gene encoding Trypsin protease. The lack of special morphological structures is one of the major limitations in the taxonomic characterization of *Pythium. Pythium* pathogenic species are also identical in appearance and ecology, resulting in confusion in the identification of these *Pythium* species by conventional approaches. A wide-ranging analysis of both morphology and molecular data is still critical for the identification of new species in the genus *Pythium*. Still, more research is required to identify and clarify the relationships among *Pythium* species.

References

Ali-Shtayeh, M.S. and Saleh, A.S. 1999. Isolation of *Pythium acanthicum, P. oligandrum,* and *P. periplocum* from soil and evaluation of their mycoparasitic activity and biocontrol efficacy against selected phytopathogenic *Pythium* species. Mycopathologia 145: 143-153.

Aliyu, T.H., Balogun, O.S. and Adesina, O.M. 2012. Effect of *Pythium aphanidermatum* on two cultivars of pepper (*Capsicum* spp). Niger. J. Agric. 8: 64-69.

Allen, T.W., Martinez A. and Burpee, L.L. 2004. Pythium blight of turfgrass. http://www.apsnet.org/education/LessonsPlantPath/pythiumblight/default.htm

Bala, K., Robideau, G.P., Levesque, C.A., de Cock, A.W.A.M., Abad, Z.G., Lodhi, A.M., Shahzad, S., Ghaffar, A. and Coffey, M.D. 2010. *PhytopythiumSindhu*. Persoonia. 24: 136-137.

Beakes, G.W., Glockling, S.L. and Sekimoto, S. 2012. The evolutionary phylogeny of the oomycete "fungi". Protoplasma 249: 3-19.

Broders, K.D., Lipps, P.E., Paul, P.A. and Dorrance, A.E. 2007. Characterization of *Pythium* spp. associated with corn and soybean seed and seedling disease in Ohio. Plant Dis. 91: 727-735.

Calvano, T.P., Blatz, P.J., Vento, T.J., Wickes, B.L., Sutton, D.A., Thompson, E.H., White, C.E., Renz, E.M. and Hospenthal, D.R. 2011. *Pythium aphanidermatum* infection following combat trauma. J. Clinical Microbiol. 49: 3710-3713.

Cram, M.M. 2003. Damping-Off. Tree Plant Notes 50: 1-5.

Crous, P.W. 2002. Damping-off. pp. 15-17. *In*: Crous, P.W. (ed.). Taxonomy and pathology of *Cylindrocladium* (Calonectria) and allied genera. The American Phytopathological Society, St. Paul, MN.

Dick, M.W. 1990. Keys to *Pythium*. University of Reading Press, Reading,UK.

Dick, M.W. 2001. The peronosporomylates. pp. 39-72. *In*: Mclaughlin, D.J., McLaughlin, E.G. and Lenke, P.A. (eds.). The Mycota VII – Part A: Systematic Evolution. Springer Verlag, Berlin.

Drechsler, C. 1952. Production of zoospores from germinating oospores of *Pythium ultimum* and *Pythium debaryanum*. B. Torrey Bot. Club 79: 431-450.

Drechsler, C. 1960. Two root-rot fungi closely related to *Pythium ultimum*. Sydowia 14(1-6).

Feng, W., Ishiguro, Y., Hotta, K., Watanabe, H., Suga, H. and Kageyama, K. 2015. Simple detection of *Pythium irregulare* using loop-mediated isothermal amplification assay. FEMS Microbiol. Lett. 362(21): 1-8.

Filer, T.H.J. and Peterson, G.W. 1975. Damping-off. *In*: Peterson, G.W. and Smith, R.S. (eds.). For. Nurs. Dis. United States. USDA Forest Service. Agric Handbook 470: 6-8.

Francis, D.M. and St-Clair, D.A. 1997. Population genetics of *Pythium ultimum*. Phytopathology 87: 454-461.

Fukuta, S., Takahashi, R., Kuroyanagi, S., Miyake, N., Nagai, H., Suzuki, H., Hashizume, F., Tsuji, T., Taguchi, H., Watanabe, H. and Kageyama, K. 2013. Detection of *Pythium aphanidermatum* in tomato using loop-mediated isothermal amplification (LAMP) with species-specific primers. Eur. J. Plant Pathol. 136: 689-701.

Fukuta, S., Takahashi, R., Kuroyanagi, S., Ishiguro, Y., Miyake, N., Nagai, H., Suzuki, H., Tsuji, T., Hashizume, F., Watanabe, H. and Kageyama, K. 2014. Development of loop-mediated isothermal amplification assay for the detection of *Pythium myriotylum*. Lett. Appl. Microbiol. 59: 49-57.

Gold, S.E. and Stanghellini, M.E. 1985. Effects of temperature on *Pythium* root rot of spinach *Spinacia oleracea* grown under hydroponic conditions. Phytopathology 75: 333-337.

Harvey, P.R., Butterworth, P.J., Hawke, B.G. and Pankhurst, C.E. 2000. Genetic variation among populations of *Pythium irregulare* in southern Australia. Plant Pathol. 49: 619-627.

Harvey, P.R., Butterworth, P.J., Hawke, B.G. and Pankhurst, C.E. 2001. Genetic and pathogenic variation among cereal, medic and sub-clover isolates of *Pythium irregulare*. Mycol. Res. 105: 85-93.

Hendrix, F.F. and Campbell, W.A. 1973. Pythiums as plant pathogens. Ann. Rev. Phytopathol. 11(1): 77-98.

Ho, H.H. 2018. The taxonomy and biology of *Phytophthora* and *Pythium*. J. Bacteriol. Mycol. 6(1): 40-45.

Hoppe, P.E. 1966. *Pythium* species still viable after 12 years in air-dried muck soil. Phytopathology 56: 1411 (Abstract).

Horst, R.K. 2013. Damping-off. Westcott's plant disease handbook. Springer Netherlands, Dordrecht 177.

Hulvey, J., Telle, S., Nigrelli, L., Lamour, K. and Thines, M. 2010. Salisapiliaceae – A new family of oomycetes from marsh grass litter of southeastern North America. Persoonia 25: 109-116.

Ivors, K.L. and Moorman, G.W. 2014. Oomycete plant pathogens in irrigation water. pp. 57-64. *In*: Hong, C.X., Moorman, G.W., Wohanka, W. and Büttner, C. (eds.). Biology, Detection, and Management of Plant Pathogens in Irrigation Water. APS Press, St Paul, MN, USA.

Jarosz, A.M. and Davelos, A.L. 1995. Effects of disease in wild plant-populations and the evolution of pathogen aggressiveness. New Phytol. 129: 371-387.

Jeger, M.J. and Pautasso, M. 2008. Comparative epidemiology of zoosporic plant pathogens. Eur. J. Plant Pathol. 122: 111-126.

Kageyama, K. 2014. Molecular taxonomy and its application to ecological studies of *Pythium* species. J. Gen. Plant Pathol. 80: 314-326.

Kirk, P.M., Cannon, P.F., Minter, D.W. and Stalpers, J.A. 2008. Ainsworth & Bisby's Dictionary of the Fungi. 10th ed. Wallingford, UK: CAB International.

Ko, W.H., Lin, M.J., Hu, C.Y. and Ann, P.J. 2010. *Aquaperonospora taiwanensis* gen. et sp. nov. in Peronosporaceae of Peronosporales. Bot. Stud. 51: 343-350.

Koike, S.T., Gladders, P. and Paulus, A.O. 2006. Vegetable Diseases: A Color Handbook. CRC Press.

Landis, T.D. 2013. Forest nursery pests: damping-off. For Nurs. Notes 2: 25-32.

Laura, E.K., Emma, C.L., Garrett Owen, W., Lesley, A., Brian, E.J., Shew, H.D. and Michael, D.B. 2013. Measuring disease severity of pythium spp. and rhizoctonia so/ani in substrates containing pine wood chips. SNA Research Conference 58: 135-142.

Le, D.P., Aitken, E.A.B. and Smith, M.K. 2015. Comparison of host range and pathogenicity of isolates of *Pythium myriotylum* and *Pythium zingiberis*. Acta Horticul. 1105: 47-54.

Lévesque, C.A., Beckenbach, K., Baillie, D.L. and Rahe, J.E. 1993. Pathogenicity and DNA restriction fragment length polymorphisms of isolates of *Pythium* spp. from glyphosate treated seedlings. Mycol. Res. 97: 307-312.

Levesque, C.A. and De Cock, A.W. 2004. Molecular phylogeny and taxonomy of the genus *Pythium.* Mycol. Res. 108: 1363-1383.

Lévesque, C.A., Brouwer, H., Cano, L., Hamilton, J.P., Holt, C., Huitema, E., Raffaele, S., Robideau, G.P., Thines, M., Win, J. and Zerillo, M.M. 2010. Genome sequence of the necrotrophic plant pathogen *Pythium ultimum* reveals original pathogenicity mechanisms and effector repertoire. Genome Biol. 11(7): R73.

Li, Q., Shen, D., Yu, J., Zhao, Y., Zhu, Y. and Dou, D. 2018. Rapid detection of *Pythium aphanidermatum* by loop-mediated isothermal amplification. J. Nanjing Agric. Univ. 41: 79-87.

Li, Y., Fan, P., Zhou, S. and Zhang, L. 2017. Loop-mediated isothermal amplification (LAMP): A novel rapid detection platform for pathogens. Micro. Pathogen. 107: 54-61.

Lumsden, R.D. and Ayers, W.A. 1975. Influence of soil environment on the germinability of constitutively dormant oospores of *Pythium ultimum*. Phytopathology 65: 1101-1107.

Lumsden, R.D., García, E.R., Lewis, J.A. and Frías, T.G.A. 1987. Suppression of damping-off caused by *Pythium* spp. In soil from indigenous Mexican chinampa agricultural system. Soil Biol. Biochem. 19: 501-508.

Martin, F.N. and Loper, J.E. 1999. Soilborne diseases caused by *Pythium* spp: Ecology, epidemiology, and prospects for biological control. Crit. Rev. Plant Sci. 18: 111-181.

Martin, F.N. 2000. Phylogenetic relationships among some *Pythium* species inferred from sequence analysis of the mitochondrially encoded cytochrome oxidase II gene. Mycologia 92: 711-727.

Martin, F.N. 2009. *Pythium* Genetics. pp. 574. *In*: Lamour, K., Kamoun, S. (eds.). Oomycete genetics and genomics: Diversity, plant and animal interactions, and toolbox. Hoboken: John Willey & Sons.

Merriam-Webster.com. 2014. Pythium. *In*: Merriam-Webster. https://www.merriam-webster.com/dictionary/pythium

Miklas, P.N., Kelly, J.D., Beebe, S.E. and Blair, M.W. 2006. Common bean breeding for resistance against biotic and abiotic stresses: From classical to MAS breeding. Euphytica 147: 105-131.

Moorman, G.W., Kang, S., Geiser, D.M. and Kim, S.H. 2002. Identification and characterization of *Pythium* species associated with greenhouse floral crops in Pennsylvania. Plant Dis. 86: 1227-1231.

Moulin, F., Lemanceau, P. and Alabouvette, C. 1994. Pathogenicity of Pythium species on cucumber in peat-sand, rockwool and hydroponics. Eur. J. Plant Pathol. 100: 3-17.

Nzungize, J.R., Lyumugabe, F., Busogoro, J.P. and Baudoin, J.P. 2012. Pythium root rot of common bean: Biology and control methods. A review. Biotechnol. Agron. Soc. Environ. 16: 405-413.

Otsyula, R.M., Buruchara, R.A., Mahuku, G. and Rubaihayo P. 2003. Inheritance and transfer of root rots (*Pythium*) resistance to bean genotypes. Afr. Crop Sci. Soc. 6: 295-298.

Palmucci, H., Wolcan, S. and Grijalba, P.E. 2011. Status of the Pythiaceae (Kingdom Stramenopila) in Argentina. I. The Genus *Pythium*. Revista de la Sociedad Argentina de Botánica 46(3-4).

Paulitz, T.C. and Adams, K. 2003. Composition and distribution of Pythium communities in wheat fields in eastern Washington state. Phytopathology 93: 867-873.

Peethambaran, C.K. and Singh, R.S. 1977. Survival of different structures of *Pythium* spp. in soil. Indian Phytopathol. 30: 347-352.

Raaijmakers, J.M., Paulitz, T.C., Steinberg, C., Alabouvette, C. and Moenne-Loccoz, Y. 2009. The rhizosphere: A playground and battlefield for soilborne pathogens and beneficial microorganisms. Plant Soil 321: 341-361.

Rusuku, G., Buruchara, R.A., Gatabazi, M., Pastor-Corrales, M.A. and Schmitthenner, A.F. 1997. Effect of crop rotation on *Pythium ultimum* and other *Pythium* species in the soil. Phytopathology 52: 27.

Schroeder, K.L., Martin, F.N., de Cock, A.W.A.M., Levesque, C.A., Spies, C.F.J., Okubara, P.A. and Paulitz, T.C. 2013. Molecular detection and quantification of *Pythium* species: Evolving taxonomy, new tools, and challenges. Plant Dis. 97: 4-20.

Schüler, C., Biala, J., Bruns, C., Gottschall, R., Ahlers, S. and Vogtmann, H. 1989. Suppression of root rot on peas, beans and beetroots caused by *Pythium ultimum* and *Rhizoctonia solani* through the amendment of growing media with composted organic household waste. J. Phytopathol. 127: 227-238.

Shen, D., Li, Q., Yu, J., Zhao, Y., Zhu, Y., Xu, H. and Dou, D. 2017. Development of a loop-mediated isothermal amplification method for the rapid detection of *Pythium ultimum*. Austral. Plant Pathol. 46: 571-576.

Spies, C.F., Mazzola, M., Botha, W.J., Langenhoven, S.D., Mostert, L. and Mcleod, A. 2011. Molecular analyses of *Pythium irregulare* isolates from grapevines in South Africa suggest that this species complex may be a single variable species. Fungal Biol. 115: 1210-1224.

Stanghellini, M.E. and Hancock, J.G. 1971. Sporangium of *Pythium ultimum* as survival structure in soil. Phytopathology 61: 157-164.

Sutton, J.C., Sopher, C.R., Owen-Going, T.N., Liu, W., Grodzinski, B., Hall, J.C. and Benchimol, R.L. 2006. Etiology and epidemiology of Pythium root rot in hydroponic crops: Current knowledge and perspectives. Summa Phytopathol. 32: 307-321.

Takahashi, R., Fukuta, S., Kuroyanagi, S., Miyake, N., Nagai, H., Kageyama, K. and Ishiguro, Y. 2014. Development and application of a loop-mediated isothermal amplification assay for rapid detection of *Pythium helicoides*. FEMS Microbiol. Lett. 355: 28-35.

Tojo, M., Hoshino, T., Herrero, M.L., Klemsdal, S.S. and Tronsmo, A.M. 2001. Occurrence of *Pythium ultimum* var. *ultimum* in a greenhouse on Spitsbergen Island, Svalbard. Euro. J. Plant Path. 107: 761-765.

Tzean, S.S. and Estey, R.H. 1981. Species of *Phytophthora* and *Pythium* as nematode destroying fungi. J. Nematol. 13: 160-163.

Uzuhashi, S., Kakishima, M. and Tojo, M. 2010. Phylogeny of the genus *Pythium* and description of new genera. Mycoscience 51: 337-365.

Van der Plaats-Niterink, J. 1981. Monograph of the genus *Pythium*. Studies in Mycology 21. Baarn, The Netherlands: Centraal Bureauvoor Schimmel Cultures, 25-39.

Wager, V.A. 1931. Diseases of plants in South Africa due to members of the pythiaceae. Dept. Agr. So. Afr. Sci. Bull. 105: 1-43.

Webster, J. and Weber, R. 2007. Introduction to Fungi. 3rd edn. Cambridge University Press, UK, 846 p.

6

Pythium Species Associated With Die-back Apple Trees and Citrus Gummosis in Tunisia

Naima Boughalleb-M'Hamdi[1]*, Najwa Benfradj[1] and Souli Mounira[2]

[1] UR 13AGR03, Department of Biological Sciences and Plant Protection, High Institute of Agronomy of Chott Mariem, 4042 Sousse, University of Sousse, Tunisia
[2] Centre Techniques des Agrumes, 8099 Zaouiet Jedidi, Beni Khalled, BP 318, Nabeul, Tunisie

Introduction

Apple and citrus are among the world's most important fruit crops and rank second and third after banana in terms of value (Savita and Avinash 2012). In Tunisia, citrus production is important (Benfradj et al. 2017), covering an area of 34,844 ha and generating about 436,429 tons in production in 2013 (FAOSTAT 2015). Apple is also an economically important crop in Tunisia, with over 27,000 ha of apple cultivated in the main apple growing areas located along the northeast and northwest of the country (Souli et al. 2014).

However, various diseases caused to citrus and apple limit their productions. In fact, diseases like apple decline (Souli et al. 2011a, b) and citrus gummosis (Benfradj et al. 2017) caused serious losses to the yield. The first report of apple decline in Tunisia was investigated by Boughalleb et al. (2006). Isolates of *Phytophthora cactorum*, obtained from roots and crowns of infected trees from Foussana and Oued Eddarb (areas of Kasserine, in Tunisia), were identified based on the morphological features. Moreover, citrus gummosis was reported in Tunisia in 1936 by Fawcett who establishes that *Phytophthora citrophthora* is the causal agent of this disease. The management strategies of this disease include the use of resistant rootstocks and the treatment of chemical fungicides. This management strategy was only able to reduce the development of these diseases but didn't stop them.

Since 2010, symptoms of apple decline and citrus gummosis have been reported in many localities in Tunisia. Thereafter, Souli et al. (2011a) and Benfradj et al. (2017) demonstrated that *Pythium* species are responsible for these above-mentioned diseases. The members of the genus *Pythium* are well known worldwide and cause serious infections in several crops (Mcleod et al. 2009, Matić et al. 2019). Intensive investigations on citrus and apple diseases revealed the role of fungus-like *Phytophthora* species as the causal agents of these two diseases. In contrast, little is known about the role of *Pythium* species in developing these diseases. Most of *Pythium* species are plant parasites and some saprophytes (Shiba et al. 2018). A number of species (355) of this genus have been described (Ho 2018). They exist as saprophytes, mutualists and parasites in soil, water, plants, fungi, insects, fish, animals and even human beings (Yu 2001, Mani et al. 2019). *Pythium*

*Corresponding author: n.boughalleb@laposte.net, n.boughalleb2017@gmail.com species are well-known hydrophilic plant pathogens inflicting serious damage on many economically important plant crops (Dick 2001). In distribution, the *Pythium* species are cosmopolitan and widely distributed throughout the world, ranging from tropical to temperate regions (Plaats-Niterink 1981). Economically, they are important as pathogens, causing serious damage to agricultural crops and turf grasses, such as soft rot of fruit, rot of roots and stems, and pre- and post-emergence of seeds and seedlings by infecting mainly juvenile or succulent tissues (Hendrix and Campbell 1973). They are rarely host-specific and often more than one species of *Pythium* may be involved in causing the disease (Ho 2009, Li et al. 2019).

Most plant pathogenic *Pythium* species don't have strict host ranges with no species-specific symptoms (Hendrix and Campbell 1973, Chen et al. 1992, Elshahawy and El-Mohamedy 2019). *Pythium* species often infect immature and undifferentiated parts of the plant, such as roots, rhizomes, emerging sprouts, and the pseudo stem of ginger, depending on the stage of their maturity (Nair 2013, Callaghan et al. 2018). Identification and characterization of *Pythium* species are often performed based on the morphological characteristics, but complications might arise due to the absence of sexual structures and the failure to induce zoosporogenesis (Nair 2013). In addition, environmental factors such as ambient temperature, media type, and age of the culture can affect the morphological and physiological characteristics and thus hinder the identification process (Hendrix and Papa 1974). On the other hand, molecular techniques are now widely used to determine the relatedness between the isolates within a species and to distinguish the difference between species. Ribosomal DNA (rDNA) of eukaryotes is arranged in tandem with repeats of specific chromosomes and is subjected to concerted evolution (Nair 2013). The ITS region of rDNA is used as a target for species-specific detection of fungi and is variable between species, but is largely conserved within species (Chen 1992, Almas et al. 2018). The ITS region consists of noncoding variable regions that are located within two rDNA repeats, and rRNA genes (Nair 2013). Molecular identification of isolated fungi was performed by amplifying and sequencing the ITS region (Al-Hawash et al. 2018).

Taxonomy of *Pythium* species

Pythium is a telluric genus, belonging to the Pythiaceae family, which also contains the genus *Phytophthora* and *Phytopythium*, responsible for serious yield losses on many crops around the world (Drenth and Goodwin 1999).

The genus *Phytophthora* was described by De Bary in 1876 for the first time after the discovery of *Phytophthora infestans*, the causal agent of potato late blight in Europe in 1840 (Waterhouse 1963). Then, Pringsheim (1858) discovered the genus *Pythium* and included it in the family of *Saprolegniaceae*. Subsequently, De Bary (1881) integrated this genus into the *Peronosporaceae* family. Then, Fisher (1892) observed that zoospores of *Phytophthora* genus emerge from the sporocyst without prior formation of the vesicles, thus making it possible to distinguish it from *Pythium* genus. In 1897, Schroter created the *Pythiaceae* family for the genus *Pythium* and *Nematosporangium*, but Butler (1907) did not agree with *Nematosporangium* as representative and included all species under the genus *Pythium*. Fitzpatrick (1923) did not accept this separation and considered that *Pythium* and *Phytophthora* could be confused for lack of distinctive characters. Tucker (1931) has developed a large study of the taxonomy of the *Phytophthora* genus with clear images of pathogenic species, a logical basis for separation from crop media and host plants. Subsequently, this genus was transferred to two different Chromista (Gunderson et al. 1987, Paquin et al. 1997) and Stramenopila kingdoms (Birch and Whisson 2001). According to Waterhouse (1974), *Pythium* species can be readily identified by the delicate, hyaline, coenocytic, freely branching flexuous hyphae (about 5 μm on average). Non-deciduous sporangia are produced only in water with variable shapes ranging from spherical, subspherical, ovate, obovate, ellipsoidal, pyriform (non-papillate or sometimes papillate without apical thickening) to lobulate and filamentous, terminal or intercalary on undifferentiated, simple, irregular or sympodially branched sporangiophores (Ho 2009). In general, most *Pythium* species grow faster than *Phytophthora* species in common agar media and have higher growth temperatures (Yu and Ma 1989). The characteristics used in the differentiation between species are the morphology and dimensions of the sporangia, oogonia, oospore, antheridia, colony characteristics, growth rate and the maximal growth temperature (Middleton 1943, Waterhouse 1967, Plaats-Niterink 1981, Dick 1990).

After the creation of *Phytophthora* genus by De Bary (1876), Waterhouse (1963) published a key to the classification for *Phytophthora* species based on morphological characters. Nowadays, the most recent identification key is the key of Erwin and Ribeiro (1996). Since then, about 100 species of *Phytophthora* have been described in the literature (Alvarez 2008, Kroon et al. 2012). This type of classification is an essential basis for the identification of this pseudo-fungus. However, it can be supplemented by molecular analyses, which are more confirmative, especially when there are several morphological characters that often make it difficult to determine species (Whisson 2010). To date, the most widely used tools for identification of various species of the genus *Pythium* are proposed by Plaats-Niterink (1981) and Dick (1990). Descriptions are still based on single isolates and therefore the extent of intraspecific variation has not been described. In 2004, Lévesque and Cock classified the *Pythium* genus, based on molecular and morphological characteristics, into 11 clades. These analyses showed the existence of a group of *Pythium* species (belonging to the K clade) which have phylogenetic characteristics distinct from the rest of *Pythium* spp. Subsequently, a phylogenetic analysis proposed by Villa et al. (2006) based on the DNA sequence of the "internal transcribed spacer", "cytochrome oxidase II" and "β-tubulin genes" regions allowed to separate these *Pythium* species (clade K) from the other species of the genus *Pythium*. Therefore, a new genus named *Phytopythium* has been created for the species of this clade (Bala 2010).

Genetic diversity of *Pythium* spp.

Genetic diversity study of *Pythium* genus is limited compared to the genus *Phytophthora* (Martin 2009). Initial studies using molecular markers to assess *Pythium* genetic diversity were based on culture collections containing only a few isolates of each species and, in many cases, the isolates originated from multiple locations and plant hosts, not from distinct populations (Martin and Kistler 1990, Chen et al. 1992, Francis et al. 1994, Dumroese and James 2005, Al-Sa'di et al. 2008).

Barr et al. (1997) were the first to examine the genetic diversity of *P. ultimum* and *Pythium irregulare*. Thus, obtained results revealed the existence of different genotypes within *P. ultimum*, suggesting an intraspecific variation. The authors also demonstrated that the use of isoenzyme and RAPD (Random Amplified Polymorphic DNA) (molecular markers) to analyze 125 isolates of *P. irregulare* separated them into two distinct groups. Garzon et al. (2005) also examined the genetic diversity of 12 *P. ultimum* isolates and found considerable genetic diversity among them. Then, analyses of Harvey et al. (2000) using RFLP markers for genetic diversity in seven populations of *P. irregulare* from cereal crops in southern Australia showed the detection of putatively heterozygous individuals within each population, implying that outcrossing also occurs within *P. irregulare*.

Later, Matsumoto et al. (2000) examined the genetic structure of *P. irregulare* using molecular markers like ITS sequences and RAPD. They identified four groups representing two distinct clusters. They found that clusters of one of those groups were closely related to *Pythium sylvaticum*. Thus, they suggested the possibility that the four groups represent two different species. Martin (2000) evaluated the phylogenetic relationship between the 24 species of *Pythium* by analyzing the sequences of the "Cytochrome oxidase gene" II. He showed that *Pythium* species were grouped into three large clades, and in a general sense, reflected the morphological characteristics of these species. Further, Haro (2009) examined 115 isolates of three *Pythium* species (*P. irregulare*, *P. sylvaticum*, and *P. ultimum*) in three forest nurseries using simple sequence repeat (SSR) and amplified fragment length polymorphism (AFLP) markers. The examination revealed distinct patterns of intraspecific variation in the three species. *P. sylvaticum* exhibited the most diversity, followed by *P. irregulare*, and substantial clonality in *P. ultimum*. The author demonstrated a significant variation among nurseries for *P. irregulare* and *P. sylvaticum*, but not *P. ultimum*. Finally, from the evidence it was proposed that certain lineages and clonal genotypes are shared among nurseries, indicating that pathogen movement among nurseries has occurred. In addition, previous studies have provided critical evidence for heterozygosis and potential outcrossing in homothallic *Pythium* species (Francis et al. 1994, Harvey et al. 2000), foundational information illustrating the presence of intraspecific variation in *Pythium* species (Francis et al. 1994, Garzon et al. 2007, Al-Sa'di et al. 2008, Lee et al. 2010, Matsumoto et al. 2000, Al-Sa'di et al. 2012), inter-populational gene flow (Harvey et al. 2000, 2001), and for a significant population structure due to location, host and/or mefenoxam sensitivity (Harvey et al. 2000, 2001, Lee et al. 2010, Al-Sa'di et al. 2012). In Tunisia, no research was done to study the genetic diversity of *Pythium* spp., the causative agents of apple decline and citrus gummosis. Despite this research, the study of *Pythium* species genetic diversity remains limited compared to other genera of the same family, such as *Phytophthora* species.

Pythium species as causal agents of citrus gummosis

Pythium species were recovered from different orchards of citrus and apple in Tunisia. A survey conducted from 2012 to 2014 by Benfradj et al. (2017) in the major growing area presented symptoms of citrus gummosis in Tunisia indicating the association of three *Pythium* species (*P. ultimum*, *P. dissotocum*, *P. aphanidermatum*) in this disease. Generally, citrus trees infected by gummosis showed symptoms of a narrow canker above the soil line on the lower trunk of the scion (Fig. 6.1A) from which the lesion expanded to affect the other parts of citrus tree; this canker is sometimes associated with a gum-exudation (Fig. 6.1B).

Figure 6.1. Symptoms of gummosis affecting the main branches of citrus trees
(A: Narrow canker; B: Gum-exudation) (Reprinted from Benfradj et al. 2017)

The identification of such species is usually performed based on morphological characters and confirmed by molecular methods (Benfradj 2017). The morphological classification showed that *Pythium* species were characterized by the presence of a yellowish colour in the aged colonies and by the presence of a marked elevation which sometimes reaches the upper surface of the Petri dish.

The difference was noted between the morphological characteristics of *Pythium* species based on colonies aspects, cardinal temperatures and pathological tests. *P. aphanidermatum* colonies are whitish and dense with aerial mycelium on PDA and V_8 media (Figs 6.2A and B), *Pythium dissotocum* colonies were characterized by whitish and dense mycelium on PDA medium (Fig. 6.2C) and *P. ultimum* was with a radiated appearance on PDA medium (Fig. 6.2E). However,

P. aphanidermatum, P. dissotocum and *P. ultimum* did not show any mycelial growth on the V$_8$ medium (Fig. 6.2D-F).

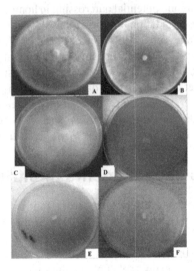

Figure 6.2. Colony morphology of *P. aphanidermatum* (A: on PDA media; B: on CMA media), *P. dissotocum* (C: on PDA media; D: on CMA media), *P. ultimum* (E: on PDA media; F: on CMA media), after six days of incubation at 25°C (Reprinted from Benfradj 2017)

For cardinal temperatures of *Pythium* species, researchers concluded that optimum temperature for *P. aphanidermatum* (35°C) was higher than *P. ultimum* (28°C) and *P. dissotocum* (25°C). Morphological aspects of *P. aphanidermatum, P. ultimum* and *P. dissotocum* were also investigated and confirmed by molecular analyses (Fig. 6.3, Table 6.1).

Table 6.1. Morphological characteristics of *Pythium* species obtained from citrus trees infected by gummosis (Benfradj et al. 2017)

Morphological characteristics	*P. aphanidermatum*	*P. dissotocum*	*P. ultimum*
Cardinal temperatures	10°C≤35°C≤40°C	5°C≤25°C≤30°C	5°C≤28°C≤34°C
Sporangia	Terminal/intercalary	Filamentous	-
Zoospores	9.2 µm-14 µm	8 µm-9 µm.	-
Main hyphae	6.02 µm-10.04 µm	4 µm-7 µm	-
Oogonia	Terminal one oospore per oogonium 19.2 µm-28.7 µm	Terminal/intercalary/ lateral/subglobose 19.5 µm-22 µm	Smooth/spherical/ terminal or rarely intercalary 18 µm-26 µm
Antheridia	Intercalary/terminal, one per oogonium, monoclinous or declinous	1-3 per oogonium, often sessile or originating very near the oogonium when monoclinous	Terminal, one or rarely two per oogonium, diclinous, monoclinous, rarely hypogenous
Oospores	Aplerotic 13.7 µm -23.3 µm	-	Spherical with a smooth, thick wall, aplerotic, globose 18.2 µm-28.4 µm
Similarity with GenBank sequences	97 % -99 %	99%	97%-100%

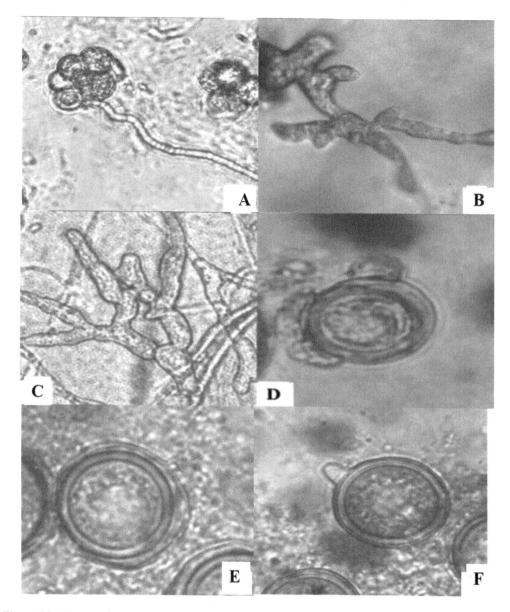

Figure 6.3. Microscopic aspects of *P. aphanidermatum* (A: Vesicle; B: Filamentous Sporangia), *P. dissotocum* (C: Filamentous sporangia; D: Oogonia with antheridia) and of *P. ultimum* (E: Aplerotic Oospore; F: monocline antheridia) (Reprinted from Benfradj 2017)

An investigation undertaken by Benfradj et al. (2017) revealed a notable difference between the severity of *Pythium* species which were isolated from citrus varieties and rootstocks. Typical symptoms of citrus gummosis were registered by the presence of a narrow canker without gum-exudation (Fig. 6.4).

Clementine was the most susceptible citrus variety; Thomson and Tangerine varieties were moderately tolerant, while Valencia and Maltese varieties were the most sensitive to *Pythium* species (Fig. 6.5A). *P. ultimum* seemed to be the most virulent (Fig. 6.5B). Inoculation of roots of rootstock 'Sour-orange' by these *Pythium* species caused damage in roots with colour varying from brown to completely black and alteration of root architecture (Fig. 6.5C).

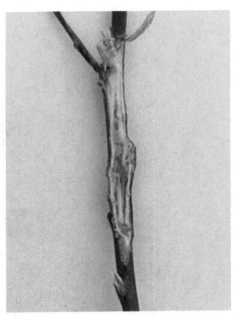

Figure 6.4. Symptoms of necrosis of branches of the 'Clementine' variety after 90 days of inoculation by *Pythium* spp. (Reprinted from Benfradj et al. 2017)

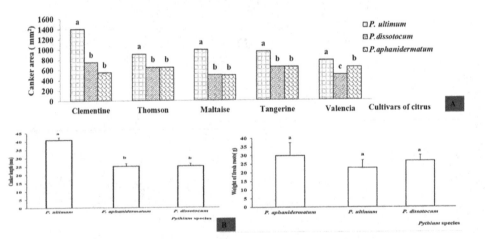

Figure 6.5. Inoculation of citrus varieties by *Pythium* species (A: *In vitro* assay; B: *In vivo* branch inoculation; C: *In vivo* roots inoculation) (Reprinted from Benfradj et al. 2017). Different letters indicate statistically significant differences.

The findings of this study improved knowledge concerning the detection and the role of *Pythium* species in citrus gummosis in Tunisia. In fact, usually, *Phytophthora* spp., especially *P. nicotianae* and *P. citrophthora,* are the most known pathogens associated with the citrus gummosis in the world (Erwin and Ribeiro 1996, Cacciola and Di San Lio 2008). The apparition of these *Pythium* species, as new causative agents of citrus gummosis in Tunisia, lead up to accentuate the damage caused by this disease. Thus, the strategy of management against this disease in Tunisia must consider the risk caused by the existence of this pathogen in citrus orchards.

Pythium species as causal agents of apple decline

In 2006-2009, Souli et al. (2011a, 2014) conducted a survey of apple orchards localized in different regions of Tunisia. Many studies performed reported the presence of symptoms of apple decline.

Symptoms of infected apple trees mainly include wilt of the young terminal shoots with wilting progressed in a downward return in any one or more branches of the tree (Fig. 6.6A). The infected area was characterized by the depression of the bark. The limit of diseased tissues was usually marked by a crack more or less deep. By removing the outer bark, the tissues were highly necrotic brown and red coloured (Fig. 6.6B).

Figure 6.6. Symptoms of apple decline (A: Branch dieback of infected apple trees observed in spring; B: Characteristic lesions caused by *Pythiaceae* on infected apple trees) (Reprinted from Souli et al. 2011a)

Based on morphological, molecular identification and pathogenicity of isolates collected from roots or soil samples of infected trees, *Pythium* species (*Pythium indigoferae, P. irregulare, P. rostratifingens, P. sterilum, P. undulatum* and *Pythium* sp.) were reported as causal agents of apple decline in Tunisia (Souli et al. 2011b, 2014).

The difference in colonies' morphology was noted between *Pythium* species. On PDA medium, *P. undulatum* and *P. irregulare* were characterized by rosaceous aspects, while *P. sterilum* and *P. indigoferae* produced petaloid ones. Morphological aspects of obtained isolates showed a morphological similarity with the five *Pythium* species and confirmed by molecular analyses (Fig. 6.7, Table 6.2).

Souli et al. (2011a) noted that the five *Pythium* species were pathogens of apple varieties (Anna, Lorka and Meski) and rootstocks of MM106. However, Anna and Lorka varieties seemed to be more resistant to *Pythium* species, *P. sterilum* was the most virulent and *P. rostratifingens* seemed to be less virulent to citrus varieties (Table 6.3).

In the same sense, Souli et al. (2011b) demonstrated that *P. indigoferae* were more virulent than *P. irregulare* onto the rootstock MM106, while *P. sterilum* isolates were the most virulent in the three tested apple varieties. Also, *P. irregulare* is also more virulent than *P. indigoferae* (Souli et al. 2011b) into varieties and rootstock apples (Figs 6.8 and 6.9).

The implication of *Pythium* species occurrences as causal agents of apple dieback and citrus gummosis diseases

Plant diseases are often thought to be caused by one species or even by a specific strain and most laboratory studies focus on single microbial strains grown in pure culture (Lamichhane and Venturi 2015). Apple dieback and citrus gummosis diseases, for a long time just *Phytophthora* species, were mentioned as causal agents worldwide.

Table 6.2. Morphological characteristics of *Pythium* species obtained from apple trees infected by dieback (Souli et al. 2014)

Morphological characteristics	P. undulatum	P. sterilum	P. indigoferae	P. irregulare
Cardinal temperatures	26-20-25°C ≤ 5-7 °C ≤ 35-40 °C	8 °C ≤23-27 °C ≤33°C	11°C≤25-28°C ≤35°C	16–23°C≤5-7 °C≤35°C
Sporangia	-	-	Inflated, forming lobed or toruloid complexes	Terminal or intercalary, globose to ellipsoidal, 19.8 ± 3.1 µm × 15.9 ± 2.6 µm
Oogonia	-	-	Spherical, smooth-walled, 19.7 ± 1.7 µm	-
Oospores	-	-	Aplerotic, smooth-walled, 18.9 ± 1.4 µm	-
Oogonia	-	-	-	Smooth walled/ irregular/terminal 19 ± 2.6 µm
Antheridia	-	-	-	Hypogynous/ monoclinous
Oospores	-	-	-	Aplerotic, spherical 18.7 ± 2.13 µm
Hyphal	-	-	-	Intercalary/ limoniform 25.3 ± 3.2
Similarity with GenBank sequences	97%	97%	98%	98%

Table 6.3. Canker area in six-year old trees of apple and one-year old MM106 rootstock six weeks after inoculation in field with three *Pythium* species (average of six replicates) (Souli et al. 2011a)

Rootstock /varieties	Canker area (10^{-6} m^2)		
	P. undulatum	P. rostratifingens	P. sterilum
MM106	653.3±18.2	558.0±4.5	620.0±14.4
Anna	237.3±11.0	273.3±5.8	248.0±6.8
Lorka	0	0	237.3±11.0
Meski	506.3±156	408.0±25.2	356.0±16.3

The existence of cortege of *Pythium* spp. as a part of the causal agents of the diseases mentioned above may induce synergistic pathogen–pathogen infections in plants among different pathogens. This synergism leads to more severe disease symptoms occurring more often than expected (Begon et al. 2006) and often leads to increased disease severity (Lamichhane and Venturi 2015). Such synergistic interactions in plants may be of crucial importance for the understanding of microbial pathogenesis and evolution and consequent development of effective disease control strategies (Lamichhane and Venturi 2015).

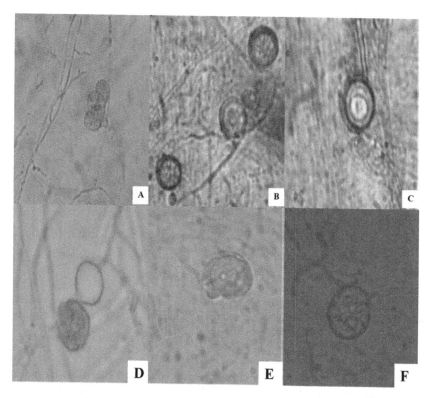

Figure 6.7. Microscopic aspects of ***Pythium indigoferae*** (A: sporangia filamentous inflated, B: oogonia and monoclinous antheridia, C: aplerotic oospore. 1 bar = 10 cm and ***Pythium irregulare*** (D: zoospore formation in a vesicle originating from sporangia, E: irregular oogonia, F: hyphal swellings. 1 bar = 10 cm) (Reprinted from Souli et al. 2011a)

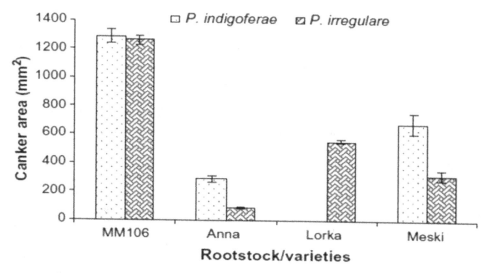

Figure 6.8. Canker area in one-year old shoots of apple and MM106 rootstock seven days after inoculation *in-vitro* with *Pythium indigoferae* and *Pythium irregulare* (average of nine replicates) (Reprinted from Souli et al. 2011b)

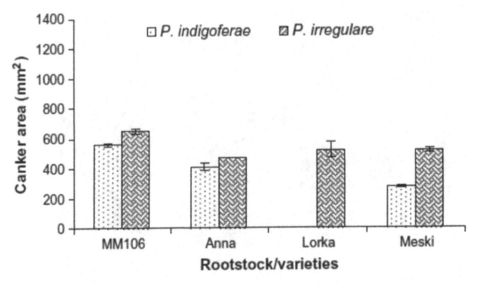

Figure 6.9. Canker area in six-year old trees of apple and one-year old MM106 rootstock six weeks after inoculation in the field with *Pythium indigoferae* and *Pythium irregulare* (average of six replicates) (Reprinted from Souli et al. 2011b)

In fact, for apple dieback and gummosis citrus diseases, chemical control is the most used technique for disease management. However, targeted chemical control strategies become limiting when a cortege of species contributes to the disease as the application of the specific substance may not necessarily result in successful disease management (Lamichhane and Venturi 2015).

It is therefore important to study the synergisms in pathogen-pathogen interactions of apple dieback and gummosis citrus diseases, as well as the underlying mechanisms to identify important links that may be manipulated to avoid the loss caused by those pathogens. According to Willocquet et al. (2002), these studies are difficult tasks since disease complexes are related to environmental conditions, cultural practices, and geography. Also, it's important to elaborate on an experimental approach leading to the identification of pathogen cortege in relation to the crop production system. In fact, diagnosis and management of both diseases take a long time, in addition to the classical isolation techniques on cultural growth media, than other more specific (e.g. immunofluorescence or PCR) or generic (e.g. morphological identification) assays used, resulting in significant yield losses (Lamichhane and Venturi 2015). Thus, new knowledge and techniques which facilitate the understanding of the total microbial species involved in plant diseases, as well as the underlying mechanisms, need to be added to these classical techniques. One of the new techniques is the studies of complex diseases from culture-independent analyses (high-throughput sequencing for example), which does not have the limitations of the classical culture-based approach, which is often lengthy and costly (Nikolaki and Tsiamis 2013). In addition, techniques such as next-generation sequencing (NGS) have enabled high-throughput analyses of complex microbial populations (Van Dijk et al. 2014). It has been revealed that microbial diversity is vastly underestimated based on classical cultivation-based techniques (Gilbert and Dupont 2011). In the last 10 years, metagenomic projects have been combined with NGS technologies boosting studies in microbial ecology at a very fast pace (Venter et al. 2004, Tringe and Hugenholtz 2008). These novel technologies are thus a powerful tool to understand the implication of two or more microbes and their contribution to plant disease occurrence.

Conclusions

Pythium species are commonly involved in apple dieback and citrus gummosis in Tunisia. This finding explains the regeneration and losses caused by the mentioned diseases in Tunisia. The occurrence of *Pythium* species could be related to the environmental conditions in soil orchards which were favourable to the development of the infection of *Pythium* species. It could also be due to the introduction and dissemination of non-native *Oomycetes* species by the international plant trade and their spread from nurseries into orchards and natural ecosystems. In addition, assumptions regarding the possible source of *Pythium* spp. pathway could be that the soil used for seedling culture has not been properly sterilized, that seedlings' roots penetrate the bottom of seedling boxes, contact infested soil, or that irrigation water obtained from rivers is contaminated. Thus, citrus gummosis and apple decline control management strategies need to take into consideration the presence of these species in order to develop a better strategy.

References

Al-Hawash, A.B., Alkooranee, J.T., Abbood, H.A., Zhang, J., Sun, J., Zhang, X. and Ma, F. 2018. Isolation and characterization of two crude oil-degrading fungi strains from Rumaila oil field Iraq. Biotechnol. Rep. 17: 104-109.

Al-Sa'di, A.M., Drenth, A., Deadman, M.L., de Cock, A.W.A.M., Al-Said, F.A. and Aitken, E.A.B. 2008. Genetic diversity, aggressiveness and metalaxyl sensitivity of *Pythium spinosum* infecting cucumber in Oman. J. Phytopathol. 156(1): 29-35.

Al-Sadi, A.M., Al-Ghaithi, A.G., Al-Balushi, Z.M. and Al-Jabri A.H. 2012. Analysis of diversity in *Pythium aphanidermatum* populations from a single greenhouse reveals phenotypic and genotypic changes over 2006 to 2011. Plant Dis. 96(6): 852-858.

Almas, S., Gibson, A.G. and Presley, S.M. 2018. Molecular detection of *Oxyspirura* larvae in arthropod intermediate hosts. Parasitol. Res. 117: 819-823.

Alvarez, L.A., Vicent, A., De la Roca, E., Bascón, J., Abad-Campos, P., Armengol, J. and García-Jiménez, J. 2008. Branch cankers on citrus trees in Spain caused by *Phytophthora citrophthora*. Plant Pathol. 57: 84-91.

Bala, K. 2010. *Phytopythium* Abad, de Cock, Bala, Robideau, Lodhi & Levesque, gen. nov. and *Phytopythium sindhum* Lodhi, Shahzad & Levesque, sp. nov. Persoonia 24: 136-137.

Barr, D.J.S., Warwick, S.I. and Desaulniers, N.L. 1997. Isozyme variation, morphology, and growth response to temperature in *Pythium irregulare*. Can. J. Botany 75(12): 2073-2081.

Benfradj, N. 2017. Identification des agents causaux de la gommose des citrus et Epidémiologie des Pythiacées incriminées dans cette maladie. PhD diss., Institut Supérieur Agronomique De Chott Meriem. Université de Sousse. 1-163.

Benfradj, N., Migliorini, D., Luchi, N., Santini, A. and Boughalleb-M'Hamdi, N. 2017. Occurrence of *Pythium* and *Phytopythium* species isolated from citrus trees infected with gummosis disease in Tunisia. Arch. of Phytopathol. PFL. 50: 5-6, 286-302.

Birch, P.R.J. and Whisson, S.C. 2001. *Phytophthora infestans* enters the genomics era. Mol. Plant Pathol. 2(5): 257-263.

Boughalleb, N., Moulahi, A. and El Mahjoub, M. 2006. Variability in pathogenicity among Tunisian isolates of *Phytophthora cactorum* as measured by their ability to cause crown rot on four apple cultivars and MM106 rootstock. J. Agron. 5(2): 321-325.

Butler, E.L. 1907. An account of the genus *Pythium* and some *chytridiaceae*. Mem. Dep. Agric. India. Bot. Ser. 1: 1-162.

Callaghan, S.E., Burgess, L.W., Ades, P., Mann, E., Morrison, A., Tesoriero, L.A. and Taylor, P.W.J. 2018. Identification and pathogenicity of *Pythium* species associated with poor growth of tomato plants in the Australian processing tomato industry. *In*: XV International Symposium on Processing Tomato 1233, pp. 147-154.

Chen, W. 1992. Restriction fragment length polymorphisms in enzymatically amplified ribosomal DNAs of three heterothallic *Pythium* species. Phytopathol. 82: 1467-1472.

Chen, W., Schneider, R.W. and Hoy, J.W. 1992. Taonomic and phylogenetic analyses of ten *Pythium* species using isozyme polymorphisms. Phytopathology – New York and Baltimore Then St Paul. 82: 1234-1244.

De Bary, A. 1876. Researches into the nature of the potato fungus, *Phytophthora infestans*. J. Roy. Agr. Soc. Engl. 2(12): 239-269.

De Bary, A. 1881. Untersuchungen uber die Peronosporeen und Saprolegnieen und die Grundlagen eines naturlichen Systems der Pilze. Abhandl. Senck. Naturforsch. Gesellsch. 12: 225-370.

Dick, M.W. 1990. Keys to *Pythium*. UK: Published by MW Dick, Reading.

Dick, M.W. 2001. Straminipilous fungi. Dordrecht: Kluwer Academic Publishers.

Drenth, A. and Goodwin, S.B. 1999. Population structure of oomycetes. pp. 195-224. *In*: Worrall, J.J. (ed.). Structure and Dynamics of Fungal Populations. Population and Community Biology Series. Kluwer Academic Publishers, Netherlands.

Dumroese, R.K. and James, R.L. 2005. Root diseases in bareroot and container nurseries of the Pacific Northwest: Epidemiology, management, and effects on outplanting performance. New Forest. 30: 185-202.

Elshahawy, I.E. and El-Mohamedy, R.S. 2019. Biological control of *Pythium* damping-off and root-rot diseases of tomato using *Trichoderma* isolates employed alone or in combination. J. Plant Pathol. 101(3): 1-12.

Erwin, D.C., Bartnicki-García, S. and Tsao, P.H. 1983. *Phytophthora* its Biology, Taxonomy, Ecology and Pathology. APS, 392 pp.

and. FAOSTAT. 2015. The Food and Agriculture Organization Corporate Statistical Database. http://www.fao.org/statistics/en (accessed May 7, 2015).

Fischer, A. 1892. Phycomycetes. Rabenhorst's Kryptogamenflora 1: 505.

Fitzpatrick, H.M. 1923. Generic concepts in the *Pythiaceae* and *Blastocladiaceae*. Mycologia. 15: 166-173.

Francis, D.M., Gehlen, M.F. and St. Clair, D.A. 1994. Genetic variation in homothallic and hyphal swelling isolates of *Pythium ultimum* var. *ultimum* and *P. ultimum* var. *sporangiferum*. Mol. Plant Microbe. Int. In. 7(6): 766-775.

Garzon, C.D., Geiser, D.M. and Moorman, G.W. 2005. Diagnosis and population analysis of *Pythium* species using AFLP fingerprinting. Plant Dis. 89: 81-89.

Garzon, C.D., Yanez, J.M. and Moorman, G.W. 2007. *Pythium cryptoirregulare*: A new species within the *P. irregulare* complex. Mycologia. 99: 291-301.

Gilbert, J.A. and Dupont, C.L. 2011. Microbial metagenomics: Beyond the genome. Annu. Rev. Mar. Sci. 3: 347-371.

Gunderson, J.H., Elwood, H., Ingold, A., Kindle, K. and Sogin, M.L. 1987. Phylogenetic relationships between Chlorophytes, Chrysophytes, and Oomycetes. P. Natl. A. Sci. 84(16): 5823-5827.

Harvey, P.R., Butterworth, P.J., Hawke, B.G. and Pankhurst, C.E. 2000. Genetic variation among populations of *Pythium irregulare* in southern Australia. Plant Pathol. 49(5): 619-627.

Harvey, P.R., Butterworth, P.J., Hawke, B.G. and Pankhurst, C.E. 2001. Genetic and pathogenic variation among cereal, medic and sub-clover isolates of *Pythium irregulare*. Mycol. Res. 105(1): 85-93.

Hendrix, F.F. and Campbell, W.A. 1973. *Pythiums* as plant pathogens. Ann. Rev. Phytopathol. 11(1): 77-98.

Hendrix, F.F. and Papa, K.E. 1974. Taxonomy and genetics of *Pythium*. APS. 1: 200-207.

Ho, H.H. 2009. The genus *Pythium* in Taiwan, China (1)—A synoptic review. Front. Biol. 4(1): 15-28.

Kroon, L.P.N.M., Brouwer, H., de Cock, A.W.A.M. and Govers, F. 2012. The genus *Phytophthora* anno 2012. Phytopathol. 102: 348-364.

Lamichhane, J.R. and Venturi, V. 2015. Synergisms between microbial pathogens in plant disease complexes: A growing trend. Front. Biol. 6: 385.

Lee, S., Garzón, C.D.   and Moorman, G.W. 2010. Genetic structure and distribution of *Pythium aphanidermatum* populations in Pennsylvania greenhouses based on analysis of AFLP and SSR markers. Mycologia 102(4): 774-784.

Lévesque, C.A. and de Cock, A.W.A.M. 2004. Molecular phylogeny and taxonomy of the genus *Pythium*. Mycol. Res. 108: 1363-1383.

Li, N., Zhou, Q., Chang, K.F., Yu, H., Hwang, S.F., Conner, R.L., Strelkove, S.E. and McLarenf, D.L. 2019. Occurrence, pathogenicity and species identification of *Pythium* causing root rot of soybean in Alberta and Manitoba, Canada. Crop Prot. 118: 36-43.

Mani, R., Vilela, R., Kettler, N., Chilvers, M.I. and Mendoza, L. 2019. Identification of *Pythium insidiosum* complex by matrix-assisted laser desorption ionization-time of flight mass spectrometry. J. Med. Microbiol. Doi: 10.1099/jmm.0.000941

Martin, F.N. and Kistler, H.C. 1990. Species-specific banding patterns of restriction endonuclease-digested mitochondrial DNA from the genus *Pythium*. Exp. Mycol. 14(1): 32-46.

Martin, F.N. 2000. Phylogenetic relationships among some *Pythium* species inferred from sequence analysis of the mitochondrially encoded cytochrome oxidase II gene. Mycologia. 92(4): 711-727.

Martin, F.N. 2009. *Pythium* Genetics. Oomycete genetics and genomics diversity. pp. 213-240. *In*: Lamour, K. and Kamoun, S. (eds.). Interactions and Research Tools. John Wiley & Sons, Hoboken, New Jersey.

Matić, S., Gilardi, G., Gisi, U., Gullino, M.L. and Garibaldi, A. 2019. Differentiation of *Pythium* spp. from vegetable crops with molecular markers and sensitivity to azoxystrobin and mefenoxam. Pest Manag. Sci. 75(2): 356-365.

Matsumoto, C., Kageyama, K., Suga, H. and Hyakumachi, M. 2000. Intraspecific DNA polymorphisms of *Pythium irregulare*. Mycol. Res. 104(11): 1333-1341.

Mcleod, A., Botha, W.J., Meitz, J.C., Spies, C.F.J., Tewoldemedhin, Y.T. and Mostert, L. 2009. Morphological and phylogenetic analyses of *Pythium* species in South Africa. Mycol. Res. 113: 933-951.

Middleton, J.T. 1943. The taxonomy, host range and geographic distribution of the genus *Pythium*. Mem. Torrey Bot. Club. 20: 1-171.

Nair, K.P. 2013. The agronomy and economy of turmeric and ginger: The invaluable medicinal spice crops. Newnes, 544 pp.

Nikolaki, S. and Tsiamis, G.G. 2013. Microbial diversity in the era of omic technologies. BioMed. Res. Int. 2013: 1-15.

Paquin, B., Laforest, M.J., Forget, L., Roewer, I., Wang, Z., Longcore, J. and Lang, B.F. 1997. The fungal mitochondrial genome project: Evolution of fungal mitochondrial genomes and their gene expression. Curr. Genet. 31: 380-395.

Plaats-Niterink, V.D. 1981. Monograph of the Genus *Pythium*. Stud. Mycol. 21: 1-239.

Pringsheim, N. 1858. Beitrage zur Morphologie und Systematik der Algen. 2. Die Saprolegnieen. Jahrb. wiss. Botan. 1: 284-306.

Savita, G.S.V. and Avinash, N. 2012. Citrus diseases caused by *Phytophthora* species. GERF Bull. Biosci. 31: 18-27.

Schroter, J. 1897. Pythiaceae. Englar and Prantl, Nat. PflFam. 1(1): 104-105.

Shiba, K., Hatta, C., Sasai, S., Tojo, M., Ohki, S.T. and Mochizuki, T. 2018. Genome sequence of a novel partitivirus identified from the oomycete *Pythium* nunn. Arch. Virol. 163(9): 2561-2563.

Souli, M., Boughalleb, N., Abad-Campos, P., Álvarez, L.A., Pérez-Sierra, A., Armengol, J. and García-Jiménez, J. 2011a. First Report of *Pythium indigoferae* and *P. irregulare* associated to apple trees decline in Tunisia. J. Phytopathol. 159(5): 352-357.

Souli, M., Boughalleb, N., Abad-Campos, P., Álvarez, L.A., Pérez-Sierra, A., Armengol, J., García-Jiménez, J. and Romdhani1, M.S. 2011b. Diversity of the *Pythium* community infecting crown and roots apple in Tunisia. Res. Plant Biol. 1(4): 16-22.

Souli, M., Abad-Campos, P., Perez-Sierra, A., Fattouch, S., Armengol, J. and Boughalleb-M'Hamdi, N. 2014. Etiology of apple tree dieback in Tunisia and abiotic factors associated with the disease. Afr. J. Microbiol. Res. 8(23): 2272-2281.

Tringe, S.G. and Hugenholtz, P. 2008. A renaissance for the pioneering 16S rRNA gene. Curr. Opin. Microbiol. 11: 442-446.

Tucker, C.M. 1931. Taxonomy of the genus *Phytophthora* de Bary. University of Missouri Agricultural Experiment Station Research Bulletin: 153.

Van Dijk, E.L., Auger, H., Jaszczyszyn, Y. and Thermes, C. 2014. Ten years of next-generation sequencing technology. Trends Gentet. 30: 418-426.

Venter, J.C., Remington, K., Heidelberg, J.F., Halpern, A.L., Rusch, D. and Eisen, J.A. 2004. Environmental genome shotgun sequencing of the Sargasso Sea. Science 304: 66-74.

Villa, N.O., Kageyama, K., Asano, T. and Suga, H. 2006. Phylogenetic relationships of *Pythium* and *Phytophthora* species based on ITS rDNA, cytochrome oxidase II and β-tubulin gene sequences. Mycologia. 98(3): 410-422.

Waterhouse, G.M. 1963. Key to the species of *Phytophthora* de Bary. Myc. Papers 92: 1-22.

Waterhouse, G.M. 1967. Key to *Pythium Pringsheim*. Mycological Papers 109: 1-15.

Waterhouse, G.M. 1968. The genus *Pythium Pringsheim*. Mycological Papers 110: 1.

Waterhouse, G.M. 1974. Other *Phytophthora* species recorded on cacao. pp. 71-79. *In*: Gregory, P.H. (ed.). *Phytophthora* – Disease of Cocoa. Longman, London.

Whisson, S.C. 2010. *Phytophthora*. *In*: Encyclopedia of Life Sciences (ELS). John Wiley & Sons Ltd., Chichester.

Willocquet, L., Savary, S., Fernandez, L., Elazegui, F.A., Castilla, N., Zhu, D., Tang, Q., Huang, S., Lin, X., Singh, H.M. and Srivastava, R.K. 2002. Structure and validation of RICEPEST, a production situation-driven, crop growth model simulating rice yield response to multiple pest injuries for tropical Asia. Ecol. Model. 153(3): 247-268.

Yu, Y.N. and Ma, G.Z. 1989. The genus *Pythium* in China. Mycosystema. 2: 1-110.

Yu, Y.N. 2001. *Pythium* and fungi. Jiangxi. Sci. 19(1): 55-60.

Pathogenic and Beneficial *Pythium* Species in China: An Updated Review

Hon H. Ho[1*] and Kamel A. Abd-Elsalam[2]

[1] Department of Biology, State University of New York, New Paltz, New York 12561, USA
[2] Plant Pathology Research Institute, Agricultural Research Center, Giza, 12619, Egypt

Introduction

Pythium species have been distributed all over the world irrespective of geographic regions. It ranges from hot to moderate (van der Plaats-Niterink 1981) and even Arctic (Hoshino et al. 1999) and Antarctic areas (Knox and Paterson 1973). They can be identified as saprophytes in the soil and water, or parasites on plants, fungi, insects, seafood, animals and humans (Yu 2001). Economically, they are significant fungal pathogens of higher plants, leading to serious harm to agricultural plants and turf grasses. It mainly causes soft rot of fruit, rot of roots and stems, and pre- and post-emergence of seeds and seedlings by infecting juvenile or succulent cells (Hendrix and Campbell 1973). Other species triggered diseases in seafood (Khulbe 1983), red algae (Takahashi et al. 1977) and mammals including humans (de Cock et al. 1987, Mendoza et al. 1996, Thianprasit et al. 1996). *Pythium* spp. can possibly be used as natural biocontrol agent for some soil-borne fungi (Jones 1995) and mosquitoes (Su 2001, Su et al. 2006), and have potential to produce beneficial metabolites which can be used in medicine and food production (Gandhi and Weete 1991, Stredansky et al. 2000). Considering all these facts, the present chapter is aimed to focus on an updated overview of all the important aspects associated with both pathogenic and beneficial *Pythium* species in landmass China.

Historical background

Prof. T.F. Yu from Plant Pathology Department at the University of Nanjing for the first time initiated investigation on plant diseases in mainland China caused by the genus *Pythium*. Researchers observed significant damping-off disease of cucumber inside the greenhouse of the Botanical Garden at the University of Nanjing. The disease was incredibly widespread in Nanjing and its vicinity. After a detailed investigation, he provided a complete description of the pathogen: *Pythium aphanidermatum* (Edson) Fitz (Yu 1934). Subsequently, he discovered that the pathogen was distributed throughout China, invading a multitude of fruit and veggies and also other important plant species like cotton and tobacco in the field and storage, leading to seedling damping-off, cottony leak and rotting of the stems, roots, fruits, and vegetable (Yu 1940, Yu et al. 1945). He also began studies on the physiology of critical pathogens and proposed procedures to regulate the condition (Yu 1934, Yu et al. 1945). Later, he isolated two species of *Pythium* in Yunnan Province, i.e. *P. spinosum* and *P.*

*Corresponding author: hoh@newpaltz.edu

ultimum that were weakly pathogenic to the root base of coffee beans (Yu 1950) and *P. spinosum* triggered dark rot of sweet potato bean (Yu 1955). Further, he classified eight distinct species of *Pythium* responsible for the seedlings of millet, including *P. aristosporum, P. arrhenomonas, P. debaryanum, P. graminicola, P. irregulare, P. monospermum, P. tardicrescens* and *P. ultimum* (Yu 1978). However, a recent era of the scholarly investigation of *Pythium* in mainland China started when Professor Y.N. Yu began to take interest in the current genus. While still at Sichuan University, he and his co-workers described the primary rot infection of pyrethrum caused by *Pythium* sp. (Ho et al. 1951). After transferring to the Institute of Microbiology, Chinese Academy of Sciences in Beijing, he introduced an outstanding research programme regarding the taxonomy and incidence of *Pythium* species in China that continued for over twenty years, resulting in publishing a monograph on the genus *Pythium* in China (Yu and Ma 1989). Although a wide range of isolates originated from natural environment, diseased or non-diseased plant parts, no pathogenicity assays were examined.

Pythium pathogenicity: An overview

Various investigators were dedicated to the study of pathogenic species of *Pythium* in China. Many species of *Pythium* were pathogenic to cereal plants and turf grass from the Family Gramineae (Poaceae) in addition to herbaceous plants and generally vegetables and fruit from families like Crucifereae (Brassicaceae), Solanaceae and Cucurbitaceae. Furthermore, the fungal pathogen, *P. spinosum* caused seedlings damping-off of millet as reported by Yu (1950). Xu and Zhang (1985) demonstrated that *P. aphanidermatum* was highly pathogenic to maize in Shandgong Province resulting in root and basal stalk rot. However, the maize stalk rot in Beijing and Zhejiang (Wu et al. 1990) and Jiangsu Province (Zhu et al. 1997) was associated with *P. inflatum* and *P. graminicola*. Wu et al. (1990) also confirmed that the causal pathogen of stalk rot of maize in China was *P. graminicola*. Shen and Zhang (1994) recognized the five most pathogenic species from Zhejiang areas associated with diseases of wheat and barley: *P. dissótocum, P. irregulare, P. polypapillatum, P. spinosum* and *P. ultimum*. Eight species of *Pythium* were recovered from springtime cropping maize seedlings in the same province which include *P. acanthophoron, P. aphanidermatum, P. debaryanum, P. gramincola, P. irregulare, P. spinosum, P. tardicrescens* and *P. ultimum* with *P. aphanidermatum* as the most virulent pathogen (Shén and Zhang 1995). Wilt disease of maize caused by *P. salinum, P. orthogonon,* and *P. spinsosum* in Guangxi, Hunan and Guangdong provinces, respectively, was reported by Zhang et al. (1996). Numerous *Pythium* spp. were responsible for the yellowing and blight of rice seedlings on dried out seedbeds, e.g. *P. aphanidermatum* and P. *catenulatum* in Jiangsu Province (Xu et al. 2001), *P. salinum, P. intermedium* and *P. acrogynum* in Nanjing region (Gao et al. 2001). Lou and Zhang (2004) isolated *P. sylvaticum* from damping-off seedlings of rice in the Hangzhou region. Wang et al. (1996) isolated ten species of *Pythium* from wheat rhizospheres in Zhejiang Province, viz. *P. adhaerens, P. aphanidermatum, P. arrhenomanes, P. catenulatum, P. diclinum, P. dissotocum, P. irregulare, P. pulchrum, P. spinosum* and *P. ultimum*. Among these, *P. adhaerens, P. catenulatum* and *P. pulchrum* were reported for the first time in Zhejiang Province. The previous *Pythium* species were pathogenic in wheat, rice, aubergine and tomato but *P. irregular* and *P. spinsoum* had been reported to be the most virulent species for both rice and wheat, respectively. Rice damping-off disease caused by *P. arrhenomanes* was screened from 2012 to 2014 inside Mongolia. This soil-borne pathogen may present a threat to the yield production of rice seedlings in the mentioned growing areas (Ling et al. 2018).

Turf grass at parks, golf places and yards was vulnerable to numerous infections. Wang et al. (2000, 2001) detected six species of *Pythium* leading to damping off and also decaying of stems and leaves of four prevalent cool-season grass species (*Festuca arundinacea, Poa pratensis, Agrostis palstriss* and *Lolium perenne*) in Zhejiang Province: *P. aphanidermatum, P. deliense, P. irregulare, P. ultimum, P. graminicola* and *P. mamillatum* and they also investigated the conditions impacting Pythium disease. *Pythium* spp. were suggested as causing diseases of cool-season lawns and turf in Shanghai location (Ma et al. 1999) along with Weifen region of Shandgong Province (Ding et al. 2005). Liu and Xie (2003) studied the most serious diseases of the frosty period turf grass in the Hexi Corridor in Gansu Province and observed that *Poa pratensis* and *L. perenne* could be harmed by *P. pythium* and *P. aphanidermatum* respectively, whereas Xue (2003) concluded that *P. graminicola* was pathogenic to both of these species of grass and *F. rubra*. Chai and He (2002) revealed that the primary turf grass illnesses during the summer season in Beijing were induced by the *Pythium* spp. which mainly infected *F. arundinacea*, while Gu et al. (2009) confirmed the pathogen of the root rot of the turf grass in the town of Lanzhou to be *P. aphanidermatum*. Eighteen species of *Pythium* were collected from Shangdong, Jiangsu, Anhui, Jiangxi and Fujian provinces in addition to the Shanghai area. The most pathogenic species were *P. ultimum*, *P. aphanidermatum*, *P. irregulare* and *P. spinosum* (Jiang et al. 1990).

P. aphanidermatum, P. dissotocum, P. kunmingense, P. spinosum and P. ultimum were found to be pathogenic to vegetable seedlings in Yinchuan, Yongning, Wuzhong, Linwu and Qingtonlin (Zhang and Jiang 1990). Zhang et al. (1990) reported that the damping-off of solanaceous and cucurbitaceous vegetables was caused by *P. spinosum, P. aphanidermatum, P. irregulare* and *P. ultimum* in the field and the disease was determined by soil temperature, rainwater and humidity. *P. spinosum* was found during the winter and springtime while *P. aphanidermatum* was collected in summer. Chinese cabbages throughout Guangdong Province suffered from a severe soft rot disease due to *P. aphanidermatum* and its pathogenicity to eighteen species of plants from seven families was confirmed (Liu 1998). Screening of diseases of watermelons in Lianing, Jilin, Heilongjiang, Hebei, Henan and Shangdong provinces showed that the fruit rot was produced by *P. aphanidermatum* and *P. debaryanum* (Wei et al. 1991), compared to the stem rot of lotus lily in Nanjing of Jiangsu Province due to *P. helicoides* (Dou et al. 2010). Four *Pythium* species had been isolated: *P. aphanidermatum, P. debaryanum, P. ultimum* and *P. irregulare* from 226 tobacco plant samples showing root neck rot symptoms in Yunnan Province, and their pathogenicity to tobacco was established, resulting not merely in root rot but seedling blight as well and the most serious pathogen was *P. aphanidermatum* (Wang 1997). *P. myriotylum* caused significant rhizome rot in ginger in Laiwu Town of Shangdong Province (Liu and Shi 2009). The tomato, a significant crop in Xinjiang Region, generally encountered root rot diseases because of *P. aphanidermatum, P. monospermum* and *P. acanthicum* (Zhao et al. 2009, Zhou et al. 2011). The pathogens causing the mung bean sprout spoilage, root rot of kidney bean and seedling rot of *Calathea* sp. were mainly *P. aphanidermatum* (Zhang et al. 2010a, b), *P. ultimum* (Wang et al. 2010a, b) and *P. myriotylum* (Oishi et al. 2010). *Pythium vexans* became a pathogen leading to the stem rot of *Dendrobium* in Yunnan Province (Tao et al. 2011), whereas *P. heterothallicum* was sometimes pathogenic to *Zea mays* (Gan et al. 2010). Furthermore, *P. elongatum* induced the soft rot of lotus lily (Liu 1992).

Moreover, several species of *Pythium* infected seedlings and the succulent tissues of vegetables, fruits, root base, leaves and stems of non-woody plants, while some species were pathogenic to woody plants. Qiu et al. (1986) identified that *P. ultimum* triggered root decay of young Chinese fir trees in Sichuan Province, inducing the yellowing of leaves and the following death after three to four years. Rhododendron foot decay was a harmful disease in Yixing of Jiangsu area, causing the wilting of seedlings and adult plants and complete loss of the crop. The causal pathogen proved to be *P. kunmingense* (Chen at al. 1997). Based on specimens collected in fifteen citrus-producing regions in Sichuan, *P. debaryanum, P. hydnosporum* and *P. hemmianum* were found leading to root decay of citrus trees and shrubs (Zhang and Huang 1994). Tea plant varieties cultivated in Hanzhou and Sichuan Province were rotted by *P. spinosum* (Lai et al. 1998, 2000). Poinsettia was vulnerable to *P. aphanidermatum* which caused root rot disease in Shanghai's Pudong New Region (Yang et al. 2003). *P. aphanidermatum* was considered among the two pathogens causing stem decay of papaya (Li et al. 2007), while *P. oligandrum* was the causal agent leading to canker disease of walnut (Shang et al. 2010). Lou et al. (2000) utilized twenty-five arbitrary primers to perform a Random Amplified Polymorphic DNA (RAPD) analysis of eight species of *Pythium* responsible for tomato seedling damping-off in the Hangzhou area. *Salvia splendens* is recorded as the host plant for *P. splendens* Han Braun (Editorial Council 1996) with no great detailed information.

Pythium species have been described as fungal pathogens for plants but they can also invade various microorganisms. The red alga, *Porphyra yezeonens* commonly cultivated in many parts of the sea coasts of provinces including Shangdong, Zhejiang and Jiangsu, was infected by *Pythium porphyrae*, the causal agent of red rot (Ma 1996, Ding and Ma 2005, Ding 2006) resulting in severe losses. One of the plant pathogens collected was identified as *Pythium chondricola* by using morphological features and molecular techniques such as sequence analysis of ITS region and cytochrome oxidase subunit 1 (Qiu et al. 2018). A single species of *P. carolinianum* was isolated from the diseased larvae of the Asian tiger mosquito (*Culex albopictus* and *C. quinquefasciatus*) in Guiyang Town of Guizhou Area in 1995 (Su et al. 2001). This knowledge is essential because of its possibilities in managing mosquito population and various follow-up studies were carried out. Huang and Su (2001) explored the mechanism of infection of larvae by fungus structures by fluorescence and scanning microscopes technique and they effectively created the cDNA library (Liu et al. 2001). Likewise, Su (2003) showed that *P. carolianianum* generated Pr-1 and Pr-2 protease processing the protein-chitin cuticle of Asian tiger mosquito and also isolated and cloned a cDNA segment of Pr1 gene. Huang and Su (2003) also examined the influence of environmental conditions on the formation of fungal spores such as zoospores, the main source for primary infection of mosquito larvae. Also, the mosquito host range for *P. carolinianum* was identified (Su et al. 2003). Yang et al. (2005a, b) developed a new molecular method to established cDNA library as a practical and prompt approach to explore for an unidentified sequence adjacent to the known to identify the full genomic sequence of the subtilisin-like protease (Pr1) gene. In addition, a novel *Pythium* species was collected from infected larvae of *Aedes albopictus* in Guiyang of Guizhou Province and recognized as *P. guiyangense* by using molecular sequence such as ITS sequence of rDNA (Su 2006).

Huang and Su (2007) carried out additional investigation into the *in vitro* development requirements of the fungus. To ensure essential safety of the ecosystem if the fungus is to be utilized on a large scale to manage mosquito inhabitants, several studies were executed to demonstrate their safety to pets (Liu and Su 2007a, b), other insect larvae (Liu et al. 2007) and crops (Zhang and Su 2008a, b). A listing of its mosquito hosts was collected (Su 2008). Scanning electron microcopy was applied to review the hyphae and asexual structures (Kong et al. 2008) and its own genetic library was produced (Hong and Su 2008). The subtilisin-like protease (Pr1) was extracted, purified and described (Wang et al. 2007a, b, Yu et al. 2007, Duan et al. 2008), while the induction circumstances for Pr2 were elucidated (Yu et al. 2008a, b). The protoplast of *P. guiyangense* was retrieved, regenerated (Zhao et al. 2008) and provided as material for genetic transformation using *Agrobacterium tumefaciens* (Zhao and Su 2008) to generate more virulent mosquito-killing strains. The impact of five prevalent pesticides on the fungus and the mosquito, *C. quinquefaciatus* in Guizhou Province was analyzed as well (Zhang and Su 2008a, b).

A new *Pythium* species termed *Pythium cedri*, isolated from *Cedrus deodara* (Pinaceae) in Jiangsu Province of China, was identified by morphological and molecular characters and supported using phylogenetic tree using ITS and COI combined sequences data (Chen 2017). Five isolates of *P. aphanidermatum* were collected from the infected rhizomes, and their pathogenicity was verified. The first record of ginger Pythium soft rot in China was caused by *P. aphanidermatum* (Li et al. 2014). Soft rot disease in *Dendrobium officinale* infected by *P. ultimum* might be related to direct penetration into the cell wall of the plant and the cellulases secreted by the pathogen (Xing et al. 2015). Three *Pythium*-like strains were isolated from the roots and stems of Chinese yam with disease symptoms. Some isolates of *P. myriotylum* and *P. ultimum* var. *ultimum* [*Globisporangium ultimum*] were identified based on the morphological characteristics, sequence analysis of fragments of the cytochrome c oxidase subunit I gene and pathogenicity tests (Zhang et al. 2018a, b). The pathogenic species of *Pythium* found so far in mainland China and the hosts for the pathogenic species are listed in Table 7.1 (adapted from Ho 2013).

Table 7.1. *Pythium* species pathogenic to different plant hosts in mainland China

Parasitic spp. on plants	Hosts
P. acanthicum	*Cucumis sativus, Lycopersicum esculentum, Zea mays*
P. acanthophoron	*Zea mays*
P. acrogynum	*Oryza sativa*
P. adhaerens	*Lycopersicum esculentum, Oryza sativa, Solanum melongena, Triticum aestivum*
P. aphanidermatum	*Agrostis palstriss, A. stolonifera, Aleuritis fordii, Amaranthus tricola, Benincasa hispida, Begonia chinensis, Begonia* spp., *Beta vulgaris, Brassica chinensis, B. oleracea* var. *capitata, B. pekinensis, B. napus, Brassica* spp., *Capsicum annuum, C. frutescens* var. *grossum, Carica papaya, Citrillus vulgaris, Chrysanthemum cineriaefolium, Chrysanthemum* spp., *Cucumis melo., C. melo.* var. *hami, C. sativus, Cucurbita maxima, Cu. moschata , Cu. moschata* var. *meloniformis, Cu. pepo, Daucus carota* var. *sativa, Diospyros kaki, Euphorbia pulcherrina, Festuca arundinacea, Gossypium herbaceum , G. hirsutum, Gossypium* spp., *Ipomaea batatas, I. reptans, Lagenaria siceraria, L. siceraria* var. *clavata , Lolium perenne , Luffa acutangula , L. cylindrical, L. leucantha, Lupinus* spp., *Lycopsersicum esculentum, Malva sylvestris, Momordica charantia, Nicotiana tabacum, Oryza sativa, Pachyrhizus tuberosus, Phaseolus vulgaris, Poa pratensis, Raphanus sativus, Secale cereale, Setaria italica, Solanum melongena, Sorghum vulgaris, Spinacia oleracea, Triticum aestivum , Zingiber officinale, Zea mays*
P. aristosporum	*Setaria glauca, Setaria italica, Setaria viridis*
P. arrhenomanes	*Capsicum annuum, Chrysanthemum* spp., *Cucumis sativus, Lycopersicum esculentum, Oryza sativa, Saccharum officinarum, Setaria italica, Solanum melongena, Sorghum vulgaris, Triticum aestivum, Zea mays*
P. catenulatum	*Daucus carota, D. carota* var. *sativa, Lycopersica esculatum, Oryza sativa, Solanum melongena, Triticum aestivum, Vigna sinensis*
P. cedri	*Cedrus deodara* (Pinaceae)
P. debaryanum	*Brassica oleracea* var. *capitata, Citrus* spp., *Cucumis melo, C. sativus, Gossypium hirsutum, Lycopersicum esculentum , Panax schinseng, Pinus massoniana, Solanum melongena, Nicotiana tabacum, Setaria italica, Sorghum vulgaris, Zea mays*
P. deliense	*Agrostis palstriss, A. stolonifer, Brassica* spp., *Capsicum annuum, Cucumis melo, C. sativus, Festuca arundinacea, Lolium perenne L., Lycopersicum esculentum , Poa pratensis, Solanum melongena*
P. diclinum	*Capsicum annuum, C. sativus, Chrysanthemum* spp., *Lycopersicum esculentum, Oryza sativa, Solanum melongena , Triticum sativum*
P. dissotocum (*=oryzeae*)	*Capsicum annuum, Chrysanthemum* spp., *Cucumis sativus, Hordeum vulgare, Glycine max, Lycopersicum esculentum, Solanum melongena, Triticum aestivum, Vicia faba, Oryza sativa*
P. echinocarpum	*Oryza sativa*
P. elongatum	*Nelumbo nucifera*
P. graminicola	*Agrostis palstriss, A. stolonifera, Capsicum annuum, Festuca arundinacea, Hordeum vulgare, Lolium perenne, Poa pratensis, Setaria italica, S. magna, Soghum vulgare, Triticum aestivum, Vicia faba, Zea mays, Zingiber officinale*
P. helicoides	*Victoria ('Longwood hybrid')*
P. hemmianum	*Citrus* spp.
P. heterothallicum	*Zea mays*

(Contd.)

Table 7.1 (*Contd.*)

Parasitic spp. on plants	Hosts
P. hydnosporum	*Citrus* spp.
P. kunmingense	*Lycopersicum esculentum, Rhododendron* spp.
P. orthogonon	*Zea mays*
P. inflatum	*Oryza sativa*
P. intermedium	*Oryza sativa, Pachyrrhizus erosus, Phaseolus vulgaris, Pisum sativum*
P. irregulare	*Agrostis palstriss, A. stolonifera, Capsicum annuum, Chrysanthemum* spp., *Cucumis sativus, Festuca arundinaceae, Hordeum vulgare, Lactuca sativa, Lagenaria siceraria, Lolium perenne, Lycopersicum esculatum, Nicotiana tobacum, Oryza sativa, Pachyrhizus erosus, Poa pratensis, Setoria italica, Solanum melongena, Sorghum vulgaris, Spinacia oleracea, Triticum aestivum, Vicia faba, Zea mays, Zingiber italica*
P. mamillatum	*Agrostis palstriss, A. stolonifera, Festuca arundinacea, Lolium perenne, Poa pratensis*
P. middletonii	*Capsicum annuum, Chrysanthemum* spp., *Cucumis sativus, Lycopersicum esculentum, Solanum melongena*
P. monospermum	*Lycopersicum esculentum, Setaria italica, Zingiber officinale*
P. myriotylum	*Calathea* sp., *Cucumis sativa, Dioscorea* spp., *Glycine max, Inula japonica, Lycopersicum esculentum, Zingiber officinale*
P. oligandrum	*Brassica pekinensis, Capsicum annuum (frutescens), Chrysanthemum* spp., *Cucumis sativus,* Family Crucifereae, *Juglans nigra, Lycopersicum esculentum, Solanum melongena*
P. orthogonon	*Zea mays*
P. paroecandrum	*Allium cepa, A. cepa var. aggregatum, A. fistulosum, A. graveolens, A. sativum, A. tuberosum, Apium graveolens, Brassica pekinensis,* Family Crucifereae
P. periplocum	*Allium sepa, A. cepa var. aggregatum, A. porrum, A. sativum, A. tuberosum*
P. perplexum	*Allium tuberosum*
P. polypapillatum	*Hordeum vulgare, Triticum aestivum*
P. pulchrum	*Lycopersicum esculentum, Oryza sativa, Solanum melongena, Triticum aestivum*
P. rostratum	*Oryza sativa, Vicia faba*
P. salinum	*Oryza sativa, Zea mays*
P. sinensis	*Capsicum frutescens, Lycopersicum esculentum*
P. spinosum	*Chrysanthemum* spp., *Capsicum* sp., *Cucumis sativus, Glycine max, Hordeum vulgare, Lycopersicum esculentum, Oryza sativa, Pachyrhizus tubersosus, Solanum melongena, Thea sinensis, Triticum aesticum, Vicia faba, Zea mays*
P. sylvaticum	*Glycine max, Gossypium , Oryza sativa*
P. tardicrescens	*Lycopersicum esculentum, Setaria italica, Zea mays*
P. torulosum	*Capsicum annuum., Chrysanthemum* spp., *Cucumis sativus, Lycopersicum esculentum, Solanum melongenae*
P. ultimum	*Agrostis palstriss, A. stolonifera, Amomum compactum, Am. krarvanh, Allium cepa, A. cepa var. aggregatum, A. fistulosum, A. porrum, A. sativum, A. tuberosum, Brassica campestris, B. juncea var. cripifolia, B. pekinensis, Calotropis procera, Capsicum annuum (frutescens), Carica papaya, Chrysanthemum* spp., *Cucumis sativus, Cunnuinghamia lanceolata, Daucus carota var. sativa, Dendrobium officinale, Dioscorea* spp. *Festuca arundinacea,*

(Contd.)

	Family Crucifereae, Fragaria ananassa, Hordeum vulgare, Leucaena leucocepha, Lolium perenne, Lycopersicum esculentum, Oryza sativa, Nicotiana tabacum, Pachyrhizus bulbosus , P. erosus, Papaver rhoeas, Phaseolus vulgaris, Pisum sativum, Poa pratensis, Portulaca pilosa, Raphanus sativa, Scutellaria baicalensis, Setaria italica, Solanum melongena, Sorghum vulgaris, Triticum aestivum, Vicia faba L., Vigna unquiculata, Zea mays
P. vexans	*Dendrobium aurantiacum, D. chrysanthum, D. chrysotoxum, D. thyrsiflorum, Dioscorea opposita, Setaria italica, Solanum tubersoum, Sorghum vulgaris, Zea mays*
Pythium unidentified species	*Agrostis* spp., *Amaranthus tricolor, Ananas sativus, Antirrhinum major, Calendula officinalis, Celosia argentea var. cristata, Chrysanthemum cineriaefolium, Citrus* spp., Family Araceae, Family Cucurbitaceae, *Festuca arundinacea, Impatiens balsamina, Lolium perenne, Manihot esculenta, Nelumbo nucifera, Nicotiana tabacum, Pachyptera alliaceae, Pelargonium hortorum, Poa pratensis, Solanum melongena, Spinacia oleraceae, Vicia faba*
Parasitic on red algae	
P. porphyrae	*Porphyra yezoensis*
Parasitic on mosquito	
P. carolinianum	*Aedes albopictus, A. elsiae, A. formosensis, A. novoniveus, Anopheles sinensis, Culex mimulus, C. minor, C. pseudovishnui, C. quinquefasciatus, C. theileri, C. tianpinggensis , C. tritaeniorhynchus*
P. guiyangense	*Aedes albopictus, A. eagypti, A. elsiae, A. formosensis, A. novoniveus, Anopheles sinensis, Culex mimulus, C. minor, C. pipiens pallens, C. pseudovishnui, C. pipiens quinquefasciatus, C. theileri, C. tianpinggensis, C. tritaeniorhynchus*
Mycoparasitic *P. oligandrum* *P. nunn*	Various *Pythium* spp. and other soil-borne fungal pathogens

Plant disease control

For the management of the plant diseases triggered by various *Pythium* species in all regions of China, it is important to have an understanding of the physiology, infection mechanism and pathogenesis of the targeted pathogens. The fungal pathogens requirements like nutritional and temperature standards have been commonly investigated along with the record of the pathogens. Chen and Lin (1949) observed the inhibition of the fungal development of *P. aphanidermatum* through anions; at the same time, Chen et al. (1989) examined the reasons stimulating the zoospore production of *P. deliense*. Moreover, Ji et al. (2000) presented a distinctive report on the ultrastructural resistance to stalk rot in maize induced by *P. inflatum* and Yang et al. (2001) proposed that the resistance mechanism to the pathogen was due to one dominant gene within the host. In the study of the pathogenesis of *P. aphanidermatum,* He et al. (1992a) showed that the zoospores aggregated at the area of root hairs and root caps of tomato and begonia and penetrated the plant cells with the aid of germ tubes. Chen et al. (1996) investigated that cucumber seedling damping-off by *Pythium* generally occurred due to the production of polygalacturonase and other cellphone wall-degrading enzymes. In another study, Xu et al. (2009a, b) discovered that the destruction of chlorophyll, carotene and other cellular changes in *Lolium perenne* was a result of the infection by *P. aphanidermatum.*

Several methods have been developed to manage plant diseases caused by *Pythium* species. For pragmatic reasons, chemical substances have been used due to the fact that they were quick performing, potent and handy to manage. Yu (1934) controlled *Pythium* diseases by acetic acid combined with oil which was able to eradicate damping-off of cucumber seedlings without causing any injuries to the seeds and seedlings. The toxicity to the pathogen was due to the acetate anions

(Chen and Lin 1949). Zinc sulfate decreased the disease progress of *P. salimum, P. orthogonon* and *P. spinosum*, the main pathogens of basal stalk rot of maize and it might be used to manage the disease effectively within the Guangxi area (Zhang 1997). More strong pesticides have been found to be inhibitory to *Pythium* species and as a result applied to manage the disease. For example, Pythium blight of turf grass can be managed by using thiophanate-methyl, mancozeb and chlorothalonil (Han and Mu 2000). Gao et al. (2001) revealed that seed treatment mixed with soil treated with three fungicides such as hymexazol, metalaxyl and diisothiocyanatomethane was effective in the control of rice blight caused by *P. salinum, P. intermedium* and *P. acrogenum*. Hymexazol and metalaxyl have been also used in the control of poinsettia root rot (Yang et al. 2003). The *P. aphanidermatum*-induced root rot disease of watermelon seedlings in Gansu Province was controlled by coating them with Vitavax FF (Wang 2004). The mixture of metalaxyl and hymexazol seemed to have a synergistic impact and was found to be extremely toxic to the mycelial progress of *P. aphanidermatum in vitro* (Zhu et al. 2010). The seed coating with a mixture of these two fungicides was verified to be useful in managing Pythium diseases of *Poa pratensis* (Wang 2010). A few fungicides like azoxystrobin, metalaxyl and difenoconazole showed high level of toxicity against *P. graminicola* and were also used to control corn stalk rot (Zhang et al. 2010a, b, c). Isothiocyanate could possibly be utilized to disinfect the soil inside the greenhouse by making use of the drip water program to control *Pythium* species (Xiao et al. 2010). In the *in vitro* assay, other chemical substances were examined for their fungicidal potentials. For instance, the development of *P. aphanidermatum* was slowed by propamocarb (Liu et al. 2007a, b, c, d) and difenoconazole (Jiang et al. 2004). Hymexazol, pyraclostrobin, bonopol and metiram were demonstrated to become toxic to *P. myriotylum*, which triggered ginger rhizome rot (Yuan et al. 2011). Many fungicides had been screened against *Pythium* spp., the causal agent of garlic decay and the mycelial progress *in vitro* was reduced by thiophanate-methyl, copper mineral sulphate, mancozeb, copper hydroxide, cymoxanil, mancobez, metiram, famoxas and cymoxanil (Zhang et al. 2008).

Although some isolates of *Pythium* established varying levels of resistance to fungicides, in particular to the greatly used metalaxyl active elements, a couple of biocontrol agents such as antagonistic bacteria, like *Pseudomonas* spp. were confirmed to be a viable alternative (Zhang et al. 1990, Lou et al. 2001, 2002). Six strains of *Rhizobacteria* that have suppressing impact on damping off disease caused by *Pythium* spp. were isolated from the rhizosphere soil of healthy tobacco crops in a severely diseased area in Yunnan (Fang 2001). *In vivo* assay including pot and field experiments in Zhejiang Province, the organic control performance of *Pseudomonas aeruginosa* against the metalaxyl-resistant *P. ultimum* and *P. spinosum* was 94.4 and 51.4%, respectively. In Shangdong area, *P. aphanidermatum* caused cucumber seedlings damping-off and the control efficiency of *Paenibacillus polymyxa* was examined under different conditions (Pan et al. 2008). The antagonistic bacteria lysed the hyphal wall causing the leakage of protoplasts and failure to form oospores. The disease management efficiency was compared with propamocarb fungicide. Moreover, it not only has impact on the germination rate of cucumber seed but also increases the emergence rate and enhances the seedling development. From the rhizosphere of ginger, *Pseudomonas fluorescens* and *Alcaligenes faecalis* were isolated and proved to be antagonistic to *P. myriotylum* which triggered the ginger rhizome rot (Wang et al. 2010a, b). *Bacillus megaterium* would produce systemic resistance in cucumber against seedling damping-off by stimulating defense-related enzymes within some plant species (Liang et al. 2011).

Some mycoparasitic fungi could possibly be used to regulate and manage different Pythium diseases. An isolate of *Trichoderma* associated with soil in Hebei Province showed promising control proficiency of 80% in safeguarding cucumber seedlings against *P. aphanidermatum* due to competition, growth within the hyphae, and secretion of compounds causing cell wall lysis of the pathogen. The result was long-lasting and it even improved the plant vigour growth of cucumber seedlings (Tian et al. 2001). An identical inhibitory mode of action was reported in the control of turf grass *Lolium perenne* in Lanzhou caused by *T. aureoviride* (Gu et al. 2011). An isolate of *T. harzianum* in Fujian Province, which minimized the fungal growth of *P. aphanidermatum* by

80% in dual-culture and the pot experiment assay, was demonstrated to have control influence on tobacco damping off much better than metalaxyl-mancozeb fungicide (Chen et al. 2009). Chen et al. (2004, 2005) revealed that *T. harzianum* controlled seedling damping off and root rots due to *P. ultimum* by stimulating the formation of plant proteomics, accounting for the induced resistance in the host plants. *In vitro* investigations demonstrated that it had been antagonistic against other plant pathogenic *Pythium* species: *P. ultimum*, *P. spinosum* and *P. irregulare* (He et al. 1992b). Seed dressing of tomatoes with oospores of *P. oligandrum* led to effectively reduced disease incidence of seed rot and damping-off from 63.6% to 79.4% caused by *P. ultimum*. Gu et al. (2011) demonstrated that *T. aureoviride* decreased *in vitro* growth of *P. aphanidermatum*, the causal agent of turfgrass root rot, with equal efficiency as the fungicide, metalaxyl and the reduction was linked to the penetration and parasitizing the hyphae of the pathogen. Furthermore, an elicitin-like protein (oligandrin) made by *P. oligandrum* could encourage systemic resistance in the host plants (Lou and Zhang 2005). Wang et al. (2007a, b) also proved that the secretion of *P. oligandrum* clearly reduced the mycelial growth of *P. sylvaticum* on agar plates . The active compound had been developed and marketed as a biofungicide: polyversum which was demonstrated to enhance the resistance against seedling blight of rice crops (Yang et al. 2007). The recent isolation of another mycoparasitic fungus, *P. nunn* (Cheng et al. 2012), is encouraging and there must be more efforts to fully utilize this species in the management of soil borne plant pathogens. There have been efforts to explore new tactics for the managing of Pythium diseases. For instance, five Chinese ectomycorrhizal fungi, *Boletus edulis*, *Suillus grevillei*, *S. luteus*, *Chromogomphus rutilus* and *Xerocomus chrysenteron*, had been found to be hostile to *P. aphanidermatum* and *P. ultimum in vitro* by secreting materials to slow down the mycelia and inhibit the forming of sporangia (Lei et al. 1995). Yu et al. (2008a, b) confirmed that the extracts of *Houttuyniae cordata* reduced the growth of *P. ultimum* and they were also used to control root rot disorder of peas. Secondary metabolites (flavones phenolics and spanonins) from *Solidago canadensis* accrued in soil could inhibit *P. ultimum* (Zhang et al. 2011).

P. aphanidermatum as a most crucial plant pathogen

Of all *Pythium* species in mainland China, 48 species are pathogenic to various plant hosts. Apart from this, one species was found pathogenic to the red alga, *Porphyra yezoensis*, and other two were reported to be pathogenic to two mosquito larvae: *P. carolinanum* and *P. guiyangense*. Moreover, *P. oligandrumm* and *P. nunn* along with some other *Pythium* species, are mycoparasitic on some plant pathogens. It is not clear if the pathogenic species in mainland China are native or exotic, being introduced from several other countries along with nursery or ornamental plants. Information on *Pythium* species, specifically from virgin forest soil, might reveal the origin of these pathogenic species already determined and the information might assist in regulating their spread. With the understanding of the negative impact of chemical fungicides on human beings and the ecosystem, it is really important to develop other non-chemical methods to control Pythium diseases. The utilization of *Trichoderma*, *P. oligandrum* and antagonistic bacteria as seed dressings and soil amendment were already successfully adopted to reduce the pathogenic *Pythium* species and also to induce resistance in the host plants. It is possible to use molecular approaches to produce more effective and stable strains of the antagonistic organisms. With additional research, antagonistic bioactive materials extracted from plants and ectomycorrhizae may become practical and useful as biofungicides.

While *P. aphanidermatum* is normally the most crucial plant-pathogenic species of *Pythium* in mainland China, it may be employed for other diverse purposes. The crude extract of the fungus in liquid medium was proved to have herbicidal activity and toxic to a number of weeds, including *Digitaria sanguinalis*, *Amaranthus retroflexus*, *Chenopodium album*, *Echinochlora crusgalli*, *Chloris virgata* and *Setaria glauca* (Zhang et al. 2005, Xu et al. 2008, Zhang et al. 2010a, b, c, d). Bioactive compound isolated from the extracts of the fungus and the outcomes of bioassay proved that it negatively affected the photosynthesis of weeds (Xu et al. 2009a, b). Zhang et al. (2010a, b, c, d) could actually isolate and determine the chemical composition of the toxin, with at least one among the components to end up being dimethyl o-phthalate. It really is hopeful that the upcoming new bioherbicide could possibly be synthesized and applied in the field. In the meantime, the toxin in *P. aphanidermatum* has been utilized to quickly screen turf grass for levels of disease resistance (Yang et al. 2008).

The finding of two mosquito-killing species such as *P. carolinianum* and *P. guiyangense* has significant benefits. Since both are generally found secure for non-target microorganisms, like animals, insects and crops, it is time to start a detailed program to use them in the management of mosquitoes which are responsible for human diseases in many countries in the world, introducing dengue fever, malaria, encephalitis, West Nile fever and yellow fever. Extra efforts are required to recognize their pathogenesis in order to eradicate mosquito larvae. Since their cDNA library has been created, it would be possible to control their very own genotypes by means of *Agrobacterium*-mediated modification or separate the violence gene which may then be inserted in other fungi which might be more appropriate for application. If mosquito larvae are destroyed by the toxins, it might be important to isolate the bioactive ingredient of the toxic substance to develop a powerful mycoinsecticide for the mosquito *Aedes aegypti*.

Conclusion

Nearly all the *Pythium* species reported from China are typically recognized as soil-borne pathogens. Unquestionably, *P. aphanidermatum* is the most destructive pathogenic species in China during the summer, with the widest plant host range, followed by *P. ultimum* and *P. irregular*. It can be used for beneficial use due to the fact that crude extracts of *P. aphanidermatum* were highly toxic to numerous weeds and hence in near future it can be used as an effective bioherbicide. *P. oligandrum* acts as both a plant pathogenic and mycoparasitic species. Oospores of the mycoparasitic *P. oligandrum* have been reported as biofungicide: polysersum WP to regulate plant diseases caused by *Pythium* spp. and other fungal pathogens. The finding of two mosquito-killing species such as *P. carolinianum* and *P. guiyangense* has significant benefits.

References

Chai, C.S. and He, W. 2002. Investigation on the main lawn and turf diseases in summer in Beijing and a laboratory fungicide selection experiment. Grassland of China 24(6): 38-42. (in Chinese)

Chen, D.H., Zhang, B.X. and Ge, Q.X. 1989. Study on the inducement of zoospore production of *Pythium deliense*. Mycosystema 8(1): 51-56.

Chen, J., Gao, H., Zhu, Y. and Wu, Y. 1996. On pathogenetic mechanism of cucumber seedling damping-off caused by *Pythium aphanidermatum*. Acta Phytopathol. Sin. 26(1): 55-61. (in Chinese).

Chen, J., Harman, G.G., Comis, A., Cheng, G.W. and Liu, H.N. 2004. The change of maize plant proteome affected by *Trichoderma harzianum* and *Pythium ultimum* I. Acta Phytopathol. Sinica 34(4): 319-328.

Chen, J., Harman, G.E., Comis, A. and Cheng, G.W. 2005. Proteins related to the biocontrol of pythium damping-off in maize with *Trichoderma harzianum* Rifai. J. Integr. Plant Biol. 47(8): 988-997.

Chen, J., LÜ, L., YE, W., Wang, Y.C. and Zheng, X.B. 2017. *Pythium cedri* sp. nov. (Pythiaceae, Pythiales) from southern China based on morphological and molecular characters. Phytotaxa 309(2): 135-142.

Chen, L.F., Wang, J.Z. and Xu, Y.G. 1997. Identification of the pathogen of Rhododendron foot rot. Acta Phytophylacica Sin. 24: 254-256. (in Chinese)

Chen, Y.H. and Lin, C.K. 1949. Physiological studies on oomycetes. 1. Inhibition of the growth of *Pythium aphanidermatum* by anions. Chinese J. Agric. 1(1): 143-152.

Chen, Z., Wu, H., Zeng, H. and Zhang, S. 2009. Identification of *Pythium aphanidermatum* isolated from tobacco field and the screening of *Trichoderma* spp. for antagonistic agent. J. Fujian Agric. Fores. University Natural Science Edition 38(1): 11-15. (in Chinese)

Cheng, C.C., Huai, W.X., Yao, Y.X. and Zhao, W.X. 2012. Morphological identification and rDNA-ITS phylogenetic analysis of *Pythium nunn*, a new record of Pythium in China. Chinese Agric. Sci. Bull. 28(18): 199-204. (in Chinese)

Chi, P.K. 1994. Fungal diseases of cultivated medicinal plants in Guangdong province. Guangdong Science and Technology Press, 106. (in Chinese)

De Cock A.W., Mendoza, L., Padhye, A.A., Ajello, L. and Kaufman, L. 1987. *Pythium insidiosum* sp. nov., the etiologic agent of pythiosis. J. Clin. Microbiol. 25(2): 344.

De Cock A.W., Lodhi, A.M., Rintoul, T.L., Bala, K., Robideau, G.P., Abad, Z.G. and Lévesque, C.A. 2015. Phytopythium: Molecular phylogeny and systematics. Persoonia: Mol. Phyl. Evol. Fungi 34: 25-39.

Ding, H. and Ma, J. 2005. Simultaneous infection by red rot and chytrid diseases in *Porphyra yezoensis* Ueda. J. Appl. Phycol. 17(1): 51-56.

Ding, H.Y. 2006. Infection by *Pythium porphyrae* in *Porphyra yezoensis* Ueda. J. Huaiyin Teachers Coll. Nat. Sci. Edition 1: 69-73. (in Chinese)

Dou, J., Tian, Y.L., Chen, Y.L., Tang, S.J. and Hu, B.S. 2010. Pythium leaf rot of Victoria 'Longwood Hybrid' and identification of the pathogen. Plant Prot. 34(5): 184-185 (in Chinese)

Duan, S.L., Yu, S. and Su, X.Q. 2008. The prokaryotic expression of a subtilisin-like protease Pr1 gene from *Pythium guiyangense* Su, a fungal pathogen of mosquitoes. J. Guiyang Med. College 33(3): 233-239.

Dunhuang, F., Meiyun, L. and Tianfei, L. 2001. Isolation of and screening for biocontrolling rhizobacteria against tobacco damping off. Xinan Nongye Daxue Xuebao China. (in Chinese)

Editorial Council, 1996. Encyclopedia of Agriculture in China: Plant Pathology Volume. 1-740. Agricultural Publishing House, Beijing. (in Chinese)

Gan, H., Chai, Z., Lou, B. and Li, J. 2010. *Pythium heterothallicum* new to China and its pathogenicity. Mycosystema 29(4): 494-501.

Gandhi, S.R. and Weete, J.D. 1991. Production of the polyunsaturated fatty acids arachidonic acid and eicosapentaenoic acid by the fungus *Pythium ultimum*. Microbiology 137(8): 1825-1830.

Gao, T.C., Ma, Y.M., Lu, Y.J. and YE, Z.Y. 2001. Identification and chemical control of pathogens causing rice blight disease on dry-raised rice seedlings. Plant Protection – Beijing 27(6): 1-3. (in Chinese)

Gu, L., Xu, B., Liang, Q. and Xue, Y. 2009. Occurrence of Pythium root rot in turf grass and identification of the pathogen. Acta Prataculturae Sin. 18(4): 75-180. (in Chinese)

Gu, L.J., Xu, B.L., Liang, Q.L. and Li, R.F. 2011. Antagonism and mechanism of action of *Trichoderma aureoviride* against *Pythium aphanidermatum* causing turfgrass root rot [J]. Acta Prataculturae Sin. 2. (in Chinese)

Han, L. and Mu, X. 2000. Disease of turfgrass and controlling in Lanzhou region. J. Beijing Forestry Univ. 22(2): 1-5. (in Chinese)

He, S.S., Zhang, B.X. and Ge, Q.X. 1992. On the antagonism by hyperparasite *Pythium oligandrum*. Acta Phytopathologica Sin. 22(1): 77-82. (in Chinese)

Hendrix, F.F. and Campbell, W.A. 1973. Pythiums as plant pathogens. Annu. Rev. Phytopathol. 11(1): 77-98.

Ho, H.H. 2013. The genus Pythium in mainland China. Mycosystema. 32(Suppl 1): 21-44.

Ho, W.C., Yu, Y.N. and Hsiang, D.Z. 1951. A preliminary study on root rot of pyrethrum. Agric. Res. China 2(2): 95-106. (in Chinese)

Hong, Y. and Su, X.Q. 2008. Construction of a genomic library of *Pythium guiyangense* Su, a mosquito-killing fungus. J. Guiyang Med. College 33(3): 248-250.

Hoshino, T., Tojo, M., Okada, G., Kanda, H., Ohgiya, S. and Ishizaki, K. 1999. A filamentous fungus, *Pythium ultimum* Trow var. *ultimum*, isolated from moribund moss colonies from Svalbard, northern islands of Norway.

Huang, J. and Su, X. 2001. Preliminary study on cytology of a new strain of mosquito-killing fungus. J. Guiyang Med. College 26(6): 484-486.

Huang, S. and Su, X. 2003. A study on effects of several environmental factors on sporulation of *Pythium carolinianum*, a fungal pathogen of mosquito larvae. J. Guiyang Med. College 28(2): 101-105. (in Chinese)

Huang, S.W. and Su, X.Q. 2007. Biological studies on *Pythium guiyangense*, a fungal pathogen of mosquito larvae. Mycosystema 26(3): 380-388.

Hulvey, J., Telle, S., Nigrelli, L., Lamour, K. and Thines, M. 2010. Salisapiliaceae – A new family of oomycetes from marsh grass litter of southeastern North America. Persoonia: Mol. Phylogeny Evol. Fungi 25: 109.

Ji, M.S., Chen, J. and Huang, G.K. 2000. Studies on the ultrastructure resistance to Pythium stalk rot in maize. J. Shenyang Agric. Univ. 31(5): 482-486.

Jiang, J.Z., Zhang, B.X. and Ge, Q.X. 1990. Studies on genus Pythium in East China I. Identification, distribution and pathogenicity of 18 *Pythium* spp. Acta Agriculturae Universitatis Zhejiangensis, 16(Suppl) 2: 206-212. (in Chinese)

Jiang, J.Z., Wu, X.M., He, F.Q., Li, X.F., Wang, C.J. and Guoji, S. 2004. Tests of toxicity of several fungicides to four pathogenic fungi on turf grass and their application. Plant Prot. 30(4): 78-80. (in Chinese)

Jones, E.E. and Deacon, J.W. 1995. Comparative physiology and behaviour of the mycoparasites *Pythium acanthophoron, P. oligandrum* and *P. mycoparasiticum*. Biocon. Sci. Technol. 5(1): 27-40.

Khulbe, R.D. 1983. Pathogenicity of some species of *Pythium pringsheim* on certain fresh water temperate fishes/pathogenität einiger arten von Pythium Pringsheim auf Süßwasserfische gemäßigter Breiten. Mycoses 26(5): 273-275.

Knox, J.S. and Paterson, R.A. 1973. The occurrence and distribution of some aquatic phycomycetes on Ross Island and the dry valleys of Victoria Land, Antarctica. Mycologia 65(2): 373-387.

Kong, X.L., Luo, R., Liu, X.L. and Su, X.Q. 2008. An observation on hyphae and asexual reproductive stages of *Pythium guiyangense* Su, a mosquito-killing fungus, with scanning electron microscope. J. Chinese Elect. Micoscopy Soc. 27(4): 331-335. (in Chinese)

Lai, C., Zhou, G., Li, F. and Lai, C. 2000. Discovery of *Pythium spinosum* on tea cutting seedlings in China and the physiological characters. Acta Phytophylacica Sin. 27(2): 117-120. (in Chinese)

Lai, C.Y., Lai, C.B. and Huang, J.X. 1998. Identification on the blight pathogen of root rot of tea cutting seedling. Guangxi Plant Prot. 4: 5-6. (in Chinese)

Li, Y., Mao, L.G., Yan, D.D., Liu, X.M., Ma, T.T., Shen, J., Liu, P.F., Li, Z., Wang, Q.X., Ouyang, C.B. and Guo, M.X. 2014. First report in China of soft rot of ginger caused by *Pythium aphanidermatum*. Plant Dis. 98(7): 101.

Li, Y.Z., Yang, M., Zhou, E.X., Wang, Z.H., Fu, C.D., Zhou, D.H., Wu, X.B. and Wu, G.L. 2007. Preliminary investigation on papaya stem base rot. Guangdong Agric. Sci. 12: 60-61. (in Chinese)

Liang, J.G., Tao, R.X., Hao, Z.N., Wang, L.P. and Zhang, X. 2011. Induction of resistance in cucumber against seedling damping-off by plant growth-promoting rhizobacteria (PGPR) *Bacillus megaterium* strain L8. Afr. J. Biotechnol. 10(36): 6920-6927. (in Chinese)

Ling, Y., Xia, J., Koji, K., Zhang, X. and Li, Z. 2018. First report of damping-off caused by *Pythium arrhenomanes* on rice in China. Plant Dis. 102(11): 2382.

Liu, A.Y. and Liu, A. 1998. Pathogenic fungi of Pythium soft rot of Chinese cabbage and its binomics. Plant Prot. 24(5): 17-19. (in Chinese)

Liu, J.R. and Xie, X.R. 2003. Study on the main diseases of cold season turfgrasses and their control techniques in Hexi corridor. Pratacultural Sci. 20(4): 48-51. (in Chinese)

Liu, P. and Su, X.Q. 2007a. A long-term test on the safety of *Pythium guiyangense* to rats. Mycostystema 26(3): 440-447. (in Chinese)

Liu, P. and Su, X. 2007b. A report of acute experiments on the safety of *Pythium guiyangense* Su to several kinds of animals. J. Guiyang Med. College 4. (in Chinese)

Liu, P., Zhou, S., Xian, D., Yang, M.F. and Su, X.Q. 2007c. An observation on the safety of *Pythium guiyangense* to *Plutellaxylo stella* larvae. J. Guiyang Med. College 32(3): 225-226. (in Chinese)

Liu, X.R., Su, X.Q., Zou, F.H. and Guo, Q. 2001. Construction of a cDNA library of *Pythium carolinianum*, a mosquito-killing fungus. Fungal Div. 7: 53-59.

Liu, Y.L., Mu, W. and Liu, F. 2007d. Studies on the control of propamocarb against *Pythium aphanidermatum* and damping-off on cucumber seedling. Modern Agrochemicals 6(1): 44. (in Chinese)

Liu, Z. 1992. Study on the rot of lotus lily. Acta Phytopathologica Sin. 22(3): 265-269. (in Chinese)

Liu, Z.W. and Shi, X.J. 2009. Identification of the pathogen of ginger *Pythium myriotylum* in Laiwu City. China Plant Prot. 7. (in Chinese)

Lou, B., Zhang, B. and Zhang, Y. 2000. DNA polymorphism of Pythium eight species of causing tomato seedlings damping-off disease in Hangzhou region. J. Zhejiang Univ. (Agri. & Life Sci.) 26(6): 607-610. (in Chinese).

Lou, B., Zhang, B., Hu, L. and Ryder, M. 2001. Resistance to metalaxyl and biological control of *Pythium* spp. Acta Phytophylacica Sin. 28(1): 55-60. (in Chinese)

Lou, B., Zhang, B., Maarten, R., Rosemary, W. and Paul, H. 2002. Biological control of seedling damping-off cucumber. Acta Phytophylacica Sin. 29(2): 109-113. (in Chinese)

Lou, B. and Zhang, B. 2004. Identification of *Pythium sylvaticum* and species-specific primers. Mycosystema 23(3): 356-365. (in Chinese)

Lou, B. and Zhang, B. 2005. Biocontrol and induction of defense responses by the nonpathogenic (Pythium) spp. Acta Phytophylacica Sini. 32(1): 93-96. (in Chinese)

Ma, J.H. 1996. A preliminary study on the red rot disease of *Porphyra yezoensis* Ueda. J Shanghai Fish Univ. 5(1): 1-7. (in Chinese)

Ma, Z., Kuai, B., Xu, C., Chen, W. and Zhang, Y. 1999. Identification and control of cool season turfgrass diseases in Shanghai area. J. Fudan Univ. Natural Sci. 38(5): 553-556. (in Chinese)

Mendoza, L., Ajello, L. and McGinnis, M.R. 1996. Infections caused by the oomycetous pathogen *Pythium insidiosum*. Infeccionescausadas por el patógenooomiceto *Pythium insidiosum*. J. Mycol. Med. 6(4): 151-164.

Oishi, M., Takahashi, M. and Kobayashi, Y. 2010. Pythium rot of *Calathea* sp. caused by *Pythium myriotylum* intercepted in import plant quarantine in Japan. Res. Bull. Pl. Prot. Service, Japan (46): 69-71. (in Japanese)

Ouyang, Y.N., Xia, L.X., Zhu, L.F., Yu, S.M. and Jin, Q.Z. 2007. Effects of polyversum, a product from *Pythium oligandrum* on growth, disease resistance and yield promotion of rice. China Rice 6: 48-51. (in Chinese)

Pan, J.J., Pan, T., Shang, Y.K., Mu, W. and Liu, F. 2008. Antifungal activity of antagonistic strain BMP-11 against *Pythium aphanidermatum* and its identification. Acta Phytophylacica Sin. 35(4): 311-316. (in Chinese)

Qiu, D.X., Li, M.C., Tan, S.B. and Duan, G.N. 1986. A preliminary study of the occurrence of China fir root rot. Scientia Silvae Sin. 22(3): 311-316. (in Chinese)

Qiu, L., Mao, Y., Tang, L., Tang, X. and Mo, Z. 2018. Characterization of *Pythium chondricola* associated with red rot disease of *Porphyra yezoensis* (Ueda) (Bangiales, Rhodophyta) from Lianyungang, China. J. Oceanol. Limnol. 1-11. https://doi.org/10.1007/s00343-019-8075-3

Shang, J., Liu, X., Pan, C., Zhao, Z. and Ma, H. 2010. Pathogen identification of walnut stalk rot disease. Scientia Silvae Sin. 46(12): 97-100. (in Chinese)

Shen, J. and Zhang, B. 1994. Pathogenic pythium species from wheat and barley in Zhejiang Province. Acta Phytopathol. Sin. 24(2): 107-112. (in Chinese)

Shen, J. and Zhang, B. 1995. Studies on pathogenic species of Pythium from spring cropping maize seedlings in Zhejiang province. Acta Phytophylacica Sin. 22(3): 265-268. (in Chinese)

Shi, X. and Xu, X. 2003. Isolating and cloning of a cDNA fragment of Pr1 gene from *Pythium carolinianum*. J. Guiyang Med. College 28(1): 1-4. (in Chinese)

Shuishan, H. and Qixin, Z.B.G. 1992. On the infection process of zoospores of *Pythium aphanidermatum* to tomato and begonia. Acta Phytopathologica Sin. (2): 13. (in Chinese)

Stredansky, M., Conti, E. and Salaris, A. 2000. Production of polyunsaturated fatty acids by *Pythium ultimum* in solid-state cultivation. Enzyme Microbial Technol. 26(2)4: 304-307.

Su, X., Zou, F., Guo, Q., Huang, J. and Chen, T.X. 2001. A report on a mosquito-killing fungus, *Pythium carolinianum*. Fungal Div. 7: 129-133.

Su, X. 2002. A preliminary study on Pr1- and Pr2-like enzyme of mosquito-killing fungi *Lagenidium giganteum* and *Pythium carolinianum*. J. Guiyang Med. College 27(5): 377-380. (in Chinese)

Su, X., Huang, S., Zhang, C. and Liu, P. 2003. A preliminary investigation on the mosquito host range of *Pythium carolinianum*. J. Guiyang Med. College 28(5): 377-379. (in Chinese)

Su, X.Q. 2008. A list of mosquito-hosts of fungal pathogen *Pythium guiyangense*. Guizhou Med. J. 32(7): 579-580. (in Chinese)

Takahashi, M., Ichitani, T. and Sasaki, M. 1977. *Pythium porphyrae* Takashashi et Sasaki sp. nov. causing red rot of marine red algae *Porphyra* spp. Trans. Mycol. Soc. Japan 18: 279-285.

Tao, Y., Zeng, F., Ho, H., Wei, J., Wu, Y., Yang, L. and He, Y. 2011. *Pythium vexans* causing stem rot of Dendrobium in Yunnan Province, China. J. Phytopathol. 159(4): 255-259.

Thianprasit, M., Chaiprasert, A. and Imwidthaya, P. 1996. Human pythiosis. Curr. Topics Med. Mycol. 7(1): 43-54.

Tian, L.S., Shi, W.L., Yang, Y.X. and Zhang, G.W. 2001. Studies on the antagonism of Trichoderma sp. against *Pythium aphanidermatum* and its application in biological control. J. Hebei Academy Sci. (2): 114-117. (in Chinese)

van der Plaats-Niterink, A.J. 1981. Monograph of the genus Pythium (Vol. 21). Baarn: Centraal Bureauvoor Schimmel Cultures.

Wang, A.Y., Lou, B.G. and Xu, T. 2007a. Inhibitory effect of the secretion of *Pythium oligandrum* on plant pathogenic fungi and the control effect against tomato gray mould. Acta Phytophylacia Sin. 34(1): 57-60. (in Chinese)

Wang, C., Sun, C. and Liu, S. 2010a. Preliminary study on pathogen of root rot of kidney bean and its control in Xinjiang. Ganhanqu Yanjiu (Arid Zone Research) 27(3): 380-384.

Wang, C.C., Du, B.H., Yao, L.T., Wang, X., Xu, Y.Y. and Ding, Y.Q. 2010b. Isolation and identification of antagonist bacteria from Ginger rhizosphere against *Pythium myriotylum* Drechsler. Microbiology/ Weishengwuxue Tongbao 37(12): 1767-1770. (in Chinese)

Wang, G.L., Ren, S.U. and Xie, X.H. 2000. Identification of pathogenic species of Pythium from cool-season grasses in Zhejiang province. Acta Phytopathologica Sin. 30(3): 286.

Wang, G.L., Ren, X.X. and Xie, X.H. 2001. Occurrence and control of Pythium damping off and leaf blight in cool-season turfgrass. Acta Phytophylacica Sin. 28(2): 129-132. (in Chinese)

Wang, J. 1997. Distributions and pathogenicity of *Pythium* spp. infecting tobacco in Yunnan Province. J. Yunnan Agric. Univ. 12(2): 97-102. (in Chinese)

Wang, J. 2004. An experiment of control effect of 40% Vitavax FF against *Pythium aphanidermatum*. Gansu nongyedaxuexuebao 39(4): 451-453.

Wang, R.X., Lei, D.W. and Su, X.Q. 2007b. A primary study on purification of subtilisin-like protease (Pr1) from *Pythium guiyangense* and its enzymological properties. Anhui Nongye Daxue Xuebao 34(4): 505. (in Chinese)

Wang, X.J. 2010. Seed coating formulations and its effects on controlling Pythium diseases of *Poa pratensis*. Pratacultural Sci. 27(11): 37-42. (in Chinese)

Wang, Z., Zhang, B. and Lou, B. 1996. On the ecology of Pythium spp. in wheat rhizosphere in Zhejiang province. Acta Phytopathol Sin. 26(1): 29-35. (in Chinese)

Wei, S.Q., Zhong, Y., Ma, Z.T. and Jiang, H. 1991. A survey on watermelon diseases in the northern China. China Fruits 1: 36-37. (in Chinese)

Wu, Q.N., Zhu, X.Y. and Wang, X.M. 1990. Isolation and the pathogenicity test of the pathogens of stalk rot of corn in China and USA. Acta Phytophylacica Sin. 17(4): 323-326. (in Chinese)

Xiao, C.K., Zhang, T., Chen, H.M., Zheng, S.H., Zhou, Z.Z., Hu, X.J., Mu, C.Q., Wang, Y.Z. and Zheng, J.Q. 2010. Studies on effect trial of soil disinfection by 20% isothiocyanate for strawberry in greenhouse. China Vegetables 21: 29-31. (in Chinese)

Xiao-Qing, S.U. 2006. A new species of Pythium isolated from mosquito larvae and its ITS region of rDNA. Mycosystema 4: 523-528.

Xing, Y.M., Li, X.D., Liu, M.M., Zhang, G., Wang, C.L. and Guo, S.X. 2015. Morphological and enzymatical characterization of the infection process of *Pythium ultimum* in *Dendrobium officinale* (orchidaceae). Cryptogamie Mycol. 36(3): 275-287.

Xu, J., Wu, C., Tong, Y. and Chen, X. 2001. Identification and pathogenicity study of fungi causing yellowing and blight of rice seedlings on dry seedbed in Jiangsu. Acta Phytopathol. Sin. 31(3): 230-235. (in Chinese)

Xu, J., Chen, H.Y., Shi, W.W., Dong, L.X. and Zhang, J.L. 2008. Biological characteristics of *Pythium aphanidermatum* and herbicidal of mutant isolates. J. Agric. Univ. Hebei 31(4): 81-86. (in Chinese)

Xu, J., Xu, W., Kang, Z., Zheng, H. and Zhang, J. 2009a. A primary study on the pathogenesis of *Pythium aphanidermatum* to Loliumperenne. Acta Prataculturae Sin. 18(4): 181-186. (in Chinese)

Xu, J., Xu, W., Zhang, L., Li, C. and Zhang, J. 2009b. Mechanism of herbicidal action of the component I from *Pythium aphanidermatum*. Front. Agric. China 3(2): 171-177. (in Chinese)

Xu, Z.T. and Zhang, C.M. 1985. On the causal organism of root and basal stalk rot of corn in Shangdong Province. Acta Phytopathol. Sin. 15(2): 103-108. (in Chinese)

Xue, F. 2003. Studies on fungal diseases of cold season turfgrass in Lanzhou. Pratacultural Sci. 20(3): 66-70. (in Chinese)

Yang, C., Hu, Y. and Li, C. 2003. Study on pathogen species of poinsettia root rot in Shanghai and drug screening. Acta Agriculturae Shanghai 19(3): 87-89. (in Chinese)

Yang, D.E., Chen, S.J., Wang, Y.G., Zhang, C.L., Li, S.R. and Wang, B. 2001. Genetic analysis of maize stalk rot (*Pythium inflatum* Mattews) resistance gene. J. Acta Phytopathol. Sin. 31(4): 315-318.

Yang, P., Xia, Y. and Su, X. 2005a. Cloning of unknown partner sequence of the subtilisin-like gene of *Pythium carolinianum* with panhandle PCR Technique. protease J. Guiyang Med. College 30(1): 4-9. (in Chinese)

Yang, P., Xia, Y.X. and Su, X.Q. 2005b. Cloning of the unknown sequence of subtilisin-like protease genein *Pythium carolinianum* by constructing mini genomic DNA library. J. Guiyang Med. College 30(2): 95-98.

Yang, Y.J., Tao, B., Li, C., Bian, X.J. and Zhang, J.L. 2008. Using the toxin of *Pythium aphanidermatum* to screen turfgrasses with disease resistance [J]. Acta Prataculturae Sin. 3. (in Chinese)

Yu, S., Duan, S.L. and Su, X.Q. 2007. Purification and characterization of a subtilisin-like protease Pr1 of *Pythium guiyangense*. J. Guiyang Med. College 32(2): 118-122. (in Chinese)

Yu, H.Z., Xiao, Z., Liu, H.H., Lei, D.Y. and Yang, X. 2008a. Control of different solvent distilled substances of *Houttuyniae cordata* against *Pythium ultimum* in pea. J. Changjiang Vegetables 3b: 80-81. (in Chinese)

Yu, S., Duan, S.L. and Su, X.Q. 2008b. A study on induction conditions for a trypsin-like protease (Pr2) of *Pythium guiyangense* Su. Journal of Guiyang Medical College 33(3): 221-224. (in Chinese)

Yu, T.F. 1934. Pythium damping off cucumber. Agric. Sin. 1(3): 1-16.

Yu, T.F. 1940. A list of important crop diseases occurring in Kiangsu Province. Lingnan Sci. J. 19: 67-68.

Yu, T.F., Chiu, W.F., Cheng, N.T. and Wu, T.T. 1945. Studies on *Pythium aphanidermatum* Edson Fitz in China. Lingnan Sci. J. 21(1-4): 45-62.

Yu, T.F. 1950. Notes on some weakly parasitic fungi associated with diseased roots of broad bean. Peking Natural History Bull. 18(4): 281-288.

Yu, T.F. 1955. Black rot of yam bean *Pachyrizus tuberosus* Spreng. Acta Phytopathol. Sin. 1(2): 177-182. (in Chinese)

Yu, T.F. 1978. Millet Diseases. Science Press, Beijing. 168 p. (in Chinese)

Yu, Y.N. and Ma, G.Z. 1989. The genus Pythium in China. Mycosystema 2: 1-110.

Yu, Y.N. 2001. Pythium and Fungi. Jiangxi Sci. 19(1): 55-60.

Yuan, J.D., Feng, K., Qi, J.S., Zhang, B., Li, L., Li, C.S. and Qu, Z.C. 2011. Toxicity determination of thirteen fungicides on Pythium causing ginger rhizome rots. Agrochemicals 50(8): 617-623. (in Chinese)

Zhang, B., Li, C.S., Li, L., Xu, Z.T. and Li, F. 2008. Control test of fungicides on Pythium root rot of garlic. Shandong Agric. Sci. 8: 023. (in Chinese)

Zhang, B.X. and Jiang, J.Z. 1990. A preliminary study of Pythium species in Yellow River irrigation area in Ningxia. Acta Agriculturae Universitatis Zhejiangensis, 16(Suppl. 2): 201-205. (in Chinese)

Zhang, B.-X., Ge, Q. X., Chen, D. H., Wang, Z. Y., and He, S. S., 1990, Biological and chemical control of root diseases on vegetable seedlings in Zhejiang province, China. pp. 181–196. *In*: D. Hornby (ed.). Biological Control of Soil-borne Plant Pathogens. C.A.B. International, Wallingford.

Zhang, C.B. and Su, X.Q. 2008a. A field evaluation on the safety of a mosquito killing fungus, *Pythium guiyangense*, to several species of crops. J. Anhui Agric. Univ. 35(3): 426-429 (in Chinese)

Zhang, C.B. and Su, X.Q. 2008b. Experimental studies on the safety of *Pythium guiyangense* to six against corn stalk rot. J. Henan Agric. Sci. (8): 90-92. (in Chinese)

Zhang, D., Min, Y., Yuan, H. and Liu, C. 2010a. Toxicity assay and field control effect of several chemicals against corn stalk rot. J. Henan Agric. Sci. (8): 90-92. (in Chinese)

Zhang, J. and Huang, Z.Y. 1994. Identification of the fungal species causing root rot in citrus. J. Southwest Agr. Univ. 16(6): 515-520. (in Chinese)

Zhang, J.L., Pang, M.H., Liu, Y.C., Zhang, L.H. and Dong, J.G. 2005. The optimum production condition of the herbicidal substance from *Pythium aphanidermatum* [J]. J. Agric. Univ. Hebei 28(4): 84-88. (in Chinese)

Zhang, L., Wu, X., Zhang, L., Kang, Y. and Liang, H. 2010b. Isolation and identification of pathogens on bean sprout spoilage and studies on its pathogenicity. J. Changjiang Vegetables 2: 71-74. (in Chinese)

Zhang, L.H., Dong, J.G. and Shi, J.G. 2010c. Liquid fermentation of *Pythium aphanidermatum* and bioassay of the metabolites. J. Agric. Univ. Hebei 33(6): 110-132. (in Chinese)

Zhang, L.H., Kang, Z.H., Jiao, X.U., Xu, W.C. and Zhang, J.L. 2010d. Isolation and structural identification of herbicidal toxin fractions produced by *Pythium aphanidermatum*. Agric. Sci. China 9(7): 995-1000.

Zhang, P.K., Wu, Q.A. and Li, S.C. 1996. Preliminary report on the investigation of occurrence and identification of the pathogen of maize wilt. Guangxi J. Agric. (3): 1-5. (in Chinese)

Zhang, P.K. 1997. Studies on the control of maize basal stalk rot. Guangxi Plant Protection 2: 7-8.

Zhang, S., Zhu, W., Wang, B., Tang, J. and Chen, X. 2011. Secondary metabolites from the invasive *Solidago canadensis* L. accumulation in soil and contribution to inhibition of soil pathogen *Pythium ultimum*. Appl. Soil Ecol. 48(3): 280-286.

Zhang, Y.L., Zhang, B., Ma, L.G., Li, C.S., Qi, K., Xu, Z.T., Qi, J.S. and Ji, Y.J. 2018a. First report of *Pythium myriotylum* causing root rot of yam in China. Plant Dis. 102(12): 2663.

Zhang, Y.L., Li, C.S., Zhang, B., Qi, J.S., Xu, Z.T. and Ma, L.G. 2018b. First report of root rot of Chinese yam caused by *Pythium ultimum* var. *ultimum* in China. Plant Dis. 102(3): 687.

Zhang, Z. and Su, X.Q. 2008. Effect of five common pesticides on *Pythium guigangense* Su and Culex quinquefaciatus Say in Guizhou. Tianjin Agric. Sciences 3: 47-50. (in Chinese)

Zhao, J., Jiang, M. and Su, X. 2008. Preparation and regeneration of protoplasts of *Pythium guiyangense* Su, a fungal pathogen of mosquitoes. J. Guiyang Med. College 33(2): 111-114. (in Chinese)

Zhao, J.N. and Su, X.Q. 2008. The genetic transformation of *Pythium guiyangense* mediated by Agrobacterium tumefaciens. Mycosystema 27(4): 594-600.

Zhao, S., Fang, X., Jiang, H. and Li, C. 2009. Identification and rDNA ITS sequence analysis of the Pythium root rot pathogens of processing tomato in Xinjiang. Acta. Phytophylacica Sin. 36(3): 219-224. (in Chinese)

Zheng, X., Qi, P. and Jiang, Z. 2001. Identification of the fungal diseases on ornamental plants (Araceae) in Guangzhou region – I.J. South China. Agric. Univ. 22(1): 51-53. (in Chinese)

Zhou, L., Ding, J.J. and Li, G.Y. 2011. Study on the biological characteristics of the major pathogen of the root-rot disease of the processing tomatoes in Xinjiang. Xinjiang Agric. Sci. 48(4): 739-743. (in Chinese)

Zhu, H., Liang, J.N., Wang, Z.M., Chen, H.D., Xi. J.P., Chou, B.H. and Wang, Q.L. 1997. Identification of pathogens causing maize stalk rot in Jiangsu. Acta. Phytophylacica Sin. 24(1): 49-54. (in Chinese)

Zhu, W.G., Hu, W.Q. and Chen, J. 2010. Seed coating formulations and its effects on controlling Pythium diseases of *Poa pratensis*. Agrochemicals 12: 920-921. (in Chinese)

The *Pythium* Complex of the Mid-North Region of Brazil

Janete Barros da Silva*, Douglas Henrique Trigueiro Silva, Francynara Pontes Rocha, Givanilso Cândido Leal, Helanny Márcia Ribeiro Trajano, Joseane Lustosa Machado, Laércio de Sousa Saraiva, Maria do Amparo de Moura Macêdo, Osiel César da Trindade Júnior, José de Ribamar de Sousa Rocha, Nayara Dannielle Costa de Sousa and Tamyres Lopes Rios

Universidade Federal do Piauí, Campus Ministro Petrônio Portella, Centro de Ciências da Natureza, Departamento de Biologia, Laboratório de Micologia, Avenida Universitária, S/N, Cep: 64049-000, Teresina - Piauí, Brasil

Introduction

The phylum Oomycota (Kingdom Straminipila) has two classes (Saprolegniomycetes and Peronosporomycetes). Saprolegniomycetes includes three orders, seven families and 33 genera, whereas Peronosporomycetes has three orders, five families and 40 genera. Further several classes, five orders, ten families and 13 genera are classified as *Incertae sedis* (Beakes et al. 2014). The phylum comprises saprobes and parasites found in aquatic and terrestrial ecosystems; its members play an important role in the decomposition of organic matter and participate in the cycling of nutrients. Important parasites attack plants, algae, fish, crustaceans, fungi, mosquito larvae, nematodes, rotifers, mammals, and even humans (Marano et al. 2008, Beakes et al. 2014).

Oomycetes are part of a diverse group of organisms known as "zoosporic fungi". Zoosporic fungi do not constitute a monophyletic group. The phylogenetic relationship between them is quite diverse (Leedale 1974, Cavalier-Smith 1981, 1987, Dick 1990, Margulis 1990, Corliss 1990, Alexopoulos et al. 1996). This group is formed by osmotrophs heterotrophic organisms and form mobile flagellate cells of asexual origin – zoospores, or sexual – the planogametes; they have some common characteristics, presenting morphological, physiological and ecological similarity with the true flagellated fungi.

Zoosporic fungi, due to the presence of flagellated spores or gametes, require water, especially during the reproduction period, to complete the life cycle and, therefore, they are also referred to as "aquatic fungi", occurring in aquatic and terrestrial ecosystems, mainly as decomposers of organic matter, where along with bacteria, they act in the mineralization of nitrogen, phosphorus, potassium and sulfur and other organic compounds' ions. In these environments, they have a direct and indirect relationship with other organisms on many levels. Some may be parasites of algae, other fungi, phanerogamic plants, fish, crustaceans, vertebrates, including humans (Christensen 1989, Hawksworth 1991, Barr 1992, Moore-Landecker 1996, Santurio et al. 1998, McMeekin and Mendonza 2000 , Steciow et al. 2012).

---

*Corresponding author: jbdsjesus272016@outlook.com

Oomycota are widely distributed and ubiquitous. Saprobic species have extensive enzymatic capacity, with the ability to degrade a wide range of vegetal and animal substrates such as cellulose (algae and plant debris), keratin (snake ecdysis, human hair and feather), chitin (exoskeleton of crustaceans and insects), lignin (dead tissue of woody plants) and sporopollenin (pollen grains). Saprotrophic species are abundantly found in decaying plant material, but their occurrence and frequency are underestimated unless specific techniques are applied for their study (Hawksworth 2004, Marano et al. 2011).

The genus *Pythium* (s.str.) was reclassified by Uzuhashi et al. (2010). In addition to maintaining the genus *Pythium* (s. Lato), four new genera were derived: *Elongasporangium*, *Globisporangium*, *Ovatisporangium* and *Pilasporangium*. The genus *Phytopythium* was separated from *Pythium* by Bala et al. (2010a).

Thus, *Pythium* species exhibit aquatic, terrestrial and saprobic habits, being mostly cosmopolitan (Hawksworth et al. 1995). As saprobes, they are capable of developing themselves under adverse conditions for the growth of other organisms, as well as being able to colonize and germinate on organic substrates (Hendrix and Campbell 1973, Amorim et al. 2011, Schroeder et al. 2013). They occur frequently in cultivated soils, in the superficial layers, and in the root region (Plaats-Niterink 1975), and with less incidence in uncultivated or acid soils (Barton 1958, Lourd et al. 1986). In favourable conditions, they become highly pathogenic, causing rot in roots, stems, fruits and pre-emergent seeds, mainly attacking young or aqueous tissues (Plaats-Niterink 1981, Alexopoulos et al. 1996, Schroeder et al. 2013).

The phylogenetic study of species and varieties of *Pythium* carried out by Lévesque and De Cock (2004), through the analysis of the ITS region of the nuclear ribosomal DNA, allowed the grouping of *Pythium* in 11 clades, enabling the reclassification of the genus with redistribution in new taxonomic groups (Villa et al. 2006, Bala et al. 2010a, b, Robideau et al. 2011, De Cock et al. 2015).

Pythium and *Phytophthora* species are highly pathogenic in plants of economic interest (Moore-Landecker 1996, Alexopoulos et al. 1996). *Pythium insidiosum* stands out as equine parasite, and other animals such as cattle, goats, dogs, cats, as well as humans. Subcutaneous ulcerative lesions in animals caused by this species have been consistently reported (Santurio et al. 1998, Leal et al. 2001, Sallis et al. 2003, Rech et al. 2004), including in humans for the first time in Brazil (Bosco et al. 2008) and in horses in Piauí State (Rocha et al. 2010).

The first research of zoosporic fungi for Northeast Brazil reported 77 species for the semiarid region, 42 being of oomycetes (Rocha 2006). In the Mid-North region, the taxonomic studies about the zoosporic mycotic were started in Piauí, in the Cerrado area, in the Sete Cidades National Park, with 76 taxa, of which 36 were of oomycetes species (Rocha 2002). In Maranhão, the first study was conducted by Sales (2009), in Timon; 17 species of oomycetes have been reported, all being first citations for the state.

The Mid-North region of Brazil, constituted by the states of Maranhão and Piauí, has little knowledge about the occurrence and distribution of oomycetes, especially for the *Pythium* Complex. Reports of diversity and their geographical distribution constitute ecological knowledge directly correlated with the biome, climate and soil type. Possibly, the present study is the first that addresses the diversity and geographic distribution of oomycetes in the region.

Research at Universidade Federal do Piauí

The information presented here comes from research carried out by the nucleus of Zoosporic Fungi studies of Universidade Federal do Piauí (UFPI), Department of Biology, in Teresina, Piauí, especially through the course of master's degree in Development and Environment, linked to the PRODEMA network.

In Piauí, the studies were developed in the counties of Caracol - Sept. 2005 to Aug. 2007, in the Serra das Confusões National Park. The Research Project about the Semi-arid Biodiversity (PPBIO)

- Piauí (Fungi) (Rocha et al. 2018); Floriano - Mar. 2006 to Jan. 2007, in the urban perimeter of the Parnaíba river (Negreiro and Rocha 2007, 2009, Rocha et al. 2014); Nazária - Mar. 2007 to Jun. 2008, in the Agricultural Field of Nazária (Pereira and Rocha 2008, 2009, 2012, Rocha et al. 2014); Piracuruca – May. 1988 to Feb. 2000, in the Sete Cidades National Park (Rocha 2001, 2006, Rocha et al. 2001, 2010). Teresina 1 - Apr. 2002 to May. 2003, in the Community Vegetable Garden of Tabuleta (Rocha et al. 2018); Teresina 2 - Jul. 2013 to Jan. 2014, in the Environmental Park Encontro dos Rios (Rocha et al. 2018, Rocha et al. 2014); Teresina 3 - Aug. 2013 to Aug. 2014, in the Poti river, in the urban perimeter of Teresina (Rocha et al. 2014, Sousa and Rocha 2017).

In Maranhão, the studies were developed in the county of Timon 1 - May. 2008 to Jun. 2009, in the settlements of Banco de Areia, Bacuri and Roncador (Sales 2009, Rocha et al. 2014). Timon 2 - Aug. 2009 to Feb. 2010, in fish farms (Costa et al. 2010). Timon 3 - Aug. 2014 to May. 2015, in the Parque Natural Municipal Lagoa do Sambico (Silva and Rocha 2017).

Some data were obtained in the specialized literature such as: Doihara and Silva 2003; Freire 2001; Mendes et al. 1998; Silva et al. 2014; Viana and Sobrinho 1998; iana et al. 2004;, Macêdo and Rocha 2017; Rocha and Macêdo 2015; Rocha et al. 2010, 2014, 2017; Trindade Júnior and Rocha 2013 and Rocha and Soares 2018.

Considering that the genus *Pythium* has undergone several reclassifications (Lévesque and De Cock 2004, Villa et al. 2006, Bala et al. 2010a, b, Uzuhashi et al. 2010) and the literature prior to these reclassifications refer only to *Pythium*, in this work the species of *Pythium* (Py.), *Phytopythium* (Phy.) and *Globisporangium* (Gl.) will be referred to as *Pythium* Complex.

Isolation techniques

The multiple baiting technique with cellulosic, chitinous and keratinous substrates was applied to isolate Oomycetes from the samples, following Milanez (1989).

Water samples were collected from the lake surface with the aid of a plastic cup and then stored in wide-mouth sterile glass flasks (100 ml Wheaton vials), with perforated plastic cover to allow oxygenation of the water. Prior to collection, nine baits were added to each flask: cellulosic substrates (corn straw, sorghum seeds, onion epidermis, cellophane and paper filter) and chitinous (termite wings) as well as keratinous (snake ecdysis and human hair), which served as organic colonization substrates.

Soil samples were collected with the aid of a sterile metal spatula, and the surface layers were removed and about 250 g of soil was collected at depth of approximately 20 cm, then packed in 500 g polyethylene bags properly labeled with the respective collection points. The collected material was transported to the Mycology Laboratory of Universidade Federal do Piauí (UFPI) for analysis.

From the water samples, about 30 ml was placed into Petri dishes (100 × 20 mm) with organic substrates. Approximately, 20 g from the soil samples was placed in Petri dishes (100 × 20 mm) with sterile distilled water to decant. After decanting, cellulosic, chitinous and keratinous substrates were added to the plates. Then the plates were incubated with the samples at room temperature (30-32 °C) for a 7-day period.

We performed microscopic analysis of the samplings in order to assess the presence of Oomycetes, which were then placed into new Petri plates with the respective substrates to be colonized (cellulosic, chitinous or keratinous). Subsequently, sterile distilled water was added to each plate. The plates were incubated for seven days for the identification of developing fungal structures. The maintenance of the strains was done with the exchange of water and adding new substrates to each plate.

Vegetative and reproductive structures were recorded by using an optical microscope (Olympus BX41 model, Tokyo, Japan), and photographed with a Nikon digital camera (Coolpix - S4100). Taxonomic descriptions of isolates were done in accordance with pertinent publications (Sparrow

1960, Johnson Jr. 1960, Scott 1961, Dick 1990, Johnson et al. 2002). Selected strains were deposited in the mycology culture collection at Universidade Federal do Piauí (UFPI).

Taxonomic studies

Beakes et al. (2014) carried out an up-to-date phylogenetic and taxonomic review of the Straminipila (Labyrinthulomycota, Hyphochytridiomycota, and Oomycota) based upon molecular sequence data, biology and evolutionary history.

The survey of species of the *Pythium* Complex for the Mid-North Region of Brazil, includes 14 species, two of which occurred exclusively in the state of Piauí. Five species common to the states of Maranhão and Piauí were reported. Currently, one taxa remains with exclusive reports for the Mid-North Region (Table 8.1), distributed in the family: *Pythiaceae* (five *Globisporangium*, five *Pythium* and four *Phytopythium*).

Currently, the Collection of Zoosporic Organisms of UFPI registers several accesses of the *Pythium* Complex, represented by 14 species: *Globisporangium echinulatum* Matthews, *Globisporangium mamillatum* Meurs, *Globisporangium perplexum* Kouyeas and Theohari, *Globisporangium proliferum* (Cornu) P.M.Kirk, *Globisporangium ultimum* (Trow) Uzuhashi, Tojo and Kakish, *Pythium aphanidermatum* Edson Fitzp, *Pythium graminicola* Subramaniam, *Pythium inflatum* Matthews, *Pythium* middletonii (Sparrow), *Pythium myriotylum* Drechsler, *Phythopythium indigoferae* Butler, *Phythopythium palingenes* Drechsler, *Phythopythium vexans* (De Bary), and *Phytophythora* sp (De Bary) (Table 8.1).

According to the catalogs of Brazilian biodiversity (Forzza et al. 2010, Flora of Brazil 2020 under preparation, 2019), 42 species of the Pythium Complex are registered throughout the country. Of these, the state of São Paulo, which today is the place where most of the studies about zoosporic organisms are concentrated, registers 30 species, then Rio de Janeiro and Piauí with 13 species each. Leaving behind only the Southeast region, the Northeast stands out with a total of 17 representatives.

Moreover, Piauí has a little more than double the total number of species registered in Pernambuco, which lists six, and is the second state of northeastern region with the highest number of occurrence records (Forzza et al. 2010, Flora of Brazil 2020 under preparation, 2019). New records of five species of *Pythium*, in São Paulo, were made by Gonçalves et al. (2016a). They follow the characteristics of the species cataloged in the Mid-North Region of Brazil. The following are the taxa, the geographic distribution in the Mid-North region and in other states of Brazil (Flora of Brazil 2020 under preparation, 2019) and the bibliographic sources.

Pythiaceae

Globisporangium

Globisporangium echinulatum (V.D. Matthews), Uzuhashi, Tojo and Kakish., *Mycoscience* 51(5): 361(2010)

Description: Zoosporangia terminal and spherical. Oogonia lateral or intercalary, sometimes catenulate, spherical, 21 μm in diameter, rarely oval, simple peduncles, oogonial wall with spiny or conical ornamentation. Antheridia present. Anteridial branches monoclinous and diclinous, single peduncle. Antheridial cells single; lateral attached, 1-2 per oogonium. Oospores aplerotic, spherical, 18.75 μm in diameter, one per oogonium and smooth wall. Fertilizer tubes not observed.

Material examined: Brazil. Piauí: Demerval Lobão, Riacho Mutum, soil samples at the point/collection S1/1, 07.VIII.2014, M.A.M. Macêdo.

Geographical distribution in Brazil: Piauí: Caracol (Rocha et al. 2018), Floriano (Negreiros 2008), José de Freitas (Pontes and Rocha 2017), Piracuruca (Rocha 2002, 2006, Rocha et al. 2001) and Teresina (Sousa and Rocha 2017, Rocha et al. 2018, Trindade Junior and Rocha 2013). It also occurs in Pernambuco and São Paulo (Milanez et al. 2007).

Table 8.1. Geographical distribution of the *Pythium* Complex occurring in the Mid-North region of Brazil

	Táxa	Maranhão	Piauí	States Other
PYTHIACEAE — Globsporângium	1. *G. echinulatum*	-	Caracol, Floriano, José de Freitas, Pedro II, Piracuruca e Teresina	Pernambuco, São Paulo
	2. *G. mamillatum*	Timon	Floriano, José de Freitas, Nazária, Pedro II, Piracuruca	Rio de Janeiro, São Paulo
	3. *G. perplexum*	-	Nazária, Teresina	(*)
	4. *G. proliferum*	Timon	Piracuruca, Teresina	São Paulo
	5. *G. ultimum*	Timon	José de Freitas, Pedro II, Piracuruca, Teresina	São Paulo
Pythium	6. *P. aphanidermatum*	-	Caracol, Floriano, Nazária, Teresina	Amazonas, Paraíba, Pernambuco, Distrito Federal. Goiás, Mato Grosso, Minas Gerais, Rio de Janeiro, São Paulo
	7. *P. graminicola*	Timon	Nazária, Teresina	Distrito Federal, São Paulo
	8. *P. inflatum*	-	Nazária	São Paulo
	9. *P. middletonii*	-	Teresina	São Paulo
	10. *P. myriotylum*	-	Nazária, Piracuruca	Distrito Federal, Goiás, São Paulo
Phytopythium	11. *Phy. indigoferae*	-	Nazária	Rio de Janeiro
	12. *Phy. paligenes*	Timon	Floriano, Pedro II, Piracuruca, Teresina	São Paulo
	13. *Phy. Vexans*	-	Caracol, Nazária, Pedro II, Piracuruca, Teresina	Pernambuco, Rio de Janeiro, São Paulo
	14. *Phytophythora sp*	-	Teresina	Ceará, Rio de Janeiro, São Paulo, Mato Grosso, Bahia

(*) Report only to the Mid North region.

Comments: The characteristics agree with the original description and resemble those described by Rocha et al. (2001). Also, it agrees with Miranda and Pires-Zottarelli (2008), Nascimento and Pires-Zottarelli (2012) and Trindade Jr. (2013). This taxon was transferred from the genus *Pythium* to *Globisporangium* (Uzuhashi et al. 2010).

Globisporangium mamillatum Meurs Uzuhashi, Tojo and Kakish., *Mycoscience* 51 (5): 362 (2010).

Description: Zoosporangia terminal or intercalary in lateral branches, globose, ovoid 20 μm long, ellipsoid or irregular, smooth or with short and irregularly distributed papillae. Oogonia terminal or intercalary, globular, 17.5-20 μm in diameter, subglobous, oval, provided with cylindrical or mammiform, irregular or bifurcated conical projections. 1 (-2) antheridia per oogonium, monoclinous or diclinous. Antheridial cells single, clavate, apical attaching to oogonium. Oospores hyaline, aplerotic, spherical, 15-17.5 μm in diameter globular and smooth-walled. Release tubes 7.5 μm length.

Material examined: Brazil. Piauí: Demerval Lobão, Riacho Mutum, soil samples at the point/ collection S4/4, 25.III.2015, M. A.M. Macêdo.

Geographical distribution in Brazil: Maranhão: Timon (Sales 2009). Piauí: Floriano (Negreiros 2008), José de Freitas (Pontes and Rocha 2017), Nazária (Pereira and Rocha 2008, 2009, 2012) and Piracuruca (Rocha 2002, 2006, Rocha et al. 2001). It also occurs in Rio de Janeiro and São Paulo (Milanez et al. 2007).

Comments: The isolates showed abundant production of oogonia. The measurements of the structures of this species are similar to those observed by Rocha et al. (2001). This taxon was transferred from the genus *Pythium* to *Globisporangium* (Uzuhashi et al. 2010).

Globisporangium perplexum H. Kouyeas and Theoh. Uzuhashi, Tojo and Kakish., *Mycoscience* 51(5): 362 (2010).

Description: Zoosporangia varied in shape, spherical (18,75-)22.5(-27.5) μm in diameter predominantly globular (20-)22.5(-25) × (25-)27.5(-. 30) μm, predominantly intercalary, terminal or sessile, occasionally in series. Oogonia terminal in lateral branches, intercalary, smooth, (20-)22,5(-25) μm in diameter. Antheridial branches monoclinous, 1 (-2) per oogonium, originated within a short distance of the oogonium. Bell-shaped anterior cells, 1 (-2) per oogonium, with broad contact with the oogonium. Oospores aplerotic, spherical, flat (16.25)17.5(-20) μm in diameter, thick wall.

Material examined: Brazil. Piauí: Demerval Lobão, Riacho Mutum, soil samples at the point/ collection S2/3, 30.I.2015, M. A.M. Macêdo.

Geographical distribution in Brazil: Piauí: Nazária (Pereira and Rocha 2008, 2009, 2012) and Teresina (Sousa and Rocha 2017).

Comments: This isolate presented similarity to the description of Plaats-Niterink (1981) and Pereira and Rocha (2008). Zoosporangia in chain and oospores with yellowish coloration were found. This should probably be the third citation for Piauí, since no other records were found. This taxon was transferred from the genus *Pythium* to *Globisporangium* (Uzuhashi et al. 2010).

Globisporangium proliferum Cornu P.P. Kirk, *Index Fungorum* 191:1. (2014).

Description: Zoosporangia terminal or intercalary, ovoid, elipsoid, globose, with internal proliferation. Oogonia terminal or lateral, spherical, 15-25 μm in diameter, oval. Sometimes with papilla. Oospores hyaline, spherical, 13-20 μm in diameter, smooth-walled. Antheridia 1-2 per oogonium, monoclinous or hypogynous, sometimes sessile, simple or branched. Antheridial cell simples, clavate, attached apically to oogonium.

Material examined: Brazil. Maranhão: Timon (Silva and Rocha 2017). Piauí: Piracuruca (Rocha 2002, 2006, Rocha et al. 2001, Milanez et al. 2007) and Teresina (Sousa and Rocha 2017). It also occurs in São Paulo.

Geographical distribution in Brazil: Piauí (Rocha et al. 2001); São Paulo (Pires-Zottarelli 1999).

Comments: Isolates produced typical spherical oogonia with oospore hyaline. It agrees with the descriptions of Plaats-Niterink (1981), PiresZottarelli (1999) and Rocha (2002) (cited as *Pythium middletonii* Sparrow). Also, according to Plaats-Niterink (1981), its pathogenicity caused disease on seedlings of soybean (*Glycine max*), tomato (*Solanum lycopersicum*), and potato (*Solanum tuberosum*). This is the first record for Maranhão State.

Globisporangium ultimum Trow Uzuhashi, Tojo and Kakish., *Mycoscience* 51(5): 363 (2010).

Description: Zoosporangia globose, subglobous; intercalary or terminal measuring 17-23 μm in diameter. Oogonium globular, smooth wall, terminal or intercalary with 22-25 μm in diameter. Antheridia monoclinous with the origin below the oogonium, rarely distant from the oogonium. Ospores single, aplerotic, globose, spherical, 17-21 μm in diameter.

Material examined: Brazil, Piauí: Teresina, Fish breeding (C1, C2, C4). 9.VII.2014, 9.IX.2014, 10.XI. 2014 and 9.III.2015, in water and soil samples at points S6/1 2014; A1/2014; S1/03 2014; S6/03 2014; A4/2 2014, A5/2 2014, A6/3 2014, A4/5 2015, L.S. Saraiva.

Geographical distribution in Brazil: Piauí: Maranhão, Timon (Sales 2009). Piauí, Caracol (Rocha et al. 2018), Floriano (Negreiros and Rocha 2008), José de Freitas (Pontes and Rocha 2017), Nazária (Pereira and Rocha 2008, 2009, 2012), Piracuruca (Rocha 2002, 2006, Rocha et al. 2001) and Teresina. It also occurs in Minas Gerais, Pernambuco, Rio de Janeiro and São Paulo.

Comments: The specimens presented characteristics different from the description of Plaats-Niterink (1981) than species with zoosporangia of (23-) 27-32 μm in diameter, and oospores with a wall thickness of 2 μm or more. Pereira (2008) characterized the species with zoosporangia between 17.5-22.5 μm in diameter and Pires-Zottarelli describes the wall of the oospore with 1.1 μm of thickness.

Pythium

Pythium aphanidermatum Edson Fitzp, 1923

Description: Zoosporangia formed by terminal complex globular, swollen; oogonium terminal, globular and smooth; antheridia sometimes terminal, ranging from 1-2 per oogonium, monoclinous or diclinous, with aplerotic oospore.

Material examined: Brazil: Piauí: Teresina: Community garden of Tabuleta, in soil sample (Trajano 2009).

Geographical distribution in Brazil: Piauí: Teresina (Trajano 2009), Nazária (Pereira and Rocha 2008, 2009), Rio de Janeiro (Milanez et al. 2007), and São Paulo (Pires-Zottarelli 1999).

Comments: It is typical fungus of the hot regions like Teresina. According to Plaats-Niterink (1981), it is a major cause of root rot, stalk, rhizome, damping-off, fruit rot, conifers, beetroot (*Beta vulgaris* L.), sugar cane (*Saccharum officinarum* L.), tomato (*Lycopersicom esculentum* M.), corn (*Zea mays* L.) and in pepper species (*Capsicum* ssp).

Pythium graminicola Subram., *Bull. Agric. Res. Inst. Pus.* 177:1 (1928).

Description: Zoosporangia terminal or intercalary with filamentous inflated complex. Oogonium 22-26 μm diâm. terminal, globose, smooth. Oospores plerotic, 20-23 μm diâm. Antheridia 1-2 per oogonium, monoclinous, diclinous with apical attachment to oogonium.

Material examined: Brazil. Maranhão: Timon, Parque Natural Municipal Lagoa Sambico, soil and water samples at collection spots, 20-X-2014, J.B. Silva (A2/2-S2/2-S4/2), 23-II-2015, J.B. Silva (S3/4-S4/4).

Geographical distribution in Brazil: Maranhão: Timon (Sales 2009, Silva and Rocha 2017). Piauí: Nazária (Pereira and Rocha 2008, 2009, 2012) and Teresina (Sousa and Rocha 2017), (Trindade Junior and Rocha 2013). It also occurs in Brasília and São Paulo (Marano et al. 2008).

Comments: Isolate produced characteristic oogonia and zoosporangia with inflated complex. The species is depicted according to the description of Plaats-Niterink (1981). It is highly pathogenic to crop plants such as corn (*Zea mays*), potato (*Solanum tuberosum*), beans (*Phaseolus vulgaris*), pineapple (*Ananas comosus*) and onion (*Allium cepa*). It is second record for Maranhão State.

Pythium inflatum V.D. Matthews, *Pythiaceae* (1931).

Description: Eucarpic thallus; mycelium with the main hyphae about 4 μm. Filamentous and inflated zoosporangia. Release of zoospores pythioid type. Oogonia globules, smooth, terminal, 25.7 μm in diameter. Antheridia terminal, 1 (-2) per oogonium, diclinous, apical attachment. Aplerotic oospores, 27.5 μm in diameter; wall thickness of 2.7 μm.

Material examined: Brazil; Piauí: Nazária (Pereira and Rocha 2008, 2009, 2012).

Comments: *Pythium inflatum* did not present significant differences in relation to the original description. This species was isolated from water sample from the Parnaíba River, from where it is obtained for irrigation of the crops. Plaats-Niterink (1981) cites this species as a parasite, being isolated from sugarcane and strawberry, causing, in the experiment, moderate infection in tomato roots.

Pythium middletonii Sparrow, *Aquatic Phycomycetes*, p. 1038 (1960).

Description: Mycelium growing radially and submerged. Hyphae up to 6 μm in diameter. Lobed appressors. Zoosporangia intercalary, in lateral branches, ellipsoid, globose, 7-37 × 15-31 μm, smooth. Zoospores 5-12 μm in diameter. Oogonium terminal, lateral, spherical, 15-25 μm in diameter, sometimes with papilla. Antheridia monoclinous or hypogynous, sometimes sessile, simple or branched. Hyaline spores, 13-20 μm in diameter. Smooth wall.

Material examined: Brazil. Piauí: Teresina, Lagoas do Norte, soil samples on sorghum seeds and onion epidermis from the lagoons of: Piçarreira do Cabrinha, Piçarreira do Lourival and Mocambinho.

Geographic distribution in Brazil: Piauí: Teresina (Trindade Junior and Rocha 2013). It also occurs in São Paulo (Milanez et al. 2007)

Comments: The description does not differ from Plaats-Niterink (1981), Pires-Zottarelli (1999) and Rocha (2002). Its pathogenicity, according to Plaats-Niterink (1981) is about soybean, tomato, potato and coniferous seedlings.

Pythium myriotylum Drechsler, *Journal of the Washington Academy of Siences* 20: 404 (1930).

Description: Colonies in CMA + PPS (Pimaricina. Penicillin. Streptomycin) with the growth of 26 mm in diameter / day. Mycelium with radial, superficial and aerial growth pattern; extensive and diffused in water / sorghum. Hyphae up to 8 μm in diameter. Appressorium clavate or lobed, in group. Zoosporangia intercalary or terminal in lateral branches, filamentous, consisting of undifferentiated parts and inflated typed or lobulated parts, of variable length and 7-11 μm wide. Long release tube, up to 100 μm or more, 2-3 μm wide. Oospores encysted 8-10 μm in diameter. Oogonium terminal or intercalary, spherical, 24-32 (27) μm in diameter, smooth. Antheridia 2-5 per oogonium, branched peduncles, often loosely involving the oogonium, declining, sometimes monoclinous, originating from several distances below the oogonium. Antheridial cells single, clavate or sickle-shaped, usually 10-18×5-8 μm, apical attached to oogonium. Oospore hyaline, spherical, 19-25 (22.5) μm in diameter. Smooth wall up to 2 μm thick.

Material examined: Brazil; Piauí: Sete Cidades National Park, J.R.S. Rocha, collected at a soil point, 28/II/2000. (SPC 1883).

Geographic distribution in Brazil: Piauí: Nazaria (Pereira and Rocha 2008, 2009, 2012), Piracuruca (Rocha et al. 2001, Rocha 2002, 2006). Causing diseases in plants - Brasília (DF) *Sclerotinia sclerotiorum* (Lib) De Bary (Gauch and Ribeiro 1998). São Paulo: Piracicaba, *Cucumis sativum* L. (Carvalho 1965); *Sechium edule* Swartz, *Daucus carota* L., *Solanum tuberosum* L., *Hibiscus esculentum* L., *Solanum melongena* L., *Phaseolus vulgaris* L Obel, *Lycopersicum esculentum* Miller, *Raphanus sativum* L. and *Curcubita pepo* L. (Cordoso et al. 1962); *Arachis hypogaea* L, *Solanum gilo* Raddi, *Glycine* sp., *Lycopersicum esculentum* Miller. (Mendes et al.1998).

Comments: It occurred as saprobe in soil samples baited with sorghum seeds and onion epidermis. Few species of *Pythium* of the group with filamentous Zoosporangia can grow 40°C. At this temperature, this isolate developed colonies CMA+p.p. with growth of 30 mm in diameter per day. The description agrees with Carvalho (1965), Middleton (1943) and Plaats-Niterink (1981).

P. myriotylum is more pathogenic at temperatures around 30°C (Gay 1969). High temperatures favour tipping in bean seedlings (*Phaseolus vulgaris* LObel) (Lumsden et al. 1975). Together with *Fusarium solani* (Mart.) Sacc., it can participate in a fungal complex and act on the peanut rot *Arachis hypogaea* L. (Frank 1972).

Phytopythium

Phytopythium indigoferae E.J. Butler P.M. Kirk, *Index Fungorum* 281: 1 (2015).

Description: Eucarpic thallus; mycelium showing main hyphae with about 8 µm. Zoosporangia branched, filaments inflated, small, with small side discharge tubes. Release of pythioid type. Oogonia terminal, globose, smooth, 20-22.5 µm in diameter, usually in connection with zoosporangia. Antheridial cells monoclinous or diclinous, terminal. Oopores aplerotic, 15-17.5 µm in diameter.

Material examined: Brazil; Piauí: Nazária (Pereira and Rocha 2008).

Geographic distribution in Brazil: Rio de Janeiro (Milanez et al. 2007, Forzza et al. 2010).

Comments: *Phytopythium indigoferae* was originally isolated from *Indigofera arrecta*, in India, being a typically saprophytic species (Plaats-Niterink 1981). It is distinguished from other species reported in this work by the direct connection between oogonia and zoosporangia, a characteristic referred to in the description from Plaats-Niterinks (1981). In the agricultural field of Nazaria, it was isolated from sandy soil, in an area of corn, beans, cassava and watermelon cultivation, being the species with the highest occurrence, in relation to the others mentioned in this study. *Phytopythium indigoferae* affects pineapple causing root rot (Andrew and Stephen 1994).

Phytopythium paligenes Drechsler Abad, de Cock, Bala, Robideau, Lohdi and Lévesque, *Persoonia* 34: 37 (2014).

Description: Thin hyphae and underdeveloped mycelium. Zoosporangia globose, terminal or spherical 20-34 µm in diameter, lateral, smooth, with internal proliferation. Oogonia terminal or intercalary, lateral or sessile, 33 µm in diameter. Antheridia 1-4 per oogonium, monoclinous. Antheridial cells simple, cylindrical with irregular contour. Oospore yellowish, aplerotic with 27 µm in diameter.

Material examined: Brazil. Piauí: Pedro II, Complexo Açude Joana, 20-XI-2015, J.L. Machado (S1/1); 25-IV-2016, J.L. Machado (S2/3); 30-V-2016, J.L. Machado (S1/4-S2/4-S8/4); 31-VIII-2016, J.L. Machado (A4/5-A5/5-S3/5-S5/5-S6/5).

Geographical distribution in Brazil: Maranhão: Timon (Sales 2009). Piauí: Floriano (Negreiros 2008); Piracuruca (Rocha et al. 2001) and Teresina (Trindade Junior and Rocha 2013, Rocha et al. 2018).

Comments: The zoosporangia of the isolate are globose and terminal, with occurrence of internal proliferation. The oogonium presented oospores and simple antheridial cells of irregular contour. The observed specimen was obtained from water and soil samples, colonizing the baits of corn straw, sorghum seed and onion epidermis, agreeing with the description reported for the country that identified the taxon in a soil sample with a saprobic way of life.

Phytopythium vexans de Bary Abad, de Cock, Bala, Robideau, Lodhi and Levesque, *Persoonia* 34: 37 (2014).

Description: Zoosporangia smooth, colourless, globose, rarely limoniform, terminal or intercalary, isolated, 17.5-22.5 µm in diameter. Oogonia terminal and solitary, 17-23 µm in diameter. Oospore colourless and smooth, spherical and aplerotic, 12.5-20 µm in diameter. Presence of an antheridium per oogonium, with antheridial cells bell-shaped, intercalary or terminal with apical attachment.

Material examined: Brazil. Piauí: Pedro II, Complexo Açude Joana, 20-XI-2015, J.L. Machado (S1/1-S2/1-S3/1); 22-II-2016, J.L. Machado (S1/2-S3/2); 31-VIII-2016, J.L. Machado (S1/5-S2/5-S3/5).

Geographical distribution in Brazil: Piauí: Caracol (Rocha et al. 2018), Nazária (Pereira and

Rocha 2008, 2009, 2012), Piracuruca (Rocha et al. 2001, Rocha 2002, 2006, Milanez et al. 2007) and Teresina (Sousa and Rocha 2017). It also occurs in Pernambuco, Rio de Janeiro and São Paulo.

Comments: The species examined presented zoosporangia, oogonia and oospores with diameter compatible with the descriptions of Plaats-Niterink (1981) and Miranda and Pires-Zottarelli (2008). The main characteristics that aid in the identification of this species are the globose zoosporangium with pityaloid release, aplerotic oospores, bell-shaped antheridial cells and monoclinous. The species can be obtained from water, soil and root of plants, and can be saprobe or parasite. In this research it was obtained from soil sample, such as saprobe in corn straw, sorghum seed and onion epidermis.

Phytophthora sp. de Bary 1876

Description: Mycelium delicate, very branched, usually monopodial. Zoosporangia usually ovoid, and generally piriform or ellipsoid, 30-50 μm of length. 20-35 μm in width. Zoospores completely formed within zoosporangia and released through papillae. Chlamydospores not observed.

Material examined: Brasil. Piauí: Teresina. Fish farms (C2). 11.V.2015, in water samples: S1/6 2015, L.S. Saraiva.

Geographic distribution in Brazil: Piauí: Teresina; North Brazil (Acre, Amazonas, Pará, Tocantins); Northeast Brazil (Alagoas, Bahia, Ceará, Maranhão, Paraíba, Pernambuco, Rio Grande do Norte, Sergipe); Centre-West Brazil (Federal District, Goiás, Mato Grosso); Southeast Brazil (Espírito Santo, Minas Gerais, Rio de Janeiro, São Paulo); South Brazil (Paraná, Rio Grande do Sul, Santa Catarina).

Comments: The cited characteristics agree with Pires-Zottarelli (1999). The specimen was identified only at the genus level, so that it will continue to be studied in order to observe the formation of sexual structures.

Conclusion

The present study contributes significantly to the knowledge of oomycetes, represented by the family Pythiaceae. Although the research is scarce about these organisms, 14 species were identified, being one of exclusiveness for the Mid-North Region of Brazil. The contribution to diversity exists as more studies are carried out. In Piauí, research began with Rocha (2001, 2002), and in Maranhão, with Sales (2009). The research group of Zoosporic Organisms of UFPI has contributed to the identification and description of oomycetes. The species, *Globisporagium perplexum*, had its first record for Piauí, in the City of Nazaria, in the studies by Pereira and Rocha (2008).

Acknowledgements

The authors thank the trainees of the Laboratory of Zoosporic Fungi of the Universidade Federal do Piauí.

References

Agrios, G.N. 2005. Plant Pathology. 5ª ed. Academic Press. St. Paul.

Alexopoulos, C.J., Mims, C.W. and Blackwell, M. 1996. Introductory Mycology. 4th ed. John Wiley and Sons, Inc., New York.

Amorim, L., Rezende, J.A.M. and Bergamin Filho, A. 2011. Manual de Fitopatologia: Princípios e Conceitos. 4ª ed. Piracicaba, São Paulo, Ceres.

Andrew, K.G. and Stephen, A.F. 1994. Pythium primer. College of Tropical Agriculture and Human Resources. University of Hawaii at Manoa.

Bala, K., Robideau, G.P., Lévesque, C.A., de Cock, A.W.A.M., Abad, Z.G., Lodhi, A.M., Shahzad, S., Ghaffar, A. and Coffey, M.D. 2010a. *Phytopythium* Abad, de Cock, Bala, Robideau, Lodhi and

Lévesque, gen. nov. and *Phytopythium sindhum* Lodhi, Shahzad and Levésque, sp. nov. Persoonia 24: 136-137.

Bala, K., Robideau, G.P., Désaulniers, N., de Cock, A.W.A.M. and Lévesque, C.A. 2010b. Taxonomy, DNA barcoding and phylogeny of three new species of *Pythium* from Canada. Persoonia 25: 2231.

Barr, D.J.S. 1992. Evolution and kingdoms of organisms from the perspective of a mycologist. Mycologia 84: 1-11.

Barton, R. 1958. Occurrence and establishment of *Pythium* in soils. Trans. Brit. Mycol. Soc. 41: 207-222.

Beakes, G.W., Honda, D. and Thines, M. 2014. Systematics of the Straminipila: Labyrinthulomycota, Hyphochytridiomycota, and Oomycota. pp. 39-97. *In*: McLaughlin, D.J. and Spatafora, J.W. (eds.). Systematics and Evolution, The Mycota, VIII: Part A, 2 ed. Springer-Verlag, Berlin, Heidelberg.

Bosco, M., Giovannetti, M. and Viti, C. 2008. Ruolo dei microrganismi nei cicli biogeochimici. pp. 1-37. *In*: Biavati, B. and Sorlini, C. (eds.). Microbiologia Agroambientale. Casa Editrice Ambrosiana Publisher, Italy.

Cavalier-Smith, T. (1981a). Eukaryote kingdoms, seven or nine? BioSystems 14: 461-481.

Cavalier-Smith, T. 1987. The origin of fungi and pseudofungi. pp. 339-353. *In*: Rayner, A.D.M., Brasier, C.M. and Moore, D. (eds.). Evolutionary Biology of the fungi. Cambridge: Cambridge University Press.

Corliss, J.O. 1990. Toward a nomenclatural protist perspective. pp. xxv-xxx. *In*: Margulis, L., Corliss, J.O., Melkonian, M. and Chapman, Robert D. (eds.). The Comet Book. Jones and Barlett, Boston, MA.

Dick, M.W. 1990. Keys to *Pythium*. College of Estate Management, Whiteknights.

Doihara, I.P. and Silva, G.S. 2003. Fungos associados a podridões em frutos de mamão no estado do Maranhão. Summa Phytopathologica 29(4): 365-366.

Florado Brasil 2020 em construção. 2019. Jardim Botânico do Rio de Janeiro. Disponível em http://floradobrasil.jbrj.gov.br. Acesso 22/III/. 2019.

Forzza, R.C., Baumgratz, J.F.A., Bicudo, C.E.M., Carvalho Jr, A.A., Costa, A., Costa, D.P., Hopkins, M., Leitman, P.M., Lohmann, L.G., Maia, L.C., Martinelli, G., Menezes, M., Morim, M.P., Coelho, M.A.N., Peixoto, A.L., Pirani, J.R., Prado, J., Queiroz, L.P., Souza, V.C., Stehmann, J.R., Sylvestre, L.S., Walter, B.M.T. and Zappi, D. 2010. Catálogo de plantas e fungos do Brasil. Vol. 1. Rio de Janeiro, Instituto de Pesquisas Jardim Botânico do Rio de Janeiro.

Freire, F.C.O. 2001. Queima de mudas do cajueiro. pp. 267-282. *In*: Luz, E.D.M.N. and Matsuoka, K. (eds.). Doenças causadas por *Phytophthora* no Brasil. Campinas, SP. Livraria Editora Rural Ltda.

Gauch, F. and Ribeiro, W.R.C. 1998. Ocorrência de espécies de Pythium potenciais micoparasitas, com oogônio equinulado, em solos de Brasília, DF. Fitolologia Brasileira 23: 176-179.

Gonçalves, D.R., De Jesus, A.L., Rocha, S.C.O., Marano, A.V. and Pires-Zottarelli, C.L.A. 2016a. New records of *Pythium* (Oomycetes, Straminipila) for South America based on morphological and molecular data. Nova Hedwigia 103(1-2): 1-12.Hawksworth, D.L., Kirk, P.M., Sutton, B.C. and Pegler, D.M. 1995. Ainsworth and Bisby's Dictionary of the Fungi. 8ª ed. Egham, International Mycological Institute

Hawksworth, D.L. 2004. Fungal diversity and its implications for genetic resource collections. Stud. Mycol. 50: 9-18.

Hendrix, F.F. and Campbell, W.A. 1973. *Pythium* as plant pathogens. Annu. Rev. Phytopathol. 11: 77-78.

Johnson Jr., T.W. 1956. The Genus *Achlya*: Morphology and Taxonomy. The University of Michigan Press. Ann Arbor.

Johnson Jr., T.W., Seymour, R.L. and Padgett, D.E. 2002. Biology and systematic of the Saprolegniaceae. Available in http://www.uncw.edu/people/padgett/book (access in 13-I-2017).

Leal, A.B.M., Leal, A.T., Santurio, J.M., Kommers, G.D. and Catto, J.B. 2001. Pitiose eqüina no Pantanal brasileiro: Aspectos clínico-patológicos de casos típicos e atípicos. Pesquisa Veterinária Brasileira 21: 151-156.

Leedale, G.F. 1974. How many are the kingdoms of organisms? Taxon 23: 261-270.

Lévesque, C.A. and de Cock, A.W.A.M. 2004. Molecular phylogeny and taxonomy of the genus *Pythium*. Mycol. Res. 108: 1363-1383.

Lourd, M., Alves, M.L.B. and Bouhot, D. 1986. Análise qualitativa e quantitativa de espécies de *Pythium* patogênicas dos solos no município de Manaus, I solos de terra firme. Fitopatologia Brasileira 11: 479.

Marano, A.V., Barrera, M.D., Steciow, M.M., Donadelli, J.L. and Saparrat, M.C.N. 2008. Frequency, abundance and distribution of zoosporic organisms from Las Canãs stream, Buenos Aires, Argentina. Mycologia 100: 691-700.

Marano, A.V., Pires-Zottarelli, C.L.A., Barrera, M.D. and Steciow, M.M. 2011. Diversity, role in decomposition, and succession of zoosporic fungi and straminipiles and submerged decaying leaves in a woodland stream. Hydrobiologia 659: 93-109.

Margulis, L. 1990. Introduction. pp. xi-xxii. *In*: Margulis, L., McKhann, H.I., Corliss, O.J., Melkonian, M. (eds.). Handbook of Protoctista. Jones and Barlett Publishers, Boston.

Mendes, M.A.S., Silva, V.L., Dianese, J.C., Ferreira, M.A.S.V., Santos, C.E.N., Gomes Neto, E. and Urben, A.F. 1998. Fungos em plantas no Brasil. Brasília, Embrapa-SPI/Embrapa-Cenargen. 569 p.

Middleton, J.T. 1943. The taxonomy, host range and geographic distribution of the genus *Pythium*. Memoirs of the Torrey Botanical Club 20: 1-171.

Milanez, A.I. 1989. Fungos de águas continentais. pp. 17-20. *In*: Fidalgo, O., Bononi, V.L. (eds.). Técnicas de coleta, preservação e herborização de material botânico. Instituto de Botânica, São Paulo.

Milanez, A.I., Pires-Zottarelli, C.L.A., Gomes, A.L. 2007. Brazilian zoosporic fungi. Winner Graph, São Paulo.

Moore-Landecker, E. 1996. Fundamentals of the Fungi. 4 ed., Prentice-Hall, New Jersey.

Nascimento, C.A. & Pires-Zottarelli, C.L.A. 2012. Diversidade de fungos zoospóricos da Reserva Biológica de Mogi Guaçu, estado de São Paulo, Brasil. Rodriguésia 63: 587-611.

Negreiros, N.C. 2008. Uso sustentável de culturas agrícolas suscetíveis a oomicetos (Oomycota) fitopatogênicos às margens do rio Parnaíba no município de Floriano, Piauí. Dissertação de Mestrado, Universidade Federal do Piauí, Teresina.

Plaats-Niterink, A.J. 1975. Species of *Pythium* in the Netherlands. J. Plant Pathol. 81: 22-37.

Plaats-Niterink, A.J. van der. 1981. Monograph of genus *Pythium*. Stud. Mycol. 21: 1-242.

Pereira, A.A. and Rocha, J.R.S. 2008. *Pythium* (Pythiaceae): Três novos registros para o nordeste do Brasil. Acta Botanica Malacitana 33: 347-350.

Pereira, A.A. and Rocha, J.R.S. 2009. Sustentabilidade na prevenção e controle de oomicetos com potencial fitopatogênico em área agrícola do Piauí, Brasil. Castro et al. (eds.). Biodiversidade e desenvolvimento no trópico ecotonal do Nordeste. – Teresina: EDUFPI, Série Desenvolvimento e Meio Ambiente, 4: 195-212.

Pereira, A.A. and Rocha, J.R.S. 2012. O gênero *Pythium* (Oomycota) no campo agrícola de Nazária, Piauí, Brasil. Rocha et al. (eds.). Sociobiodiversidade no meio norte brasileiro. Teresina: EDUFPI, Série Desenvolvimento e Meio Ambiente, 7: 127-138.

Pires-Zottarelli, C.L.A. 1999. Fungos zoospóricos dos Vales dos Rios Moji e Pilões, Região de Cubatão, SP. Tese de doutorado, Universidade Estadual Paulista, Rio Claro, São Paulo.

Rech, R.R., Graça, D.L. and Barros, C.S.L. 2004. Pitiose em um cão: Relato de caso e diagnósticos diferenciais. Clínica Veterinária 50: 68-72.

Robideau, G.P., de Cock, A.W.A.M., Coffey, M.D., Voglmayr, H., Brouwer, H., Bala, K., Chitty, D.W., Désaulniers, N., Eggertson, Q.A., Gachon, C.M.M., Hu, C.H., Küpper, F.C., Rintoul, T.L., Sarhan, E., Verstappen, E.C.P., Zhang, Y., Bonants, P.J.M., Ristaino, J.B. and Lévesque, A.C. 2011. DNA barcoding of oomycetes with cytochrome c oxidase subunit I and internal transcribed spacer. Mol. Ecol. Resour. 11: 1002-1011.

Rocha, J.R.S., Milanez, A.I. and Pires-Zottarelli, C.L.A. 2001. O gênero *Pythium* (Oomycota) em áreas de cerrado no Parque Nacional de Sete Cidades, Piauí, Brasil. Hoehnea 28: 209-230.

Rocha, J.R.S. 2002. Fungos zoospóricos em área de cerrado no Parque Nacional de Sete Cidades, Piauí, Brasil. Tese de Doutorado, Universidade de São Paulo, São Paulo.

Rocha, J.R.S. 2006. Filos Chytridiomycota e Oomycota. Giulietti, A.M., (Ed.). Diversidade e caracterização dos fungos do semi-árido. Recife: Associação Plantas do Nordeste/Instituto do Milênio do Semi-Árido. Vol. II, Chap. 4, 75-95.

Rocha, J.R.S., Silva, V.S., Santos, S.L., Dias, L.P., Rodrigues, E.P., Batista Filho, D.M., Feitosa Junior, F.S. and Barbosa, R.D. 2010. Pitiose cutânea equina. Primeiro relato de caso no Piauí. Revista do Conselho Federal de Medicina Veterinária 15: 24-27.

Rocha, J.R.S., Sousa, N.D.C., Negreiros, N.C., Santos, L.A., Pereira, A.A., Sales, P.C.L. and Trindade Júnior, O.C. 2014. The genus *Pythiogeton* (Pythiogetonaceae) in Brazil. Mycosphere 5: 623-634.

Rocha, J.R.S. and Macêdo, M.A.M. 2015. First record of *Brevilegnia longicaulis* Johnson (Saprolegniales) in Brazil. Curr. Res. Environm. Appl. Mycol. 5(2): 78-81.

Rocha, J.R.S., Rocha, F.P. and Machado, J.L. 2017. O gênero *Myzocytiopsis* (Oomycota) no Estado do Piauí: Novos registros para o Brasil. Gaia Scientia 11(1): 98-115.

Rocha, M. and Pires-Zottarelli, C.L.A. 2002. Chytridiomycota e Oomycota da Represa do Guarapiranga. São Paulo, SP. Acta Botânica Brasilica 16: 287-309.

Rocha, M. 2004. Micota zoospórica de Lagos, com diferentes trofias do Parque Estadual das Fontes do Ipiranga (PEFI), São Paulo, SP. Dissertação de Mestrado, Universidade de São Paulo, São Paulo.

Rocha, J.R.S., Machado, J.L., Silva, J.B., Trindade Júnior, O.C., Santos, L.A., Rodrigues, E.P. and Cronemberger, Á.A. 2018. O gênero Olpidiopsis (Oomycota) no Nordeste do Brasil. Rodriguesia 69: 2035-2053.

Sales, P.C.L. 2009. Potabilidade da água e presença de oomicetos (Oomycota) em poços freáticos nos povoados Banco de Areia, Bacuri e Roncador no município de Timon Maranhão. Dissertação de Mestrado, Universidade Federal do Piauí, Teresina.

Sallis, E.S.V., Pereira, D.I.B. and Raffi, M.B. 2003. Pitiose cutânea em eqüinos: 14 casos, 2001. Ciência Rural 33: 899-903.

Santurio, J.M., Monteiro, A.B. and Leal, A.T. 1998. Cutaneous Pyhtiosis insidiosi in calves from the Pantanal region of Brazil. Mycopathologia 141: 123-125.

Saraiva, L.S. 2016. Diversidade de oomicetos (Oomycota), potencial patogênico e manejo na piscicultura em Teresina-Pi. (Dissertação de Mestrado). Universidade Federal do Piauí. Piauí.

Schroeder, K.L., Martin, F.N., de Cock, A.W., Lévesque, C.A., Spies, C.F. Okubara, P.A. and Paulitz, T.C. 2013. Molecular detection and quantification of *Pythium* species: Evolving taxonomy, new tools, and challenges. Plant Dis. 97(1): 4-20.

Scott, W.W. 1961. A monograph of the genus *Aphanomyces*. Virginia Agricultural Esperiments Station. Tech. Bull. 151: 1-95.

Silva, G.S., Rêgo, A.S. and Leite, R.R. 2014. Doenças da vinagreira no Estado do Maranhão. Summa phytopathol. vol. 40 no.4 Botucatu Oct./Dec. 2014.

Silva, J.B. and Rocha, J.R.S. 2017. Oomycets (Oomycota) from Maranhão State, Brazil. Hoehnea 44(3): 394-406.

Sousa, S.B., Rocha, J.R.S., Lucena, R.F.P. and Barros, R.F.M. 2017. Uso de macrofungos em região de Caatinga no Nordeste do Brasil. Gaia Scientia 11: 6.

Sparrow, F.K. 1960. The aquatic Phycomycetes. Ann Arbor, University of Michigan Press 2: 1181.

Species Fungorum. 2016. Search Species Fungorum. Available in http://www.speciesfungorum.org/ (access 23-III-2019).

Steciow, M.M., Milanez, A.I., Pires-Zottarelli, C.L.A., Marano, A.V., Letcher, P.M. and Vélez, C.G. 2012. Zoosporic true fungi, heterotrophic straminipiles and plasmodiophorids: Status of knowledge in South America. Darwiniana 50: 25-32.

Trajano, H.M.R. 2009. Resistência de espécies de pimenta (*Capsicum* sp.) a *Pythium* sp. (Dissertação Mestrado). Universidade Federal do Piauí. Piauí.

Trajano, H.M.R., Rocha, J.R.S. and Lopes, A.C.A. 2009. Resistência genética de pimentas (*Capsicum* spp.) a fungos fitopatógenos (*Pythium* spp.) provenientes do semiárido piauiense e de Teresina e sua aplicabilidade. Lopes et al. (eds.). Sustentabilidade do semiarido. Teresina: EDUFPI, Série Desenvolvimento e Meio Ambiente 3: 419-435.

Trindade-Júnior, O.C. and Rocha, J.R.S. 2013. *Brevilegnia linearis* Coker (Saprolegniales, Oomycota): Um novo registro para or Brasil. Pesquisas, Botânica 64: 341-345.

Uzuhashi, S., Tojo, M. and Kakishima, M. 2010. Phylogeny of the genus *Pythium* and description of new genera. Mycoscience 51: 337-365.

Viana, F.M.P. and Sobrinho, C.A. 1998. Fitomoléstias identificadas na microrregião do litoral piauiense: 1988-1997. EMBRAPA Comunicado Técnico 74: 1-4.

Viana, F.M.P., Freire, F.C.O., Araújo, J.R.G. and Pessoa, M.N.G. 2004. Influência da variedade da copa na incidência da gomose-de-*Phytophthora* em porta-enxerto de limoeiro "cravo" no Estado do Piauí. Fitopatologia Brasileira 29(1): 103.

Villa, O.N., Kageyama, K., Asano, T. and Suga, H. 2006. Phylogenetic relationships of *Pythium* and *Phytophthora* species based on ITS rDNA, cytochrome oxidase II and b-tubulin gene sequences. Mycologia 98(3): 410-422.

Pythium spp. on Vegetable Crops: Research Progress and Major Challenges

Pratibha Sharma[1*], Prashant P. Jambhulkar[2], Raja, M.[1] and Shaily Javeria[3]

[1] Department of Plant Pathology, SKN College of Agriculture, Jobner - 303329, Jaipur, Rajasthan
[2] Agricultural Research Station (MPUAT, Udaipur), Banswara, Rajasthan
[3] Division of Seed Science and Technology, ICAR - IARI, New Delhi

Introduction

The genus *Pythium* established by Pringsheim in 1858 and placed in the family Saprolegniaceae is multiform, physiologically distinct with a wide host range. The first described species are *P. entophytum* Pringsheim and *P. monospermum* Pringsheim. Later, *P. monospermum* was reported as the type species and *P. entophytum* was transferred to the genus Lagenidium. These species were included by Schroter (1897) in a new family, the Pythiaceae, which was placed in the Saprolegniales, which was later included in order Peronosporales. The monographs on *Pythium* spp. were written by Butler (1907) followed by Matthews (1931), taxonomy of *Pythium* spp. by Sideris (1931, 1932), taxonomy and geographic distribution by Middleton (1943), and *Pythium* spp. from Argentina Frezzi (1956). Waterhouse in 1967 and 1968 described further 89 specific and sub-specific taxa with the original descriptions and figures.

Pythium spp. are soil borne pathogens and cosmopolitan in nature. It attacks juvenile or succulent tissues of young seedlings, thus considered a primitive parasite. Symptoms of *Pythium* infection on chrysanthemum were prominently seen in nursery beds, greenhouse and row crops and the mortality is sudden and fast (Cox 1969). Abawi et al. (2006) reported that besides being root pathogens, they also cause severe foliage blight and root necrosis in beans. Similar reports were made by Schwartz et al. (2007) that *Pythium* spp. cause a combination of root rot disease complex symptoms which reduces the yield. Their presence can be observed in soil and water and have distribution worldwide with wide host range. *Pythium* spp. is a disastrous pathogen causing losses of seeds, pre-emergence and post emergence damping off, rots of seedling, roots or basal stalks, fruit and vegetable decay during cultivation, storage, transport or at market (Sheikh 2010). Several vegetable crops are affected by soil-borne pathogens and it causes serious losses in both glass house and field conditions from emergence of seedlings. Due to poor root aeration, over irrigation or excess moisture, root damage during transplantation and alteration in root temperature are important triggers for *Pythium* spp. outbreaks.

Pythium species overwinters as mycelium in plant left over debris or oospores in the soil. They germinate in conducive microclimatic conditions and produce zoosporangia. The zoosporangium produces several zoospores which are encysted and cause infection to the germinating plant seedling

*Corresponding author: psharma032003@yahoo.co.in

below the soil surface or at the crown region (the joining point of root and stem), above ground level of the seedlings adjoining to the soil surface. The mycelium of the fungus grows within infected tissues and damages the plants. The sporangia develop on the infected site of the plants leading to secondary infection under favourable environmental conditions. Under adverse conditions, the fungus turns to sexual reproduction sets and produces oospores, which is a source of survival under the unfavourable conditions. Besides *Pythium* species, other fungi like *Rhizoctinia solani* and *Fusarium* species also survive in the soil as resting spores in the form of sclerotia and chlamydospores, respectively. Under favourable environmental conditions, hyphae of *R. solani* grow from these sclerotia and infect the seedlings. In case of *Fusarium* spp., chlamydospores germinate and produce sporodochia, which in turn produce micro and macroconidia. These conidia germinate into germ tube and cause infection to the seedlings. *Pythium* spp. causing serious losses in productivity of numerous kinds of vegetable crops in field, polyhouse and hydroponic systems around the world including cabbage and cauliflower, chilly, cucumber, ginger, spinach, tomato and brinjal predominantly are caused by *Pythium aphanidermatum*, *Pythium dissotocum*, members of *Pythium* spp. group F, and *Pythium ultimum* var. *ultimum*. This chapter is an effort to review the major symptoms, epidemiology and management of major vegetable crops.

Damping-off of cabbage and cauliflower

Disease occurrence and distribution: Cabbage (*Brassica oleracea* var. *capitata* L) and cauliflower (*Brassica oleracea* var. *botrytis* L) are very commonly grown vegetables. The damping-off disease is of worldwide occurrence, primarily in those regions having moderate temperatures and high soil moisture. In India, it is prevalent throughout the country causing one or the other important diseases (Chauhan et al. 2000, Gupta and Shyam 1996). In nursery, the pathogen often causes more than 50 per cent seedling mortality. In some regions of Himachal Pradesh, the incidence was more than 50 per cent during March-April (Gupta and Shyam 1996). The incidence was more in hybrid Royal Saluis than the varieties Snowball 16 and PSBK 1. In Kullu valley, up to 60 per cent loss in cauliflower has been reported (Kapoor 1999).

Causal organism: The seedlings of cabbage and cauliflower suffer due to the attack of *Pythium* spp., *Fusarium* spp., and *Rhizoctonia solani* at nursery stage. The species reported by various workers are *Pythium ultimum* Trow. var. *ultimum* and *Pythium aphanidermatum* (Edson) Fitzpatrick. *Rhizoctonia solani* produces typical wire stem symptoms. This fungus produces brown hyphae that are upto 12 μm in diameter, has dolipore septa, with right angle branching and a cross wall in the adjacent branch. Since this fungus does not produce asexual spores, the perfect stage develops white, thin, hymenial layer near the collar region which comprises basidia and hyaline basidiospores that measure 6-14 × 4-8 μm. This disease is highly favoured by the temperature in the range of 20 to 25°C and relative humidity from 95 to 100% (Sharma and Sharma 1985). The fungus is polyphagous in nature and has been recorded on a number of hosts from various parts of the world. It attacks 287 genera belonging to dicotyledonous and monocotyledonous families of Angiosperms and also three each of Gymnosperms and Pteridophytes (Ahuja 1976).

Symptoms: The poor emergence of seed and sudden falling off the seedlings are the major symptoms of damping off. This nursery disease is favoured by pathogen inoculum and favourable environmental conditions. The wire stem stage is different to damping-off where light to dark brown areas appear on the stem extending 1-2 cm below the soil line (Gupta and Shyam 1996). The lesions girdle the stem, which become sunken and at later stage turn in to tough and woody structure. The stem area becomes wiry in appearance, relatively thinner than the healthy stem known as wire stem (Walker 1952). The lower leaves loose turgidity and give wilting appearance and the affected plants die prematurely. Plants contracting infection at later stages give unthrifty, stunted appearance and produce small sized curds. The infection is prominent at the beginning of head formation, when the lower leaves develop dark brown, oval lesions. The lesions turn soft and watery due to secondary infection. The pathogen, *Rhizoctonia solani,* with *Pythium* spp. causes damping off in nursery, and after transplanting it inflicts wire stem disease that outrightly kills young seedlings in main field (Fig. 9.1).

Figure 9.1. Pre-emergence damping-off of cauliflower seedlings (Photo courtesy: Raghuveer Singh, ICAR Research Complex, HEN region, Baser)

Epidemiology and favourable conditions: The soil borne nature of *Pythium* spp. allows the pathogen to survive in soil or plant debris for prolonged periods. The seedlings get affected before emergence leading to poor germination. Post emergence the seedlings that topple down or die are said to "damp-off". The severity of the disease depends on the amount of pathogen in the soil and environmental conditions. Cool, cloudy weather, high humidity, wet soils, compacted soil, and overcrowding especially favour development of damping-off. Fields with unequal levels of nitrogen, phosphorus and potassium, soil pH ranging between 7.4 and 8.5 and having sandy-silt loam texture predispose the plants to the attack by pathogen (Sharma and Chaudhary 1984). Application of ammoniacal form of nitrogen decreases the disease incidence as compared to nitrate form (Chernyaeva and Bisuas 1980). The seedling stage up to 20 days is more prone to attack of *R. solani*as compared to 35 days and 50 days old plants (Chauhan et al. 2000). Sandy soil favours maximum wire stem incidence while clay soil the least (Chauhan et al. 2000). The fungus hibernates and perpetuates in the soil when seedlings topple down and perish. The primary source of inoculum of *R. solani* is in the form of mycelium or sclerotia favoured by high humidity and temperature. The sclerotia are resistant to drought and cope large fluctuations in temperature. It is present in most soils and once established, remains there indefinitely. It spreads through rain, irrigation water, agricultural tools etc.

Disease management

Cultural practices

Use of healthy seed, change of nursery site every year, disinfection of the plant bed either with soil solarization or by fumigants like formaldehyde, deep ploughing during summer months, shallow seeding to encourage rapid emergence and selection of the healthy seedlings at the time of transplanting should be followed to keep the disease under check. The disease reduced appreciably by treating the nursery soil with powders of commercial cellulose, rice stubbles or water hyacinth biomass in combination with NH_4NO_3, which was found to be most effective (Kundu and Nandi 1985). Total microbial population also increased in amended soil. It is observed that when C : N ratio increases in soil, bacterial and actinomycetes increase and the fungal population decreases.

Biological control

Applications of bioagents either as seed treatment or soil application have been competent to manage the disease incidence. A spore mixture of *Streptomyces arenae* and. *S. chibaensis* was found to be effective against the disease besides improving shoot length and dry mass of seedlings (Kundu and Nandi 1984). Seed treatment and soil application of *Trichoderma* spp. suppress damping-off symptoms and increase the growth of plants (Roy et al. 1998). *Pseudomonas* strains release antibiotic and, produce siderophore and, therefore, seed treatment helps in improving germination and growth of seedlings, yield and chlorophyll content of leaves. Similarly, reduction in disease incidence was observed by Bhagat and Pan (2008) by seed and soil application of *T. harzianum*.

Integrated management

The nursery from the ground level should be raised to 10 cm above to help good drainage. Soil solarisation of nursery soils for 2-3 weeks by using 45 micron thick polythene sheet helps in suppression of pathogen population. Treatment of nursery soil with 10-15 g of powder formulation of *Trichoderma* spp. along with neem cake at 50 g/m² impregnated is very effective in raising healthy nursery. Seed treatment with *Trichoderma* spp. @ 4 g/kg, seedling dip for 30 min with *Trichoderma* spp. @ 10 g/litre helps to manage rots in the nursery. Soil drenching with captan 75 WP @ 2.5 gm/ litre, adoption of wide spacing of 60 × 50 cm, growing of Indian mustard as trap crop after every 25 rows of cabbage (One row of mustard sown 15 days before and second 25 days after planting of cabbage; first and last row should be of mustard) are recommended to develop damping-off free nursery (Sardana et al. 2019).

Damping-off of chilli

Occurrence and distribution: Damping-off of Chilli (*Capsicum annuum* L.) caused by *Pythium aphanidermatum* (Edson) Fitz. is a major limitation to chilli production causing 62% seedling mortality and 90% of plant death both at pre or post-emergence stage in the nursery fields and greenhouse (Majeed et al. 2018).

Damping-off disease causes heavy losses to the growers. It is responsible for 90 per cent of plant death either at pre or post-emergence damping-off in the nurseries, fields (Sowmini 1961) and greenhouse (Jarvis 1992). This disease is considered as an important limiting factor in successful cultivation of crop plants throughout the world.

Losses: The crop is very important both at green and ripened stage; therefore, there are huge economic losses due to damage to seed and seedlings. To replenish the losses, there is another indirect cost of replanting which is generally delayed (Horst 2013). Five decades back, Sowmini (1961) reported *Pythium* spp. is responsible for 90% of plant death either as pre- or post-emergence damping-off of chilli in nurseries or fields. Damping-off disease in chilli, caused by *Pythium aphanidermatum,* produces about 60% losses at seedling stage in both nursery and field level (Jadhav and Ambadkar 2007). Manoranjitham et al. (2001) reported 60% and Ramamoorthy et al. (2002) reported 62% mortality of seedlings affected by damping-off. This pathogen is quite general and unspecific in its host range capable of causing serious loss under field and greenhouse condition affecting newly emerging seedlings by rapidly infecting germinating seeds, therefore creating obstruction in management (Ellis et al. 1999).

Causal organism: The pathogens associated with this disease are either seed or soil-borne or both. Species of *Pythium, Phytophthora, Fusarium* and *Sclerotium* attack the plants in juvenile stage. In chilli crop *Pythium ultimum* Trow. Fitz. and *Pythium aphanidermatum* (Edson) are reported to cause damping-off. Among them, *P. aphanidermatum* is more in habitats like hydroponic culture and low-lying areas, while *P. ultimum* generally occurs relatively in dry areas (Zagade et al. 2013). Soil moisture near saturation stimulates mycelial growth as well as asexual reproduction. Propagules causing infections include vegetative mycelium, germinating sporangia, zoospores and oospores. Some propagules, particularly zoospores, may come in contact with chilli plants by chemotaxis. Infective propagules are capable of direct penetration, but wounds enhance penetration and infection.

Symptoms: Damping-off occurs typically in two stages, i.e. the pre-emergence and the post-emergence phase. In the pre-emergence phase, the seedlings are toppled and get killed just before they reach the soil surface. The young radical and plumule are killed which later cause complete rotting of the seedlings. The post-emergence phase is characterized by the infection of the young, juvenile tissues of the collar just near the ground level (Agrios 2005).

The main characteristic symptom of the disease is pre-emergence damping-off, i.e. rotting of the seeds and seedlings before actual emergence from the soil and post-emergence damping-off which is severe when the seedlings are in cotyledon stage. The infected tissues become soft and water soaked resulting in toppling over of the entire plant on the soil surface (Singh 1990). On inoculation, symptoms were noticed in the inoculated seedlings within 48h as unhealthy appearance; later, it turns as dark water soaked discolouration observed on lower leaves, which gradually increased upwards. The seedling stem also produce water-soaked lesions which later turns to brown discolouration (Fig. 9.2). Stem portion becomes soft and weak, which later toppled down and damped-off within a week resulting in death of the seedlings.

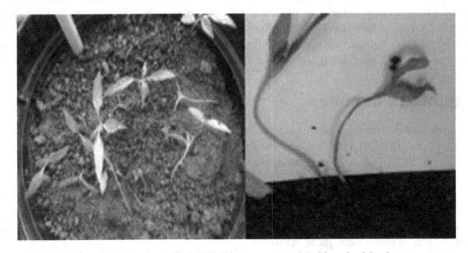

Figure 9.2. Damping-off of chilli (Photo courtesy: Marbiangdor Mawlong, Nagaland University, Medziphema)

Epidemiology and favourable conditions: Damping-off of chillies is soil borne in nature and monoculture practice makes the crop susceptible to disease. The susceptible cultivars help to spread the disease. The disease becomes severe at high temperature and humidity (Gold and Stanghellini 1985). The adult plants are more resistant to damping-off. The favourable factors are moist soil, poor drainage with 90-100% relative humidity and soil temperature around 20°C. The disease is further aggravated by high density of seedlings and poor drainage, low lying areas, clayey soils, flat beds with low content of sand and continuous cultivation in the same field.

Disease management

Cultural practices

The timely application of management strategies helps the nursery disease management (Sullivan 2004). The selection of healthy nursery not infected earlier with soil borne pathogens and transplanting of seedlings in a well-drained soil are the best cultural practices for keeping away the damping-off. Cultural practices to reduce damping-off include avoidance of nursery sowing in the same bed year after year and frequent heavy irrigations. Soils with low clay content were more conducive to the disease (Wokocha 1987). Most successful applications of solarization have been achieved as a pre plant treatment but post-plant treatments have also proved effective. Soil solarization involves covering soils with clear, thin polyethylene sheets in mid-summer, generally for 4 to 6 weeks, before sowing the host crop. (Stapelton and Devay 1984). During solarization, the plastic sheets trap solar energy and prevent heat losses caused by evaporation and conservation, thus creating a greenhouse effect. In Mediterranean type climates, temperatures as high as or higher than 60°C are achieved in the surface layers of soil, generally representing an increase of 10 to 12°C than the uncovered soil. Reports show that many nematode and fungal pathogens are killed by temperatures within the range of 45°C to 55°C (Pullman et al. 1981, Porter and Merriaman 1983). Consequently, the principal effect of solarization is to raise soil temperatures to levels which are lethal to soil borne pathogens. The use of resistant cultivars can be the most important weapon for damping-off disease control. However, desirable cultivars with resistance to important soil borne diseases are not always available.

Chemical control

The soil drenching with metalaxyl M + mancozeb at 1.2% concentration was found to be very effective in reducing the seedling mortality to an extent of 27.20% as reported by Palakshappa et al. (2010). They reported that metalaxyl M + mancozeb (4% + 64%) WP @1.20% was found effective and recorded 20.80 per cent seedling mortality (Palakshappa et al. 2010).

Bhat and Shrivastava (2003) evaluated fourteen fungicides against *P. aphanidermatum*, at 250, 500 and 1000 ppm concentrations. They found that Emisan, captan, mancozeb, carbendazim 12% + mancozeb 63% WP, propiconazole and RIL F 004 completely inhibited mycelial growth of *P. aphanidermatum* at all concentrations, while Blitox and Calixin was effective at 500 and 1000 ppm concentrations with 100 per cent inhibition. Jiskani et al. (2007) evaluated fungicides such as Benlate, Topsin-M, Derosal and copper oxychloride which were applied as soil drench. Seed germination was significantly increased with Topsin-M applied as soil drench after 15 days of sowing (18.50) followed by benlate and copper oxychloride (14.00). Zagade et al. (2012) indicated that seed treatment with a combination of metalaxyl + *T. hamatum* was effective when compared with the evaluated fungicides. Pandey et al. (2016) evaluated *in vitro* bioefficacy of three fungicides viz. Ridomil, carbendazim and Thiram (each @ 50, 100, 150 and 200 ppm) against *P. aphanidermatum*; reported Ridomil was the most effective with maximum average mycelial growth inhibition (99.81%) followed by carbendazim (73.05%).

Biological control

Management of damping-off in chilli caused by *P. aphanidermatum* involving cultural practices and toxic chemicals have both advantages and disadvantages. The use of plant growth promoting rhizobacteria for biocontrol of plant diseases is currently one of the major areas of research in biology. There are various mechanisms involved for the success of biocontrol agents namely antibiosis or mycoparasitism, competition for nutrients in soil or induction of defence responses in host plant (Saxena et al. 2013). Abeysinghe (2009) reported that a combination of bacterial and fungal compatible strains of *Bacillus subtilis* CA32 and *Trichoderma harzianum* RU01 were successful in suppressing damping-off disease. *Pseudomonas* spp. in combination with other biocontrol agents or applied singly have been tested for their biocontrol potential against many diseases as well as for the damping-off of chilli (Jain et al. 2014, Muthukumar et al. 2010b).

Integrated disease management

The control of diseases is efficient when integrated measures are adopted and started from early during land preparation. The integration of cultural, soil solarization, bioagents and chemical methods are very effective for the management of damping-off. Champawat and Sharma (2003) evaluated soil solarization, bioagents, organic amendments, fungicides and cultural methods in chilli crop against *P. aphanidermatum*. Bioagents used were *T. viride* and T. *harzianum* as a seed dresser and soil inoculation @ 4 g/kg seed and 500 g in 50 kg FYM/ha, respectively. *Pseudomonas fluorescens* was applied as seed treatment @10 g/kg seed and 2.5 kg in 50 kg FYM/ha as soil inoculation, while *Azatobacter chrococcum* as seed treatment @15 g/kg seed and soil inoculation in 50 kg FYM/ha. Neem cake was applied @150 kg/ha as soil amendment. Fungicides viz. Captan was applied as seed dresser @ 2.5 g/kg seed which was found effective in chilli. Soil solarization with transparent polyethylene mulch (25 μm) during MayJune, coupled with incorporation of bioagents like *Trichoderma* spp. and seed treatment with fungicides, was found effective in increasing seed germination, seedling vigour and reduced disease severity. Muthukumar et al. (2010a) tested talc based formulation of *T. viride, P. fluorescens* and Zimmu leaf extract individually or in consortia for their efficacy against *P. aphanidermatum* under pot culture condition. A combined application of *T. viride* + *P. fluorescens* + Zimmu leaf extract as a seed treatment resulted in a significant reduction of the pre-emergence and post-emergence damping-off incidence to 8.3 and 17 per cent, respectively, in chilli.

Damping-off, seedling and root rots of cucurbits

Disease occurrence and distribution: Damping-off and root rot have been reported from nursery beds and young seedlings of some of the most commonly grown cucurbits mainly cucumber (*Cucumis sativus* L), melon (*Cucumis melo* L), bottle gourd (*Lagenaria siceraria* (Mol) Standi), and watermelon (*Citrullus lanatus* (Thumb) Matsum and Nakai). In greenhouses, cucumber damping-off disease is widespread. These diseases have increased rapidly under protected cultivation.

Causal organism: Cucurbit damping-off and seedling root rots have been reported to be caused by various species of *Pythium* like *P. aphanidermatum* (Edson) Fitzp, *P. irregulare* Buisman, *P. ultimum* Trow., *P. spinosum* Swada, *Rhizoctonia solani* Kuhn, *P. capsici* Leon. and *Phytophthora drechsleri* Tucker, which occur independently and also in combination with the species of the *Pythium, Fusarium* spp. including *F. oxysporum* f. sp. *cucumerinum* Owen that frequently causes rotting of seeds and also damping-off.

Symptoms: Water-soaked lesions occurring at soil level are the first visible symptoms of damping-off in cucurbitaceous family. The disease spreads and ultimately the seedling wilts and decays. Seedlings may die in the soil itself before emergence or may be attacked when the cotyledonary leaves are opening. After emergence of the seedlings from the soil surface, water-soaked lesions appear at the collar region and seedlings droop after 2 to 3 days which is symptomatic of post-emergence damping-off (Fig. 9.3(a)). Often, plants that have survived damping-off may show symptoms of root rot. Root rot affected roots show a watery grey appearing cottony leak, sometimes called Pythium fruit rot or Pythium cottony leak, and is a common disease of cucurbits which appears in the form of white cottony growth of the pathogen on the fruit (Fig. 9.3(b)).

Epidemiology and favourable conditions: Damping-off of cucurbits is usually most severe in conditions having high soil moisture, high plant population, lack of aeration, cool, damp and cloudy weather. In addition, root rot is also favoured by deep planting. Severe incidence of *P. aphanidermatum* was found in watermelons and muskmelon which were sown in cool, moist soil which appears to form crusts around or over the hypocotyls. Soon after one week of sowing or transplanting, seedlings get more prone to damping-off. Unsterilized soil in green houses is more susceptible to damping-off fungi and heavy irrigation also exacerbates damping-off. Pythium fruit rot is enhanced due to moist soil at growing time.

Disease management

Cultural practices

Soil solarization and soil enrichment with organic matter or oil cakes and/or biofumigation

Figure 9.3(a). Damping-off of cucumber at seedling stage

Color version at the end of the book

Figure 9.3(b). Cottony leak of cucumber fruits

Color version at the end of the book

have a great potential to prevent incidence of damping-off. Deadman et al. (2006) reported that biofumigation with cabbage residue incorporated at the top 20 cm of soil at the rate of 5 kg m⁻ and solarization enhanced crop growth. Effects on pathogen inoculum levels, disease severity and plant growth parameters were greater during summer seasons than in winter. Simple cultural practices like sanitation measures, irrigation schedule, raised beds and good quality seeds help to keep the disease away.

Chemical control

Seed treatment with Captan or Thiram @ 2.5 to 3 g/kg seed is found effective to control incidence of damping-off disease. Drenching with the mixture of mancozeb (0.25%) and carbendazim (0.1%) to the plants after seedling emergence controls the disease significantly.

Biological control

Application of biocontrol agents through seed treatment and soil application in the nursery bed before sowing is a good option to control the disease. Sharifi et al. (2007) reported that actinomycetes (*Streptomyces*) isolates inhibited the growth of the *P. aphanidermatum* in culture as well as under greenhouse conditions against damping-off disease in melon. Cucumber seeds treated with spores of *Serratia marcescens* N4-5 and ethanol extracts of the bacterial cultures provided significant suppression of damping-off disease caused by *P. ultimum*. Besides cucumber, the spores of this bacterium also suppressed damping-off caused by *P. ultimum* on cantaloupe, musk melon, and pumpkin (Robert et al. 2007). Another antagonistic bacterium *Bacillus pumilus* strain SQR-N43 was able to control damping-off disease of cucumber caused by *R. solani*. The application of bioagent stabilizes the population and increases the active form of the antagonist (Huang et al. 2012). Robert et al. (2016) treated cucumber seeds with ethanol extract of *Serratia marcesens* N4-5 and found to control damping-off caused by *P. ultimum*. They also reported that N4-5 extract was found compatible with two Trichoderma isolates and can be used as a seed treatment in combination with in-furrow application of *Trichoderma* isolates. A significant reduction (up to 70%) by *T. harzianum* was also reported under field conditions by Al-Ameiri (2014).

Soil biofumigation by incorporation of mustard crop in soil at flowering stage and soil cover with plastic for ten days is commonly used by farmers for managing many soil borne problems including cucumber damping-off. Soil amendment with canola residue was also found very efficient as reported by Moccellin et al. (2017). Amanizadeh et al. (2011) reported Trichomix-HV a consortial bioformulation, a mixture of *Trichoderma harzianum* strain T969 and the fungicides Metalaxyl and Metalaxyl MZ when applied into soil medium at a cfu of 10^7 conidia ml/L and was found to be effective in controlling damping-off disease caused by *P. aphanidermatum, P. ultimum* and *P. irregular* in polyhouses. Chen et al. (2015) developed a container medium with spent blewit peat compost (SBPC), consisting 50% (v/v) SBMC, 0.3% (w/v) lime and 50% (v/v) BVB for suppressing damping-off in cucumber seedlings. The SBPC enhanced plant growth promotion of cucumber seedlings and reduced disease incidence. Microbial population of SBMC compost showed that it consists of 20 microbes. Among these, *Bacillus aryabhattai* CB13 effectively suppressed *P. aphanidermatum* in the steamed SBPC medium. Steamed SBPC treated cucumber plants show 58% disease incidence but there was suppression up to 4% in the bio-formulated container medium. This new bioformulation showed high potential to manage damping-off in commercial nurseries.

Soft rot of ginger

Occurrence and distribution: Ginger (*Zingiber officinale* Roscoe) Pythium soft rot was first reported by Butler (1907) in India. Presently, this disease is prevalent throughout ginger growing countries of the world (Dohroo 2005).

Causal Organism: Pythium soft rot (PSR) on ginger was first observed in India by Butler (1907) caused by *Pythium gracile*. Confusion of species existed till Dohroo (2005) considered *P. gracile* as a synonym for *P. aphanidermatum*. On ginger, *P. aphanidermatum* was recorded in China, India and Japan, but when assessed for pathogenicity on ginger in Japan in a pot trial, PSR symptoms were not observed (Ichitani and Shinsu 1980). However, Krywienczyk and Dorworth (1980) did serological assay and consolidated the two species under the name of *P. aphanidermatum*. Further morphological comparisons of a number of isolates identified as *P. butleri* and *P. aphanidermatum* carried out by Plaats-Niterink (1981) failed to identify characters that would distinctly differentiate the two species. The consolidation of these two species was further confirmed by PCR-RFLP (restriction fragment length polymorphism) analysis of the ITS (internal transcribed spacer) regions (Wang and White 1997). Rajan et al. (2002) reported PSR in Sikkim, India to be caused by a group of fungal pathogens *viz. Pithier* sp., *Fusarium* sp. and *R. solanaceaum* and *Pratylenchus coffeae* increased the severity of infection along with *F. oxysporum*. Eleven species of Pythium were listed by Dohroo (2005) as causal organisms from different regions of the world, but the species most commonly associated with the disease are *P. aphanidermatum* and *P. myriotylum*.

Epidemiology and favourable conditions: Many species of oomycete *Pythium* have been associated with soft rot of ginger but *P. aphanidermatum* is most prevalent and widely associated with soft rot (Dake and Edison 1989). Oospores of *P. aphanidermatum* survive up to 40-42 weeks in soil Kulkarni (2011). *Pythium* is soil borne and its infection begins at collar region of pseudostem and later spreads to rhizome and thus cause complete decay of inner tissues (Selvan et al. 2002). Low temperature favours *P. vexans* for infection, but for *P. aphanidermatum* and *P. myriotylum* to cause infection optimum temperature is 34°C and 18.5°C in storage pits while relative humidity is 60-80% during storage (Dohroo 2001). Higher rhizome rot disease incidence may be attributed to heavy rains and improper drainage. Further, Kulkarni (2006) reported that monocropping also aggravates occurrence of soft rot. Disease progress is more in the month of July-August, which coincides with heavy rains and humid weather (Suryanarayana et al. 2005).

Losses: Growing ginger year after year on the same piece of land has immensely predisposed the crop to soft rot of ginger resulting in 30-70% yield losses during cropping season and storage as well (Rana and Arya 1991). The major losses in ginger occur at post-harvest stage when the precious harvest is lost due to negligence in crop production and storage in underground pits and open heaps at various stages. Dohroo (2005) reported losses of 50–90% due to soft rot of ginger sometimes occurring in major production areas such as the tropical regions of India. The post-harvest losses are affected by various biotic and abiotic factors. During storage, the soft rhizomes are affected by fungal infection (50%) within 3 to 4 weeks (Sharma et al. 2017).

Symptoms: Symptoms develop at any stage of crop growth, but are most commonly seen when the crop is at grand growth stage during summer and autumn. Infection in mature plant takes place through roots or via the collar region (Fig. 9.4(a)), with the first aboveground symptoms being leaf yellowing and the collapse of affected shoots (Fig. 9.4(b)). The disease develops in the form of water-soaked lesions on the developing rhizome (Fig. 9.4(c)), which later spreads causing rots rapidly under favourable environmental conditions, and is eventually destroyed (Dohroo 2005). The disease generally spreads when seed is cut from rhizomes harvested from wet or poorly drained areas and is stored under moist conditions (Pegg and Stirling 1994). Soft rot of ginger caused by *Pythium* spp. is a major limiting factor in ginger production (Dohroo 1991). Many species of *Pythium* have

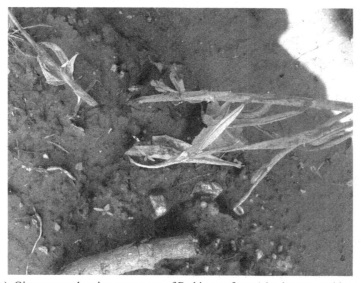

Figure 9.4(a). Ginger crop showing symptoms of Pythium soft rot (also known as rhizome rot of ginger)

Figure 9.4(b). Wilting and yellowing symptoms of Pythium soft rot of ginger

Figure 9.4(c). Rhizome rot of ginger. The dark brown lesions at growth buds of the ginger rhizome

been reported to cause the disease; however, *Pythium aphanidermatum* is the most predominant (Joshi and Sharma 1982, Srivastava et al. 1998).

Disease management

Cultural practices

PSR spreads very fast since the causal organism has a wide host range. The crop rotation is one of the most effective solution to this problem. Mathur et al. (2002) were able to manage the occurrence of PSR caused by *P. myriotylum* by the combined application of soil solarization and by captan seed dressing +Phorate soil application. However, Harvey and Lawrence (2008) suggested that crop rotations could change the prevalence of *Pythium* spp. populations within a field, by reducing the inoculum. *Pythium* spp. can overwinter in the soil for years once introduced (Hoppe 1966), and therefore, integration of management strategies is the right approach (Mathur et al. 2002, Smith and Abbas 2011). Smith and Abbas (2011) also suggested an integrated programme including cultural practices and a strict quarantine package of practices. Solarization at 47°C for 30 min under 200 mm polyethylene sheet has been found to be effective for ginger seed (Lokesh et al. 2012), though soil solarization for longer periods (up to four weeks) has also been reported by Christensen and Thinggaard (1999) and Deadman et al. (2006). In recent years, numerous environment friendly control measures against PSR have been investigated. In a four-year programme of summer and winter crop rotations of ginger for microbial indicators, Rames et al. (2013) concluded that the population of *Pythium myrotylum* species was reduced in fumigated or fallow plot as compared to the continuous cultivation of pasture grass (*Digitaria eriantha* subsp. *pentzii*). The increase in soil carbon levels and wet conditions which support growth and yield of ginger helps in reducing soft rot (Smith et al. 2011, Stirling et al. 2012). Soil fumigation with 1,3 dichloropropene (Telone®) or irradiation of soils suppresses beneficial microbes which allows PSR to be higher in pot trials (Stirling et al. 2010). However, Stirling et al. (2012) indicated that whenever conditions for *P. myriotytlum* development are ideal, such as high temperature and saturated soils, approaches to retain the presence of antagonist soil microorganisms were not as effective at controlling the pathogen. Stirling et al. (2012) showed that application of organic manure once a while was not sufficient to allow maintenance of soil C levels after three years of cultivation, indicating that total C dropped from 54 g/kg to 39 g/kg which was not significantly different from that found in a bare fallow field. Thus, regular applications of green biomass and organic amendments to the soil are required to maintain optimum organic C in soil C (Kibblewhite et al. 2008, Smith et al. 2011). Sharma et al. (2010) reported PSR incidence was less than 15% in the soil with 2.25% greater organic carbon content depicting a significant negative relationship between organic carbon content and PSR incidence. Higher microbial populations were reported in a minimum tillage led soils as compared to conventional tillage practice by Rames et al. (2013). However, the soils in minimum till fields also tended to set hard; thereby, ginger spread, establishment and proliferation of rhizomes were poor following direct mechanical drilling of the seed (Smith et al. 2011). To solve this problem, Smith et al. (2011) proposed crop rotation for beds under minimum tillage and should be rotary-hoed before planting seed ginger to create better till for germination and growth of the crop.

Plant extract or botanicals

Application of neem (*Azadirachta indica*) seed powder and punarnava (*Boerhavia diffusa*) leaves during land preparation has been suggested to be useful in reducing disease intensity to 89% (Gupta et al. 2013). Application of neem cake at 0.5% mass/mass soil was found to suppress *P. aphanidermatum*. This was due to the increased number of beneficial microorganisms not directly inhibiting *P. aphanidermatum*, which suppressed disease (Abbasi et al. 2005). Lawson (1996) reported inhibitory effect of garlic extract against *P. aphanidermatum*. Similarly, Block et al. (1992) and Shashidhara et al. (2008) reported garlic and onion extracts to be effective against bacterial and fungal species. The significant inhibitory zone with the application of garlic extract on *P. aphanidermatum* was linked to the presence of higher amount of sulphur containing compounds like allicin. Similarly, the inhibitory efficacy of onion extract may be due to its reactive, volatile, odour producing and lachrymal properties. Whitemore and Naidu (2000) opined that the inhibitory effect of allicin was due to the presence of S-methyl-L-cysteine sulphoxide. Soil application of bioformulation of *Trichoderma harzianum* and seed treatment with onion and garlic were observed to be the best in limiting the soft rot of ginger; besides, it showed significant improvement in plant growth and yield promoting parameters (Dohroo et al. 2012). Bhat (2000) has also reported the significant control of soil-borne pathogen including *Pythium* spp. by employing the extract of botanical for seed treatment. The results proved that seed treatment with the extract of botanicals (onion and garlic) @ 25% and soil application of *T. harzianum* were useful in managing the soft rot disease of ginger besides being eco-friendly in the integrated disease management system.

Biological control

Dohroo (2001) reported successful control of soft rot of ginger by using *T. viride* for seed treatment. *Trichoderma* spp. have been reported to be a successful biocontrol agent (Sharma et al. 2014).

The efficacy of *T. harzianum* can be attributed to its faster multiplication in FYM as has also been opined by Danielson and Davey (1973). The *in vitro* studies reported that the non-volatile and volatile compounds produced by *T. viride* inhibit the growth of *P. myriotylum* isolated from ginger (Rathore et al. 1992). Similar observations were made by Shanmugam et al. (2013) for *T. harzianum* and *T. saturnisporum* which also showed strong inhibitory effect against *P. splendens*. Along with *Trichoderma* spp., several rhizobacteria were also antagonistic and showed significant antagonism of *P. myriotytlum* growth in dual culture assays (Bhai et al. 2005).

Soil inoculation with *T. harzianum* colonized wheat grains (9.9 × 10[8] CFU/g) reduced the disease to just 3% as compared to a control treatment without *T. harzianum* at 81% (Rajan et al. 2002). Ram et al. (2000) reported that seed ginger coated with *T. harzianum* was effective in reducing PSR incidence. Disinfestation of ginger seed surface was done by 1% HOCl for 5 min further soaked in *Trichoderma* spp. talc-based formulation (6 × 10[7] CFU/ml) and followed by three applications of talc-based formulation (3 × 10[6] CFU/g) to the soil at 15-day-intervals from time of planting increased the disease suppression. The application of combination of different microbes has also been useful in disease management. Gupta et al. (2010) reported that rhizome soaking in *T. harzianum* + *Glomus mosseae* + fluorescent Pseudomonad strain G4 reduced infection of *P. splendens* to 10%. Interestingly, seed dipping in a suspension of *P. fluorescens* or *Bacillus* sp. (10[8] CFU/ml) for one hour combined with inoculation with *Glomus* sp. at time of planting showed absolute suppression of PSR (Bhai et al. 2005). Moreover, it is well-known that antagonists may also play a significant role in plant growth promotion. The soil application with biocontrol agents are very effective in growth and yield of ginger (Bhai et al. 2005, Gupta et al. 2010, Shanmugam et al. 2013).

Variable and inconsistent results have been observed in the management of ginger rot in field as compared to the pot studies. Singh (2011) reported that *T. harzianum* was able to suppress PSR to the same level as Ridomil® MZ. Shanmugam et al. (2013) also reported that the efficacies of biocontrol agents enhanced in combination as compared to solo application.

Chemical control

The periodic drenching of fungicides Metiram 55% +Pyraclostrobin 5% was found very effective in reducing loss with *P. aphanidermatum* (Dohroo et al. 2012). The treatment of seed ginger (a section of the rhizome used as planting material) with Ridomil® MZ (metalaxyl + mancozeb) at 6.25 g/L along with the soil treatment with Thimet (phorate) and Ridomil MZ at 10 L/3 m[2] plot at 60 days after sowing has been reported to give the best control of *P. myriotylum* on an experimental ginger field in southern Rajasthan, India. The enhanced performance of solarized plots is achieved by using a thick transparent polythene film over the soil surface for 20 days (Mathur et al. 2002). Rajan et al. (2002) in a pot trial reported no disease in the combination of Fytolan (copper oxychloride) 0.2% + Ridomil 500 ppm + Bavistin (carbendazim) 0.2% + Thimet as compared to single application, though each one of these was also effective in reducing rhizome rot and root rot caused by *Pythium* sp., *Fusarium* sp. and *R. solanaceaum*. Seed dipping with Ridomil MZ at a rate of 1.25 g/L could increase survival of rhizomes by about 30% as compared to hot water treatment at 51°C for 30 minutes in a field naturally infested with *P. aphanidermatum* in Raigarh, India (Singh 2011).

In another pot trial, Smith and Abbas (2011) reported that by coating ginger seed with Ridomil (metalaxyl), Maxim® XL (fludioxonil) and Proplant® (propyl carbamate hydrochloride) gives a better control of PSR caused by *P. myriotylum*. The large scale application of seed dipping causes contamination of seed and, therefore, alternatives have been tried in Australia. Therefore, farmers in Australia are now advised to use disease-free certified seed ginger that is allowed to air dry until the cut surfaces are fully suberized. Instead of dipping, Lokesh et al. (2012) reported that seed ginger can be solarized at 47°C under 200 μm polyethylene sheets for 30 min for *Pythium* spp. disinfestation.

Host plant resistance

Nair and Thomas (2007) in a study on the characterization and expression of resistance gene candidates (RGCs) accessions of the wild species of genus Zingiber, *Z. zerumbet* and *Z. cernuum* and three accessions of *Z. officinale*: cv. kuruppampady, cv. Rio-de-Janeiro could identify 42 RGCs belonging to non-toll-interleukin receptor nucleotide binding site leucine rich repeat (TIR NBS-LRR). Similarly, Kavitha and Thomas (2008a), through mRNA differential display in *Z. zerumbet*, investigated a putative soft rot resistant species. Wild *Zingiber* species exhibit both sexual and asexual reproduction, and therefore offer the potential of introducing variability into the cultivated crop including potential resistance traits (Kavitha and Thomas 2008c). Nair and Thomas (2013) isolated and characterized ZzR1 gene from *Z. zerumbet*, a very useful gene pool to elucidate resistance mechanisms. Genetic transformation of ginger can be an important tool for developing resistant cultivar since improving edible ginger cultivars by conventional breeding was not successful.

Induced resistance

Induced resistance is triggered in plants by prior application of biotic or abiotic elicitors. Induced resistance is an alternative to genetic resistance by upregulating induced resistance mechanisms. Karmakar et al. (2003) demonstrated systemic resistance in highly susceptible cultivar Varada, Suprabha and Maran against *P. aphanidermatum* by soaking rhizomes for one hour 5mM of salicylic acid (SA) and its analogs DL-aminobutyric acid (BABA), and 2,1,3 benzothiadiazole (BTH) resulted in significant reduction in subsequent PSR. The upregulation of resistance proteins even after eight weeks suggests the application of abiotic or biotic elicitors. Similar observations were made for highly susceptible ginger cultivar 'Varada' by Kavitha and Thomas (2008b). The treated non-susceptible variety *Z. zerumbet* showed no difference in the response to these signaling molecules (Kavitha and Thomas 2008b). Kiran et al. (2012) reported that in *Z. zerumbet* genes responsible for a hypersensitive response and jasmonic acid (JA), SA, and ethylene (ET) biosynthesis were themselves markedly upregulated but down regulated in *Z. officinale* against *P. myriotylum*. The application pre-inoculated with avirulent strains of *P. aphanidermatum* was reported by Ghosh et al. (2006) to reduce disease severity and there was an increase in expression of five particular defence proteins. Ghosh and Purkayastha (2003) reported that by dipping seed rhizome in 10% leaf extract of *Acalypha indica* or 10% root extract of *Boerhaavia diffusa* followed by three foliar sprays of the suspension at seven days intervals, as reported by Pandey et al. (2010), could lower PSR incidence.

Damping-off tomato and brinjal

Disease occurrence and distribution: Damping-off is a severe disease of solanaceous crops including tomato (*Lycopersicon esculentum* Mill.) and eggplant (*Solanum melongena* L.) in nursery. Several soil borne pathogens attack seed, seedling and transplant and thus cause disease in plants. The disease is not only responsible for the poor seed germination and stand of seedlings, but also carry the pathogens into the main field and affect the crop. Shyam (1991) reported that diseased seedlings are one of the major sources for rapid and efficient spread of disease in the field.

Losses: Damping-off disease spreads in nurseries and transplanted plants from initial infection points, causing large areas in which usually all the seedlings are killed. Many researchers reported severe losses caused by this disease. Gattani and Kaul (1951) reported 72.3 per cent damping-off disease in tomato. Similarly, Pandey and Dubey (1997) recorded tomato damping-off (*P. aphanidermatum*) up to 100 per cent, while this disease was reported to cause about 60% losses at seedling stage in both nursery and field (Manoranjitham et al. 2000, Jadhav and Ambadkar 2007).

Symptoms: Affected plants generally occur in scattered patches in nursery or in lower parts of the slope fields. In even slope fields, diseased plants usually occur randomly in the fields. Damping-off may occur as pre- or post-emergence. In pre-emergence damping-off, target of the pathogen is seed and on infection it fails to grow after sowing. On attack of the pathogen, seeds turn soft, mushy and brown, and later decompose as a result of seed infection. In post-emergence damping-off, the seed germinate from underneath the soil, but dies soon. The affected portions, mainly roots, hypocotyls and crown of the plant, become pale brown, soft, water soaked and weaker than non-affected part. Infected stems collapse (Fig. 9.5). Root rot and collar rot makes the plant stunted. Severity of the symptoms usually changes with the stage of the crop development. Pathogen takes time to establish in the host and thus infection by the pathogen(s) may occur much later after plant emerges; in this case, the infection is usually not severe but the overall plant growth and yield production may be affected. Roots are the most severely affected part of the plant and may show wilt or rot in warm or windy weather. Severe root rotting results in nutrient deficiency symptoms in the plant since nutrients may not be sufficiently available from the soil up through the plant. Since number of pathogens are responsible for identical symptoms, identification of respective pathogen after isolation is needed to confirm diagnosis.

Figure 9.5. Damping-off of tomato seedlings (Photo courtesy: Susanta Banik,
Nagaland University, Medziphema)

Causal organism: Damping-off disease is caused by number of *Pythium* spp. such as *Pythium aphanidermatum, P. butleri, P. inflatum, Phytophthora parasitica, Fusarium oxyspoprum, F. solani, Phomopsis vexans, Colletotrichum capsici, Rhizoctonia solani* and *Sclerotium rolfsii*. In *Pythium* spp., *P. aphanidermatum* and *P. butleri* are common species and mycelium of *P. aphanidermatum* is white, well developed and woolly but not as abundant or vigorous as of *P. butleri* and less profuse. The hyphae is 2.8 to 7.5 μm in diameter, sporangiophore produces swollen, toruloid, branched sporangia. Encysted zoospores are 6.7-12.0 μm in size. Spherical, smooth oogonia are formed terminally on later hyphae and antheridia are monoclinous, intercalary or terminal. On each oogonia, 1-2 antheridia attach. Oospores are single aplerotic, moderately thick walled (17-19 μm).

Epidemiology and favourable conditions: Seedlings during the initial three weeks are susceptible under the following predisposing factors: heavily infested soil or medium growth, overwatering or poor drainage, overcrowding of the seedling with poor ventilation, excessive doses of nitrogen, variable environmental conditions such as rainy, cloudy weather that result in prolonged soil moisture, and low sunlight that prevents drying. These pathogenic fungi cause disease in a wide range of plants. These pathogens belong to similar genera of fungi with different environmental requirements for the disease development. For example, cool and wet climate favours *Pythium* spp. and *Phytophthora* spp.. Within *Pythium* spp., *P. ultimum* proliferate well under the low temperatures, while *P. aphanidermatum* is active in high temperatures. Overwintering spores of these fungi can survive well in soil for a long period in the non-availability of host plants. According to Tripathi and Grover (1976), favourable temperature for pre-emergence damping-off is 20-25°C, while post-emergence damping off was more severe at 40°C. High soil moisture and soil temperature accelerate development of the disease (Shyam 1991). These fungi colonize on the various plant debris. Clay soil favoured the damping off diseases incidence caused by *P. aphanidermatum* as compared to sandy soil (Muthukumar et al. 2009). Irrigation system in greenhouses contributes in fast spread of the *Pythium* spp. They form oospores and chlamydospores on decaying of plant roots which later survive for long period under adverse conditions and lead to subsequent infection (Tsitsigiannis et al. 2008). The disease incidence is enhanced, if nematodes are already present in nursery soil, which helps to cause injury to the roots for entry of the pathogen.

Disease management

Cultural practices

Soil solarization of irrigated soil for 30 days decreased pre and post occurrence of damping off disease in tomato (Hazarika 2004). Nursery sowing in the same bed year after year with frequent heavy irrigation should be avoided (Shyam 1991). Healthy tomato seedling should be raised in the nursery bed with the field soil added with sand and farm yard manure (Rana et al. 1989).

Biological control

Intana and Chamswarng (2007) reported antifungal metabolites from four strains of *Bacillus* which had the ability to inhibit mycelial growth *of P. aphanidermatum*, causing damping-off of tomato. Muthukumar et al. (2010b) reported that among ten endophytic isolates of *Pseudomonas fluorescens*, EBS 20 produced the largest inhibition zone and inhibited growth of *P. aphanidermatum* mycelium. This isolate also produced more amounts of salicylic acid, siderophore and HCN. Seed treatment with native biological consortium *Pseudomonas* and *Azotobacter* species exhibited an enhanced level of protection compared to the native individual antagonistic bacteria to control pre- and post-emergence damping-off tomato (Champawat and Sharma 2003, Thiruvudainambi et al. 2004, Dutta and Dutta 2007, Pandey and Simon 2010). *Pseudomonas fluorescens* not only reduces damping-off caused by *Pythium* spp. but also increased plant growth development in tomato and brinjal (Pandey and Simon 2010). Apart from this, it also induced defence enzymes involved in the phenyl propanoid pathway which collectively contributed to enhance pathogen resistance. Seed and soil treatment with *Bacillus subtilis* Bs2 also effectively reduced the damping off incidence and increased shoot and root length (Pandey and Simon 2010). *T. harzianum* and *T. viride* formulations applied as seed treatment reduced damping-off in tomato and brinjal in the field and enhanced the plant biomass under field conditions (Hazarika et al. 2000, Jayraj et al. 2006, Kumar and Rao 2006, Poi 2008, Basharat et al. 2011). Seed and soil treatments with *T. viride* and *T. harzianum* alone or in combination with FYM and *Gliocladium virens* reduced the damping off and root rot diseases (Hazarika et al. 2000, Basharat et al. 2011). Anandh and Prakasam (2001) reported that activated formulation of

T. harzianum had the highest population in the soil and the least disease incidence. Seed treatment with *Murraya koenigii* (20% extract) showed better results in managing damping-off (Hasan and Devi 2005). Inoculation of mycorrhizal fungus *Glomus fasciculatum* either simultaneously or two weeks before the inoculation with *P. aphanidermatum* reduced damping-off and increased plant height in comparison to control. Florence (2011} evaluated *in vitro* bioefficacy of bioagents against *P. aphanidermatum*, causing damping off of tomato by dual culture technique and reported *T. viride* most effective with maximum inhibition (91.12%) of the test pathogen, followed by *T. viride* (89.21%), *T. viride* (NBA11) (88.22%) and *T. harzianum* (NBA11) (86.42%). However, *B. megaterium* and *P. fluorescens* were least effective with comparatively minimum mycelial inhibition of 41.85 and 70.05 per cent, respectively.

Hooda et al. (2011) reported that neem cake extract, *Lantana camara* leaf extract, cow urine, *Sapium* sp. *Ligustrum nepalensis, Utrica parviflora, Eucalyptus* sp. and azadirachtin were found very effective for the management of pre-emergence damping-off, while *Thuja compacta* and *Curcuma longa* were also found effective to control post-emergence damping-off.

Plant extract and botanicals

Pandey and Dubey (1994) reported that essential oil extracted from the leaves of 30 angiospermic plants depicted antifungal activity against *P. aphanidermatum* and *P. debaryanum*. Of these, essential oil from *Ocimum sanctum, Hyptis suaveolens* and *Murraya koenigii* showed strong antifungal activity against *P. aphanidermatum* and *P. debaryanum* causing damping-off of tomato. Seed treatment of tomatoes with Lippia oil exhibited 88.9% and 71.3% reduction in disease when challenge inoculated with *P. aphanidermatum* and *P. debaryanum*, respectively (Kishore and Dubey 2002). Gholve et al. (2014) evaluated botanicals against *P. aphanidermatum* which cause damping-off in brinjal and reported that garlic was found very effective and recorded highest mean mycelial growth inhibition of 94.83%. The second and third best botanicals which were found effective were Adulsa (75.53%) and Datura (60.65%).

Chemical control

Many researchers worked on seed treatment and soil sterilization with fungicides for management of this disease in the nursery. Shukla et al. (1977) reported efficacy of soil fumigation with metham, formalin and methyl bromide to reduce the population of soil borne fungi. Pre-sowing application of dazomet (Basamid G @ 30 g/m2) was found effective in the management of damping off caused by *P. aphanidermatum* (Patel and Patel 1998). Various fungicides such as cuprous oxide, MEMC, copper oxychloride, carbendazim, thiram carbendazim + thiram and carboxin have been recommended for the seed treatment to control the disease in the vegetables (Shukla et al. 1977, Vyas et al. 1981, Shyam 1991, Kaushal and Sugha 1995). Rahman and Bhattiprolu (2005) studied that seed treatment with metalaxyl 64% +mancozeb 64% @ 2 g/kg seed alone or along with soil application of metalaxyl 64% + mancozeb 64% @ 0.2% twice, first immediately after germination and second at 20 days after germination, resulted in the effective control of damping-off disease in the solanaceous crop nurseries.

Integrated management of disease

Soil solarization with introduction of biocontrol agent *P. fluorescens* (PfT-8) and *T. harzianum* reduced the rhizospheric population of *Pythium* spp. by 86% and 82%, respectively. Combination of soil solarization (30-40 days), organic amendments (FYM and poultry manure) and seed treatment with *T. viride* was found to be the most effective approach for reducing pre and post emergence of damping-off in tomato. Consortial application of *P. fluorescens* (PfT-8) as seed treatment and soil application along with soil application of organic amendments such as FYM of poultry manure is needed which reduces disease incidence to 90%, improves plant health and increases rhizospheric population of bioagent (Jayraj et al. 2007).

Damping-off and foliar web blight disease in spinach

Distribution and occurrence: Spinach (*Spinacia oleracea* L.) is a leafy green vegetable crop native to central and western Asia. Damping-off and seedling blights are the major diseases of this crop. Many fungal pathogens like *Pythium* spp. and *R. solani*, and perhaps others such as *Fusarium solani,* are reported to cause this disease.

Cool to warm soil temperatures, excess soil moisture, and delayed seedling emergence are congenial conditions for occurrence of damping-off. Soil dwelling pathogens are associated with damping-off and can survive in soil as dormant oospores or sclerotia for extended period of time, pathogenically on alternate hosts and weeds, and saprophytically on crop remains. Lack of sanitation spreads *Pythium* spp. and *Rhizoctonia* spp., which usually disseminate through irrigation water, contaminated soil on equipment, and movement from infected plant materials.

Causal organism: *Pythium aphanidermatum* and *P. dissotocum* have been reported as damping-off and seedling pathogen of spinach, while *P. aphanidermatum* is the cause of web blight of spinach (Liu et al. 2018). However, foliar web blight of spinach is a unique foliar disease first time observed in Arizona (Liu et al. 2018).

Epidemiology and favourable conditions: The primary source of infection is in soil. The conidia acts as secondary source of infection which disseminates through rain splash or wind. The predisposing factors for disease development are high relative humidity, high soil moisture due to water logged soil, cloudiness and low temperatures below 24°C for few days. Overcrowded seedlings, dampness due to high rainfall, poor drainage and excess of soil solutes create congenial microclimate and thus hamper plant growth and increased incidence of damping-off disease. Hot conditions (27-42°C) under sprinkler irrigation and high density plantings (>8.0 million seeds/ha) are commonly associated with web blight of spinach.

Symptoms: Damping-off disease in spinach occurs before or after crop emergence. The seed coat of the spinach seed is rotten in pre emergence damping off. Radical and cotyledons may appear brown and soft soon after germination but fail to emerge. Usually, water-soaked and greasy lesions are seen on hypocotyls and roots after emergence when infected with *Pythium* spp. causing plants to collapse and die. Post-emergence damping-off caused by *Pythium* spp. or *R. solani* begins as elliptical shaped, brown to black sunken cankers on radicals and hypocotyls with a sharp margin between infected and healthy tissue.

Water-soaked or brown to black root colour is seen on infected plants. The upper portion and the tip of the taproot is surrounded by necrotic lesions. When the disease appears in severe form, the roots get rotted and seedlings are killed. The factors important for disease severity are susceptible cultivar, soil texture, irrigation, and pathogen populations. Damping- off is more severe in clay or poorly draining soils with a continuous spinach production year after year. While all stages of spinach can be infected by root rot pathogens, young seedlings are more susceptible. Web blight symptoms appear as water-soaked foliage and rapid collapse of young plants, with luxuriant, white, cottony mycelia on plant leaves.

Disease management

The inoculum levels of *P. ultimum* provide consistent level of disease pressure in spinach causing pre- and post-emergence damping-off. Application of fungicides, botanicals, or biocontrol agents as drench treatment was found to increase disease severity because extra liquid added to the potting medium compared to seed treatment since *P. ultimum* thrives well under wet conditions (Hendrix and Campbell 1973). Continuous cultivation of spinach in soil having more wetness favours damping-off. The cool, wet weather, saturated or compacted soils favour *Pythium* spp. and *R. solani,* which survive in soil for long time. Tea compost significantly reduces damping-off disease (Scheuerell and Mahaffee 2004).

Two proprietary organic disinfectant namely GTG I and GTG II (latter also containing *T. harzianum* T22) were able to manage the disease equivalent to conventional fungicide, Mefenoxam against *P. ultimum* (Cummings et al. 2009). This suggests the organic disinfectant in GTG I and GTG II was the main ingredient that reduced damping-off from *P. ultimum*. Islam (2010) used *Lysobacter* sp. SB-K88 as biocontrol agent to suppress damping off in spinach caused by *Aphanomyces cochlioides* and *Pythium* spp. The SB-K88 effectively inhibited growth and decomposed AC-5 mycelia and suppressed the release of zoospores from the hyphae. After studying the mechanism, Islam (2010) inferred that *Lysobacter* sp. SB-K88 suppresses damping-off diseases through exerting multifaceted antagonistic effects against the Peronosporomycetes. Conditions favourable to these pathogens can result in up to 100% mortality and yield loss if non-treated seeds are planted into infested soils.

Control measures include planting spinach seed in field sites that are suppressive to the pathogens but have conducive growing conditions for spinach and use of fungicide seed treatments such as metalaxyl (Bayer Crop Science, Research Triangle Park, NC), which is effective at controlling *Pythium* spp. but not *R. solani*. Seed treatment with fungicides has been the primary control measure against many damping-off pathogens of spinach (Foss and Jones 2005).

Challenges and future direction

Pythium spp., also known as "water molds," belong to Oomycetes, and are among the most devastating plant pathogens. They are taxonomically distinct from other important plant pathogens. *Pythium* spp. are reported to attack the vegetable crops with other soil borne pathogens *Rhizoctonia* spp., *Fusarium* spp. and *Verticillium* spp., which always warrant a plant pathologist to confirm the species status and pathogenicity on different crops. Although soft rot in some countries is commonly associated with bacterial wilt and Fusarium yellows, thus considered a disease complex, it seems that certain species by themselves are responsible for soft rots. This has raised challenges to plant pathologists for research and farmers in field. Pythium soft rots can be complex, so further work is necessary to determine interactions among species that cause PSR together with other soil borne pathogens and pests of ginger. Morphological identification is very difficult due to high variability in the species. With the advanced technologies, identification of species can be more accurate and reliable by sequencing various gene loci such as ITS, LSU, SSU, CoxI, CoxII and β-tubulin. Other approaches, namely RFLP, RAPD, SSR, q-PCR, ELISA and SDS-PAGE, have been well proven as diagnostic tools in identifying *Pythium* spp. The recent development of LAMP for *P. aphanidermatum* and *P. myriotylum* provides a new hope for a quick and accurate means to detect pathogenic species in the field. But these techniques face a problem in differentiating closely related species like *P. myriotylum* and *P. zingiberis*. With the new era of sequencing technology (next generation sequencing), the whole genome sequence of single species can be easily and accurately obtained. Currently, several species have been fully sequenced and deposited in genbank. Hopefully, these freely available valuable resources will be useful in finding a better gene region for differentiation of closely related species or pathotypes.

Plant pathologists need to find sustainable solutions for disease management and, therefore, integration of cultural, chemical, biological, host resistance and induced resistance was found to be effective. Application of single methodology has not proved to be successful in disease management. Chemical management gives promising results, but continuous application of fungicides, like metalaxyl resistance or any other conventional fungicide and phytotoxicity at harvest, leads to environmental pollution causing human health hazards. Thus, there is an urgent need to replace alternative fungicides with lesser residue. Management of oomycetes generally requires different group of fungicides than what will protect against *Fusarium* spp. and *Rhizoctonia* spp. To get ecologically sustainable solutions, great efforts are required to find potential strains of biocontrol agents. It can be an economically viable alternative that can be implemented at farmers' fields. Biocontrol scientists are very optimistic with some important fungal and bacterial antagonists like

Trichoderma spp., *Pseudomonas* spp. and *Bacillus* spp. strains, which have been proved to suppress the establishment of pathogen. A great deal of understanding of ecology, genetics and epidemiology of *Pythium* spp. associated with these diseases is required to overcome constraints in biological management. These drawbacks are due to differential distribution of strains of biocontrol agents and pathogenic *Pythium* spp. Improving soil health by applications of biocontrol agents, organic matter, minimum tillage, and the use of suitable crop rotations should reduce the inoculum densities of pathogenic species as well as their suppression by increasing competition through enrichment in numbers and diversity of beneficial microorganisms in the soil. At the same time, such healthy soils should promote vegetable crops, allowing the plant to cope better with both biotic and abiotic stress. It will also be important to use certified clean seed and applications of chemicals or plant extracts as part of an integrated programme. There should be a strong and good quarantine practice to avoid cross infections. Option of resistant varieties is also very challenging since we need to have seedling resistance, which can be possible with the advanced techniques.

References

Abawi, G.S., Ludwig, J.W. and Gugino, B.K. 2006. Bean root rot evaluation protocols currently used in New York. Ann. Rep. Bean Improv. Coop. 49: 83-84.

Abbasi, P.A., Riga, E., Conn, K.L. and Lazarovits, G. 2005. Effect of neem cake soil amendment on reduction of damping-off severity and population densities of plant-parasitic nematodes and soilborne plant pathogens. Can. J. Plant Pathol. 27: 38-45.

Abeysinghe, S. 2009. Effect of combined use of *Bacillus subtilis* CA32 and *Trichoderma harzianum* RUOI on biological control of *Rhizoctonia solani* on *Solanum melongena* and *Capsicum annuum*. Plant Pathol. J. 8: 9-16.

Agrios, G.N. 2005. Plant Pathology, Fifth Edition. Academic Press Inc., San Diego, CA.

Ahuja, S.C. 1976. Banded leaf and sheath blight of maize. Ph.D. Thesis, Postgraduate School, IARI, New Delhi.

Al-Ameri, N.S. 2014. Control of cucumber damping-off in the field by the bio-agent *Trichoderma harzianum*. Int. J. Agric. For. 4(2): 112-117.

Anandh, K. and Prakasam, V. 2001. Effect of activated *Trichoderma* formulations in managing damping off disease of tomato. South Indian Hort. 49(special): 275-277.

Anonymous, 2017. Horticultural Statistics at a glance. Government of India, Ministry of Agriculture and Farmers Welfare.

Basharat, N., Simon, S., Das, S. and Soma, R.V. 2011. Comparative efficacy of *Trichoderma viride* and *T. harzianum* in management of *Pythium aphanidermatum* and *Rhizoctonia solani* causing root-rot and damping-off diseases. J. Plant Dis. Sci. 6: 60-62.

Bhagat, S. and Pan, S. 2008. Biological management of root and collar rot of cauliflower (*Rhizoctonia solani*) by a talc-based formulation of *Trichoderma harzianum* Rifai. J. Biol. Cont. 22: 483-486.

Bhai, R.S., Kishore, V.K., Kumar, A., Anandaraj, M. and Eapen, S.J. 2005. Screening of rhizobacterial isolates against soft rot disease of ginger (*Zingiber officinale* Rosc.). J. Spices Arom. Crops. 14(2): 130-136.

Bhat, M.N. 2000. *In vitro* evaluation of some leaf extracts against *Pythium* species causing soft rot of ginger in Sikkim. Plant Dis. Res. 15(1): 97-100.

Bhat, M.N. and Shrivastava, L.S. 2003. Evaluation of some fungicides and neem formulations against six soil borne pathogens and three *Trichoderma* spp. *in vitro*. Plant Dis. Res. 18(1): 56-59.

Block, E., Naganathan, S., Putman, and Zhao, S.H. 1992. Allium chemistry: HPLC analysis of thiosulphate from onion, garlic, wild garlic, Scallion, shellol, elephant garlic and chinese chive: Unique highly allyl to methyl ratio in some garlic samples. J. Agri. Food Chem. 40: 2418-2423.

Butler, E.J. 1907. An account of the genus *Pythium* and some *Chytridiaceae*. Memoirs of the Department of Agriculture in India. Botanical Series 1(5): 1-160.

Champawat, R.S. and Sharma, R.S. 2003. Integrated management of nursery diseases in brinjal, chilli, cabbage and onion. Indian J. Mycol. Plant Pathol. 33: 290-291.

Chauhan, R.S., Maheshwari, S.K. and Gandhi, S.K. 2000. Effect of soil type and plant age on stem rot disease (*Rhizoctonia solani*) of cauliflower (*Brassica oleracea* var. *botrytis*). Agri. Sci. Digest. 20: 58-59.

Chen, J.T., Lin, M.J. and Huang, J.W. 2015. Efficacy of spent blewit mushroom compost and *Bacillus aryabhattai* combination on control of *Pythium* damping-off in cucumber. J. Agri Sci. 153(7): 1257-1266.

Chernyaeva, I.L. and Bisuas, M.C. 1980. Effect of different forms of nitrogen fertilizers and organic additives on *Rhizoctonia solani* Kuhn, the pathogen of black leg of cabbage seedlings. Mikologiya Fitopatol. 14: 453.

Christensen, L.K. and Thinggaard, K. 1999. Solarization of greenhouse soil for prevention of Pythium root rot in organically grown cucumber. J. Plant Pathol. 81(2): 137-144.

Cox, R.S. 1969. Control of Pythium wilt of chrysanthemum in south Florida. AGRIS. 53: 912- 913.

Cummings, J.A., Miles, C.A. and Toit, L.J. 2009. Greenhouse evaluation of seed and drench treatments for organic management of soilborne pathogens of spinach. Plant Dis. 93: 1281-1292.

Dake, G.N. and Edison, S. 1989. Association of pathogens with soft rot of ginger in Kerala. Indian Phytopathol. 42: 116-119.

Danielson, R.M. and Davey, C.B. 1973. Non-nutritional factors affecting the growth of Trichoderma in culture. Soil Biol. Biochem. 5: 495-504.

Deadman, M., Hasani, H.A. and Sa'di, A.A. 2006. Solarization and biofumigation reduce *Pythium aphanidermatum* induced damping off and enhance vegetative growth of greenhouse cucumber in Oman. J. Plant Pathol. 88(3): 335-337.

Dohroo, N.P. 1991. New record of bacterial wilt of ginger in Himachal Pradesh. Presented at symposium on "Pathological Problems of Economic Crop Plants and their Management." Held at CPRI, Shimla.

Dohroo, N.P. 2001. Etiology and management of storage rot of ginger in Himachal Pradesh. Indian Phytopathol. 54(1): 49-54.Dohroo, N.P. 2005. Diseases of ginger. pp. 305-340. *In*: Ravindran, P.N and Babu, K.N. (eds.). Ginger, the Genus *Zingiber.* CRC Press, Boca Raton.

Dohroo, N.P., Kansal, S., Mehta, P. and Ahluwalia, N. 2012. Evaluation of eco-friendly disease management practices against soft rot of ginger caused by *Pythium aphanidermatum*. Plant Dis. Res. 27(1): 1-5.

Dutta, S. and Dutta, D. 2007. Evaluation of biocontrol potentiality of native plant growth promoting bacteria against *Rhizoctonia solani* mediated damping off disease of tomato. J. Mycopathol. Res. 45: 201-206.

Ellis, R.J., Timms-Wilson, T.M., Beringer, J.E., Rhodes, D., Renwick, A., Stevenson, L. and Bailey, M.J. 1999. Ecological basis for biocontrol of damping-off disease caused by *Pseudomonas fluorescens* 54/96. J. App. Microbiol. 87: 454-463.

Florence, U. 2011. Management of tomato damping off caused by *Pythium aphanidermatum* (Edson) Fitzp. through integrated approach. M.Sc. thesis submitted to University of Agricultural Sciences, Bangalore.

Forter, I. and Merriman, P.R. 1985. Evaluation of soil solarization for control of root diseases of row crops in Victoria. Plant Pathol. 34(1): 108-118.

Foss, C.R. and Jones, L.J. 2005. USDA Crop Profile for Spinach Seed in Washington. USDA Pest Management Centers. http://cipm.ncsu.edu/cropprofiles/docs/waspinachseed.html.

Frezzi, M.J. 1956. Especies de Pythium fit opatogenas identificadusen la Republica Argentina. Revista de investigacienesagricolas. Buenos Aires 10: 113-241.

Gattani, M.L. and Kaul, T.N. 1951. Damping off of tomato seedlings, its cause and control. Indian Phytopathol. 4: 156-161.

Gholve, V.M., Tatikundalwar, V.R., Suryawanshi, A.P. and Dey, U. 2014. Effect of fungicides, plant extracts/botanicals and bioagents against damping off in brinjal. Afr. J. Microbiol. Res. 8(30): 2835-2848.

Ghosh, R. and Purkayastha, R.P. 2003. Molecular diagnosis and induced systemic protection against rhizome rot disease of ginger caused by *Pythium aphanidermatum*. Curr. Sci. India 85: 1782-1787.

Ghosh, R., Datta, M. and Purkayastha, R.P. 2006. Intraspecific strains of *Pythium aphanidermatum* induced disease resistance in ginger and response of host proteins. Indian J. Exp. Biol. 44: 68-72.

Gold, S.E. and Stanghellini, M.E. 1985. Effects of temperature on *Pythium* root rot of spinach *Spinacia oleracea* grown under hydroponic conditions. Phytopathol.75: 333-337.

Gupta, M., Dohroo, N.P., Gangta, V. and Shanmugam, V. 2010. Effect of microbial inoculants on rhizome disease and growth parameters of ginger. Indian Phytopathol. 63: 438-441.

Gupta, S.K. and Shyam, K.R. 1994. Anti-sporulant activity of fungicides on downy mildew (*Peronospora parasitica*) of cabbage (*Brassica oleracea*var. *capitata* var. *capitata*). Indian J. Agric. Sci. 64: 891-893.

Gupta, S.L., Paijwar, M.S. and Rizvi, G. 2013. Biological management of rot disease of ginger (*Zingiber officinales* Rosc.). Trends Biosci. 6(3): 302.

Harvey, P. and Lawrence, L. 2008. Managing Pythium root disease complexes to improve productivity of crop rotations. Outlooks Pest Manage. 19: 127-129.

Hassan, M.G. and Devi, L.S. 2005. Management of pre and post emergence diseases of tomato (*Lycopersicon esculentum* Mill.) by botanicals. J. Mycopathol. Res. 43(1): 129-131.

Hazarika, D.K., Sharma, P., Paramanick, T., Hazarika, K. and Phookan, A.K. 2000. Biological management of tomato damping-off caused by **Pythium aphanidermatum**. Indian J. Plant Pathol. 18: 36-39.

Hazarika, B.N. 2004. Effect of soil solarization with polythene mulch in tomato nursery. Adva. Plant Sci. 17: 525-527.

Hendrix, F.F. and Campbell, W.A. 1973. Pythium as plant pathogens. Annu. Rev. Phytopathol. 11: 77-98.

Hooda, K.S., Joshi, D., Dhar, S. and Bhatt, J.C. 2011. Management of damping-off of tomato with botanicals and bio-products in North Western Himalayas. Indian J. Horti. 68: 19- 223.

Hoppe, P.E. 1966. *Pythium* species still viable after 12 years in air-dried muck soil. Phytopathol. 56: 1411.

Horst, R.K. 2013. Damping-off. Westcott's Plant Disease Handbook. Springer Netherlands, Dordrecht, p. 177.

Huang, H., Yu, Y., Gao, Z., Zhang, Y., Li, C., Xu, X., Jin, H., Yan, W., Ma, R., Zhu, J. and Shen, X. 2012. Discovery and optimization of 1,3,4-trisubstituted-pyrazolone derivatives as novel, potent, and nonsteroidalfarnesoidXreceptor (FXR) selective antagonists. J. Med. Chem. 55: 7037-7053.

Ichitani, T. and Amemiya, J. 1980. *Pythium gracile* isolated from the foci of granular dermatitis in the horse (*Equus caballus*). Trans. Mycol. Soc. Japan. 21: 263-265.

Intana, W. and Chamswarng, C. 2007. Control of Chinese-kale damping-off caused by *Pythium aphanidermatum* by antifungal metabolites of *Trichoderma virens* Songklanakarin. J. Sci. Tech. 29: 919-927.

Islam, M.T. 2010. Mode of antagonism of a biocontrol bacterium *Lysobacter* sp. SB-K88 toward a damping-off pathogen *Aphanomyces cochlioides*. World J. Microbiol. Biotechnol. 26: 629-637.

Jadhav, V.T. and Ambadkar, C.V. 2007. Effect of *Trichoderma* spp. on seedling emergence and seedling mortality of tomato, chilli and brinjal. J. Plant Dis. Sci. 2: 190-192.

Jain, R.K., Singh, P.K., Vaishampayan, P.K. and Parihar, S. 2014. Management of damping off (*Pythium aphanidermatum*) in chilli by *P. fluorescens*. Int. J. Agric. Environ. Biotechnol. 7(1): 83-86.

Jarvis, W.R. 1992. Managing disease in greenhouse crops. American Phytopathological Society Press, St. Poul, Minn.

Jayaraj, J., Radhakrishnan, N.V. and Velazhahan, R. 2006. Development of formulations of *Trichoderma harzianum* strain M1 for control of damping-off of tomato caused by *Pythium aphanidermatum*. Arch. Phytopathol. Plant Protect. 39: 1-8.

Jayaraj, J., Parthasarathi, T. and Radhakrishnan, N.V. 2007. Characterization of a *Pseudomonas fluorescens* strain from tomato rhizosphere and its use for integrated of management tomato damping-off. Bio. Control. 52: 683-702.

Jiskani, M.M., Pathan, M.A., Wangan, K.H., Imran, M. and Abro, H. 2007. Different fungicides such as benlatetopsin - M. derosal and Copper oxychloride were applied as soil drench. Pak. J. Bot. 39(7): 2749-2754.

Joshi, L.K. and Sharma, N.D. 1982. Diseases of ginger and turmeric. pp. 260. *In*: Nair, M.K., Premkumar, T., Ravindran, P.N. and Sarma, Y.R. (eds.). National Seminar on Ginger and Turmeric Central Plantation Crops Research Institute, Kasaragod (Kerala), India.

Kapoor, K.S. 1999. Fungal and bacterial diseases of crucifers. pp. 184-215. *In*: Verma, L.R. and Sharma, R.C. (eds.). Diseases of Horticultural Crops: Vegetables, Ornamentals and Mushrooms. Indus Publishing Co., New Delhi.

Karmakar, N.C., Ghosh, R. and Purkayastha, R.P. 2003. Plant defence activators inducing systemic resistance in *Zingiber officinale* Rosc. Against *Pythium aphanidermatum* (Edson) *Fitz.* Indian J. Biotech. 2: 591-595.

Kaushal, N. and Sugha, S.K. 1995. Role of *Phomopsis vexans* in damping off of seedlings in eggplant and its control. Indian J. Mycol. Plant Pathol. 25: 189-191.

Kavitha, P.G. and Thomas, G. 2008a. Defence transcriptome profiling of *Zingiber zerumbet* (L.) Smith by mRNA differential display. J. Biosciences 33: 81-90.

Kavitha, P.G and Thomas, G. 2008b. Expression analysis of defense-related genes in *Zingiber* (Zingiberaceae) species with different levels of compatibility to the soft rot pathogen *Pythium aphanidermatum*. Plant Cell Rep. 27: 1767-1776.

Kavitha, P.G and Thomas, G. 2008c. Population genetic structure of the clonal plant *Zingiber zerumbet* (L.) Smith (Zingiberaceae), a wild relative of cultivated ginger, and its response to *Pythium aphanidermatum*. Euphytica. 160: 89-100.

Kibblewhite, M.G., Ritz, K. and Swift, M.J. 2008. Soil health in agricultural systems. Philos. T. R. Soc. B. 363: 685-701.

Kiran, A.G., Augustine, L., Varghese, S. and Thomas, G. 2012. Reactive oxygen species sequestration, hypersensitive response and salicylic acid signalling together hold a central role in *Pythium* resistance in *Zingiber zerumbet* (Zingiberaceae), 6th International Zingiberaceae Symposium Calicut University Campus, Calicut, Kerala, India, pp. 32-33.

Kishore, N. and Dubey, N.K. 2002. Fungitoxic potency of some essential oils in management of damping-off diseases in soil infested with *Pythium aphanidermatum* and *Pythium debaryanum*. Indian J. For. 25: 463-468.

Krywienczyk, J. and Dorworth, C.E. 1980. Serological relationships of some fungi of the genus *Pythium*. Can. J. Bot. 58: 1412-1417.

Kulkarni, S. 2006. Investigations on the etiology and integrated management of rhizome rot of ginger and turmeric in northern Karnataka. Paper presented in XVII workshop on All India Coordinated Research Project on Spices. Indian Institute of spices Research, Calicut, Kerala, p. 56.

Kulkarni, S. 2011. Etiology, epidemiology and integrated management of rhizome rot of ginger in Karnataka. Indian Phytopathol. 64(2): 113-119.

Kundu, P.K. and Nandi, B. 1984. Control of cauliflower damping off by using antagonist coated seeds. Pedobiologia 27: 43-48.

Kundu, P.K. and Nandi, B. 1985. Control of *Rhizoctonia* disease of cauliflower by competitive inhibition of the pathogen using organic amendments in soil. Plant and Soil 83: 357-362.

Lawson, L.D. 1996. Garlic: A review of its medicinal effects and indicated active compounds. pp. 176-209. *In*: Lawson, C.D. and Bauer, R. (eds.). Phytomedicines of Europe: Their Chemistry and Biological Activity. ASC Press, Washington DC.

Liu, B., Feng, C., Matheron, M.E. and Correll, J.C. 2018. Characterization of foliar web blight of spinach caused by *Pythium aphanidermatum* in desert southwest of United States. Plant Disease 102: 608-612.

Lokesh, M.S., Patil, S.V., Gurumurthy, S.B., Palakshappa, M.G. and Anandaraj, M. 2012. Solarization and antagonistic organisms for management of rhizome rot of ginger in Karnataka. Int. J. Plant Protect. 5: 195-200.

Majeed, M., Mir, G.H., Hassan, M., Mohuiddin, F.A., Paswal, S. and Farooq, S. 2018. Damping off in chilli and its biological management – A Review. Int. J. Curr. Microbiol. App. Sci. 7(04): 2175-2185.

Malthews, V.D. 1931. Studies on the genus Pythium. University of North Carolina Press, pp. 136.

Manoranjitham, S.K., Prakasham, V. and Rajappan, K. 2001. Biocontrol of damping off of tomato caused by *Pythium aphanidermatum*. Indian Phytopathol. 54(1): 59-61.

Mathur, K., Ram, D., Poonia, J. and Lodha, B.C. 2002. Integration of soil solarization and pesticides for management of rhizome rot of ginger. Indian Phytophathol. 55: 345-347.

Middleton, J.T. 1943. The taxonomy, host range and geographic distribution of the genus *Pythium*. Mem. Torrey Bot. Club 20: 1-171.

Moccellin, R., dos Santos, I., Heck, D.W., Malagi, G. and Giaretta, R.D. 2017. Control of cucumber damping-off caused by *Pythium aphanidermatum* using canola residues. Trop. Plant Pathol. 42: 291.

Muthukumar, A., Eswaran, A. and Sanjeev Kumar, K. 2009. Effect of different soil types on the incidence of chilli damping-off incited by *Pythium aphanidermatum*. Agric. Sci. Dig. 29(3): 215-217.

Muthukumar, A., Eswaran, A. and Sangeetha, G. 2010a. Induction of systemic resistance by mixtures of fungal and endophytic bacterial isolates against *Pythium aphanidermatum*. Acta Physiol. Pl. 33: 1933-1944.

Muthukumar, A., Nakkeeran, S., Eswaran, A. and Sangeetha, G. 2010b. *In vitro* efficacy of bacterial endophytes against the chilli damping-off pathogen *Pythium aphanidermatum*. Phytopathologia Mediterranea 49(2): 179-186.

Nair, R.A. and Thomas, G. 2007. Isolation, characterization and expression studies of resistance gene candidates (RGCs) from *Zingiber* spp. Theor. Appl. Genet. 116: 123-134.

Nair, R.A. and Thomas, G. 2013. Molecular characterization of ZzR1 resistance gene from *Zingiber zerumbet* with potential for imparting *Pythium aphanidermatum* resistance in ginger. Gene. 516: 58-65.

Palakshappa, M.G., Lokesh, M.S. and Parameshwarappa, K.G. 2010. Efficacy of Ridomil Gold (metalaxyl M + mancozeb (4%+64% WP) against chilli damping off caused by *Pythium aphanidermatum*, Karnataka. J. Agric. Sci. 23(3): 445-446.

Pandey, A. and Simon, L.S. 2010. Bacterial antagonists for the management of damping off of brinjal. Nat. Acad. Sci. Lett. 33(11/12): 365-368.

Pandey, A.K., Awasthi, L.P., Srivastava, J.P. and Sharma, N.K. 2010. Management of rhizome rot disease of ginger (*Zingiber officinale* Rosc.). J. Phytol. 2: 18-20.

Pandey, M., Ahmad, S. and Khan, K.Z. 2016. *In vitro* evaluation of some antagonists and plant extracts against *Pythium aphanidermatum* causes damping off of chilli. The Bioscan 11(2): 879-884.

Pandey, V.N. and Dubey, N.K. 1994. Antifungal potential of leaves and essential oils from higher plants against soil phytopathogens. Soil Biol. Biochem. 26: 1417-1421.

Pandey, V.N. and Dubey, N.K. 1997. Antifungal potentiality of some higher plants against *Pythium* species causing damping-off in tomato. J. Nat. Acad. Sci. Lett. 20(5-6): 68-70.

Patel, H.R. and Patel, B.N. 1998. Evaluation of dazomet (Basamid G) against rove beetles, damping-off, weeds and nematodes in bidi tobacco nursery. Indian J. Mycol. Plant Pathol. 28(2): 134-139.

Pegg, K.G. and Stirling, G.R. 1994. Ginger. pp. 55-57. *In*: Persley, D. (ed.). Diseases of Vegetable Crops. Department of Primary Industries, Queensland.

Plaats-Niterink, A.J.V.D. 1981. Monograph of the genus *Pythium*. Stud. Mycol. 21: 1-244.

Poi, S.C. 2008. Effect of talc based biocontrol agents on damping-off disease in chilli and brinjal. Environ. Ecol. 26(4): 1492-1493.

Pringsheim, N. 1858. Beitraegezur morphologie und systematik algae. 1. Die Saprolegnieen. J. Wiss. Bot. 1: 284-306.

Pullman, G.S., DeVay, J.E., Garber, R.H. and Weinhold, A.R. 1981. Soil solarization: Effects on Verticillium wilt of cotton and soilborne populations of *Verticillium dahliae*, *Pythium* spp., *Rhizoctonia solani*, and *Thielaviopsis basicola*. Phytopathol. 71(9): 954-959.

Rahman, M.A. and Bhattiprolu, S.L. 2005. Efficacy of fungicides and mycorrhizal fungi for the control of damping-off disease in nurseries of tomato, chilli and brinjal crops. Karnataka J. Agric. Sci. 18(2): 401-406.

Rajan, R.P., Gupta, S.R., Sarma, Y.R. and Jackson, G.V.H. 2002. Disease of ginger and their control with *Trichoderma harzianum*. Indian Phytopathol. 55: 173-177.

Ram, D., Mathur, K., Lodha, B.C. and Webster, J. 2000. Evaluation of resident biocontrol agents as seed treatments against ginger rhizome rot. Indian Phytopathol. 53(4): 450-454.

Ramamoorthy, V., Raghuchander, T. and Samiyappan, R. 2002. Enhancing resistant of tomato and hot pepper to *Pythium* diseases by seed treatment with fluorescent pseudomonas. Eur. J. Plant Pathol. 108: 429-441.

Rames, E.K., Smith, M.K., Hamill, S.D. and De Faveri, J. 2013. Mircrobial indicators related to yield and disease and changes in soil microbial community structure with ginger farm management practices. Australas. Plant Pathol. 42: 685-692.

Rana, K.S. and Arya, P.S. 1991. Rhizome rot and yellow disease of ginger in Himachal Pradesh. Indian J. Plant Path. 21: 242-245.

Rana, R., Gangopadhyay, S. and Banerjee, M.K. 1989. Incidence of damping off of tomato in different types of nursery beds. Plant Dis. Res. 4(2): 179.

Rathore, V.R.S., Mathur, K. and Lodha, B.C. 1992. Activity of volatile and non-volatile substances produced by *Trichoderma viride* on ginger rhizome rot pathogens. Indian Phytopathol. 45: 253-254.

Roberts, D.P., McKenna, L.F., Lakshman, D.K., Meyer, S.L.F., Kong, H., de Souza, J.T., Lydon, J., Baker, C.J. and Chung, S. 2007. Suppression of damping-off of cucumber caused by *Pythium ultimum* with live cells and extracts of *Serratia marcescens*. Soil Biol. Biochem. 39: 2275-2288.

Roberts, D.P., Lakshman, D.K., McKenna, L.F., Emche, S.E., Maul, J.E. and Bauchen, G. 2016. Seed treatment with ethanol extract of *Serratia marcescens* is compatible with *Trichoderma* isolates for control of damping off of cucumber caused by *Pythium ultimum*. Plant Dis. 100: 1278-1287.

Roy, S.K., Das, B.C. and Bora, L.C. 1998. Non pesticidal management of damping off of cabbage caused by *Rhizoctonia solani* Kuhn. J. Agric. Sci. Soc. North East India 11: 127-130.

Sardana, H.R., Bhat, M.N., Choudhary, M. and Vaibhav, V. 2019. Integrated Management Strategies for Cabbage and Cauliflower. Extension folder, NCIPM, New Delhi.

Saxena, A., Mishra, S., Raghuwanshi, R. and Singh, H.B. 2013. Biocontrol agents: Basics to biotechnological applications in sustainable agriculture. pp. 141-164. *In*: Tiwari, S.P., Sharma, R. and Gaur, R. (eds.). Recent Advances in Microbiology, Vol. 2. Nova Publishers, U.S.A.

Scheuerell, S. and Mahaffee, W. 2004. Compost tea as a container medium drench for suppressing seedling damping-off caused by *Pythium ultimum*. Phytopathol. 94: 1156- 1163.

Schröter, J. 1897. Saprolegnineae III Pythiaceae. Engler&Prantl Nat PflFam. 1: 104-105.

Schwartz, H.F., Gent, D.H., Gary, D.F. and Harveson, R.M. 2007. Dry bean, *Pythium* wilt and root rots. High plains IPM Guide, a cooperative effort of the University of Wyoming, University of Nebraska, Colorado State University and Montana State University.

Selvan, M.T., Thomas, K.G. and Manoj Kumar, K. 2002. Ginger (*Zingiber officinale* Rosc.). pp. 110-131. *In*: Singh, H.P., Sivaraman, K. and Selvan, M.T. (eds.). Indian Spices – Production and Utilization. Coconut Development Board, India.

Shanmugam, V., Gupta, S. and Dohroo, N.P. 2013. Selection of a compatible biocontrol strain mixture based on co-cultivation to control rhizome rot of ginger. Crop Prot. 43: 119-127.

Sharifi, M., Ghorbanli, M. and Ebrahimzadeh, H. 2007. Improved growth of salinity-stressed soybean after inoculation with pre-treated mycorrhizal fungi. J. Plant Physiol. 164: 1144-1151.

Sharma, B.R., Dutta, S., Roy, S., Debnath, A. and De Roy, M. 2010. The effects of soil physico- chemical properties on rhizome rot and wilt disease complex incidence of ginger under hill agro-climatic region of West Bengal. Plant Pathol. J. 26(2): 198-202.

Sharma, P., Sharma, M., Raja, M. and Shanmugam, V. 2014. Status of Trichoderma research in India: A review. Indian Phytopathol. 67(1): 1-19.

Sharma, R.C. and Sharma, S.L. 1985. Correlation of meteorological factors with stalk rot of cauliflower caused by *Sclerotinia sclerotiorum*. Indian J. Ecol. 12: 327-330.

Sharma, S., Dohroo, N.P., Shanmugam, V., Phurailatpam, S., Thakur, N. and Yadav, A.N. 2017. Integrated disease management of storage rot of ginger (*Zingiber officinale*) caused by *Fusarium* sp. in Himachal Pradesh, India. Int. J. Curr. Microbiol. App. Sci. 6(12): 3580-3592.

Sharma, Y. and Chaudhary, K.C.B. 1984. Epidemiology of cauliflower wilt in different soils in Jammu. Indian J. Plant Pathol. 2: 78-79.

Shashidhara, S., Lokesh, M.S., Lingaraju, S. and Palakshappa, M.G. 2008. *In vitro* evaluation of microbial antagonists, botanicals and fungicides against *Phytophthora capsici* Leon, the causal agent of foot rot of black pepper. Karnataka J. Agri. Sci. 21: 527-531.

Sheikh, H.A. 2010. Two pathogenic species of Pythium: *P. aphanidermatum* and *P. diclinum* from a wheat field. Saudi J. Biol. Sci. 17(4): 347-352.

Shukla, P., Singh, R.P. and Prasad, M. 1977. Studies on damping off of tomato in different types of nursery beds. Indian J. Mycol. Plant Pathol. 7(2): 157-158.

Shyam, K.R. 1991. Important diseases of summer and winter vegetable crops and their management. pp. 62-69. *In*: Jain et al. (eds.). Horticultural Techniques and Postharvest Management Directorate of Extension Education, UHF, Nauni.

Sideris, C.P. 1931. Taxonomic studies in the family Pythiaceae. I: Nematosporangium. Mycologia 23: 252-295.

Sideris, C.P. 1932. Taxonomic studies in the family Pythiaceae. II: Pythium. Mycologia 24(1): 14-61.

Singh, A.K. 2011. Management of rhizome rot caused by *Pythium, Fusarium* and *Ralstonia* spp. in ginger (*Zingiber officinale*) under natural field conditions. Indian J. Agric. Sci. 81: 268-270.

Singh, R.S. 1990. Plant Diseases, 6th ed. Oxford and IBH Pub. Co., New Delhi, pp. 615.

Smith, M. and Abbas, R. 2011. Controlling *Pythium* and Associated Pests in Ginger, RIRDC Publication No. 11/128, Canberra.

Smith, M.K., Smith, J.P. and Stirling, G.R. 2011. Integration of minimum tillage, crop rotation and organic amendments into a ginger farming system: Impacts on yield and soil borne diseases. Soil Till. Res. 114: 108-116.

Sowmini, R. 1961. Studies of phycomycetes in agricultural soils with special reference to Pythiaceae. M.Sc. (Agri.) Thesis, University of Madras, pp. 160.

Srivastava, L.S., Gupta, S.R., Mahota, U.P. and Neopani, B. 1998. Microorganisms associated with ginger in Sikkim. J. Hill Res. 11: 120-122.

Stapleton, J.J. and De Vay, J.E. 1984. Thermal components of soil solarization as related to changes in soil and root microflora and increased plant growth response. Phytopathol. 74(3): 255-259.

Stirling, G.R., Smith, J.P., Hamill, S.D. and Smith, M.K. 2010. Identifying and developing soils that are suppressive to Pythium rhizome rot of ginger. p. 89. *In*: Sterling, G.R. (ed.). Proceedings of the Sixth Australasian Soil Borne Diseases Symposium, Twin Waters, Queensland.

Stirling, G.R., Smith, M.K., Smith, J.P., Stirling, A.M. and Hamill, S.D. 2012. Organic inputs, tillage and rotation practices influence soil health and suppressiveness to soil borne pests and pathogens of ginger. Austral. Plant Pathol. 41: 99-112.

Sullivan, P. 2004. Sustainable Management of Soil Borne Plant Diseases. ATTRA, National Sustainable Agriculture Information Service, California, U.S.A.

Suryanarayana, V., Kulkarni, S. and Lokesha, N.M. 2005. Soft rot of ginger – a threatened scenario in Malnad tract of Karnataka. In the National Seminar on 'Emerging Trends in Plant Pathology and Their Social Relevance.

Thiruvudainambi, S., Meena, B., Kanthaswamy, V., Srinivasan, K. and Veeraraghavathatham, D. 2004. Biological management of damping off disease in tomato. South Indian Horti. 52(1/6): 368-370.

Tsitsigiannis, D.I., Antoniou, P.P., Tjams, S.E. and Paplomatas, E.J. 2008. Major diseases of tomato, pepper, and eggplant in greenhouse. Eur. J. Plant Sci. Biotechnol. 2(S1): 106-124.

Vyas, S.C., Andotra, P.S. and Joshi, L.K. 1981. Effect of systemic fungicides on control of root rot on vegetables caused by *Rhizoctonia bataticola* and plant growth. Pesticides 15(11): 22-24.

Walker, J.C. 1952. Diseases of Vegetable Crops. McGraw Hill Book Company, New York, p. 529.

Wang, P.H. and White, J.G. 1997. Molecular characterization of *Pythium* species based on RFLP analysis of the internal transcribed spacer region of ribosomal DNA. Physiol. Mol. Plant Pathol. 51: 129-143.

Waterhouse, G.M. 1967. Key to Pythium Pringsheim. Mycol. Pap. 109: 1-15.

Waterhouse, G.M. 1968. The genus Pythium. Diagnosis (or descriptions) and figures from the original papers. Mycol. Pap. 109: 1-50.

Webster, J. and Weber, R.W.S. 2007. Introduction to Fungi. 3rd edn. UK: Cambridge University Press, p. 846.

Whitemore, B.B and Naidu, A. 2000. Thiosulphate. pp. 265-380. *In*: Naidu, A.S. (ed.). Natural Food Antimicrobial System. CRC Press, Boca Raton, FL.

Wokocha, R.C. 1987. Effect of soil type on the damping-off of tomato seedlings caused by *Sclerotium rolfsii* in the Nigerian savanna. Plant and Soil 98(3): 443-444.

Zagade, S.N., Deshpande, G.D., Gawade, D.B., Atnoorkar, A.A. and Pawar, S.V. 2012. Biocontrol agents and fungicides for management of damping off in chilli. World J. Agri. Sci. 8(6): 590-597.

Zagade, S.N., Deshpande, G.D., Gawade, D.B. and Atnoorkar, A.A. 2013. Studies on pre- and post-emergence damping off of chilli caused by *Pythium ultimum*. Indian J. Plant Protect. 41(4): 332-337.

Zamanizadeh, H.R., Hatami, N., Aminaee, M.M. and Rakhshandehroo, F. 2011. Application of biofungicides in control of damping disease off in greenhouse crops as a possible substitute to synthetic fungicides. Int. J. Environ. Sci. Technol. 8: 129.

10

Host Plants and Specificity of the Genus *Pythium*

Kamel A. Abd-Elsalam[1,2*] and Amal-Asran[1]

[1] Plant Pathology Research Institute, Agricultural Research Center (ARC), Egypt
[2] Unit of Excellence in Nano-Molecular Plant Pathology Research Center – Plant Pathology Research Institute, 9 Gamaa St., 12619 Giza, Egypt

Introduction

The various members belonging to the genus *Pythium* are common plant pathogens showing a wide range of host species like plants, mammals, fungi and algae (Webster and Weber 2007). This genus contains over 355 identified species with a broad host range and is living in a diverse global and aquatic environment (Dick 2001, Webster and Weber 2007, Ho 2018). Middleton (1943) expects that approximately 60 species of *Pythium* can infect the higher plants, both cultivated and in forest soils. *P. ultimum* infected over 150 plant species belonging to different plant species (Middleton 1943, Hendrix and Campbell 1973). The genus *Pythium* is a causal organism for seed rots, seedling damping-off, and root rots in a wide range of plants, particularly in greenhouse production. Also, some of the *Pythium* species are reported as causal pathogens for post-harvest diseases such as soft rots of fleshy vegetable and fruit and other plant parts which are in contact with soil in the field, in storage, in transit, and at the market (Agrios 2005). *P. aphanidermatum* and *P. ultimum,* with a broad host range, did not reveal a special response to the plant host. Broad-leaf plants and grasses are susceptible to various species of *Pythium*. *Pythium* affects most of the economically important crops such as wheat, mustard, and forest pine trees and grasses (Hendrix and Campbell 1973). Moreover, it also causes important economic losses to cucumber, tomato, chilli, cotton, leguminous crops and trees like citrus, peach, pear and apple. Soft fleshy plants such as squash, cabbage, beans and potatoes also may be rotted by certain species of *Pythium*. Apart from this, *Pythium* is regularly reported to cause unlimited problems in greenhouses, seed beds and nurseries. For example, *P. myriotylum* and *P. aphanidermatum* are both able to infect different plants such as tomatoes, cucurbits, peanuts, or ornamentals, besides grasses like wheat, oats, rye, corn, or turf grasses. Solanaceous weeds and *Crotalaria* spp., many weedy grasses, and other weeds may be hosts of *Pythium* spp. and assist as an alternative host as a source of inoculum next season (Kucharek and Mitchell 2000).

Many of the new *Pythium* host plants, recently described and previously known diseases from new hosts, have been included in the seventh edition of *Westcott's Plant Disease Handbook*, 2008 (Horst 2008) and *Field Manual of Diseases in Trees and Shrubs* (Horst 2013). Frezzi (1956, 1977) and Palmucci et al. (2011) thoroughly reviewed the findings of the fungus Genus *Pythium* specified in cultivated plant hosts in Argentina. The number of species of the current genus varies based on their geographical distribution, hosts affected, and symptoms. The most significant number of

*Corresponding author: kamelabdelsalam@gmail.com

plant hosts is damaged by *P. ultimum* and *P. debaryanum*, then followed by *P. irregulare* and *P. aphanidermatum* (Palmucci et al. 2011). The diversity of *Pythium* species within the world suggests that maybe a wider sort of species is still not reviewed well (Martin 2009). This chapter is mainly focused on the present status of host range for the genus *Pythium*, symptoms and their specificity.

Pythium symptoms

The diseases initiated by various *Pythium* species were divided into two types, i.e. soil-borne diseases that affect plant parts in touch with or in the soil (roots, lower stem, seeds, tubers, and fleshy fruits) and foliar diseases that affect above parts of plant including leaves, young stems, and fruits (Agrios 2005). *Pythium* species infect the roots of mature plants, causing necrotic lesions on root tips, fine feeder roots and, less often, on tap roots. *Pythium* infections are generally limited to meristematic root tips, root epidermis, the cortex of roots, and fruits, but simple infections develop when the pathogen transfers inside plant tissue and spreads into the vascular structure. This includes fungal pathogens that cause the root rot of several plant species in the hydroponic lifestyle (Watanabe 2008). There are generally many different types of symptoms induced by some *Pythium* pathogenic species such as seedling damping-off, root rot, wilt and Pythium blight.

Seedling damping-off disease

The infected seedlings appear spindly and small, and the first true leaf is often short on the family's Poaceae, warped and cupped. *Pythium* is frequently disseminated almost evenly in soils. This means that in the absence of a protective treatment for all plants affected by *Pythium* to a similar extent and that unless severe, the infection progresses undetected. Therefore, detection and diagnosis of *Pythium* root rot based on foliar symptoms are positive and when the disease severity is moderate to serious, it is often unidentified similar to *Rhizoctonia* damage (Horst 2013). The effects of *Pythium* root rot are often under-assessed. *Pythium*-infected root system seems small with a limited number of lateral and adventitious roots. Symptoms of appearance include soft, yellow and light brown colouration in infected areas, particularly near the tips. They often lose finer-feeder roots, their exterior layers rotting away and exposing the interior vascular tissues to produce the appearance of *Rhizoctonia*-like 'spear points'. Various species are known to cause seedling damping-off, e.g. *P. debaryanum* and, perhaps more frequently, *P. ultimum*. *P. aphanidermatum* is associated with stem rot and damping-off of 53 plant hosts (Fig. 10.1). We showed a list of the main plant hosts susceptible to several *Pythium* species in Table 10.1.

Rot root diseases

Rot root disease symptoms produced by the genus *Pythium* cause severe losses in certain crops in different growing areas, which has led to an important work dedicated to the management of diseases by using bio-control agents and eco-friendly treatments (Frank and Joyce 1999). There are several symptoms of root rot disease that farmers can easily identify if their crops are infected with *Pythium*. Yellowing of the lower leaves, plants take on a grey/green appearance with leaves that flag slightly during the heat of the day, and stunting are the most detectable in foliar symptoms that plants exhibit when they are being infected by the current pathogen. The degree of these symptoms differs mainly based on disease severity of the infected plants. Most of the time, mainly in the first stages, it is difficult to investigate symptoms on the top growth. *Pythium* typically attacks the root tips first followed by infection upward in the root system. Roots infected with *Pythium* are generally soft, pulpy, and might appear to be a darker shade of brown. The outer overlaying of the foundation, the cortex, typically rotted and slides off without problems whilst pulled, leaving the string-like vascular bundles in the back of this is normally called sloughing of the roots. *Pythium* is an opportunistic pathogen that generally invades the plants through broken or stressed tissues.

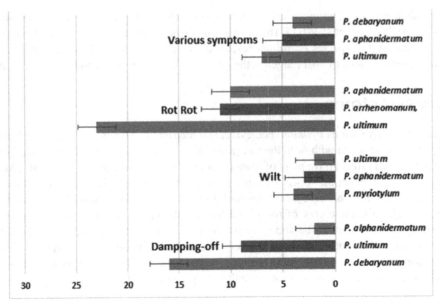

Figure 10.1. Three major *Pythium* species ranking caused various type of symptoms such as damping-off, wilt, root rot and various symptoms on different plant species

Color version at the end of the book

Table 10.1. List of the main plant hosts susceptible to several *Pythium* species, the causal organisms of seedlings disease damping-off

Diseases' symptoms	Host name	Pythium species
Damping-off	Alfaalfa (*Medicago sativa*)	*P. aphanidermatum*
Damping-off	Amaranthus (Love-Lies-Bleeding, Princes-Feather, Joseph-Coat, Spiny Amaranth)	*P. debaryanum*
Damping-off	Barberry (*Berberis*)	*P. debaryanum*
Damping-off	Bean, Kidney, Lima (*Phaseolus vulgaris, P. limensis*)	*P. ultimum; P. debaryanum*
Damping-off	Beet (*Beta vulgaris*)	*Pythium* spp., *P. aphanidermatum*
Damping-off	Buffaloberry (Shepherdia)	*P. ultimum*
Damping-off	Butterfly Flower (Schizanthus)	*P. ultimum*
Damping-off	Cabbage (*Brassica oleracea*)	*Pythium* spp.
Damping-off	Carrot (*Daucus carota* var. *sativa*)	*Pythium* spp.
Damping-off	Catalpa	*P. ultimum*
Damping-off	Celery, Celeriac (*Apium graveolens*)	*P. mastophorum, Pythium* spp.
Damping-off	Citrus Fruits (Grapefruit, Lemon, Lime, Orange)	*Pythium* spp.
Damping-off	Clarkia	*P. debaryanum*
Damping-off	Coleus	*Pythium* spp.
Damping-off	Cockscomb (*Celosia cristata*)	*P. debaryanum*
Damping-off	Delphinium (Larkspur)	*Pythium* spp.
Damping-off	Desert-Willow (Chilopsis)	*P. ultimum*
Damping-off	Eggplant (*Solanum melongena*)	*P. debaryanum, P. catenulatum*
Damping-off	Elm (Ulmus)	*Pythium* sp.
Damping-off	Fern, Brake (Pteris)	*P. intermedium*
Damping-off	Flame bottletree (*Brachychiton acerifolius*)	*P. debaryanum*

(Contd.)

Table 10.1. (*Contd.*)

Diseases' symptoms	Host name	Pythium species
Damping-off	Ginseng (Panax)	*P. debaryanum*
Damping-off	Gypsophila (Babys-Breath)	*P. debaryanum*
Damping-off	Impatiens (Garden Balsam, Sultan)	*Pythium* sp.
Damping-off	Kochia (Summer-Cypress)	*P. debaryanum*
Damping-off	Lettuce (Lactuca)	*Pythium* spp.
Damping-off	Lily (Lilium)	*P. debaryanum*
Damping-off	Lobelia (Cardinal-Flower, Blue Lobelia)	*P. debaryanum*
Damping-off	Locust (Robinia)	*Pythium* spp.
Damping-off	Melon, Muskmelon, Cantaloupe, Cassabra (*Cucumis melo*)	*P. debaryanum*
Damping-off	Onion (*Allium cepa*)	*Pythium* spp.
Damping-off	Orchids	*Pythium* spp.
Damping-off	Osage-Orange (Maclura)	*P. ultimum*
Damping-off	Papaya (*Carica papaya*)	*P. aphanidermatum, P. debaryanum*
Damping-off	Parrotfeather (*Myriophyllum*)	*Pythium* spp.
Damping-off	Peanut (*Arachis*)	*P. myriotylum,*
Damping-off	Pepper (*Capsicum*)	*Pythium* spp.
Damping-off	Pepper-Grass, Garden Cress (*Lepidium*)	*P. debaryanum*
Damping-off	Pine (Pinus)	*Pythium* spp.
Damping-off	Pyrethrum (*Chrysanthemum cinerariifolium, C. coccineum*)	*Pythium* spp.
Damping-off	Radish (Raphanus)	*P. debaryanum*
Damping-off	Rhubarb (Rheum)	*Pythium* spp.
Damping-off	Sage (Salvia) (Includes Blue, Clary, Sauceleaf, Scarlet, Black ornamental forms)	*P. debaryanum*
Damping-off	Snapdragon (*Antirrhinum*)	*Pythium* spp.
Damping-off	Soybean (*Glycine max*)	*Pythium* spp.
Damping-off	Spinach (*Spinacea*)	*Pythium* spp.
Damping-off	Spruce (*Picea*)	*P. ultimum*
Damping-off	Squash and Pumpkin (Cucurbita)	*P. debaryanum*
Damping-off	Stock (Matthiola)	*Pythium* spp.
Damping-off	Sweet Alyssum (Lobularia)	*P. ultimum*
Damping-off	Sweet Pea and Perennial Pea (Lathyrus)	*Pythium* spp.
Damping-off	Swiss Chard (*Beta vulgaris* var. *cicla*)	*P. aphanidermatum*
Damping-off	Tomato	*Pythium* spp., *P. catenulatum*
Damping-off	Turnip (*Brassica rapa*)	*P. ultimum*
Damping-off	Watermelon (*Citrullus*)	*Pythium* spp.
Damping-off	White mulberry (*Morus alba*)	*P. aphanidermatum*
Damping-off	Wild Rice (*Zizania*)	*P. torulosum*

Flowers which are under pressure due to excessive soluble salt accumulations, negative drainage of the developing medium, and over-watering are mainly susceptible to *Pythium* infestations. Plant roots harmed by bugs such as fungus gnat larvae (*Bradysia* spp.), can provide an entry site for the pathogens via the feeding wounds. In many examples, the damage important for root infections (along with insect feeding, extraordinarily dry conditions, and extra salt degrees) occurs earlier to make the moist and cool conditions more favourable to this pathogen. *Pythium* is frequently known as a water mould because moist conditions are more suitable to infect and reduce plant host growth. Root damage is the most important factor for root infections (along with insect feeding, dry conditions, and extra salt degrees) happens earlier to create the moist and cool conditions more favourable for this pathogen. Some *Pythium* species caused yellowing symptoms of the lower leaves and plants turn grey, and also stunting is one of the most recognizable above-ground symptoms. The degree of these symptoms varies on the severity of the infection. Commonly, within the early stages, symptoms of top growth are not discovered. In African countries, plant pathologists isolated the following species from bean plants suffering from root rot diseases: *P. nodosum, P. echinulatum, P. pachycaule, P. oligandrum, P. acanthicum, P. chamaehyphon, P. folliculosum, P. indigoferae, P. lutarium, P. macrosporum, P. myriotylum, P. paroecandrum, P. torulosum, P. vexans, P. zingiberis, P. graminicola, P. spinosum, P. ultimum, P. arrhenomanes, P. catenulatum, P. diclinum, P. dissotocum, P. rostratum, P. salpingophorum* and *P. deliense* (Buruchara et al. 2007, Nzungize et al. 2011). Similarly, *P. macrosporum, P. oligandrum, P. spinosum* isolated from sorghum, *P. glomeratum* from potato, *P. arrhenomanes, P. ultimum* and *P. heterothallicum* were recovered from maize, peas and sweet potato, respectively (Gichuru 2008, Nzungize et al. 2012). Because of the wide range of hosts and distribution, the lack of visible symptoms and the wide number of species, it seems that the reductions in yield due to this pathogen have been under-reported and not fully understood (Martin 2009). The most important *Pythium* species causing the root rot are *P. ultimum* var. *ultimum, P. ultimum* var. *sporangiifereum, P. aristosporum, P. volutum, P. torulosum, P. irregulare, P. sylvaticum,* etc. (Chamswarng and Cook 1985). There are at least 14 *Pythium* species that were recognized to cause seedling blight and root rot: *P. acanthicum, P. adhaerens, P. aphanidermatum, P. arrhenomanes, P. graminicola, P. irregulare, P. paroecandrum, P. pulchrum, P. rostratum, P. splendens, P. tardicrescens, P. ultimum* and *P. vexans* (Chen 1999). A list of the main host plants susceptible to several *Pythium* species and the causal organisms of wilt disease have been presented in Table 10.2.

Table 10.2. List of the main plant hosts susceptible to several *Pythium* species, the causal organisms of wilt disease

Diseases' symptoms	Host name	Pythium species
Wilt	Bean	*P. myriotylum*
Wilt	Cucumber	*P. ultimum*
Wilt	Crownvetch	*P. myriotylum*
Wilt	Kangaroo Paw (*Anigozanthos*)	*P. myriotylum*
Wilt	Lettuce	*P. tracheiphilum, P. uncinulatum*
Wilt	Nicotiana	*P. aphanidermatum*
Wilt	Muskmelon (*Cucumis melo*)	*P. splendens*
Wilt	Kiwi (Actinidia)	*P. ultimum*
Wilt	Lettuce (Lactuca)	*Pythium* spp., *P. tracheiphilum*
Wilt	Nicotiana (Flowering Tobacco)	*P. aphanidermatum*
Wilt	Olive *(Olea europaea)*	*Pythium* spp.
Wilt	Palm, Coconut (Cocos)	*Pythium* spp.
Wilt	Peanut	*P. myriotylum*
Wilt	Sweet Pepper (*Capsicum annuum*)	*P. aphanidermatum*

Pythium-wilt and blight

Pythium-wilt is one of the most serious diseases of lettuce since it causes a broad wilting and destruction of lettuce foliage. The symptoms of these diseases can be jumbled with other problems such as root rot or damping-off (Horst 2008). As the disease progresses, it wilted outer leaves during the warmer seasons and eventually turned yellow before becoming brown and dead. In advanced stages, the entire foliar canopy can likewise wilt; such plants are not harvestable. First, the pathogen invades the small feeder roots of the lettuce, causing them to be soft and brown. In the late stage disease development, the taproot is shown to be darkly discoloured and the entire root system can be rotted. Pythium blight, sometimes seen as a cottony blight, is one of the most damaging turfgrass diseases. Several species of *Pythium* (water-fungi) are responsible for inducing the disease. The disease can "explode" in just a few days if conditions are right. *P. aphanidermatum* is one of the vital fungal pathogens of cottony blight in *Solanum melongena* (Eggplant) (Horst 2013). Cottony blight symptoms appeared on different hosts which included melon, muskmelon, cantaloupe, and Cucumis *melo* (Cassabra) caused by *P. periplocum* and *P. aphanidermatum*, respectively. We present a list of the main plant hosts susceptible to several *Pythium* species causing rot root in Table 10.3.

Table 10.3. List of the main plant hosts susceptible to several *Pythium* species, the causal organisms of rot root in ninety four plant species

Diseases' symptoms	Host name	Pythium species
Rot, Root	Abelia	*Pythium* sp
Rot, Root	Ageratum	*P. mamillatum*
Rot, Root	Aloe	*P. ultimum*
Rot, Root	Artichoke, Globe (*Cynara scolymus*)	*P. aphanidermatum*
Rot, Foot	Aster (Callistephus)	*P. ultimum*
Rot, Root	Avocado (*Persea americana*)	*Pythium* spp.
Rot, Root	Azalea (Rhododendron)	*Pythium* spp.
Rot, Root	Basil (*Ocimum*)	*P. irregulare*
Rot, Root	Bean, Kidney, Lima (*Phaseolus vulgaris, P. limensis*)	*P. aristorum, P. catenulatum, P. dissotocum, Pythium* spp.
Rot, Root	Beet (*Beta vulgaris*)	*P. deliense*
Rot, Root	Begonia	*Pythium* sp.
Rot, Root	Brachiaria (*Broadleaf signalgrass*)	*P. arrhenomanes*
Rot, Root	Cactus, Prickly Pear (*Opuntia*)	*P. debaryanum*
Rot, Root	Calceolaria (Slipperwort)	*P. mastophorum, P. ultimum*
Rot, Root	Calendula (Pot Marigold)	*P. ultimum*
Rot, Root	Candytuft (Iberis)	*P. oligandrum*
Rot, Root	Cape-Marigold (Dimorphotheca)	*P. ultimum*
Rot, Root	Carnation (*Dianthus caryophyllus*)	*Pythium* sp.
Rot, Root	Centaurea	*Pythium* sp.
Rot, Root	Chamaecyparis	*P. ultimum*
Rot, Root, Damping-off	Chick-Pea, Garbanzo (*Cicer*)	*P. ultimum*
Rot, Root, Leaf	Chinese Evergreen (*Aglaonema*)	*P. splendens*
Rot, Root	Chrysanthemum (*Dendranthema grandiflora*)	*Pythium* sp., *P. ultimum*

(Contd.)

Table 10.3. (*Contd.*)

Diseases' symptoms	Host name	Pythium species
Rot, Root	Cotton	*Pythium* spp.
Rot, Root	Collinsia (Blue-Lips, Blue-Eyed Mary)	*P. mamillatum*
Rot, Root	Columbine (Aquilegia)	*P. mamillatum*
Rot, Root	Corn, Sweet (*Zea mays* var. *saccharata*)	*Pythium* sp.
Rot, Root	Crassula	*Pythium* sp.
Rot, Stem, Soft root	Cucumber (*Cucumis sativus*)	*Pythium* spp.
Rot, Root	Delphinium (Larkspur)	*P. aphanidermatum, P. ultimum, P. vexans*
Rot, Root	Dianthus (Garden Pinks)	*P. ultimum*
Rot, Root	Dieffenbachia	*P. splendens*
Rot, Root	Digitaria (Large Crabgrass)	*P. arrhenomanes*
Rot, Root	Dogwood, Flowering (*Cornus florida*)	*Pythium* sp.
Rot, Root	Echinochloa (Barnyardgrass)	*P. arrhenomanes*
Rot, Root	Elephants-Ear (Colocasia)	*P. debaryanum*
Rot, Root	Eleusine (*Goosegrass*)	*P. arrhenomanes*
Rot, Root	Elm (*Ulmus*)	*P. debaryanum*
Rot, Root	English Daisy (*Bellis perennis*)	*P. mastophorum*
Rot, Root	Fuchsia	*P. rostratum, P. ultimum*
Rot, Root	Gaillardia (Blanket Flower)	*P. ultimum*
Rot, Root	Geranium (*Pelargonium*)	*P. aphanidermatum, P. dissotocum, P. ultimum, P. heterothallicum, P. irregulare, P. myriotylum*
Rot, Root	Gerbera (Transvaal Daisy)	*P. irregulare*
Rot, Root	Gloxinia (Sinningia)	*P. ultimum*
Rot, Root	Godetia	*P. ultimum*
Rot, Root	Caper Spurge (Euphorbia)	*P. aphanidermatum*
Rot, Root	Grasses, Lawn, Turf	*P. debaryanum, P. arrhenomanes, P. ultimum*
Rot, Root	Guayule (Parthenium)	*P. aphanidermatum*
Rot, Root	Heuchera (Alum-Root, Coral-Bells)	*P. hypogynum, P. ultimum.*
Rot, Root	Ice Plant (Carpobrotus)	*P. aphanidermatum*
Rot, Root	Itchgrass (Rottobellia)	*P. arrhenomanes*
Rot, Root	Johnsongrass (Sorghum)	*P. arrhenomanes, P. ultimum*
Rot, Root	Lentil (Lens)	*P. irregulare*
Rot, Root	Lettuce (Lactuca)	*P. myriotylum*
Rot, Root	Lily (Lilium)	*Pythium* spp.
Rot, Root	Lisianthus (*Eustoma grandiflora*)	*P. irregulare*
Rot, Root	Locust (Robinia)	*P. myriotylum*
Rot, Root	Lupine (Lupinus)	*P. ultimum, P. paroecandrum*

(*Contd.*)

Rot, Root	Marigold (Tagetes)	*P. ultimum*
Rot, Root	Maize	*P. arrhenomanes*
Rot, Root	Nephthytis	*P. splendens*
Rot, Root	Nicotiana (Flowering Tobacco)	*P. myriotylum*
Rot, Root	Nutsedge (*Cyperus rotunders*)	*P. arrhenomanes*
Rot, Black	Orchids	*P. ultimum, P. splendens*
Rot, Stem, Root	Parrotfeather (Myriophyllum)	*P. carolinianum*
Rot, Root	Peanut (Arachis)	*Pythium* sp.
Rot, Root	Peperomia	*P. splendens*
Rot, Root	Pepper (Capsicum)	*P. aphanidermatum, P. myriotylum, P. helicoides, P. splendens, P. arrhenomanes, P. catenulatum, P. irregulare, P. graminicola*
Rot, Root	Philodendron	*P. splendens*
Rot, Root	Pitcher-Plant (Sarracenia)	*P. graminicola*
Rot, Root	Potato	*P. glomeratum*
Rot, Root	Poinsettia (*Euphorbia pulcherrima*)	*P. debaryanum, P. perniciosum, P. ultimum;*
Rot, Root	Primrose (*Primula*)	*P. irregulare*
Rot, Root	Prunella (Self-Heal, Heal-All)	*P. palingenes; P. polytylum*
Rot, Root	Radish (Raphanus)	*P. aphanidermatum*
Rot, Stem	Ranunculus (Buttercup, Crowfoot)	*Pythium* sp.
Rot, Root	Sempervivum (Houseleek)	*Pythium* sp.
Rot, Root	Shasta Daisy (Leucanthemum × Superbum)	*Pythium* sp.
Rot, Root	Snakeweed (*Polygonum*)	*P. helicoides*
Rot, Root	Sorghum	*P. macrosporum, P. oligandrum, P. spinosum*
Rot, Root	Soybean (*Glycine max*)	*P. aphanidermatum, P. dissotocum*
Rot, Root	Spurge Caper (*Euphorbia lathyris*)	*P. aphanidermatum*
Rot, Root	Strawberry (Fragaria)	*Pythium* sp.
Rot, Root	Sunflower (Helianthus)	*P. debaryanum*
Rot, Root	Sweet Pea and Perennial Pea (Lathyrus)	*Pythium* spp.
Rot, Root	Sweet Potato	*P. heterothallicum*
Rot, Root	Sweet William (*Dianthus barbatus*)	*P. ultimum*
Rot, Root	Switchgrass (Panicum)	*P. arrhenomanes*
Rot, Root	Tomato	*P. myriotylum P. arrhenomanes*
Rot, Stem, Fruit	Tomato	*P. myriotylum*
Rot, Root	Tuberose (Polianthes)	*P. debaryanum*
Rot, Root	Turnip (Brassica rapa)	*Pythium* sp.
Rot, Root	Water-Cress (*Nasturtium officinale*)	*P. debaryanum*
Rot, Leaf and Stem	Water-Lily (Nymphaea)	*Pythium* spp.

Diverse symptoms

Pythium is generally considered an important pathogen as it destroys juvenile or succulent tissues of initial seedlings. The fungus is responsible for the fast rot of the seeds and for emerging radicles inflicting seed deterioration. Different types of symptoms caused by *Pythium* spp. including crown rot, seedling rot, leaf blight, root rot, cottony blight, watery leak, blight, cottony, fruit rot, bulb rot, and seedlings stunting can cause pre-emergence damping off. They are additionally correlated with root pruning that leads to decreased plant vigour of more mature seedlings. Even though this results in a reduction in plant growth and yield and requires implementation of control measures, those pathogens commonly are not deadly to mature crops (Martin 2009). A list of the main plant hosts susceptible to several infections caused by *Pythium* species, the causal organisms of crown rot, leaf rot, stem rot, blight, cottony leak, watery rot, blackleg, blossom, and another type of symptoms on some plant species has been mentioned in Table 10.4.

Table 10.4. List of the main plant hosts susceptible to several *Pythium* species, the causal organisms of rot crown, rot leaf, rot stem, blight, cottony leaky, watery leak, blackleg, blossom, and other type of symptoms on some plant species

Diseases' symptoms	Host name	Pythium species
Rot, Root and Crown	African Violet (*Saintpaulia*)	*P. ultimum*
Rot, Seedling Root	Amaryllis (*Hippeastrum*)	*P. debaryanum*
Rot, Soft	Bean	*P. aphanidermatum*
Rot, Stem	Cabbage (*Brassica oleracea*)	*P. debaryanum*
Rot, Leaf	Cat-tail (Typha)	*P. helicoides*
Rot, Root, Damping-off	Chick-Pea, Garbanzo (Cicer)	*P. ultimum*
Rot, Root, Leaf	Chinese Evergreen (Aglaonema)	*P. splendens*
Blight, Root	Clover (*Trifolium incarnatum, T. pratense, T. repens, T. stoloniferum, T. subterraneum, T. vesiculosum*)	*P. ultimum, P. irregulare*
Rot, Cottony Leak.	Eggplant (*Solanum melongena*)	*P. aphanidermatum*
Rot, Blackleg, Stem and Cutting Rot.	Geranium (*Pelargonium*)	*P. debaryanum, P. mamillatum, P. splendens, P. ultimum, P. vexans*
Blight, Cottony	Grasses, Lawn, Turf	*P. aphanidermatum*
Rot Crown	Impatiens (Garden Balsam, Sultan)	*P. ultimum*
Rot, Root Cottony Leak	Melon, Muskmelon, Cantaloupe, Cassabra (*Cucumis melo*)	*P. periplocum, P. aphanidermatum*
Rot, Watery Leak	Potato (*Solanum tuberosum*)	*P. debaryanum, Pythium* spp.
Rot, Blossom-End; Root	Squash and Pumpkin (Cucurbita)	*P. aphanidermatum*
Rot, Fruit	Squash and Pumpkin (Cucurbita)	*P. ultimum*
Blight, Aerial	Tomato	*P. myriotylum*
Rot, Bulb	Tulip (*Tulipa*)	*P. ultimum*
Rot, Damping-off, Stunting	Wheat	*P. abappressorium*

Various species are known to cause seedling damping-off, e.g. *P. debaryanum*, and perhaps more frequently, *P. ultimum* and *P. aphanidermatum* are associated with stem rot and damping-off of 53 plant hosts. While the most frequent species to cause wilt symptoms include: *P. myriotylum*, *P. aphanidermatum*, and *P. ultimum*, respectively, the causal organisms responsible for Pythium rot are usually *P. ultimum*, *P. arrhenomanes*, and *P. aphanidermatum* in 94 plant hosts.

Host specificity

A lot of phytopathogenic fungi are greatly host-specific. In most examples, host-unique interactions are usually developed at the time of infection on host plants (Borah et al. 2018). While a few *Pythium* species are modern within their spread with a large range of host variables, every single species nevertheless offers slightly partial pathogenicity. Similarly, each Pythium species offers an intraspecific difference, promoting that only a few invading pathogen strains of the species will certainly become pathogenic (Garzon et al. 2005). The degree of host specificity of fungal pathogen varies based on pathogen species, isolates, and plant species (Zhou and Hyde 2001, Agrios 2005). There are many examples of fungal plant pathogens occurring on one host (Shivas and Hyde 1997). Some soil-borne pathogens, dominated by species of *Pythium, Phytophthora, Fusarium,* and *Rhizoctonia* are significant natural pathogens of tropical trees. The host specificity is no longer being screened in the Janzen–Connell (JC) hypothesis (Connell 1971). They argued for prevalent frequency and distance-dependent mortality of seeds and seedlings by damping-off disease (Dalling et al. 1998, Hood et al. 2004, Augspurger and Wilkinson 2007). Past investigations of the comparative pathogenicity of *Pythium* species were constrained to temperate crop species (Abad et al. 1994, Larkin et al. 1995), herbs (Mills and Bever 1998) and trees (Packer and Clay 2000, Reinhart et al. 2005). Mills and Bever (1998) reported four *Pythium* species infecting four different host species; however, every species caused a distinctive extent of reduction in plant mass and root : shoot ratios of a kind species. Some of the novel isolates of *Pythium* species were found to be pathogenic to crop species. These findings suggest that not all local dispersal events or invasions of a new habitat by a pathogen will lead to finding an appropriate host. Likewise, each species has an intraspecific variant (Garzon et al. 2005), indicating that in reality, a few invading pathogen strains of a species may be pathogenic.

Pythium species are universal in their host range and many species can infect different hosts (e.g. *P. ultimum* with 719 host records) because a few species including *P. graminicola* and *P. arrhenomanes* are primary pathogens of graminous hosts (Middleton 1943, van der Plaats-Niterink 1981, Schroeder et al. 2013). Deep and Lipps (1996) stated that *P. arrhenomanes* was the primary cause of *Pythium* root rot in corn. Eleven *Pythium* spp. like *P. attrantheridium, P. dissotocum, P. echinulatum, P. graminicola, P. inflatum, P. irregulare, P. helicoides, P. sylvaticum, P. torulosum, P. ultimum* var. *ultimum, P. ultimum* var. *sporangiiferum* and *P. dissotocum* were found to be most pathogenic to corn and soybeans (Broders et al. 2007). The principal *Pythium* species recognized to cause root rot in wheat are *P. aristospurum, P. arrhenomanes, P. graminicola, P. ultimum, P. heterothallic* and *P. torulosum* (Paulitz 2010). *Pythium* spp. and *P. spinosum* have been reported as the most virulent species on soybean and corn, even as in rice *Pythium* spp. and *P. spinosum* were the most virulent species. *P. irregulare* and *P. spinosum* isolates varied in rates of virulence in soybean and corn (Urrea Romero 2015). The specific examples of host specificity for the genus *Pythium* are a damping-off disease of seeds and seedlings (Augspurger and Wilkinson 2007). *Pythium* species differ in their pathogenicity and inflicting differential mortality of potential host species. The host specificity of soil pathogens causing damping-off disease in tropical forests is presently unknown. *P. polymastum* has never been associated with root rot in vegetable producing areas in California; however, the pathogen was described as a causal organism of damping-off of broccoli and cauliflower seedlings in greenhouses in California in 1998 (Aegerter and Davis 1998). Vanterpool (1974) defined the host-specificity of *P. polymastum* for the family of Brassicaceae.

The lifestyle of a novel plant family-specific root pathogen consisting of *P. brassicum* shows that rotation with non-*Brassicaceae* plants and control of weeds belonging to this plant family may be effective in disease control (Stanghellini et al. 2014). Several studies emphasize the important role of specialized phytopathogens in structuring and keeping the species composition of plant communities (Connell 1971, Mills and Bever 1998, Wardle et al. 2004). So far, no study has explored the role of unspecified soil pathogens in determining plant distribution. The host specificity of soil pathogens inflicting damping-off disease in tropical forests is unknown.

Pathogenicity tests of interaction specificity from two experiments, with three host genotypes (Belgium, Louisiana, and Pennsylvania) and 37 *Pythium* isolates (10 Europe and 27 USA), explained that *Pythium* pathogens from *Prunus serotina* in its native range were non-specific host genotypes, while *Pythium* pathogens from its non-native range vary with plant host (Reinhart et al. 2005). Buetof and Bruelheide (2011) determined whether non-specialized pathogens can affect the growth and performance of wild plant species. *P. ultimum* is among these pathogens that lack its host specificity (Buetof and Bruelheide 2011). This may lead to the conclusion that unspecified pathogens have the capacity to change the distribution of untamed plant species, those naturally appearing in less humid climates. Benítez et al. (2013) presented features of the complexity of plant diseases and host-pathogen interactions essential for understanding pathogen regulation of plant species diversity. It requires a combination of molecular techniques and statistical models to match the scale of complexity of these challenges. The effect of the present-pathogen interplay is measured via the level of host specialization of the host and interactions with the ecosystem. Changes in host survival are employed to forecast pathogen-mediated effects on plant society changes. Interactions

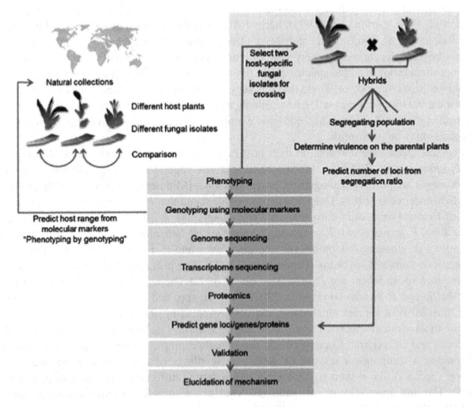

Figure 10.2. Diagrammatical illustration of comparative approaches applied to determine molecular host-specificity determinants. Methods beginning with a set of natural fungal isolates (yellow historical past) are contrasted with strategies starting with segregating populations of defined strains (grey historical past). Both starting substances use the same collection of studying procedures (Reprinted from Borah et al. 2018).

throughout pathogen, host, and environmental conditions provide details on the mechanisms leading to the fluctuations in host survival (Borah et al. 2018). In the new method, the full pool of wild isolates is used for phenotypic and molecular studies to hyperlink genomic markers to phenotypic developments (Fig. 10.2). A well-defined pathosystem is the basis of any follow-up comparative analysis. From this pool of data, two diverse methods are surveyed. In one method, two isolates with dissimilar host-specificity are crossed or hybridized, producing a segregating population with different abilities of virulence to infect one or the other host plant. In the second method, the complete pool of normal isolates is used for phenotypic and molecular assays to link genomic markers to phenotypic characters (Borah et al. 2018). Confronting the complexity of this system, i.e. multi-host pathogens to co-contamination, will help the sphere to continue to develop from attributing plant mortality to a black box of pathogens to a further knowledge of the ecology and biology of the synergies between pathogens, hosts, and environment, and the contribution of these to the renovation of diversity (Fig. 10.3).

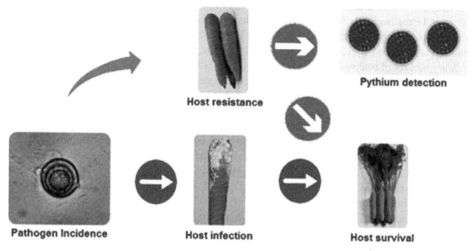

Figure 10.3. Pathogen-host agri-ecosystem contributions to host survival and pathogen detection. The current figure includes results from pathogen-host environs interactions. Green and blue arrows constitute pathogen and host biology characteristics, respectively, that generate contributions to disease, disease development, and plant host survival

Conclusion

Many *Pythium* species have very wide host plants and have also been discovered in most agricultural soils. Various *Pythium* species causing diseases like pre-emergence killing, damping-off, root rot, and wilt showed diverse signs and symptoms. *Pythium* spp. tends to be very generalistic and unspecific in their host range because of a large number of pathogenic species in the genus *Pythium*, their extensive plant hosts and various distribution, the lack of described seen symptoms of the disease and morphological characterization for different species. Additionally, many differences in virulence and host range have been reported among various *Pythium* spp.; however, knowledge about the mechanisms responsible for those observations and the genetics that manage them are still unknown. Consequently, more studies are required to discover the correct number of various plant hosts infected with *Pythium* species, in addition to screening their host range specificity. Additional surveys would be needed to add more plant species host ranges and also to evaluate the diversity of the *Pythium* population in several crop growing areas.

References

Abad, Z.G., Shew, H.D. and Lucas, L.T. 1994. Characterization and pathogenicity of *Pythium* species isolated from turfgrass with symptoms of root and crown rot in North Carolina. Phytopathology 84: 913–921.

Aegerter, D.J. and Davis, R.M. 1998. First report of *Pythium polymastum* on broccoli and cauliflower in California. Plant Dis. 82: 1282.

Agrios, G.N. 2005. Plant Pathology, 5th edn. Amsterdam: Elsevier.

Augspurger, C.K. and Wilkinson, H.T. 2007. Host specificity of pathogenic *Pythium* species: Implications for tree species diversity. Biotropica 39(6): 702-708.

Benítez, M.S., Hersh, M.H., Vilgalys, R. and Clark, J.S. 2013. Pathogen regulation of plant diversity via effective specialization. Trends Ecol. Evol. 28(12): 705-711.

Borah, N., Albarouki, E. and Schirawski, J. 2018. Comparative methods for molecular determination of host-specificity factors in plant-pathogenic fungi. Int. J. Mol. Sci. 19(3): 863.

Broders, K.D., Lipps, P.E., Paul, P.A. and Dorrance, A.E. 2007. Characterization of *Pythium* spp. associated with corn and soybean seedling disease in Ohio. Plant Dis. 91: 727-735.

Buetof, A. and Bruelheide, H. 2011. Effects of an unspecialized soil pathogen on congeneric plant species with different geographic distributions. Preslia 1; 83(2): 205-217.

Buruchara, R.A., Mahuku, G., Mukalazi, J. and Levesque, A. 2007. *Pythium* species associated with *Pythium* root rot of beans (*Phaseolus vulgaris* L.) in Eastern Africa. Cali, Colombia: CIAT, 42-53.

Chamswarng, C. and Cook, R.J. 1985. Identification and comparative pathogenicity of *Pythium* species from wheat roots and wheat-field soils in the Pacific Northwest. Phytopathology 75: 821-827.

Chen, W. 1999. *Pythium* root rot. pp. 11-12. *In*: Dodd, J.L. and White, D.G. (eds.). Compendium of Corn Diseases. Third edition. The American Phytopathological Society, St. Paul, MN.

Connell, J.H. 1971. On the role of natural enemies in preventing competitive exclusion in some marine animals and in rain forest trees. pp. 298-312. *In*: Den Boer, P.J. and Gradwell, G. (eds.). Dynamics of Populations. Cent. Agric. Publ. Doc., Wageningen.

Dalling, J.W., Swaine, M.D. and Garwood, N.C. 1998. Dispersal patterns and seed bank dynamics of pioneer trees in moist tropical forest. Ecology 79: 564-578.

Deep, I.W. and Lipps, P.E. 1996. Recovery of *Pythium arrhenomanes* and its virulence to corn. Crop Prot. 15: 85-90.

Dick, M.W. 2001. The peronosporomylates. pp. 39-72. *In*: Mclaughlin, D.J., McLaughlin, E.G. and Lenke, P.A. (eds.). The Mycota, VII: Part A. Systematic Evolution. Springer Verlag, Berlin.

Frank, N.M. and Joyce, E.L. 1999. Soil borne plant disease caused by *Pythium* spp.: Ecology, epidemiology and prospects for biological control. Critical Rev. Plant Sci. 18(2): 111-181.

Frezzi, M.J. 1956. Especies de *Pythium* fitopatógenas identificadas en la República Argentina. Revista Invest. Agric. 10: 113-241.

Frezzi, M.J. 1977. Especies delgénero *Pythium* and *Phytophthora* fitopatógenas identificadas en Argentina. Boletín Serie Didáctica 2. (Eds.) Inst. de Cs Agronómicas. UNCb.

Garzon, C.D., Geiser, D.M. and Moorman, G.W. 2005. Diagnosis and population analysis of *Pythium* species using AFPL fingerprinting. Plant Dis. 89: 81-89.

Gichuru, G.V. 2008. Influence of farming systems and crop host varieties on *Pythium* root rots epidemics in a highland agroecology of South Western Uganda. PhD thesis. Makerere University, Kampala, Uganda.

Hendrix, J.J. and Campbell, W.A. 1973. *Pythium* as plant pathogens. Annu. Rev. Phytopathol. 11: 77-98.

Ho, H.H. 2018. The taxonomy and biology of *Phytophthora* and *Pythium*. J. Bacteriol. Mycol. 6(1): 40-45.

Hood, L.A., Swaine, M.D. and Mason, P.A. 2004. The influence of spatial patterns of damping-off disease and arbuscular mycorrhizal colonization on tree seedling establishment in Ghanaian tropical forest soil. J. Ecol. 92: 816-823.

Horst, R.K. 2008. Westcott's Plant Disease Handbook, Seventh Edition, Springer-Verlag Berlin Heidelberg, 1349 p.

Horst, R.K. 2013. Field Manual of Diseases on Trees and Shrubs. DOI 10.1007/978-94-007-5980-0_28, © Springer, Dordrecht, 196 p.

Horst, R.K. 2013. Field Manual of Diseases on Garden and Greenhouse Flowers. Springer, New York.

Larkin, R.P., English, J.T. and Mihail, J.D. 1995. Identification, distribution, and comparative pathogenicity of *Pythium* spp. associated with alfalfa seedlings. Soil Biol. Biochem. 27: 357-364.

Martin, F. 2009. Pythium genetics. pp. 213-239. *In*: Kurt Lamour and Sophien Kamoun (eds.). Oomycete Genetics and Genomics: Diversity, Interactions, and Research Tools. John Wiley & Sons.

Middleton, J.T. 1943. The taxonomy, host range and geographic distribution of the genus Pythium. Mem. Torrey Bot. Club 20(1): 1-171.

Mills, K.E. and Bever, J.D. 1998. Maintenance of diversity within plant communities: Soil pathogens as agents of negative feedback. Ecology 79: 1595-1601.

Nzungize, J., Gepts, P., Buruchara, R., Buah, S., Ragama, P., Busogoro, J.P. and Baudoin, J.P. 2011. Pathogenic and molecular characterization of *Pythium* species inducing root rot symptoms of common bean in Rwanda. Afr. J. Microbiol. Res. 5(10): 1169-1181.

Nzungize, J.R., Lyumugabe, F., Busogoro, J.-P. and Baudoin, J.-P. 2012. *Pythium* root rot of common bean: Biology and control methods. A review. Biotechnol. Agron. Soc. Environ. 16(3): 405-413.

Packer, A. and Clay, K. 2000. Soil pathogens and spatial patterns of seedling mortality in a temperate tree. Nature 404: 278-281.

Palmucci, H., Wolcan, S. and Grijalba, P.E. 2011. Status of the Pythiaceae (Kingdom Stramenopila) in Argentina. I. The Genus *Pythium*. Bol. Soc. Argent. Bot. 46(3-4): 197-211.

Paulitz, T.C. 2010. *Pythium* root rot. pp. 45-47. *In*: W.W. Bockus, R.L. Bowden, R.M. Hunger, W.L. Morrill, T.D. Murray and R.W. Smiley (eds.). Compendium of Wheat Diseases, 3rd ed. APS Press. St. Paul. MN.

Reinhart, K.O., Royo, A.A., Van der Putten, W.H. and Clay, K. 2005. Soil feedback and pathogen activity in *Prunus serotina* throughout its native range. J. Ecol. 93: 890-898.

Reinhart, K.O., Van der Putten, W.H., Tytgat, T. and Clay, K. 2011. Variation in specificity of soil-borne pathogens from a plant's native range versus its non native range. Int. J. Ecol. 2011.??

Schroeder, K.L., Martin, F.N., de Cock, A.W.A.M., Lévesque, C.A., Spies, C.F.J., Okubara, P.A. and Paulitz, T.C. 2013. Molecular detection and quantification of *Pythium* species: Evolving taxonomy, new tools, challenges. Plant Dis. 97: 4-20.

Shivas, R.G. and Hyde, K.D. 1997. Biodiversity of plant pathogenic fungi in the tropics. pp. 47-56. *In*: Hyde, K.D. (ed.). Biodiversity of tropical microfungi. Hong Kong University Press, Hong Kong, China.

Stanghellini, M.E., Mohammadi, M., Förster, H. and Adaskaveg, J.E., 2014. *Pythium brassicum* sp. nov.: A novel plant family-specific root pathogen. Plant Dis. 98(12): 1619-1625.

Urrea Romero, K.E., 2015. *Pythium*: Characterization of Resistance in Soybean and Population Diversity. Theses and Dissertations. 1272. University of Arkansas, Fayetteville, http://scholarworks.uark.edu/etd/1272

van der Plaats-Niterink, A.J. 1981. Monograph of the genus *Pythium*. Stud. Mycol. No. 21. Vanterpool, T.C. 1974. *Pythium polymastum* pathogenic on oilseed rape and other crucifers. Can. J. Bot. 52: 1205-1208.

Wardle, D.A., Bardgett, R.D., Klironomos, J.N., Setala, H., Van der Putten, W.H. and Wall, D.H. 2004. Ecological linkages between aboveground and belowground biota. Science 304: 1629-1633.

Webster, J. and Weber, R.W.S. 2007 Introduction to Fungi, Third edition. Cambridge University Press, 975 p.

Zhang, B.Q. and Yang, X.B. 2000. Pathogenicity of *Pythium* populations from corn-soybean rotation fields. Plant Dis. 84: 94-99.

Zhou, D.O. and Hyde, K.D. 2001. Host-specificity, host-exclusivity and host-recurrence in saprobic fungi. Mycol. Res. 105: 1449-1457.

PART II
Identity and Taxonomy

Taxonomic Challenges in the Genus *Pythium*

Reza Mostowfizadeh-Ghalamfarsa* and Fatemeh Salmaninezhad

Department of Plant Protection, Shiraz University, Shiraz, Iran

Introduction

Pythium Pringsh. species are among the most devastating plant pathogens, which can be found from tropical to even cold regions worldwide. However, many other species of this cosmopolitan genus have been reported as saprobes, and animal, algal and even other fungal putative parasites (Kageyama 2014). As a result, identification and taxonomy of these species is a matter of importance. Classification of *Pythium* species has always given rise to many questions due to the difficulties in isolation, accurate identification and characterization of the species.

Species of *Pythium* are classified in the order *Oomycota,* class *Oomycetes*. *Oomycetes* belong to the kingdom *Stramenopila* (*Chromista),* a considerably diverse group including brown algae and planktonic diatoms. They are non-photosynthetic eukaryotic microorganisms and have mycelium containing cellulose and glucans but have no cross walls except to separate living (cytoplasmic) parts of hyphae from old part from which the cytoplasm has been withdrawn (Agrios 2005). *Oomycetes* are a diverse group of eukaryotes living in terrestrial, limnic and marine habitats worldwide (Diehl et al. 2017). These organisms include several destructive plant pathogens (Baldauf et al. 2000, Dick 2001, Cavalier-Smith and Chao 2006). *Oomycetes* have previously been classified under the kingdom Fungi with which they share several similar traits such as mycelial growth and osmotrophic nutrition. Diploid life cycle, gametangial meiosis before fertilization, formation of oospores as sexual spores, motile spore formation, and form of energy storage as mycolaminarins are the main differences between oomycetes and the true fungi. Two major lineages have been reported for oomycetes: Saprolegniomycetidae and Peronosporomycetidae. Saprolegnialean species have been reported as animal pathogens infecting fish, fish eggs, amphibians and crustaceans, as well as plant and algal pathogens. Some non-pathogenic saprobes have also been reported (Johnson et al. 2002, Densmore and Green 2007, Thines and Kamoun 2010, Pelizza et al. 2011, Caballero and Tisserat 2016). Peronosporales are known as agriculturally relevant group and the most aggressive pathogens of crops and ornamental plants, as well as forest trees (Jung and Burgess 2009, Bennett et al. 2017). This order includes *Pythium* species.

In the early 1900s, association of *Pythium* species with plant root diseases was discovered. As a result, it soon became clear that these fungi were important plant pathogens. The genus comprises species that include plant pathogens (e.g. *P. aphanidermatum* (Edson) Fitzp and *P. ultimum* Trow), invertebrate pathogens (e.g. *P. guiyangense* X. Q. Su), algal pathogens (e.g. *P. porphyrae* M. Takah

---

*Corresponding author: rmostofi@shirazu.ac.ir

and M. Sasaki and *P. chondricola* de Cock), mycoparasites (e.g. *P. acanthicum* Drechsler, *P. nunn* Liftsh., Stangh. and R. E. D. Baker, *P. oligandrum* Drechsler and *P. periplocum* Drechsler), mammal pathogens (e.g. *P. insidiosum* de Cock, L. Mend., A. A. Padhye, Ajello and Kaufman) and saprophytes (e.g. *P. intermedium* de Bary) (Drechsler 1943, van der Plaats-Niterink 1981, Ali-Shtayeh 1985, Lodha and Webster 1990, Donaldson and Deacon 1993, Ribeiro and Butler 1995, Hwang et al. 2009, Uzuhashi et al. 2010, Carmona et al. 2017, Klochkova et al. 2017, McCarthy and Fitzpatrick 2017, Majeed et al. 2018). Many *Pythium* species are soil-borne plant pathogens that cause serious economic loss on a wide variety of hosts, while others are more limited in host and geographic range or affect plants only under special environmental conditions (van der Plaats-Niterink 1981, Czeczuga et al. 2005, Kawamura et al. 2005, Hwang et al. 2009, Li et al. 2010, Weiland et al. 2012, Ho 2013, Mostowfizadeh-Ghalamfarsa 2015). These species are among the most prevalent and most important causes of seed rot, seedling damping-off, and root rot of all types of plants, and also of soft rots of fleshy fruits in contact with the soil (Martin and Looper 2010, Mufanda et al. 2017, Rojas et al. 2017). Phytopathogen species mostly have broad host ranges. However, some have been reported to affect certain species. For example, *P. ultimum* can cause damping-off on almost 719 plant species while *P. arrhenomanes* Drechsler is restricted to the plant family, *Poaceae* (Rizvi and Yang 1996, Richard et al. 2005, Schroeder et al. 2013, Giroux 2017). New *Pythium* species are being discovered as pathologists investigate soil organisms associated with plant growth problems.

More than 250 *Pythium* species have been described (Hyde et al. 2014) and it is not surprising to see descriptions of new species once in a while. Identification and characterization of *Pythium* species have been considered as one of the major issues among taxonomists since its first description. In order to identify *Pythium* species accurately, one has to concentrate on various aspects regarding morphological, morphometric, physiological, molecular and phylogenetic characteristics. Considering these aspects, identification of *Pythium* species has become a great challenge to taxonomists. This chapter focuses on the morphological and molecular identification of the *Pythium* species as well as the phylogenetic status and taxonomic challenges of these significant oomycetes.

A history of identification

The genus *Pythium* was established by Pringsheim in 1858. It was first located in the family *Saprolegniaceae* and *Pythium monospermum* Pringsh. was considered as the type species. By the end of 19th century, taxonomic details of the genus had been elucidated and many new species had been described. Schröter (1897) had established its relationship with other oomycetes and the genus was placed in a new family, *Pythiaceae*. Species of *Pythium* were merely interesting to taxonomic mycologists until it was revealed that these oomycetes cause root disease in many plants. The taxonomy of *Pythium* was mainly based on morphological features of sporangia, oogonia, oospores, and antheridia (Waterhouse 1968, van der Plaats-Niterink 1981).

Based on morphological features of *Pythium* species, some identification keys have been generated so far. An extensive publication was conducted by Sideris (1931, 1932) on his taxonomic studies of *Pythium* spp. isolated from pineapple and other plants in Hawaii. However, his proposed classification system had vague and indefinite species separations which was not accepted by other mycologists, and later researchers realized that he had described the same species under different names (e.g. Hendrix and Campbell 1973, Dick 1990, Zitnick-Anderson 2013). Meanwhile, the taxonomic position of known *Pythium* species was refined and a reasonable key was provided by Matthews (1931). During the same period, a number of new species were fully described with illustrations and remarkable examples (Drechsler 1940, 1946). With more species being described, Matthews' key has been treated as old-fashioned. Middleton (1943) assembled an extensive monograph of the genus *Pythium* based on morphology, host ranges, and illustrations of the more important species. His key to the species reflected the importance of morphological features and offered a good picture of relationships based on broad groupings. However, this monograph was not ideal for distinguishing species because of some minor variation to separate species from each other. Middleton's studies were limited to investigation of single isolates of most species, or were dependent upon his interpretation of published descriptions. Hence, checking the range of variation within or between species in many instances was impossible. With the exception of these flaws, and excluding species described since 1940, this monograph remained as the chief taxonomic reference on the genus till 1973. Later on, a good taxonomic monograph for 17 species from Argentina was written, which was a well-illustrated study and in general followed Middleton's concept of the genus (Frezzi 1956). Waterhouse (1967, 1968) generated a key to the genus in 1967 based on the original descriptions and illustrations of all described species and her key, as well as all other keys, worked well with species with well-defined morphological features. These monographs were also difficult to use especially with abnormality forms, or those without a full complement of taxonomic characters.

Using sporangial morphology and sexual organs characteristics, Dick (1990) generated the most recent key of the genus *Pythium*. Although Dick (1990) has offered a reasonable key for identification of *Pythium* species, the most comprehensively used identification key to date is the Monograph of the Genus *Pythium*, which was organized by van der Plaats-Niterink in 1981 (Lévesque and de Cock 2004, Kageyama 2014, Majeed et al. 2018). However, with more species being described, monographs are required to persistently update and add new information to the old ones (Grijalba et al. 2017). With the advent of molecular tools, and based on phylogenetic analysis of some housekeeping nuclear (e.g. Internal transcribed spacer region of rDNA, and β-tubulin) as well as cytoplasmic (e.g. cytochrome c oxidase subunit 1, and 2) genes (Lévesque and de Cock 2004, Villa et al. 2006, Robideau et al. 2011), new species are still continuously being discovered and described.

Identification of *Pythium* species: Morphological approach

Pythium species exhibit unique distinctive features such as their sexual and asexual structures. Asexual structures include mycelium, sporangia, zoospores, vesicle, and chlamydospores (Fig. 11.1). Main hyphae, similar to other oomycetes, are usually coenocytic. *Pythium* sporangia are often spherical or filamentous (Middleton 1943). With direct germination of sporangia, vesicles are produced on the apex of discharged tubes from sporangia. As protoplasm accumulates in vesicle, zoospore formation takes place. However, some species, such as *P. debaryanum* R. Hesse, do not produce zoospores (Middleton 1943, Zitnick-Anderson 2013, Salmaninezhad and Mostowfizadeh-Ghalamfarsa 2017, Zitnick-Anderson et al. 2017).

Sporangia are the most diverse structures in *Pythium* species, which can be ovoid, obovoid, globose, filamentous, pyriform, limoniform, and can be positioned intercalary, terminally, or proliferating (Middleton 1943, van der Plaats-Niterink 1981). Each shape shares some intermediate form. For instance, filamentous sporangia might appear as inflated lobed branches (such as those in *P. aphanidermatum*) or as unbranched (van der Plaats-Niterink 1981). Strictly filamentous sporangium recognition might be very difficult since it can't be distinguished from main hypha (Fig. 11.1) (van der Plaats-Niterink 1981). A very good example of a species with strictly filamentous sporangia is *P. porphyrae* (Hwang et al. 2009).

Although zoospores of *Pythium* species differ in shape, they have identical components (Middleton 1943). The number of zoospores produced in a single vesicle might range from two to 128 (van der Plaats-Niterink 1981). Zoospores shape have generally been reported as bean or pear-shaped with two lateral flagella, anterior tinsel and a posterior whiplash type. (Middleton 1943).

The sexual structures include the oogonium, antheridia, and oospores. Sexual structures usually produce under specific conditions such as darkness (Middleton 1943, Dick 1990). Identification of sexual structures is much easier than asexual ones. This is mainly due to their specific visible features (Middleton 1943, van der Plaats-Niterink 1981). van der Plaats-Niterink (1981) has reported that the oogonium has less variation in shape than that observed for sporangia. However, new *Pythium* species show various kinds of oogonia, ranging from globose to limoniform and some other shapes. *Pythium plurisporium* Abad, Shew, Grand and L. T. Lucas is a good example of amorphous oogonium (Abad et al. 1995). The oogonial wall can either be smooth or have ornamentation of various kinds (Fig. 11.1) (van der Plaats-Niterink 1981). Oogonia can be positioned intercalary or terminally. Antheridia are usually recognizable after the oogonia maturity (Middleton 1943) and they either branch from the same hypha as the oogonium (monoclinous) or from a different hypha (diclinous). Antheridia can originate from the oogonial stalk (hypogynous) or can be attached anywhere on the oogonium (paragynous) (Figs 11.1 and 11.2) (van der Plaats-Niterink 1981) and they may have apical, apical branched, campanulate or broad attachment to oogonium. Antheridia can be helical, terminal or intercalary and their stalk can be absent, short or long (Middleton 1943, Zitnick-Anderson 2013). A recent study on new *Pythium* species revealed that another type of antheridium, amphigynous type, also exist (Abdul Baten et al. 2015); this type of antheridium completely surrounds the oogonial stalk. The morphology and origin of antheridia are very valuable for species identification since they are very specific.

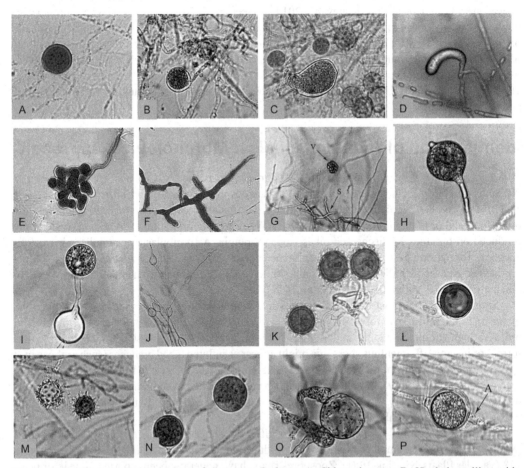

Figure 11.1. Morphological structures of the genus *Pythium*: A. Chlamydospore; B. Hyphal swelling with regular shape; C. Hyphal swelling with irregular shape; D. Appressorium; E. Filamentous, inflated sporangium; F. Filamentous, inflated, dendroid sporangium; G. Strictly filamentous sporangium with vesicle; H. Globose sporangium; I. Vesicle located at the apex of sporangium's discharged tube; J. Proliferation in sporangia; K. Plerotic oospores; L. Aplerotic oospore; M. Ornamented oogonia; N. Smooth-walled oogonia; O. Oogonium with paragynous antheridia; P. Oospore with hypogynous antheridium

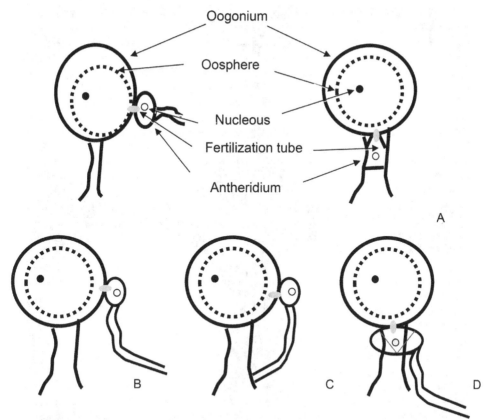

Figure 11.2. Morphology of gametangia in *Pythium* species and different types of antheridial attachment to oogonia: A. (From left to right) Paragynous antheridium, hypogynous antheridium; B. Paragynous, diclinous antheridium; C. Paragynous, monoclinous antheridium; D. Amphigynous antheridium

Fertilization occurs when an antheridium attaches to an oogonium, and oospore formation takes place. The term zygote might be used for oospore in some papers. During fertilization, gametangial meiosis takes place and the fertilized oogonium results in a diploid zygote or oospore. Usually one oospore develops; however, a single oogonium might consist more than one oospore in some species, such as *P. plurisporium* and *P. multisporum* Poitras (Abad et al. 1995). The oospore wall might be smooth or reticulate, thin or thick, although thick wall is more common in *Pythium* species. Oospore contents, which are comprised of granular or opaque protoplasm, are also valuable in the species-level identification (Middleton 1943). Plerotic and aplerotic are two common terms for oospores with completely filled protoplasm and not completely filled with protoplasm, respectively (Fig. 11.1) (Middleton 1943). van der Plaats-Niterink (1981) reported that the oospores might also have either one or multiple refringent bodies within the cavity.

Pythium species were considered as homothallic because Sparrow's research showed that in all of his isolates, sexual structures were produced abundantly despite diclinous antheridial existence (Sparrow 1931). In contrast, Campbell and Hendrix (1967) with similar studies of hyphal tipping and dual cultures on water agar observed the production of both sexual structures only in the contact zone between two compatible partners. Hence, they concluded that *Pythium* species can also be heterothallic as well and may need a compatible mating type for sexual reproduction.

Some other factors have been considered important for identification of *Pythium* species. These factors include colony morphology and growth rate (Kageyama 2014). van der Plaats-Niterink (1981) has classified *Pythium* growth habit as chrysanthemum, radiate, rosette, intermediate, or without any obvious pattern (Fig. 11.3). Also, aerial mycelium can vary in pigmentation from

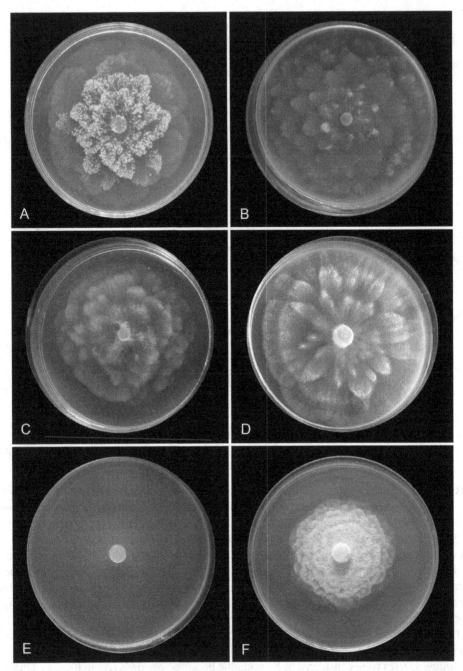

Figure 11.3. Different types of growth patterns of *Pythium* species on solid media. A. Rosette pattern; B. Chrysanthemum pattern; C. Median pattern (Rosette and Chrysanthemum); D. Radiate pattern; E. No pattern; F. Median pattern (Rosette and Radiate)

white, pale yellow, to even greyish purple/lilac. Growth rate is another important factor, separating isolates from each other. *Pythium* species, in general, usually grow faster than true fungi. However, there are some species with low growth rate such as *P. rostratum* Butler and *P. sulcatum* Pratt and Mitch. (Kageyama 2014). Moreover, cardinal temperature is another characteristic that taxonomists frequently conduct in order to describe a new species (Kageyama 2014).

Identification of *Pythium* species: Molecular approach

In the early 1980s, the first molecular research performed on *Pythium* was begun (Lipps 1980). Molecular identification of oomycete species was achieved through the application of polymerase chain reaction (PCR) with primers for amplifying housekeeping genes which reflected the species diversity existing within the *Pythium* genus. Primers used in molecular identification of *Pythium* species, include regions within the ribosomal DNA such as the internal transcribed spacer (ITS), elongation factor 1α, β-tubulin (*Btub*), large subunit (LSU), and small subunit (SSU) of rRNA gene, and also NADH dehydrogenase subunit I, cytochrome c oxidase subunit 1 (*cox*1) and cytochrome c oxidase subunit 2 (*cox*2) located in the mitochondria (Dick et al. 1999, Villa et al. 2006, Belbahri et al. 2008, Morajelo et al. 2008, Bala et al. 2010, Hyde et al. 2014, Tangphatsornruang et al. 2016).

The main region that has been utilized to study oomycetes is nuclear rDNA. The D2 domain of the large nuclear ribosomal subunit (LSU) seems to be the first region which was analyzed for identification of *Pythium* species (Briard et al. 1995). However, it was later discovered that LSU lacks sufficient resolution between species (Robideau et al. 2011). As a result, it is believed that LSU is better suited for genus- and family-level relationships' studies in oomycetes (Voglmayr and Riethmüller 2006, Robideau et al. 2011, Hyde et al. 2014). Using the small nuclear ribosomal subunit primers, Dick et al. (1999) sequenced 14 species of the oomycetes, concluding that divergences within *Phytophthora* and *Pythium* lines exist. Although the genus *Pythium* seemed to be paraphyletic in their study, they left the established genus as it was. de Cock and Lévesque (2004) have also successfully used the large subunit of the ribosomal DNA containing the highly divergent regions D1-D3, as a molecular marker for *Pythium* species identification.

The ITS region of rDNA is considered as the most extensively used for identification and systematic studies in mycology, mainly of the universal amplification of a highly variable region through all taxa, such as oomycetes (Lévesque and de Cock 2004). This amplification is due to the development of PCR primers by White et al. (1990). Analysis of the non-coding regions (i.e. ITS1 and ITS2), which were found between coding regions (i.e. 18S, 28S and 5.8S), have been conducted to study the intragenic relationships within *Pythium* species and intergenic relationships among *Pythium* and *Phytophthora* species (Matsumoto et al. 1999, Cooke et al. 2000, Matsumoto et al. 2000, Villa et al. 2006). In 1999, a phylogenetic analysis of 30 known *Pythium* species revealed that the ITS region could be a great target in designing species-specific primers to identify and detect *Pythium* species (Matsumoto et al. 1999). Wang et al. (2003) designed species-specific primers based on ITS1 region of 34 *Pythium* species, among which five primer pairs were applicable for their corresponding species. Several other studies have proven that the ITS region is useful to other investigation such as genetic variation and phylogeny of *Pythium* species, and to distinguish *Pythium* flora in greenhouse crops (Moorman et al. 2002, Schurko et al. 2003, Kageyama et al. 2007, Bala et al. 2010, Kageyama 2014).

DNA barcoding seems to be a rapid and reliable tool for the identification of species, which enables researchers to unravel the diversity and distribution of oomycetes (Schroeder et al. 2013). The ITS of rDNA and cytochrome c oxidase subunit 1 (*cox*1) of mtDNA were suggested as standard DNA barcode markers for oomycetes by Robideau et al. (2011). These authors believed that because *cox*1 is a protein-coding region, its sequence alignment might be easier and accurate species identification might be possible. However, several problems have arisen since these loci's choice. The lack of backbone studies for generating a comprehensive ITS database for some genera and insufficient variability in ITS for phylogenetic distinction in closely related species were the most important issues regarding this choice. Due to these limitations, additional genes, *cox*1 and *cox*2, have been proposed as phylogenetic markers within oomycetes (Cook et al. 2001, Hudspeth et al. 2003). The *cox*1 locus has been used mainly in studies of *Pythium* and its sister taxon *Phytophthora* (de Cock and Lévesque 2004, Kroon et al. 2004, Bala et al. 2010, Robideau et al. 2011, Schroeder et al. 2013), while the *cox*2 locus has been extensively applied in phylogenetic studies of downy mildews (Testen et al. 2014), and also in *Pythium* spp. (Martin 2000, Villa et al. 2006, Senda et al. 2009, Uzuhashi et al. 2010), *Phytophthora* spp. (Martin and Tooley 2003, Villa et al. 2006), white blister rusts (Mirzaee et al. 2013) and basal *Oomycetes* (Hulvey et al. 2010, Choi et al. 2015). Since the amplification of *cox*1 locus varies with different lineages and sample ages of various oomycetes except *Pythium* and *Phytophthora* species, its application to other oomycete groups is limited (Thines 2007, Thines et al. 2009). It has been suggested that *cox*1 shares a similar intraspecific variation with ITS; however, ITS provides greater interspecific variation comparing with *cox*1 (Robideau et al. 2011). Recent researches using molecular markers have demonstrated that the sequencing results constructed phylogenetic trees with 99% confidence level. This indicates that the molecular markers are able to effectively classify and identify *Pythium* species (de Cock and Lévesque 2004, Bala et al. 2010, Schroeder et al. 2013). The identification of a single isolate might be carried out via BLAST (https://blast.ncbi.nlm.nih.gov/Blast.cgi) search of the obtained sequences. High homology of a given sequence with verified species' sequences suggests the existence of a same species. Researchers use DNA databases such as DDBJ, EMBL and GenBank to do a nucleotide sequence search. All these databases share all registered sequences. While using public databases for identification, it is of great importance that all of the reference sequences be checked and see if they come from phylogenetic studies in valid publications; otherwise, it might lead the identification to errors. There are also some available specific databases for molecular identification of oomycetes, such as Phytophthora Database (http://www.phytophthoradb.org/index.php) and the Phytophthora-ID website (http://phytophthora-id.org/) (Park et al. 2008, Grünwald et al. 2011). However, no *Pythium* database for specific identification has been constructed yet.

Phylogenetic trees are considered as an illustration of the relationship between *Pythium* species. The trees also serve as a backbone for creating DNA barcodes for identification of *Pythium* species. These trees can result from the combination of morphological and molecular studies of *Pythium* species. As a consequence, both morphological and the DNA barcoding are critical in accurate identification of *Pythium* species (de Cock and Lévesque 2004, Robideau et al. 2011, Schroeder et al. 2013).

Phylogeny of the genus *Pythium*

Advances in molecular technologies and studies on the genus *Pythium* gradually led to advances in phylogenetic of *Pythium* species (Schroeder et al. 2013, Voglmayr et al. 2014). Exerting various nuclear and mitochondrial regions, researchers have divided *Pythium* species into different groups (Hyde et al. 2014). In addition, the genus *Pythium* itself has been split into some other genera (Uzuhashi et al. 2010, 2016).

Applying partial sequencing of the mitochondrial cytochrome c oxidase subunit 2 gene (*cox*2), Martin (2000) studied the phylogenetic relationships between 24 species of the genus *Pythium*. Based on sporangial shape, three phylogenetic groups were formed. The first group contained species with filamentous sporangia, whereas the second and third groups comprised globose sporangia and formed two distinctive clades. Some species including *P. pulchrum* Minden, *P. rostratum* Butler and *P. nunn* did not cluster with any other clades. He presented six clusters in the genus *Pythium*, namely, I, II, IIIA, IIIB, IIIC and IIID. It was concluded that there was a correlation between sporangial shapes and clustering of species; however, an absolute correlation was not observed (Martin 2000).

Using ITS and LSU data as well as morphological characteristics, Lévesque and de Cock (2004) concluded that there were 11 clades (A – K) within the species of the genus *Pythium* (Lévesque and de Cock 2004). Unlike oogonial ornamentation and heterothallism, sporangial shape correlated very well with the major clades described by Leveque and de Cock (2004). They concluded that the clusters I, II, IIIA, IIIB, IIIC and IIID of Martin (2000) represent the clusters E – G, I, D, B, C and A in their study, respectively (Lévesque and de Cock 2004). Three *Pythium* species, which did not cluster with any other clades in Martin (2000) study, were located into clades E and J. *Pythium rostratum* and *P. pulchrum* clustered into clade E whereas *P. nunn* was located into clade J (Lévesque and de Cock 2004). Clade H members had not been identified in Martin's analyses (2000). These authors also stated that the species within clade K were appropriately classified in *Pythium*. This surprising conclusion has been reported previously, that the species in the clade K of *Pythium* have more characters in common with those of *Phytophthora* species (Briard et al. 1995, Panabières et al. 1997, Dick 2001). As a result, it was concluded that *Pythium* was a polyphyletic genus due to clade K, which was located between all *Pythium* species and the mostly non-marine *Phytophthora* species (Lévesque and de Cock 2004, Bala et al. 2010). Major morphological clustering of the genus *Pythium* is supported by molecular phylogenies, which separates *Pythium* species into two groups: a monophyletic group consisting of clades A to D and a monophyletic group formed by clades E to J *sensu* Lévesque and de Cock (2004). This classification is based on sporangial morphology. The first group is mainly characterized by the formation of filamentous sporangia while the second group forms globose sporangia. However, clade K is an exception, which has been reported to be phylogenetically distinct from the other 10 clades (Lévesque and de Cock 2004).

Based on molecular and morphological studies, clade K members reassigned as genus *Phytopythium* Abad, De Cock, Bala, Robideau, A.M. Lodhi and Lévesque (Bala 2010), based on its similarity to both *Pythium* and *Phytophthora* species. *Phytopythium* species have intermediate characteristics between *Pythium* and *Phytophthora* species, morphologically (Bala 2010, Marano et al. 2014, de Cock et al. 2015, Jesus et al. 2016, Ascunce et al. 2017). They have been reported to develop internal proliferation of sporangium like *Phytophthora* species do but their sporangia produce vesicles in a way which resembles *Pythium* species (Abdul-Baten et al. 2014, Marano et al. 2014, de Cock et al. 2015, Jesus et al. 2016). Based on nuclear (LSU, ITS and *Btub*) and mitochondrial (*cox*1 and *cox*2) genes, it has been proved that former clade K of *Pythium* is a sister group of the monophyletic *Phytophthora*-downy mildew clade (Briard et al. 1995, Villa et al. 2006, Belbahri et al. 2008, Uzuhashi et al. 2010, Robideau et al. 2011). However, no such results were true based on small subunit (SSU) of rDNA (de Cock et al. 2015). Hence, clade K of *Pythium* is regarded as *Phytopythium* so far; nevertheless, its evolutionary relationship with other members of *Peronosporales* should be resolved. It was confirmed that the genus *Phytopythium* is more related to *Phytophthora* and *Halophytophthora* (Briard et al. 1995, Villa et al. 2006, Belbahri et al. 2008, Uzuhashi et al. 2010, Abdul Baten et al. 2014, de Cock et al. 2015). Consequently, placement of this genus on the basal line of *Peronosporales* lineage suggested that zoospore formation of this clade is an ancestral feature, meaning that vesicle formation may have lost in the evolution (Uzuhashi et al. 2010).

Identification of *Pythium* species via molecular techniques and phylogenetic analyses might provide several evidences to split this genus (Ho et al. 2012). Uzuhashi et al. (2010) divided *Pythium sensu lato* into five major clades (1 to 5) based on LSU of rDNA D1/D2 and *cox*2 genes. They believe that *Pythium* species belonging to the clades E to J *sensu* Lévesque and de Cock (2004) should be transferred into two new genera, *Globisporangium* Uzuhashi, Tojo and Kakish (clade 4) and *Elongisporangium* Uzuhashi, Tojo and Kakish (clade 5), and other species located in clades A to D should remain as *Pythium* (clade 3), based on molecular and morphological traits. Besides, they mentioned that although all clade 4 members, which contained clades E, F, G, I and J were closely related, it was obvious that the members did not cluster as a monophyletic group (Uzuhashi et al. 2010). However, no phylogenetic support for any rearrangement of these clades was reported (de Cock et al. 2015). Reassignment of these clades has not been widely used. In other words, some papers might site clades A to J as *Pythium sensu lato* (Ascunce et al. 2017), while others might site them as *Pythium sensu stricto* (for clades A to D) and new genera (i.e. *Globisporangium, Elongisporangium, Ovatisporangium* Uzuhashi, Tojo and Kakish and *Pilasporangium* Uzuhashi, Tojo and Kakish) (Uzuhashi et al. 2016).

The genus *Ovatisporangium* was considered as a synonym of *Phytopythium* as it was recommended earlier (Uzuhashi et al. 2010). *Ovatisporangium* is referred to those species which mostly have ovoid sporangia with apical or lateral papilla, all of which produce discharged tubes and vesicles. This genus, which was located to clade 1 *sensu* Uzuhashi et al. (2010), corresponds with the species belonging to clade K *sensu* Lévesque and de Cock (2004). *Globisporangium* is referred to the species which have globose hyphal swellings or sporangia. It also corresponds with the species belonging to clades E-G, I and J, which were presented by Lévesque and de Cock (2004). Members of this genus have also been reported as important pathogens of various plants. For example, *G. oryzicola* Uzuhashi and Tojo has been reported as a rice pathogen (Uzuhashi et al. 2016). *Pilasporangium* refers to those species with globose and nonproliferating sporangia without any zoospore formation, which includes only one species (i.e. *P. apinafurcum* Uzuhashi and Tojo) (Uzuhashi et al. 2010). Its phylogenetic position has not been included in other studies (i.e. Lévesque and de Cock 2004, Robideau et al. 2011). Moreover, this species has been reported to be saprobe and parasite in aquatic and dry lands (Uzuhashi et al. 2010). The genus *Elongisporangium* produces very large elongated and clavate sporangia, which is not a common characteristic of *Pythium* species. It corresponds well with the species in the clade H *sensu* Lévesque and de Cock 2004. Uzuhashi et al. (2010) concluded that sporangial shape must be regarded as an important taxonomic criterion. Therefore, *Pythium* species were restricted to those producing filamentous sporangia. *Elongisporangium* species contain clavate to elongate sporangia, *Pilasporangium* species have non-proliferating sporangia and their secondary hyphae are branched complexly, and *Globisporangium* species have globose, proliferating sporangia (Uzuhashi et al. 2010, 2016).

Ribosomal small and large subunits have strongly suggested that *Pythium* has paraphyletic nature (Schroeder et al. 2013). Phylogeny of the genus *Pythium* based on ITS sequences reveals that divergence occurs within this genus according to sporangia types (Bala et al. 2010). Researches have exposed that the globose type is likely to be ancestral because both outgroup species and the species in the outmost *Pythium* cluster produce globose sporangia (Bala et al. 2010). Clade D *sensu* Lévesque and de Cock (2004) with similar ITS sequences include both globose and filamentous types of sporangia (Bala et al. 2010). Clade definition is also affected by taxonomic characteristics such as homothallism and heterothallism, and oogonium ornamentation. The existence of oogonium ornamentation in six groups within clades F and G is an example to prove this claim (Bala et al. 2010). However, many authors have concluded that there is no correlation between evolutionary patterns and phylogenetic analysis based on morphological features (Martin and Tooley 2003, de Cock and Lévesque 2004, Tambong et al. 2006, Bala et al. 2010).

Challenges in morphological identification of the genus *Pythium*

Understanding the precise biology of *Pythium* species and characterizing the evolutionary relationships among them necessitates the accurate identification of the isolates. Although molecular tools have facilitated the identification process, application of morphological traits is still inevitable. Classifying the species of *Pythium* has always given rise to problems due to various reasons, such as difficulty in isolating certain species and the lack of identification data for species (van der Plaats-Niterink 1981, Zitnick-Anderson 2013). Another problem many researchers have had since *Pythium* was first described has been the identification of the morphological features of the various species (Bala et al. 2010). As a consequence, identification of *Pythium* species based only on morphological characteristics is quite difficult. This is mainly due to the low number of typical isolates, non-homothallic species, sexually sterile isolates, the existence of similar morphological features among different species groups and considerable fluctuations in sexual and asexual structures' size and shape within species (Hendrix and Papa 1974, Ali-Shtayeh 1985, Martin and Kistler 1990, Klassen et al. 1996, Barr et al. 1997, Kageyama 2014). While morphological classification of the genus *Pythium* has been used as a strong tool to plant pathologists traditionally, it has been proven that several morphological species are polyphyletic assemblages (Villa et al. 2006).

Phylogenetic studies have reported that the diversity of oomycetes has been mostly underestimated (Broders et al. 2007). Researchers' knowledge of oomycete diversity is mainly biased to the economically relevant plant pathogens, and little is known about saprobic or marine species (Hulvey et al. 2010, Nigrelli and Thines 2013). Therefore, the number of oomycetes is still only about one-tenth of the expected species number (Hawksworth 2001). Recovering both pathogen and saprophyte *Pythium* species is one of the major concerns of taxonomists (Broders et al. 2007). *Pythium* species are generally difficult to isolate from soil or plant material by the usual nutrient agar techniques because these media favour the more competitive, ubiquitous saprophytes. While these fungi can be readily isolated from plant material using water agar or corn meal agar, they were usually isolated only infrequently prior to the development of selective media (Kageyama 2014).

Pythium species can be grouped according to their growth habit. However, since multiple strains of an individual species may vary and species with unequal morphological characteristics might have similar growth habit, specific identification of species based only on growth habit would be by no means of the method of specific species identification (Zitnick-Anderson 2013).

Limited number of morphological features suitable to identify and describe *Pythium* species is one of the major concerns of researchers (Lévesque and de Cock 2004). Many studies have been conducted on the morphology of various *Pythium* species. Although morphology is the most valuable criterion for oomycete identification and diversity, morphological species identification is time-consuming and very difficult. However, accurate taxonomic identification of *Pythium* species based only on their morphological traits has always been problematic to taxonomists, which is mainly because of the absence of definite structures (Bala et al. 2010). For instance, *P. aphanidermatum* and *P. deliense* Meurs are two *Pythium* species exhibiting very similar characteristics (McLeod et al. 2009). Moreover, variations have been observed even between the isolate of a single species, which led to inconclusive results. Ornamentation of oogonia and heterothallism has been traditionally used to separate *Pythium* species. These characteristics might be acquired or lost easily through evolution (Matsumoto et al. 1999, Martin 2000, Lévesque and de Cock 2004). Besides, such features did not correlate with major clades of *Pythium* species (Uzuhashi et al. 2010).

Many *Pythium* species can have multiple variations of a specific morphological feature within a single species, which is called pleomorphism (van der Plaats-Niterink 1981, Zitnick-Anderson 2013). For instance, pleomorphism had been observed in *P. deliense,* in which the antheridia can be in the monoclinous, diclinous, intercalary, or terminal positions (Van der Pläats-Niterink 1981). Other examples of this phenomenon are: *P. adhaerens* Sparrow with both terminal and intercalary oogonia, *P. anadrum* Drechsler with both monoclinous and diclinous antheridia as well as unisporous and multisporous oogonia, *P. catenulatum* V. D. Matthews with both terminal and intercalary oogonia as well as monoclinous, diclinous, clavate and crook-necked antheridia, *P. hydnosporum* (Mont.) J. Schröt. and *P. mastophorum* Drechsler with both plerotic and aplerotic oospore, *P. hypogynum* Middleton with both terminal and intercalary sporangia, and *P. multisporum* with subglobose, globose, oblong and limoniform sporangia as well as both monoclinous and diclinous antheridia (van der Plaats-Niterink 1981). Although rare, the shape of the oogonium can be smooth or ornamented in some *Pythium* species such as *P. heteroogonium* Mostowfizadeh-Ghalamfarsa and Salmaninezhad, *P. irregulare* Buisman*,* and *P. carbonicum* B. Paul (Middleton 1943, van der Plaats-Niterink 1981, Salmaninezhad and Mostowfizadeh-Ghalamfarsa 2019). Moreover, some morphological features may change during different environmental conditions. For example, high temperatures (above 33°C) may result in abnormal oogonium production, decrease in the number of ornamented oogonia and increase in the number of diclinous antheridia in *P. irregulare* (van der Plaats-Niterink 1981, Mostowfizadeh-Ghalamfarsa 2016). For these reasons identifying *Pythium* species based on morphological traits has been a persistent problem for even the most experienced mycologists (de Cock and Lévesque 2004).

Species complex is another major concern of taxonomists. For instance, with examining more than 10,000 *Pythium* isolates from 2100 soil samples, Hendrix and Campbell (1970) concluded that *P.*

debaryanum Hesse *sensu* Middleton and *P. irregulare*, the most abundant species in the literature till 1973, indeed formed a species complex with distinct features only at the extremes but which merged at a median point. Other species likewise proved better subjects for species complexes than pointless exercises attempting to determine which of the several names to use. Besides, environmental and laboratory conditions might have considerable influence on morphology. This has been shown by Biesbrock and Hendrix (1967) that how morphological features of the oogonia, antheridia, and sporangia varied with differences in temperature, light, and nutrient media in a taxonomic study of *P. irregulare.*

Working with *Pythium* spp. isolates, one might encounter some obstacles during the course of species identification, most of which are in a practical aspect. Some of these complications were related to the biology of these microorganisms and some of them had something to do with the scientific tools and software availability. Most of the *Pythium* spp. isolates need sterols to produce sexual or asexual reproduction organs, such as zoosporangia and oospores in artificial media. It is possible to add some phytosterols in the form of sterol-rich plant materials such as hemp seed extract or pure sterols, for instance, β-sitosterol, into media (Mostowfizadeh-Ghalamfarsa 2015). However, for some isolates, it takes a long time to produce any sexual or asexual organs and some of them never produce anything but a mycelial mat of coenocytic hyphae. Additionally, some sexual organs, such as antheridia may disappear soon after fertilization. This makes the culture observation a time-consuming and laborious task. As it was discussed previously, some *Pythium* species produce more than one type of sporangia or antheridia. To avoid any confusion related to this pleomorphism, the cultures must be absolutely pure.

There is a high level of morphological overlapping among convergent species, especially phylogenetically related ones, which makes them an identification challenge. It is not that easy to find a compatible mating type for a heterothallic *Pythium* isolates to stimulate oospore production. Therefore, the identification must be solely based on asexual morphology, which is error-prone. The biology of *Pythium* species is not the only issue in the course of identification and there are several software-based concerns which have something to do with the accessible tools for the identification. The most comprehensive *Pythium* species identification key available (van der Plaats-Niterink 1981) only covers 120 out of 250 reported species. Additionally, the latest identification key (Dick 1990) is almost 25 years old. On the other hand, there are also no descriptive sheets, no web-based database and no molecular barcode metadatabase for *Pythium* species. If a molecular identification is the approach of choice, species-specific primers are designed for only around 20 species (http://sppadbase.ipp.cnr.it/), most of which are developed for plant pathogens. There were some attempts for generating web-based interactive keys based on Lucid Builder platform (Moorman et al. 2014). Nevertheless, it is a modification of van der Plaats-Niterink (1981) identification key and not all the species in the original key are included.

Challenges in molecular identification of the genus *Pythium*

The ITS region of the rDNA has exclusively been used for *Pythium* species identification. This is mainly due to the availability of a vast number of sequences in public databases. Besides, it is easy to be amplified for DNA sequencing in most oomycete species with the application of universal PCR primers (Robideau et al. 2011). Also, interspecific variation level could be observed using ITS. Recent studies have proven that the ITS regions of 5.8S gene varied from 750 to 1050 bp, which is much longer than the usual 300 to 700 bp of Eumycota. Lévesque and de Cock (2004) stated that the longer length would present more technical challenges for PCR products' direct sequencing. But it also might provide more characters for analysis (Lévesque and de Cock 2004). Because of the apparent absence of functional limitations on this untranslated region of rDNA, large amounts of indels (i.e. insertions and deletions), which could be even observed within a single strain, might prevent the alignment of ITS sequences (Robideau et al. 2011). Unclear, heterogeneous sequences can be obtained for the rDNA ITS region. This is especially true for heterothallic species due to the diploid nature of *Pythium* species and situation of rDNA sequences on several multiple copied chromosomes (Martin 1995, Kageyama et al. 2007, Kageyama 2014). A possible solution to this matter might be the cloning of PCR products before performing sequencing.

Because of the low level of sequence divergence among phylogenetically related species and intraspecific variation, ITS region would not be the ultimate choice of some species (Schroeder et al. 2013). For this reason, some alternative nuclear (e.g. β-tubulin, and LSU of the rDNA) and mitochondrial (e.g. *cox1, cox2*, and NADH dehydrogenase subunit 1) loci have been proposed (Martin 2000, Lévesque and de Cock 2004, Villa et al. 2006, Morajelo et al. 2008, Robideau et al. 2011). Although using these regions might help for more accurate identification, this might prompt researchers with less experiences in oomycete taxonomy to sequence genes that might not resolve some species (e.g. LSU) or for which there is limited reference data available (e.g. NADH dehydrogenase subunit 1). Application of *cox1* and *cox2* as mitochondrial markers should be with regard to their maternal evolution, due to which they might produce incongruent phylogenies (Hyde et al. 2014). Although *Btub* region can be amplified and sequenced in most of *Pythium* species and can be easily used in concatenated datasets, it is not a powerful marker to resolve species-level phylogenies (Hyde et al. 2014).

Some *Pythium* species might need more than one locus (e.g. both ITS and *cox1*) to be evaluated for accurate identification (Robideau et al. 2011, Hyde et al. 2014). However, there have been several reports concluding that some species can't be differentiated even with the comparison of both ITS and *cox1* sequences (Robideau et al. 2011, Hyde et al. 2014). There are many examples of such species including *P. myriotylum* Drechsler and *P. zingiberis* Takah, *P. aristosporum* Vanterp and *P. arrhenomanes, P. amasculinum* Yu, *P. hydnosporum, P. lycopersicum* G. Karaca, Teped. and B. Paul and *P. ornamentum, P. folliculosum* B. Paul and *P. torulosum* Coker and P. Patt., *P. conidiophorum* Jokl and *P. salpingophorum* Drechsler, *P. debaryanum* and *P. viniferum* Paul, *P. irregulare, P. cryptoirregulare* Garzón, Yánez and G.W. Moorman and *P. cylindrosporum* Paul, *P. acrogynum* Yu and *P. hydpogynum* Middleton, *P. erinaceus* Robertson and *P. ornacarpum* Paul, *P. minus* Ali-Shtayeh and *P. pleroticum* Takesi Itô, and *P. dimorphum* Hendrix and Cambp. and *P. undulatum* Petersen (Robideau et al. 2011). Some would suggest that such species might have the chance to be formally synonymized using multiple hypervariable genes in further investigations (Hyde et al. 2014). Furthermore, such approach might be useful to resolve species complexes such as *P. irregulare, P. paroecandrum* Drechsler, *P. cylindrosproum* Paul, *P. cryptoirregulare* and *P. mamillatum* Meurs (Barr et al. 1997, Matsumoto et al. 2000, Spies et al. 2011, Hyde et al. 2014). As a consequence, it might be possible to use a single locus for identification; however, it is crucial to examine two or more loci before identifying or describing a new species (Schroeder et al. 2013). The possible existence of multiple forms within the same isolate is a matter of importance, especially regarding rDNA region. This phenomenon has been reported for both IGS and ITS region of *Pythium* (Buchko and Klassen 1990, Klassen and Buchko 1990, Martin 1990, 1995, Uzuhashi et al. 2009, Schroeder et al. 2013).

Despite the fact that rDNA has been widely used in phylogenetic studies, it should not be concluded that the evolution of one gene might represent the evolution of the entire genome (Shen 2001, Villa et al. 2006). As a result, separate sampling of many possible independent genes and their comparison with each other to study whether they support each other or not is important (Villa et al. 2006, Robideau et al. 2011). This approach, based on multiple gene genealogy concordance (Taylor et al. 2000), is receiving much attention nowadays. For example, using multiple markers (ITS, *cox2* and *Btub*), Villa et al. (2006) confirmed the intermediate evolutionary position of clade K species between *Pythium* and *Phytophthora*. They also suggested that clade H species occupy an intermediate position (Villa et al. 2006). This conclusion was, however, in contrast with the suggested position of this clade nestled among clades E, F, G, I, and J by Lévesque and de Cock (2004). Although Hyde et al. (2014) used multimarker phylogeny (18S-ITS-28S, *cox2* and β-tubulin) of 152 *Pythium* species, they could not provide sufficient support for the evolutionary association of clade H with any of the other recognized groups within *Pythium* (Hyde et al. 2014).

Some researches have revealed that species with different kinds of morphological characteristics may have similar DNA sequences, which suggests that those species are actually related (de Cock and Lévesque 2004). For instance, both *P. perplexans* and *P. mastophorum* are located in the same clade, meaning molecularly they are very similar. However, from the point of view of morphological features, *P. mastophorum* has ornamented oogonium whereas *P. perplexans* has smooth wall oogonium (de Cock and Lévesque 2004). Submission of a *Pythium* species into GenBank database for molecular identification might result in matching with multiple species (Schroeder et al. 2013). This is mainly due to the submission of erroneous data entered into GenBank or the regions being similar.

Some other molecular techniques have also been employed for the identification of *Pythium* species. Kageyama et al. (2002) once detected *P. sulcatum* (a slow-growing species) from soils and plant tissues using serological methods. Although serological methods can be used for the identification of some microorganisms such as bacteria, due to the production of a species-specific antibodies' difficulty, the availability of these methods for *Pythium* detection is limited (Kageyama 2014).

All these taxonomic challenges may contribute to frequent errors in accurate identification of species especially for those lacking adequate experience working with this genus (Lévesque and de Cock 2004). Although molecular techniques have significantly assisted in the identification of unknown *Pythium* species, morphological features are still essential in supporting the identifications defined by molecular techniques (Zitnick-Anderson 2013). On the other hand, in order to describe a new species, *Pythium* taxonomy based on morphological characteristics has been increasingly supplemented with molecular techniques (Paul 2003).

Possible strategies for resolving taxonomic challenges in the genus *Pythium*

Since pathologists encounter *Pythium* species only in culture, when attempting identification, laboratory environment requires careful standardization for comparative studies (Benfradj et al. 2017). Considering both morphological and molecular identification methods and their advantages and defects, it is extremely recommended to use both morphological and molecular methods for an accurate identification of *Pythium* species (Benfradj et al. 2017, Naznin et al. 2017).

It has been suggested that a researcher should take extreme cautions using DNA databases for the identification of a single species. For instance, one should verify that the entire sequence (not a part of it) has high homology with other verified sequences in a database, when a given sequence shows high homology with one in the database (Kageyama 2014). Besides, confirmation of a verified species in a database is a matter of importance, especially when the list of homologous sequences shows more than one species. In general, it is believed that if a sequence was used in the DNA barcoding of Robideau et al. (2011) (i.e. sequences with "voucher" in front of the isolate name), it is more reliable (Kageyama 2014). Moreover, checking the known sequence variability in any species showing homology with the query sequence is a crucial step in molecular species identification. The phylogenetic analysis would be helpful to show the possible relationships between a given query isolate and its related species in a database.

Based on multiple gene genealogy studies, the application of SSU and LSU nuclear rRNA genes for generic level phylogenies within *Pythium sensu lato*, the application of ITS and *cox2* for sub-generic, inter- and intra-specific level phylogenies, and the application of ITS and *cox1* for non-phylogenetic species identification are recommended as genetic markers for accurate identification of *Pythium* (Villa et al. 2006, Uzuhashi et al. 2010, Robideau et al. 2011, Hyde et al. 2014, Uzuhashi et al. 2016). Resolving the systematic status of the *Pythium* genus and concordance between taxonomists on keeping *Pythium* as a single genus or splitting the genus into four separate genera, would deeply help those who are involved in *Pythium* identification, especially plant pathologists, and prevent any confusion.

The online Lucid Key for *Pythium* species (https://keyserver.lucidcentral.org/key-server/player.jsp?keyId=121) (Moorman et al. 2014), which does not contain any species described since 1981, needs to be updated. Finally, constructing a molecular identification database, at least for pathogenic species, to foster accurate and faster identification of the species would be an important step for tackling the challenges of *Pythium* taxonomy.

Conclusion

The genus *Pythium* includes a number of recognized species with wide distributions and host ranges (Majeed et al. 2018). Identification of *Pythium* species traditionally has been conducted based on morphological features such as sporangia, oogonia, and antheridia, the type and size of oospores, homothallism *vs.* heterothallism, growth habit, and growth rate in culture media. These features can also vary under different cultural conditions and many species show similar morphological characters. Some of these characteristics can also change or be acquired or lost easily. The ITS region of the nuclear rDNA has been established to be variable at the family, genus, and species level for *Pythium* (Ashwathi et al. 2017). Application of the ITS region seems to be the most popular choice of many researchers working with this genus (Ashwathi et al. 2017). However, it is recommended to use more than one gene to describe a new species (Robideau et al. 2011) and application of both ITS and *cox1* genes might be researchers' ultimate choice of species-specific identification (Robideau et al. 2011, Hyde et al. 2014). Convergent morphological features as well as molecular similarities existing between two distinct *Pythium* species urge taxonomists to adopt both morphological and molecular identification for accurate species identification.

References

Abad, G., Shew, H.D., Grand, L.F. and Lucas, L.T. 1995. A new species of *Pythium* producing multiple oospores isolated from bentgrass in North Carolina. Mycologia 87: 896-901.

Abdul Baten, M.D., Asano, T., Motohashi, K., Ishiguro, Y., Ziaur Rahman, M., Inaba, S., Suga, H. and Kageyama, K. 2014. Phylogenetic relationships among *Phytopythium* species, and re-evaluation of *Phytopythium fagopyri* comb. nov. recovered from damped-off buckwheat seedlings in Japan. Mycol. Prog. 13: 1145-1156.

Abdul Baten, M.D., Mingzhu, L., Motohashi, K., Ishiguro, Y., Rahman, M.Z., Suga, H. and Kageyama, K. 2015. Two new species, *Phytopythium iriomtense* sp. nov. and *P. aichiense* sp. nov., isolated from river water and water purification sludge in Japan. Mycol. Prog. 14: 2-13.

Agrios, G.N. 2005. Plant diseases caused by fungi. pp. 385-614. *In*: Agrios, G.N. (ed.). Plant Pathology. Elsevier Science, The Netherlands.

Ali-Shtayeh, M.S. 1985. Value of oogonial and oospore dimentions in *Pythium* species differentiation. T. Brit. Mycol. Soc. 85: 781-784.

Ascunce, M.S., Huguet-Tapia, J.C., Ortiz-Urquiza, A., Keyhani, N.O., Braun, E.L. and Goss, E.M. 2017. Phylogenomic analysis supports multiple instances of polyphyly in the oomycete peronosporalean lineage. Mol. Phylogenet. Evol. 114: 199-211.

Ashwathi, S., Ushamalini, C., Parhasaranthy, S. and Nakkeeran, S. 2017. Morphological, pathogenic and molecular characterization of *Pythium aphanidermatum*: A casual pathogen of coriander damping-off in India. Pharma Innovation J. 6: 44-48.

Bala, K. 2010. *Phytopythium* Abad, de Cock, Bala, Robideau, Lodhi and Levesque, gen. nov. and Phytopythium sindhum Lodhi, Shahzad and Levesque, sp. nov. Persoonia 24: 136-137.

Bala, K., Robideau, G.P., Désaulniers, N., de Cock, A.W.A.M. and Lévesque, C.A. 2010. Taxonomy, DNA barcoding and phylogeny of three new species of *Pythium* from Canada. Persoonia 25: 22-31.

Baldauf, S.L., Roger, A.J., Wenk-Siefert, I. and Doolittle, W.F. 2000. A kingdom-level phylogeny of eukaryotes based on combined protein data. Science 290: 972-977.

Barr, D.J.S., Warwick, S.I. and Desaulners, N.L. 1997. Isozyme variation in heterothallic species and related asexual isolates of *Pythium. Can. J. Bot.* 75: 1927-1935.

Belbahri, L., McLeod, A., Paul, B., Calmin, G., Morajelo, E., Spies, C.F., Botha, W.J., Clemente, A., Descals, E., Sanchez-Hernandez, E. and Lefort, F. 2008. Intraspecific and within-isolate sequence variation in the ITS rRNA gene region of *Pythium mercuriale* sp. nov. (Pythiaceae). FEMS Microbiol. Lett. 284: 17-27.

Benfradj, B., Migliorinim D., Luchi, N., Santini, A. and Boughalleb-MHamdi, N. 2017. Occurrence of *Pythium* and *Phytophthora* species isolated from citrus trees infected with gummosis disease in Tunisia. Arch. Phytopathol. Plant Protect. 50: 286-302.

Bennett, R.M., Nam, B., Dedeles, G.R. and Thines, M. 2017. *Phytopythium leanoi* sp. nov. and *Phytopythium dogmae* sp. nov., *Phytopythium* species associated with mangrove leaf litter from the Philippines. Acta Mycologia 52: 1103-1116.

Biesbrock, J.A. and Hendrix, F.F. 1967. A taxonomic study of *Pythium irregulare* and related species. Mycologia 59: 943-952.

Briard, M., Dutertre, M., Rouxel, F. and Brygoo, Y. 1995. Ribosomal RNA sequence divergence within the Pythiaceae. Mycol. Res. 99: 1119-1127.

Broders, K.D., Lipps, P.E., Paul, P.A. and Dorrance, A.E. 2007. Characterization of *Pythium* spp. associated with corn and soybean seed and seedling disease in Ohio. Plant Dis. 91: 727-735.

Buchko, J. and Klassen, G.R. 1990. Detection of length heterogeneity in the ribosomal DNA of *Pythium ultimum* by PCR amplification of the intergenic region. Curr. Genet. 18: 203-205.

Caballero, J.R.I. and Tisserat, N.A. 2016. Transcriptome and secretome of two *Pythium* species during infection and saprophytic growth. Physiol. Mol. Plant Pathol. 99: 41-54.

Campbell, W.A. and Hendrix, F.F. 1967. A new heterothallic *Pythium* from Southern United States. Mycologia 59: 274-278.

Carmona, M.A., Sautua, F.J., Grijalba, P.E., Cassina, M. and Pérez-Hernández, O. 2017. Effect of potassium and manganese phosphites in the control of *Pythium* damping-off in soybean: A feasible alternative to fungicide seed treatment. Pest Manag. Sci. 74: 366-374.

Cavalier-Smith, T. and Chao, E.E. 2006. Phylogeny and megasystematics of phagotrophic heterokonts (kingdom Chromista). J. Mol. Evol. 62: 388-420.

Choi, Y.J., Beakes, G., Glockling, S., Kruse, J., Nam, B., Nigrelli, L., Ploch, S., Shin, H., Shivas, R.G., Telle, S., Volgmayr, H. and Thines, M. 2015. Towards a universal barcode of oomycetes – A comparison of the *cox*1 and *cox*2 loci. Mol. Ecol. Resour. 15: 1275-1288.

Cook, K.L., Hudspeth, D.S.S. and Hudspeth, M.E.S. 2001. A cox2 phylogeny of representative marine peronosporomycetes (Oomycetes). Nova Hedwigia, Beiheft 122: 231-243.

Cooke, D.E.L., Drenth, A., Duncan, J.M., Eagels, G. and Brasier, C.M. 2000. A molecular phylogeny of *Phytophthora* and related oomycetes. Fungal Genet. Biol. 30: 13-32.

Czeczuga, B., Mazalska, B., Godlewska, A. and Muszynska, E. 2005. Aquatic fungi growing on dead fragments of submerged plants. Limnologica 35: 283-297.

de Cock, A.W.A.M. and Lévesque, C.A. 2004. New species of *Pythium* and *Phytophthora*. Stud. Mycol. 50: 481-487.

de Cock, A.W.A.M., Lodhi, A.M., Rintoul, T.L., Bala, K., Robideau, G.P., Abad, Z.G., Coffey, M.D. and Lévesque, C.A. 2015. *Phytopythium*: Molecular phylogeny and systematics. Persoonia 34: 25-39.

Densmore, C.L. and Green, D.E. 2007. Diseases of amphibians. Ilar J. 43: 235-254.

Dick, M.W. 1990. Keys to Pythium. University of Reading, Reading, UK.

Dick, M.W., Vick, M.C., Gibbings, J.G., Hedderson, T.A. and Lastra, C.C.L. 1999. 18S rDNA for species of *Leptolegnia* and other Peronosporomycetes: Justification for the subclass taxa Saprolegniomycetidae and Peronosporomycetidae and division of the Saprolegniaceae *sensu lato* into the Leptolegniaceae and Saprolegniaceae. Mycol. Res. 103: 1119-1125.

Dick, M.W. 2001. Straminipilous Fungi. Kluwer Academic Publishers, The Netherlands.

Diehl, N., Kim, G.H. and Zuccarello, G. 2017. A pathogen of New Zealand *Pyropia plicata* (Bangiales, Rhodophyta), *Pythium porphyrae* (Oomycota). Alga 32: 29-39.

Donaldson, S.P. and Deacon, J. 1993. Effects of amino acids and sugars on zoospore taxis, encystment and cyst germination in *Pythium aphanidermatum* (Edson) Fitzp., *P. catenulatum* Matthews and *P. dissotocum* Drechs. New Phytol. 123: 289-295.

Drechsler, C. 1940. Three species of *Pythium* associated with root rot. Phytopathology 30: 189-213.

Drechsler, C. 1943. Two species of *Pythium* occurring in southern states. Phytopathology 33: 261-299.

Drechsler, C. 1946. Several species of *Pythium* peculiar in their sexual development. Phytopathology 36: 781-864.

Frezzi, M.J. 1956. Especies de Pythium fitopatogenas identificadas en la Reublica Argentina. J. Agric. Res. 10: 113-241.

Giroux, E. 2017. Using RNA-Seq to identify oospore wall specific Carbohydrate-Active Enzyme (CAZyme) coding genes of *Pythium ultimum* var. ultimum, an oomycete plant pathogen. Doctoral dissertation, Carleton University, Ottawa.

Grijalba, P.E., Palmucci, H.E. and Guillin, E. 2017. Identification and characterization of *Pythium graminicola,* causal agent of kikuyu yellows in Argentina. Trop. Plant Pathol. 42: 284-290.

Grünwald, N.J., Martin, F.N., Larsen, M.M., Sullivan, C.M., Press, C.M., Coffey, M.D., Hansen, E.M. and Parke, J.L. 2011. Phytophthora-ID.org: A sequence-based *Phytophthora* identification tool. Plant Dis. 95: 337-342.

Hawksworth, D.L. 2001. The magnitude of fungal diversity: The 1.5 million species estimate revisited. Mycol. Res. 105: 1422-1432.

Hendrix, F.F. and Campbell, W.A. 1973. Pythiums as plant pathogens. Ann. Rev. Phytopathol. 11: 77-98.

Hendrix, F.F. Jr. and Campbell, W.A. 1970. Distribution of *Phytophthora* and *Pythium* species in soils in the continental United States. Can. J. Bot. 48: 377-384.

Hendrix, F.F. Jr. and Papa, K.E. 1974. Taxonomy and genetics of *Pythium. Proc. Am. Phytopathol. Soc.* 1: 200-207.

Ho, H.H., Chen, X.X., Zeng, H.C. and Zheng, F.C. 2012. The occurarence distribution of *Pythium* species in Hainan island of south China. Bot. Stud. 53: 525–534.

Ho, H.H. 2013. The genus *Pythium* in mainland China. Mycosystema 32: 20-44.

Hudspeth, D.S.S., Stenger, D. and Hudspeth, M.E.S. 2003. A *cox*2 phylogenetic hypothesis for the downy mildew white rusts. Fungal Divers. 13: 47-57.

Hulvey, J., Telle, S., Nigrelli, L., Lamour, K. and Thines, M. 2010. Salisapiliaceae – A new family of oomycetes from marsh grass litter of southeastern North America. Persoonia 25: 109-116.

Hwang, E.K., Park, C.S. and Kakinuma, M. 2009. Physiochemical responses of *Pythium porphyrae* (Oomycota), the causative organism of red rot disease in Porphyra to acidification. Aquac. Res. 40: 1777-1784.

Hyde, K.D., Nilsson, R.H., Alias, S.A., Ariyawansa, H.A., Blair, J.E., Cai, L., de Cock, A.W.A.M., Dissanayake, A.J., Glockling, S.L., Goonasekara, I.D., Gorczak, M., Hahn, M., Jayawardena, R.S., van Kan, J.A.L., Laurence, M.H., Levesque, C.A., Li, X., Liu, J., Maharachchikumbura, S.S.N., Manamgoda, D.S., Martin, F.N., McKenzie, E.H.C., McTaggart, A.R., Mortimer, P.E., Nair, P.V.R., Pawlowska, J., Rintoul, T.L., Shivas, R.G., Spies, C.F.J., Summerell, B.A., Taylor, P.W.J., Terhem, R.B., Udayanga, D., Vaghefi, N., Walther, G., Wilk, M., Wrzosek, M., Xu, J. and Yan, J. 2014. One stop shop – Backbones trees for important pathogenic genera: I 2014. Fungal Divers. 67: 21-125.

Jesus, A.L., Gonçalves, D.R., Rocha, S.C.O., Marano, A.V., Jerônimo, G.H., De Souza, J.I., Boro, M.C. and Pires-Zottarelli, C.L.A. 2016. Morphological and phylogenetic analyses of three *Phytopythium* species (Peronosporales, Oomycota) from Brazil. Cryptogamie Mycol. 37: 117-128.

Johnson, T., Seymour, R. and Padgett, D. 2002. Biology and Systematics of the Saprolegniaceae. University of North Carolina at Wilmington, Wilmington, NC.

Jung, T. and Burgess, T.I. 2009. Re-evaluation of *Phytophthora citricola* isolates from multiple woody hosts in Europe and North America reveals a new species, *Phytophthora plurivora* sp. nov. Persoonia 22: 95-110.

Kageyama, K., Kobayashi, M., Tomita, M., Kubota, N., Suga, H. and Hyakumachi, M. 2002. Production and evaluation of monoclonal antibodies for the detection of *Pythium sulcatum* in soil. J. Phytopathol. 150: 97-104.

Kageyama, K., Senda, M., Asano, T., Suga, H. and Ishiguro, K. 2007. Intra-isolate heterogeneity of the ITS region of rDNA in *Pythium helicoides*. Mycol. Res. 111: 416-423.

Kageyama, K. 2014. Molecular taxonomy and its application to ecological studies of *Pythium* species. J. Gen. Plant Pathol. 80: 314-326.

Kawamura, Y., Yokoo, K., Tojo, M. and Hishiike, M. 2005. Distribution of *Pythium porphyrae*, the causal agent of red rot disease of *Porphyrae* spp., in the Ariake sea, Japan. Plant Dis. 89: 1041-1047.

Klassen, G.R. and Buchko, J. 1990. Subrepeat structure of the intergenic region in the ribosomal DNA of the oomycetous fungus *Pythium ultimum*. Curr. Genet. 17: 125-127.

Klassen, G.R., Balcerzak, M. and de Cock, A.W.A.M. 1996. 5S ribosomal RNA gene spacers as species-specific probes for eight species of *Pythium*. Phytopathology 86: 581-587.

Klochkova, T.A., Jung, S. and Kim, G.H. 2017. Host range and salinity tolerance of *Pythium porphyrae* may indicate its terrestrial origin. J. Appl. Phycol. 29: 371-379.

Kroon, L.P.N.M., Bakker, F.T., Bosch, G.B., Bonants, P.J.M. and Fliera, W.G. 2004. Phylogenetic analysis of *Phytophthora* species based on mitochondrial and nuclear DNA sequences. Fungal Genet. Biol. 41: 766-782.

Lévesque, C.A. and de Cock, A.W.A.M. 2004. Molecular phylogeny and taxonomy of the genus *Pythium*. Mycol. Res. 108: 1363-1383.

Li, W., Zhang, T., Tang, X. and Wang, B. 2010. Oomycetes and fungi: Important parasites on marine algae. Acta. Oceanol. Sin. 29: 74-81.

Lipps, P.E. 1980. A new species of *Pythium* isolated from wheat beneath snow in Washington. Mycologia 72: 1127-1133.

Lodha, B.C. and Webster, J. 1990. *Pythium acanthophoron* a mycoparasite, rediscovered in India and Britain. Mycol. Res. 94: 1006-1008.

Majeed, M., Mir, G.H., Mohuiddin, F.A., Hassan, M., Paswal, S. and Farooq, S. 2018. *In vitro* efficacy and population dynamics of fungal and bacterial antagonists against chilli damping off. Int. J. Curr. Microbiol. Appl. Sci. 7: 3024-3030.

Marano, A.V., Jesus, A.L., De Souza, J.I., Leão, E.M., James, T.Y., Jerônimo, G.H., de Cock, A.W.A.M. and Pires-Zottarelli, C.L.A. 2014. A new combination in *Phytopythium*: *P. kandeliae* (Oomycetes, Straminipila). Mycosphere 5: 510-522.

Martin, F.N. 1990. Taxonomic classification of asexual isolates of *Pythium ultimum* based on cultural characteristics and mitochondrial DNA restriction patterns. Exp. Mycol. 14: 47-56.

Martin, F.N. and Kistler, H.C. 1990. Species-specific banding patterns of restriction endonuclease-digested mitochondrial DNA from the genus *Pythium*. Exp. Mycol. 14: 32-46.

Martin, F.N. 1995. Electrophoretic karyotype polymorphisms in the genus *Pythium*. Mycologia 87: 333-353.

Martin, F.N. 2000. Phylogenetic relationships among some *Pythium* species inferred from sequence analysis of the mitochondrially encoded cytochrome oxidase II gene. Mycologia 92: 711-727.

Martin, F.N. and Tooley, P.W. 2003. Phylogenetic relationships among *Phytophthora* species inferred from sequence analysis of mitochondrially encodes cytochrome oxidase I and II genes. Mycologia 95: 269-284.

Martin, F.N. and Looper, J.E. 2010. Soilborne plant diseases caused by *Pythium* spp.: ecology, epidemiology and prospects for biological control. Crit. Rev. Plant Sci. 18: 111-181.

Matsumoto, C., Kageyama, K., Suga, H. and Hyakumachi, M. 1999. Phylogenetic relationships of *Pythium* species based on ITS and 5.8S sequences of ribosomal DNA. Mycoscience 40: 321-331.

Matsumoto, C., Kageyama, K., Haruhisa, S.U.G.A. and Hyakumachi, M. 2000. Intraspecific DNA polymorphisms of *Pythium irregulare*. Mycol. Res. 104: 1333-1341.

Matthews, V.D. 1931. Studies on the genus *Pythium*. University of North Carolina Press, Chapel Hill.

McCarthy, C.G.P. and Fitzpatrick, D.A. 2017. Phylogenomic reconstruction of the oomycete phylogeny derived from 37 genomes. mSphere 2: e00095-17.

McLeod, A., Botha, W.J., Meitz, J.C., Spies, C.F.J., Tewoldemedhin, Y.T. and Mostert, L. 2009. Morphological and phylogenetic analysis of *Pythium* species in South Africa. Mycol. Res. 113: 933-951.

Middleton, J.T. 1943. The Taxonomy, Host Range and Geographic Distribution of the Genus Pythium. Lancaster Press Inc, Lancaster, PA, USA.

Mirzaee, M.R., Ploch, S., Runge, F., Telle, S., Nigrelli, L. and Thines, M. 2013. A new presumably widespread species of *Albugo* parasitic to *Strigosella* spp. (Brassicaceae). Mycol. Prog. 12: 45-52.

Moorman, G.W., Kang, S., Geiser, D.M. and Kim, S.H. 2002. Identification and characterization of *Pythium* species associated with greenhouse floral crops in Pynnsylvania. Plant Dis. 86: 1227-1231.

Moorman, G.W., May, S. and Ayers, K.M. 2014. The key to *Pythium* species. Online: http://keys. lucidcentral.org.key-server.player.jsp?keyId=121 (Accessed: 28 October 2018).

Morajelo, E., Clemente, A., Descals, E., Belbahri, L., Calmin, G., Lefort, F., Spies, C.F.J. and McLeod, A. 2008. *Pythium recalcitrans* sp. nov. revealed by multigene phylogenetic analysis. Mycologia 100: 310-319.

Mostowfizadeh-Ghalamfarsa, R. 2015. The current status of *Pythium* species in Iran: Challenges in taxonomy. Mycol. Iran. 2: 79-87.

Mostowfizadeh-Ghalamfarsa, R. 2016. *Pythium* species in Iran. Shiraz, Iran: Shiraz University Press.

Mufanda, F., Muzhinji, N., Sigobodhla, T., Marunda, M., Chinheya, C.C. and Dimbi, S. 2017. Characterization of *Pythium* spp. associated with root rot of tobacco seedlings produced using the float tray system in Zimbabwe. J. Phytopathol. 165: 737-745.

Naznin, T., Hossain, M.J., Nasrin, T., Hossain, Z. and Sarowar, M.N. 2017. Molecular characterization reveals the presence of plant pathogenic *Pythium* spp. around Bangladesh Agricultural University Campus, Mymensingh, Bangladesh. Int. J. Agric. Res. 1: 1–7.

Nigrelli, L. and Thines, M. 2013. Tropical oomycetes in the German Bight Climate warming or overlooked diversity? Fungal Ecol. 6: 152-160.

Panabières, F., Ponchet, M., Allasia, V., Cardin, L. and Ricci, P. 1997. Characterization of border species among Pythiaceae: Several *Pythium* isolates produce elicitins, typical proteins from *Phytophthora* spp. Mycol. Res. 101: 1459-1468.

Park, J., Park, B., Veeraraghavan, N., Jung, K., Lee, Y.H., Blair, J.E., Geiser, D.M., Isard, S., Mansfield, M.A., Nikolaeva, E., Park, S., Russo, J., Kim, S.H., Greene, M., Ivors, K.L., Balci, Y., Peiman, M., Erwin, D.C., Coffey, M.D., Rossman, A., Farr, D., Grunwald, N.J., Luster, D.G., Schrandt, J., Martin, F., Rebeiro, O.K., Makalowska, I. and Kang, S. 2008. *Phytphthora* database: A forensic database supporting the identification and monitoring of *Phytophthora.* Plant Dis. 92: 966-972.

Paul, B. 2003. *Pythium glomeratum,*: aA new species isolated from agricultural soil taken in north-eastern France, its ITS region and its comparison with related species. FEMS Microbiology Letters 255: 47-52.

Pelizza, S.A., Cabello, M.N., Tranchida, M.C., Scrosetti, A.C. and Bisaro, V. 2011. Screening for a culture medium yielding optimal colony growth, zoospore yield and infectivity of different isolates of *Leptolegnia chapmanii* (Straminipila: Peronospromycetes). Ann. Microbiol. 61: 991-997.

Ribeiro, W.R. and Butler, E.E. 1995. Comparison of the mycoparasites *Pythium periplocum, P. acanthicum* and *P. oligandrum.* Mycol. Res. 99: 963-968.

Richard, W.S., Peter, H.D. and Bruce, B.C. 2005. Compendium of Turfgrass Diseases. St. Paul, USA: APS Press.

Rizvi, S.S.A. and Yang, X.B. 1996. Fungi associated with soybean seedling disease in Iowa. Plant Dis. 80: 57-60.

Robideau, G.P., de Cock, A.W.A.M., Coffey, M.D., Voglmayr, H., Brouwer, H., Bala, K., Chitty, D.W., Desaulniers, N., Eggertson, Q.A., Gachon, C.M.M., Hu, C., Kupper, F.C., Rintoul, T.L., Sarhan, E., Verstappen, E.C.P., Zhang, Y., Bonants, P.J.M., Ristaino, J.B. and Lévesque, A.C. 2011. DNA barcoding of oomycetes with cytochrome c oxidase subunit I and internal transcribed spacer. Mol. Ecol. Resour. 11: 1002-1011.

Rojas, J.A., Jacobs, J.L., Napieralski, S., Karaj, B., Bradley, C.A., Chase, T., Esker, P. D., Giesler, L.J., Jardine, D.J., Malvick, D.K., Markell, S.G., Nelson, B.D., Robertson, A.E., Rupe, J.C., Smith, D.L. Sweets, L.E., Tenuta, A.U., Wise, K.A. and Chilvers, M.I. 2017. Oomycete species associated with soybean seedlings in north America – Part II: Diversity and ecology in realtion to environmental and edaphic factors. Phytopathology 107: 293-304.

Salmaninezhad, F. and Mostowfizadeh-Ghalamfarsa, R. 2017. Taxonomy, phylogeny and pathogenicity of *Pythium* species in rice paddy fields of Fars Province. Iran. J. Plant Pathol. 53: 31-50.

Salmaninezhad, F. and Mostowfizadeh-Ghalamfarsa, R. 2019. Three new *Pythium* species from rice paddy fields. Mycologia. In Press.

Schroeder, K.L., Martin, F.N., de Cock, A.W.A.M., Lévesque, A., Spies, C.F., Okubara, P.A. and Paulitz, T.C. 2013. Molecular detection and quantification of *Pythium* species: Evolving taxonomy, new tools and challenges. Plant Dis. 97: 4-20.

Schröter, J. 1897. Pythiaceae. Engler & Prantl, Nat PflFam, 1: 104-105.

Schurko, A.M., Mendoza, L., Lévesque, C.A., Désaulniers, N.L., de Cock, A.W.A.M. and Klassen, G.R. 2003. A molecular phylogeny of *Pythium insidiosum*. Mycol. Res. 107: 537-544.

Senda, M., Kageyama, K., Suga, H. and Lévesque, C.A. 2009. Two new species of *Pythium, P. senticosum* and *P. takayamanum*, isolated from cool-temperature forest soil in Japan. Mycologia 101: 439-448.

Shen, Q.S. 2001. Molecular phylogenetic analysis of *Grifola frondosa* (Maitake) and related species and the influence of selected nutrient supplements on mushroom yield. Doctoral dissertation. The Pennsylvania State University Graduate School, USA.

Sideris, C.P. 1931. Taxonomic studies in the family Pythiaceae. 1. *Nematosporangium*. Mycologia 23: 252-295.

Sideris, C.P. 1932. Taxonomic studies in the family Pythiaceace. 2. *Pythium*. Mycologia 24: 14-61.

Sparrow, F.K. 1931. The classification of *Pythium*. Science 73: 41-42.

Spies, C.F.J., Mazzola, M., Botha, W.J., Langenhoven, S.D., Mostert, L. and McLeod, A. 2011. Molecular analyses of *Pythium irregulare* isolates from grapevines in South Africa suggest a single variable species. Fungal Biol. 115: 1210-1224.

Tambong, J.T., de Cock, A.W.A.M., Tinker, N.A. and Lévesque, C.A. 2006. Oligonucleotide array for identification and detection of *Pythium* species. Appl. Environ. Microbiol. 72: 2691-2706.

Tangphatsornruang, S., Ruang-areerate, P., Sangsrakru, D., Rujirawat, T., Lohnoo, T., Kittichotirat, W., Patumcharoenpol, P., Grenville-Briggs, J.G. and Krajaejun, T. 2016. Comparative mitochondrial genome analysis of *Pythium insidiosum* and related oomycete species provides new insights into genetic variation and phylogenetic relationships. Gene 575: 34-41.

Taylor, J.W., Jacobson, D.J., Kroken, S., Kasuga, T., Geiser, D.M., Hibbett, D.S. and Fisher, M.C. 2000. Phylogenetic species recognition and species concepts in fungi. Fungal Genet. Biol. 31(1): 21-32.

Testen, A.L., Jimenez-Gasco, M.D., Ochoa, J.B. and Backman, P.A. 2014. Molecular detection of *Peronospora variabilis* in quinoa seeds and phylogeny of the quinoa downy mildew pathogen in South America and the United States. Phytopathology 104: 379-386.

Thines, M. 2007. Characterization and phylogeny of repeated elements giving rise to exceptional length of ITS2 in several downy mildew genera (*Peronosporaceae*). Fungal Genet. Biol. 44: 199-207.

Thines, M., Choi, Y.J., Kemen, E., Ploch, S., Holub, E.B., Shin, H.D. and Jones, J.D.G. 2009. A new species of *Albugo* parasitic to *Arabidopsis thaliana* reveals new evolutionary patterns in white blister rusts (*Albuginaceae*). Persoonia 22: 123-128.

Thines, M. and Kamoun, S. 2010. Oomycete-plant coevolution: Recent advances and future prospects. Curr. Opin. Plant Biol. 13: 427-433.

Uzuhashi, S., Tojo, M. and Kakishima, M. 2009. Structure and organization of the rDNA intergenic spacer region in *Pythium ultimum*. Mycoscience 50: 224-232.

Uzuhashi, S., Tojo, M. and Kakishima, M. 2010. Phylogeny of the genus *Pythium* and description of new genera. Mycoscience 51: 337-365.

Uzuhashi, S., Hata, K., Matsuura, S. and Tojo, M. 2016. *Globisporangium oryzicola* sp. nov., causing poor seedling establishment of directly seeded rice. Antonie van Leeuwenhoek, 110: 543–552.

van der Plaats-Niterink, A.J. 1981. Monograph of the Genus *Pythium*. Studies in Mycology. No. 21. Centraal Bureauvoor Schimmel Cultures, The Netherlands.

Villa, N.O., Kageyama, K., Asano, T. and Suga, H. 2006. Phylogenetic relationships of *Pythium* and *Phytophthora* species based on ITA rDNA, cytochrome oxidase II and β-tubuline gene sequences. Mycologia 98: 410-422.

Voglmayr, H. and Riethmuller, A. 2006. Phylogenetic relationships of *Albugo* species (white blister rusts) based on LSU rDNA sequence and oospore data. Mycol. Res. 110: 75-85.

Voglmayr, H., Montes-Borrego, M. and Landa, B.B. 2014. Disentangling *Peronospora* on *Papaver*: Phylogenetics, taxonomy, nomenclature and host range of downy mildew of opium poppy (*Papaver somniferum*) and related species. PloS One 9: e96838.

Wang, P.H., Wang, Y.T. and White, J.G. 2003. Species-specific PCR primers for *Pythium* developed from ribosomal ITS1 region. Lett. Appl. Microbiol. 37: 127-132.

Waterhouse, G.M. 1967. Key to Pythium *Pringsheim*. Mycological Paper 109. CAB International Mycological Institute, UK.

Waterhouse, G.M. 1968. The genus Pythium *Pringsheim.* Mycological Paper 110. CAB International Mycological Institute, UK.

Weiland, J.E., Beck, B.R. and Davis, A. 2012. Pathogenicity and virulence of *Pythium* species obtained from forest nursery soils on Douglas-Fir seedlings. Plant Dis. 97: 744-748.

White, T.J., Bruns, T., Lee, S. and Taylor, J. 1990. Amplification and direct sequencing of fungal ribosomal RNA genes for phylogenetics. pp. 315-322. *In*: Innis, M.A., Gelfand, D.H., Sninsky, J.J. and White T.J. (eds.). PCR Protocols: A Guide to Methods and Applications. Academic Press, San Diego, USA.

Zitnick-Anderson, K.K. 2013. Characterization and identification of *Pythium* on soybean in North Dakota. Doctoral dissertation. North Dakota State University, North Dakota, USA.

Zitnick-Anderson, K.K., Norland, J.E., Mendoza, L.E.R., Fortuna, A. and Nelson, B.D. 2017. Probability models based on soil properties for predicting presence-absence of *Pythium* in soybean roots. Microb. Ecol. 74: 550-560.

Diagnosis of *Pythium* by Classical and Molecular Approaches

Shivannegowda Mahadevakumar[1*] **and Kandikere Ramaiah Sridhar**[2]

[1] Department of Studies in Botany, University of Mysore, Manasagangotri, Mysore 570 006, Karnataka, India
[2] Department of Biosciences, Mangalore University, Mangalagangotri, Mangalore 574 199, Karnataka, India

Introduction

Pathogenic fungi are the causal agents of many detrimental diseases in plants and economically important crops resulting in considerable loss worldwide. They infect a wide range of plant species or are confined to one or a few host species. Some of them are obligate parasites which require live host to grow and reproduce, but most of them being saprophytes survive in the soil, water and air without association with live host. Isolates of pathogenic fungi could be differentiated by morphological characteristics, host range (formae speciales), aggressiveness (pathotypes or races) or capability to establish stable vegetative heterokaryons by fusion of genetically different strains (Capote et al. 2012). Detection as well as accurate identification of plant pathogens is the most essential criteria to control or to initiate preventive or curative measures. Special attention should be given to early diagnosis of pathogens in seeds, mother plants and propagative plant material to prevent introduction and dissemination of pathogens to unaffected regions (Capote et al. 2012). To fulfill such criteria, rapid, sensitive and precise methods of detection and identification of fungal pathogens are increasingly necessary for making decision to control diseases. Traditionally, the most prevalent techniques employed to identify plant pathogens are dependent upon culture-based morphological approaches. Such methods are time-consuming, laborious and require extensive knowledge of the classical taxonomy of pathogen. Another limitation of early diagnosis includes difficulty in culturing some pathogens *in vitro* (Goud and Termorshuizen 2003). To overcome such limitations, to improve the accuracy and reliability, it is obvious to depend on the molecular approaches. A variety of molecular methods have been employed in detection, identification and quantification of several plant pathogenic fungi. Molecular methods are immensely valuable in understanding the genetic variability of pathogenic populations and also for the description of new fungal species/varieties. Overall, molecular methods are more rapid, specific, sensitive and accurate in diagnosis of plant pathogens. Plant pathogens like *Phytophthora*, *Pythium*, obligate biotrophic pathogens and soil inhabitants require special criteria for diagnosis and identification to enforce preventive measures. The genus *Pythium* is a well-known pathogen, which causes several diseases in plants as well as animals. Isolation and identification of these members results in array of complexities; hence, a wide range of molecular techniques have been followed for identification and diagnosis. Therefore, the main rationale of this chapter is to offer an overview of classical and

*Corresponding author: mahadevakumars@gmail.com

molecular techniques necessary to evaluate various *Pythium* species with emphasis on their merits and demerits.

Pythium

The genus *Pythium* was established by Pringshein in 1858 (*Pythium monospermum* Pringsh.) by placing in the family Saprolegniaceae as type species. As and when more information accumulated by the end of 19th century, taxonomic features of the genus were clarified by description of many new species. The genus was transferred to a new family Pythiaceae (Schröter 1897). This Oomycete genus is widely distributed throughout the world and usually occurs in fresh waters, marine waters and terrestrial habitats. *Pythium* currently includes plant pathogens (*Pythium aphanidermatum* and *Pythium gracile*), animal pathogens (*Pythium insidiosum*), algal pathogens (*Pythium porphyrae*), mycoparasites (*Pythium oligandrum*) and saprophytes (Kageyama 2014). *Pythium* is also one of the most important groups among soil-borne plant pathogens, and occurs in almost all agricultural soils capable to attack roots by causing root rots of thousands of host species leading to reduced crop quality as well as yield. Most *Pythium* species are necrotrophs by infecting tender tissues and they are commonly referred to as 'common cold' due to its chronic nature and ubiquitous distribution (Cook and Veseth 1991, Schroder et al. 2013).

Accurate diagnosis of *Pythium* infected root is often confused with those of root rots caused by *Aphanomyces, Chalara, Cylindrocladium, Fusarium, Phytophthora* and *Rhizoctonia*. Explicit identification of the *Pythium* species in root samples is very important as they differ in their host range as well as temperature regimes. *Pythium* commonly causes pre- and post-emergent damping-off in seedlings. Such pre-emergent damping-off results in failure of seed germination or the radicle decay. Post-emergent damping-off refers to the collapse of seedlings as a result of attack at the soil line or from an infection that starts at the root tip, which moves upwards the soil line. The mycelia on root surface are hyaline and very fine filaments without holding soil particles. Diagnosis ideally requires precise microscopic examination and isolation from infected tissue onto a selective artificial media. The fine vegetative threads without septa develop thick-walled oospores in *Pythium*, which are clearly visible in infected root tissues. In the absence of a suitable host, *Pythium* survives in the soil by production of oospores, chlamydospores and also as mycelial fragments, which are activated under the suitable environmental conditions. In favourable conditions, *Pythium* enters a constant phase of pathogen development by formation of sporangia, which release numerous tiny swimming zoospores.

Phytophthora, Pythium and Phytopythium

The genera *Phytophthora* and *Pythium* include many economically important species, which have been placed in the Kingdom Chromista or Strominophila, which is distinct from other true fungi. Morphologically, *Phytophthora* and *Pythium* are very similar in possessing coenocytic, hyaline and freely branching mycelia and oogonia usually with single oospore. But, the difference between them lies in the mode of zoospore differentiation and their discharge pattern. In *Phytophthora*, the zoospores are differentiated within the sporangium and on maturation release zoospores in an evanescent vesicle at the sporangial apex. In *Pythium*, the protoplast of a sporangium is transferred usually through an exit tube to a thin vesicle outside the sporangium where zoospores are differentiated and released upon rupture of the vesicle. Many species of *Phytophthora* are destructive pathogens, especially in dicotyledonous woody trees, shrubs and herbaceous plants. *Pythium* attacks primarily monocotyledonous herbaceous plants, whereas some of them are also capable of causing diseases in fishes, red algae and mammals including humans. The relationships of the genus *Pythium* with other Oomycete genera have been well established and its systematic position and relationship to other fungi remain unchanged (Hendrix and Papa 1974). However, the nomenclature *of Pythium* is in a state of instability owing to molecular evaluations that revealed new concerns with their names.

Recently, Lévesque and De Cook (2004) worked on phylogenetic analysis involving 116 species and varieties of *Pythium* employing parsimony as well as phonetic analysis of ITS region of nuclear ribosomal DNA, D1, D2 and D3 regions of the adjacent large subunit nuclear ribosomal DNA. Nearly, half the *Pythium* strains revealed a phylogeny congruent with ITS data. A total of 40 ex-type specimens and 20 newly described species were included in their study. The parsimony analysis generated two distinct clades: i) the first clade represented by *Pythium* species with filamentous sporangia; ii) the second clade represented by globose sporangia (Clade-K). On the other hand, Uzuhashi et al. (2010) proposed splitting *Pythium* into four genera primarily based on sporangium morphology: *Elongisporangium, Globisporangium, Ovatisporangium* and *Pilasporangium*.

Villa et al. (2006) performed phylogenic analysis of *Pythium* and *Phytophthora* based on the sequences of three genomic regions such as rDNA ITS, β-tubulin and COX-II genes. The results showed that *Pythium* species in clade K were more closely related to *Phytophthora* species than to other *Pythium* species. Finally, Bala et al. (2010) erected a new genus *Phytopythium* from the *Pythium* clade K as described by Lévesque and De Cook (2004) based on maximum likelihood and Bayesian phylogenetic analysis of the nuclear ribosomal DNA (LSU and SSU), and mitochondrial DNA cytochrome oxidase subunit 1 (COX-I). Statistical analyses of pair-wise distances also strongly supported the *Phytopythium* as a separate phylogenetic entity. The *Phytopythium* is morphologically intermediate between the genera *Phytophthora* and *Pythium*. It is unique in having papillate, internally proliferating sporangia and cylindrical or lobate antheridia (de Cock et al. 2015). The genus *Phytopythium* currently includes 15 species (*P. boreale, P. carbonicum, P. chamaehyphon, P. citrinum, P. cucurbitacearum, P. delawarii, P. helicoides, P. indigoferae, P. litrale, P. megacarpum, P. montanum, P. oedochilum, P. ostracods, P. sterinum* and *P. vexans*) (Levesque and de Cock 2004, Bala et al. 2010, de Cock et al. 2015).

Diagnosticians need to be alert while changing the name as the biology, host range and other related aspects do not change more drastically. Many debates were held to whether a given 'species' has a great deal of diversity or it is actually an artificial grouping of several species. *Pythium irregulare* is one of the examples for such dilemma. Some experts believe that the species is very diverse in its morphology, while others recommend splitting of *P. irregulare* into *Pythium cryptoirregulare, P. cylindorsporium, P. irregulare* and *P. regulare*.

Isolation and recognition

Tsao (1970) was the first to develop a selective medium for the cultivation of *Pythium* under controlled conditions, which consists of pimaricin, pentachloronitrobenzene (PCNB) and vancomycin in CMA. Subsequently, the selective media for growth and development of *Pyhium* spp. have been modified several times. Ali-Shtayeh et al. (1986) developed the VP3 medium (consisting vancomycin, penicillin, pimaricin and PCNB) and Jeffers and Martin (1986) developed the PARP medium (consisting pimaricin, ampicillin, rifampicin and PCNB). Due to carcinogenic nature of PCNB, Morita and Tojo (2007) developed a new selective medium called NARM (consisting nystatin, ampicillin, aifampicin and miconazole) free from PCNB (Kageyama 2014). Ali-Shtayeh et al. (1986) used NARM in 'surface soil dilution method' to isolate *Pythium* spp. from soil samples on 0.1% (w/v) water agar derived from the soil suspension, which is plated onto the surface of the selective medium. Isolation bias has been seen in isolating *Pythium* from soil samples because survival structures of several species could be dormant or require nutrients from the host (Kageyama and Nelson 2003). For instance, *P. myriotylum* is very difficult to isolate from infested soils or plant samples, but it could be isolated using baiting techniques (Kageyama and Ui 1982, 1983, Wang and Chang 2003, Wang et al. 2003a). In addition, slow-growing species like *P. sulcatum* is known to cause cavity spot and blotted root rot in carrot are difficult to isolate because of competition with fast-growing species. In such instances, baiting methods are necessary for isolation (Kageyama et al. 1996). The baiting methods employ trapping materials such as host plant tissues and nutrient-rich substrates. Watanabe (1981, 1989) surveyed the *Pythium* in agricultural fields all over Japan using cucumber seed traps and used this baiting technique for isolation of *Pythium* spp. Watanabe et al. (2008) further demonstrated that turf grass leaf blades and perilla seeds strongly attract zoospores of *P. aphanidermatum*, *P. helicoides*, and *P. myriotylum*, and thus used these traps for monitoring the pathogens in hydroponic cultures. Membrane filters can also be used to isolate *Pythium* (and *Phytophthora*) spp. directly from water samples (Hong et al. 2002). Suzuki et al. (2013) successfully employed this method for monitoring *P. aphanidermatum* and *P. myriotylum* in tomato hydroponic cultures.

Classical identification

Identification of pathogens can be carried out either by classical methods or by advanced techniques. Conventional methods of disease diagnosis include culturing of pathogens on semi-defined media, observation of colony characteristics, sporulating structures and fruit bodies. Owing to advances in instrumentation and molecular biology, the schemes facilitate diagnosis of plant pathogens more rapidly. Advanced methods often employed in diagnosis of plant pathogens include: immunological methods, DNA/RNA probe technology and polymerase chain reaction (PCR) by amplification of nucleic acid sequences. These techniques have potential advantages over conventional diagnostic methods especially in rapidness, accuracy, quantification and could be followed by the personnel who are not specialized in taxonomy (Ward et al. 2004). More importantly, these techniques also assist in detection of non-culturable pathogens. Furthermore, molecular techniques are useful in revealing new diseases with unknown etiology. Comparative analysis of genomic sequences allows the phylogenetic reconstruction of the pathogen relationships at different taxonomic levels.

Molecular diagnosis of pathogens has undergone major advances over the last four decades, which has helped in development and application of these techniques for various purposes. The advent of antibody-based detection, especially monoclonal antibodies (MA) and enzyme-linked immune sorbent assay (ELISA) were the important breakthrough. Application of these approaches was the turning point especially in virology and bacteriology owing to rapidness in diagnosis. Thereafter, the DNA-based technologies especially PCR were revolutionary as they augmented and amplified the original target DNA by several million-folds.

Although various selective media are available for the isolation and identification of *Pythium* spp., they are time-consuming and demand high expertise. Whenever the selective media are used for isolation of *Pythium*, there will be many colonies representing more number of *Pythium* spp., which needs immediate transfer to fresh media for growth to ascertain morphological and cultural features. Further, identification of *Pythium* spp. up to the species level is difficult exclusively on morphological characteristics. Other factors which complicate morphological identification include: i) the high number of non-typical isolates; ii) the heterothallism in some species; iii) sexually sterile isolates; iv) variations in the size and shape of structures within species; v) similar morphology among different groups of species (Hendrix and Papa 1974, Ali-Shtayeh 1985, Martin and Kistler 1990, Klassen et al. 1996, Barr et al. 1997). To circumscribe such difficulties, simple, rapid and accurate methods of detection are essential. Although serological methods are popular in detection of *Pythium*, the inherent problem is production of species-specific antibodies.

Serological diagnosis

Ever since the introduction of ELISA for the detection of plant virus infection (Clark and Adams 1977), serological assays have gradually become valuable support for plant certification schemes. Now it is used to diagnose several plant pathogens including bacteria, fungi, and phytoplasmas. The major advantages of this technique are that it enables the handling of a large numbers of samples and require only one antiserum preparation against the pathogen in a single mammalian species (Ahoonmanesh et al. 1990). However, the major drawback of serological technique lies with requirement of different antibody-enzyme conjugate for each pathogen to be detected and it is highly strain-specific and thus sometimes fails to detect closely related strains of the same fungus (Koenig 1978). Even though the molecular techniques are highly sensitive and capable of distinguishing many species of *Pythium* and other phytopathogenic fungi, the techniques are not quantitative and also not designed for identification of pathogens directly from the environmental samples (e.g. soil or plant) (Li et al. 2010). Serological assays on the other hand have been demonstrated to be highly specific, sensitive, simple, rapid, cost-effective and could be automated for large-scale applications. Recent investigations also suggest that plants exude defense proteins in response to pathogenic fungal secretions during host-pathogen interactions (Gupta et al. 2015). Study of these polypeptides could be utilized to develop a rapid and early detection method for fungal infection based on immunological assays. Polyclonal antibodies (PA) have been used previously to detect multiple isolates of pathogens (Gautam et al. 1999, Biazon et al. 2006, Fleurat-Lessard et al. 2010).

The usefulness of commercial ELISA for detection and quantification of *Pythium* spp. in plants has been demonstrated about three decades ago (MacDonald et al. 1990, Shane 1991). Several species of *Pythium* were tested and diagnosed through ELISA techniques. *Pythium ultimum* is a ubiquitous soil fungus that causes seedling blight and root rot throughout the world. Culturing methods are widely used to quantify inoculum levels of *P. ultimum* in soil (Stanghellini and Hancock 1970, Martin 1992). *Pythium ultimum* and other members of the genus *Pythium* could also be detected in plant tissues by isolation on a wide variety of semi-defined agar media (Martin 1992), but obtaining pure cultures and identifying them up to species level is cumbersome (van der Plaats-Niterink 1981). Another complication in the detection and identification of *P. ultimum* by culturing is that some isolates are asexual, producing hyphal swellings without diagnostic sexual structures (van der Plaats-Niterink 1981).

Recently, the PCR using *P. ultimum*-specific primers was reported to be a potential   method of choice (Kageyama et al. 1997). However, Yuen et al. (1998) are of the opinion that none of the current detection techniques allow for rapid and specific quantification of *P. ultimum* biomass in plant tissues. Therefore, species-specific detection by serological techniques gained momentum and several species have been diagnosed serologically. A double-antibody sandwich indirect ELISA was developed by Yuen et al. (1998) for the detection and quantification of *P. ultimum*. A polyclonal antibody produced against cell walls of *P. ultimum* was employed as the capture antibody, while *P. ultimum*-specific monoclonal antibody (i.e. MAb E5) was used for recognition. The MAb E5 was also developed specifically for identification of *P. ultimum* (Yuen et al. 1993).

The *P. ultimum* was detected by ELISA in roots of bean, cabbage and sugar beet seedlings grown in pathogen-infested soil. The ELISA optical density readings for infected bean and sugar beet root samples were highly correlated ($r > 0.9$) with infection levels determined by culturing the samples on water agar medium (Yuen et al. 1998). The specificity of MAb E5 was demonstrated in two studies involving 21 isolates of *P. ultimum* and over 60 isolates of other fungal species (Yuen et al. 1993, Avila et al. 1995). Successful ELISA detection of *P. ultimum* in plant tissues using MAb E5 was possible only when the antibody was employed in an indirect ELISA format, in which plant tissue extracts were adsorbed onto the wells of microtiter plates (Yuen et al. 1993, Avila et al. 1995). The MAb E5 proved that *P. irregulare* infected samples were detected and those with no *Pythium* infection did not react in the ELISA. The ELISA method developed was highly sensitive and the fungus was detected in culture extracts diluted up to a ratio of 1:5,000,000 and in roots with less than one infection per 100 cm root length. Similarly, the PA against *Pythium aphanidermatum* proteins were developed for the detection of rhizome rot in ginger using serological assays. Under the optimal experimental conditions, detection limit of *P. aphanidermatum* by indirect ELISA was 10 µg/ml with a linear working range from 5-100 µg/ml ($R^2 = 0.994$) (Ray et al. 2018). Identification of *Pythium insidiosum*, a causal agent of pythiosis, relies on cultural characteristics, serological diagnosis and molecular assay. Serodiagnosis of *P. insidiosum* requires culture filtrate antigen extracted from the pathogen and exo-1,3-beta-glucanase gene (PinsEXO1) encodes a 74kDa immune-reactive protein, which could be recognized by sera from pythiosis patient but not from healthy individuals. Keeratijarut et al. (2013) developed a peptide ELISA to detect antibodies against *P. insidiosum* based on predicted antigenic determinants of exo-1,3-beta-glucanase by developing antigenic determinant (epitope) of PinsEXO1 to serodiagnosis of pythiosis based on peptide ELISA.

The major advantage of serology-based detection is that these assays are very rapid and specially developed serology-based field kits are available in the market for the detection of *Pythium*. However, it is important to consider that presence of *Pythium* need not necessarily indicate that it is the primary cause for symptoms in hosts. *Pythium* often infects dead and decaying plant tissues following infection by other pathogens (e.g. *Chalara* or *Phytophthora*). Thus, serology-based detection warrants that infected plants should still be assessed by a specialized diagnostic laboratory even after a positive result from field-based kits.

Molecular markers

With the advent of biochemical and molecular techniques, classical methods were replaced by modern molecular tools such as molecular markers to study fungal populations. Molecular techniques used for identification and assessment of the genetic diversity of various *Pythium* species includes: RFLP, AFLP, SSR, ISSR, SCAR, ITS and PCR-RFLP (Liu and Whittier 1995, Harvey et al. 2000, 2001, Lian and Hogetsu 2002, Garzon et al. 2005, 2007, Al-Sadi et al. 2007, 2008a, b, 2012, Lee and Moorman 2008). These techniques help researchers for assessment of DNA in different fields such as phylogenetic studies to determine the genetic relationships, genetic mapping, gene localization, gene transformation and diagnosis of diseases. Microsatellites or simple sequence repeats are tandemly repeated motifs of 1-6 bases found in eukaryotic genomes (Sawaya et al. 2013) including fungi (Tautz 1989). As they were first described (Litt and Luty 1989, Tautz 1989, Weber and May 1989), microsatellites have been widely used as tools in studies of genetic variation in natural populations. In addition, they are codominant, multiallelic, highly polymorphic and require only small amounts of DNA for PCR analysis. Microsatellites could be amplified using primer pairs complementary to their flanking regions and fragment length polymorphism will be detected by gel electrophoresis. To amplify microsatellite loci by PCR, primers must be developed using the flanking sequences. Sequence information for primer design can be obtained by constructing and screening a small insert plasmid library (Hughes and Queller 1993) or by enriching a DNA library for microsatellites (Karagyozov et al. 1993). However, both the methods are labour-intensive and also expensive. Therefore, several alternative strategies, including modifications of the RAPD method, PCR isolation of microsatellite arrays and primer extension enrichment protocols have been devised (Zane et al. 2002).

Yin-Ling et al. (2009) described a new method for the isolation of microsatellite markers from *P. helicoides* genome. They adopted a strategy combining dual-suppression PCR and thermal asymmetric interlaced PCR to determine sequences flanking microsatellite regions in *P. helicoides* and verified the advantages of those markers for population analyses of *P. helicoides*. Employing this approach, they further developed a simple method for designing PCR primers to amplify microsatellite markers from *P. helicoides*.

Application of a single amplified fragment length polymorphism (AFLP) for understanding the spatial distribution of *Pythium* species was reported by many researchers. Garzon et al. (2005) detailed the existence of little genetic variation in a sample of 23 isolates from eight countries belonging to four continents with AFLP primer combination. Although the RAPD and AFLP analyses were rapid and convenient procedures, they can only detect dominant markers. Thus, these methods cannot be used to identify heterozygotes in diploid species like *P. helicoides*. Furthermore, the RAPD analysis is poorly reproducible between laboratories and researchers. The RFLP analysis, although very informative, is time consuming owing to requirement of Southern blots (Herrero and Klemsdal 1998, Kageyama et al. 1998, Matsumoto et al. 1999, 2000).

Inter Simple Sequence Repeat (ISSR) assay has been applied in genetic studies since 1994 (Zietkiewicz et al. 1994, Vasseur et al. 2005, Amini et al. 2014). The ISSR primers are modified SSR primers based on sequences surrounding the microsatellite region. This marker showed higher polymorphism than RAPD primers owing to repeated regions in the genome (Es-selman et al. 1999). Repetitive motifs in the genomes of the *Pythium* group F isolates were demonstrated using this marker (Vasserur et al. 2005). Figure 12.1 schematically represents various methods which could be adapted for identification of *Pythium* spp.

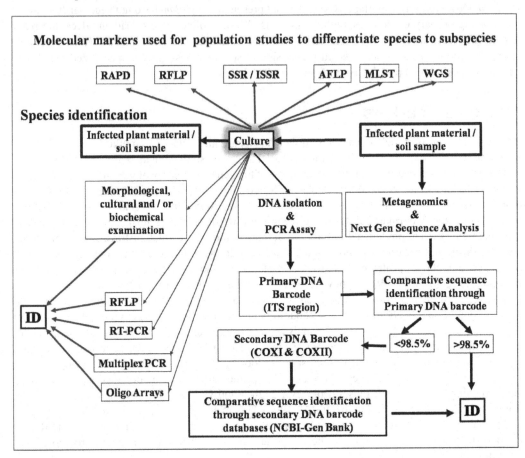

Figure 12.1. Suggested schematic representation of molecular methods used for identification and differentiation of *Pythium* species (this pragmatic approach on top highlights the molecular markers routinely used for identification and differentiation; lower panel consists of molecular platforms available for the identification)

Copy Number of Marker and Quantification

The officially approved COX-I barcode and the proposed ITS barcode are present in multiple copies in a genome. This has significant advantages during amplifying DNA from difficult templates and also for direct detection of species from environmental samples. However, the copy number of the ITS locus has not been determined for *Pythium*; thus, there is indirect evidence to suggest that there may be variations in copy numbers at an intra- and inter-specific levels (Martin 2009). Spies et al. (2011) also suggested that the situation is similar for *P. vexans*. Standard curves for a qPCR assay based on the ITS region were different, which are dependent on the phylogenetic sub-grouping of the species. The estimates of DNA concentration based on Ct differed by more than two orders of magnitude among the *P. vexans* sub-groups. With a similar ITS assay for *P. irregulare* and *P.*

ultimum, the intra-specific variability was generally less than one order of magnitude (Spies et al. 2011). This denotes that it is important to consider the possibility of multiple gene copies for developing assays (array or PCR-based) that will be used to quantify the presence of a pathogen and to perform appropriate tests to determine if this influences the accuracy of the results. In a soil quantification assay for *Verticillium dahliae*, Bilodeau et al. (2011) estimated copy number differences in the IGS rDNA for a geographically diverse collection of isolates ranging from 24-73 copies per genome. Similarly, it is possible to apply such approach for *Pythium* cultures collected from wide geographical regions.

Single nucleotide polymorphism

In most of the instances, closely related pathogens having a wide range of hosts or pathogenicity frequently differ in a single or a few base pairs in the target genes universally used for identification. Therefore, single nucleotide polymorphisms (SNP) are useful in any routine diagnostic assay (Martin 2008, Delmotte et al. 2008). The SNPs were used as genetic markers to differentiate various plant pathogenic fungi causing disease in economically important crop plants. A SNP-based diagnostic method has been developed to detect and differentiate isolates of *Phytophthora* species from Europe and United States (*P. ramorum*) based on the DNA sequence differences of mitochondrial COX-I gene (Kroon et al. 2004). Similarly, for variations in DNA sequences of microsatellite flanking regions of *Phytophthora infestans,* SNP genetic marker was employed for typing the pathogen (Abbott et al. 2010).

Identification and differentiation of pathogenic and non-pathogenic *Pythium* based on SNP have been reported by various researchers. Notable contribution of using SNP for identification and differentiation for *P. myriotylum* comes from Gómez-Alpízar et al. (2011) and Le et al. (2017). Gómez-Alpízar et al. (2011) reported a technique, where SNP was employed to develop a PCR-RFLP for detection of *P. myriotylum* (causal agent of cocoyam root rot disease). Sequences of the ITS-rDNA containing ITS1 and ITS2 of *P. myriotylum* isolates from cocoyam and other hosts were aligned and they generated a restriction map to analyze the genetic diversity based on PCR-RFLP (Gómez-Alpízar et al. 2011). The rDNA-ITS alignment revealed a new SNP that involves thymine/cytosine downstream to previously reported SNP guanine/adenine between isolates of *P. myriotylum* which are pathogenic to cocoyam and non-pathogenic strains. Le et al. (2017) developed a PCR-RFLP based on SNP typing for the detection of isolates of *Pythium myriotylum*, causal agent of *Pythium* soft rot (PSR) disease in capsicum and ginger in Australia under different regimes. In their report, the whole genome data from four *P. myriotylum* isolates were recovered from three hosts, one from *Pythium zingiberis* isolate derived and analyzed for sequence diversity based on SNP. They noted that a higher number of true and unique SNP occurred in *P. myriotylum* isolates obtained from ginger with symptoms of PSR from Australia than other *P. myriotylum* isolates. Overall, more numbers of SNP were discovered in the mitochondrial genome than those in the nuclear genome. Among them, a single substitution from the cytosine (C) to the thymine (T) in the partially sequenced COX-II gene of 14 representatives of PSR *P. myriotylum* isolates was within a restriction site of HinP1I enzyme, which was applied for the development of PCR-RFLP detection and identification of the isolates without sequencing. The PCR-RFLP was sensitive enough to detect PSR causing *P. myriotylum* strains from artificially infected ginger without isolation of pure cultures (Le et al. 2017). The SNP analysis is greatly helpful in rapid and early detection of the pathogen that could be of great advantage in certifying planting materials as disease-free, enhancing sustainable management practices and limiting economic losses.

DNA array for detection

The DNA arrays are used to detect the presence of specific sequences. Known DNA sequences are fixed on to a supporting material like glass slide and/or membrane. Sample DNA is applied to the array and hybridization occurs if a sequence that is complementary to one on the array is present in the sample. The hybridized DNA is detected using fluorescent dyes. Usually, the PCR is used to amplify and label the targeted region. The DNA arrays are useful for detecting multiple plant pathogens in environmental samples (Lévesque et al. 1998). Recently, Tambong et al. (2006) developed an oligonucleotide array for the identification and detection of *Pythium* species. The array contains 172 oligonucleotides which are complementary to specific diagnostic regions of the rDNA ITS region and could be used to detect more than 100 different *Pythium* species. Hybridization patterns will be distinct for each species and the oligonucleotides for each species will also be distinct.

PCR and DNA barcode

Limitations of conventional diagnoses compelled researchers to consider advantages of PCR-based methods for accurate identification and detection of *Pythium* spp. There are various commercially available conventional PCR from 1990s and they have been used for specific identification purpose of plant pathogens. Initiatives are also underway to identify the DNA sequences that could differentiate all plants, animals and fungi. Such sequences are called DNA barcodes with reference to the barcodes applied to food packages and other goods. Robideau et al. (2011) recommended that sequences from the COX-I gene and the rDNA ITS region could be used for DNA barcoding of Oomycetes. The COX-I gene is more variable than the rDNA ITS region, but already a wealth of rDNA ITS sequence data is available. Thus, these two regions of DNA will be required for identifying any given isolate more authentically.

Primers for *Pythium*

Conventional keys for identification of *Pythium* species are mainly designed on morphological and biological characteristics (Waterhouse 1967, Plaats-Niterink 1981). The characters considered to formulate the key are highly variable and a large number of *Pythium* species have been described (Hendrix and Campbell 1973). The overlapping characters frequently make identification more complex.

Some studies have reported specific probes and primers related to identification of *Pythium*. Martin (1991) used selected DNA fragments from mtDNA of *Pythium oligandrum* and *P. sylvaticum* as probes for identification. Klassen et al. (1996) used parts of the 5S-rRNA gene spacers to make probes specific for eight species of *Pythium* (*Pythium acanthicum*, *P. anadurm*, *P. intermedium*, *P. macrosporum*, *P. mastophorum*, *P. odanogenense*, *P. sylvaticum* and *P. ultimum*). Lévesque et al. (1994, 1998) suggested the use of ITS1 region of rDNA as a species-specific probe for *Pythium acanthicum*, *P. aphanidermatum*, *Phytophthora cinnamomi* and *P. ultimum*. Chen (1994) used specific primers for identification and detection of *Pythium arrhenomanes*. The organization of ITS region typically consisting 18S-rRNA representing SSU and 28SrRNA on the other side flanked by LSU and in the central placed 5.8SrRNA region along with location of primer sites are presented in Fig. 12.2. Figure 12.3 presents COX-I and COX-II gene cluster in *Pythium*.

In most instances, the ITS region is widely used for PCR detection. Different workers have designed a wide range of primers to amplify specific region in ITS-rDNA. However, universal primers developed by White et al. (1990) were highly successful in amplifying the specific region in most of the fungi. Although other DNA sequences are informative as well as specific, the ITS region has been widely employed for designing the primer. Nevertheless, the universal primers were not so effective in amplifying the rDNA region of *Pythium*. Therefore, other combinations of primers have been designed and tested (Tojo et al. 1998, Paul 2002, Singh et al. 2013). For about 26 different species of *Pythium*, species-specific primers were developed and standardized in conventional PCR methods. They are widely used for identification/confirmation of species and are presented along

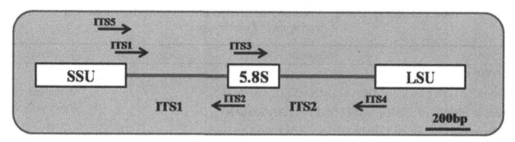

Figure 12.2. Schematic structure of ITS gene region and location of primer sites in fungi (organization of ITS region typically consisting of 18S-rRNA representing SSU, 28SrRNA on the other side flanked by LSU and in the centre 5.8SrRNA region along with location of primer sites are represented)

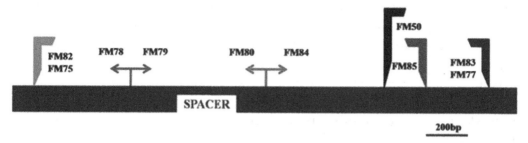

Figure 12.3. Organization of the mitochondrially encoded COX-I and COX-II gene clusters in *Pythium* and location of the PCR primers used for template amplification and sequencing reactions (refer Table 12.1 for details of primer sequence)

with their source of DNA (host or culture), primer sequence designation and reference sources in Table 12.1.

Although there are several advantages for selecting ITS-rDNA sequences for phylogenetic analysis, use of DNA coding sequences for protein combined with rDNA sequences is advisable to further validate phylogenetic studies. Among the genes encoding proteins, very few reports are available on use of COX-II gene for phylogenetic studies (Villa et al. 2006). COX-II is the component of the respiratory chain that catalyzes the reduction of oxygen to water. Subunits 1-3 form the functional core of the enzyme complex, whereas subunit 2 transfers the electrons from cytochrome c. Unlike the ITS regions, COX-II is a mitochondrial-encoded gene, which is generally considered to be more variable than the nuclear DNA. Therefore, COX-I and COX-II genes are employed to analyze phylogenetic relationships of *Pythium* as well as *Phytophtora* spp. (Martin 2000, Martin and Tooley 2003). A list of primers (multi-locus) for various genes targeted and used for molecular identification and phylogenetic analysis of *Pythium* spp. are presented in Table 12.2.

Multiplex PCR

Molecular methods for diagnosis and identification of *Pythium* species have been developed using species-specific primers (Kageyama et al. 1997, Godfrey et al. 2003, Wang et al. 2003a, b) and for each species, separate PCR-primers are necessary for amplification. Although species-specific primers were developed for accurate identification of *Pythium* spp., they are very difficult to identify other than the target organism; as a result, it demands more time as well as reagents. In principle, multiplex PCR uses multiple primer pairs and allows simultaneous detection of multiple species of *Pythium*. In order to reduce the time and labour, Asano et al. (2010) developed a multiplex PCR system, which has been successfully employed to detect five species of *Pythium* (*P. aphanidermatum*, *P. arrhenomanes*, *P. graminicola*, *P. torulosum* and *P. vanterpoolii*), while Ishiguro et al. (2013) developed another PCR protocol for detection of three more *Pythium* species (*P. aphanidermatum*, *P. helicoides* and *P. myriotylum*).

Table 12.1. List of primers designed and tested for COX-I, COX-II, ITS, Tubulin and Actin genes of *Pythium* spp.

Primer	Sequence	Expected size	Reference
COX-I			
Forward: FM75	5-CCTTGGCAATTAGGATTTCAAGAT-3		Martin and
Reverse: FM82	5-TTGGCAATTAGGTTTTCAAGATCC-3		Tooley (2003)
Reverse: FM77	5-CACCAATAAAGAATAACCAAAAATG-3		
Reverse: FM83	5-CTCCAATAAAAAATAACCAAAAAT-3		
FM78	5-ACAAATTTCACTACATTGTC-3	Sequencing primers for COX-II	
FM79	5-GGACAATGTAGTGAAATTTGT-3		
FM80	5-AATATCTTTATGATTTGTTGAAA-3		
FM84	5- TTTAATTTTTAGTGCTTTTGC-3	Sequencing primers for COX-I	
FM85	5-AACTTGACTAATAATACCAAA-3		
FM50	5-GTTTACTGTTGGTTTAGAT-3		
COX-II			
Forward: FM58	5-CCACAAATTTCACTACATTGA-3 (21bp)	563 bp	Martin and
Reverse: FM66	5'-TAGGATTTCAAGATCCTGC-3 (19bp)		Tooley (2003)
ITS			
Forward: ITS1	5-TCCGTAGGTGAACCTGCGG-3 (19bp)	700-900 bp	White et al.
Reverse: ITS4	5-TCCTCCGCTTATTGATATGC-3 (20bp)		(1990)
β-Tubulin			
Forward: BT5	5-GTATCATGTGCACGTACTCGG-3	658 bp	Villa et al.
Reverse: BT6	5- CAAGAAAGCCTTACGACGGA-3		(2006)
Actin			
5FWDACT	GTATGTGCAAGGCCGGTTT		Villa et al.
MIDREVACT	ATGAGGCAGACCTAGCCACCAAG		(2006)

RT-PCR and qPCR

Although conventional PCR-based techniques are very sensitive and capable of breaking up many species of *Pythium*, they are neither quantitative nor practical for identification directly from environmental samples (Schroeder et al. 2006). Among the PCR techniques developed in the pace of modern technology, a rapid, precise and effective PCR technique used to detect and quantify from environmental samples is the RT-PCR or qPCR technique. It is a combination of fluorescence detection and PCR platforms to monitor the amplification on each cycle in a thermo cycler. It is also referred as qPCR technique due to its capability in estimation of the amount of target DNA amplified. The RT-PCR is based on the use of fluorogenic probes (hybridization probes or exonuclease probe like TaqMan) (Nitsche et al. 1999) or dyes (SYBRR GreenI) (Morrison et al. 1998), which relies on measuring the intensity of a fluorescent signal that is proportional to the amount of DNA generated during the PCR amplification leading to accurate quantification (Wittwer et al. 1997). The SYBRR GreenI has an additional advantage that it does not demand the design of specific complementary probes to target the DNA. Thus, this is a method of choice to detect and quantify fungi (Schnerr et al. 2001, Filion et al. 2003a, b). The amount of target DNA present in a sample is inversely proportional to the fractional cycle number (Ct) at which the amplicon is first detected. Li et al. (2010) developed another rapid, precise and effective RT-PCR technique for estimating the population densities of *P. intermedium* from soils using species-specific primers.

Table 12.2. Species-specific ITS based primers developed for the diagnosis of
Pythium spp. using conventional PCR technique

Pythium species	Primer	Sequence (5'-3')	Host and source	Reference
P. acanthicum	Pac1 ITS2*	GTGCCTCGTCTTGTTGAAAG GCTGCGTTCTTCATCGATGC	-/ Culture	Wang et al. (2003a)
P. aphanidermatum	Pa1/ITS2	TCCACGTGAACCGTTGAAATC	-	Wang et al. (2003a)
P. aphanidermatum	AsPyF AsAPH2Brev	CTGTTCTTTCCTTGAGGTG GCGCGTTGTTCACAATAAATTGC	Turf grass	Asano et al. (2010)
P. arrhenomanes	Par1/ITS2	AAGTGTAGTTAATTCTGTACGCTGC	-	Wang et al. (2003a)
P. arrhenomanes	AsPyF AsARRR	CTGTTCTTTCCTTGAGGTG CGTCCAAGAGCAATAACCACTC	Turf grass	Asano et al. (2010)
P. graminicola	AsGRAF AsPyR	GGGCTGCATGTATGTGTAGT ATTCTGCAATTCGCATTAC	Turf grass	Asano et al. (2010)
P. heterothallicum	Phe1/ITS2	TTTGTATGAGATCAGCTGAT	-	Wang et al. (2003a)
P. inflatum	Pinf1/ITS2	AAGGTGGGCGCATGTATGTG	-	Wang et al. (2003a)
	Pint1/ITS2	ATACTGCTGGCGGGTGCGAG	-	Wang et al. (2003a)
P. intermedium	PINTERf PINTERr	ATGCAGAGGCTGAACGAA CTGTATTCATAGCCGAAACGA	Carrot/Soil	Klemsdal et al. (2008)
P. irregulare	Pir1/ITS2	AGCGGCGGGTGCTGTTGCAG	Culture	Wang et al. (2003a)
P. myriotylum	Pmy5/ITS2	GCTGCTGTTATGGCGGACT	Ginger	Wang et al. (2003b)
P. oligandrum	P.OLIG.F P.OLIG.R04	CTGTGCTTCGTCGCAAGACT CTTTAAAAAGACAGCGCGAGA	Mushroom	Godfrey et al. (2003)
P. oligandrum	Po1/ITS2	TGCGTCTATTTTGGATGCGG	-	Wang et al. (2003)
P. porphyrae	PP1 PP2	TGTGTTCTGTGCTCCTCTCG CCCAAATTGGTGTTGCCTCC	Seaweed	Park (2006)
P. rostratum	Pro1/ITS2	TAGTGTAGCTTTTGTTGCGC	-	Wang et al. (2003)
P. splendens	Pspl1/ITS2	GAAGGTCGGAGTAAAATCTGGC	-	Wang et al. (2003)
P. spinosum	Pspi1/ITS2	TGTGTGTTGTGATCGTGCCT	-	Wang et al. (2003)
P. sulcatum	Psu3/ITS2	CACGTGAACCGTAATAATCA	-	Wang et al. (2003)
P. sulcatum	PSULCf PSULCr	GCCGCTTTATTGTGGTCT TCTTCTTTACCCCACAAGTGA	Carrot	Klemsdal et al. (2008)
P. sylvaticum	PSYLf PSYLr	CGCTGTGGTTGGTATATTTGT GCCAATTGCACAAGTACAAA	Carrot	Klemsdal et al. (2008)

(Contd.)

Table 12.2. (*Contd.*)

Pythium species	Primer	Sequence (5'-3')	Host and source	Reference
P. torulosum	AsPyF	CTGTTCTTTCCTTGAGGTG	Turf grass	Asano et al. (2010)
	AsTOR6	CGCCTGCCGAAACAGACTAG		
P. ultimum	K1	ACGAAGGTTGGTCTGTTG	Cucumber and sugar beet	Kageyama et al. (1997)
	K3	TCTCTACGCAACTAAATGC		
P. ultimum	Pu1F1	GACGAAGGTTGGTCTGTTG	Potato	Cullum et al. (2007)
	Pu2R1	CAGAAAAAGAAAGGCAAGTTTG		
P. vanterpoolii	AsVANF	GGTGGATAGCGGCGTATTT	Turf grass	Asano et al. (2010)
	AsPyR	ATTCTGCAATTCGCATTAC		
P. violae	PVIOLf	ATGTGTGTGTGCGGGACT	Carrot	Klemsdal et al. (2008)
	PVIOLr	CCACTCCCCAAAGAGAGAAGT		
P. vipa	PVIPAf	CAGCGGTTGGTATATTCGTT	Carrot	Klemsdal et al. (2008)
	PVIPAr	AAAAAGAAGTGCACAAATAGATGA		

Generally, the SYBRR GreenI is used as a fluorescent dye and it binds to double stranded DNA in each annealing and extension steps. However, recently this dye has been replaced by TaqMan probe. The TaqMan probe is capable of distinguishing SNP within the probe binding site, conferring a higher degree of target specificity as well as the ability to multiplex amplifications. The RT-PCR is very sensitive, high throughput (>350 samples/run), employs gel electrophoresis without use of ethidium bromide and the melting temperature could be used as an amplicon-specific character. Schroeder et al. (2013) used RT-PCR for the quantitative detection of 14 *Pythium* species from plant tissues, soil, and water. This technique is highly useful for ecological, epidemiological and diagnostic studies. The *P. intermedium* from cool-temperate natural forests was quantified through RT-PCR to demonstrate the pathogen's uneven distribution (Li et al. 2010). Schroeder et al. (2006) used species-specific primer sets with RT-PCR to analyze the distribution of nine *Pythium* spp. in wheat field soils. The RT-PCR has been successfully employed to quantify a total of 13 plant-pathogenic *Pythium* spp. by different researchers. Details of *Pythium* species quantified or detected using RT-PCR along with details of primer sequence and host are presented in Table 12.3.

RT-PCR for environmental samples

The RT-PCR is a highly useful technique specifically developed for ecological, epidemiological studies and diagnostic purposes. The RT-PCR studies have shown an uneven distribution of *P. intermedium* in cool-temperate natural forests (Li et al. 2010) and distribution of nine *Pythium* species in wheat field soil samples using species-specific primers (Schroeder et al. 2006). However, there are two technical issues to employ RT-PCR methods to detect pathogens efficiently in environmental samples: i) the first issue is to design the specific primer sets; the ITS-rDNA region will be of practical target since species-specific sequences have evolutionally accumulated in this region; further, the ITS-rDNA sequences of almost all *Pythium* species are available in the DNA databases. In addition, the genome contains multiple copies of the rDNA (Martin 1995); therefore, the PCR detection of these sequences is more sensitive than the single copy genes; ii) the second issue is the method of extracting DNA from environmental samples. A general protocol for detecting a wide range of soil microorganisms in various soils is still unavailable. This lacunae is partly due to those factors that hinder detection of thick cell walls of survival/reproductive structures, which are resistant to breaking; strong adherence of microorganisms to soil particles or organic matter; and inhibitors in the soil especially the humic acid, which hinders the PCR amplification of microbial DNA (Kageyama 2014).

Kageyama et al. (2003a, b) developed an integrated method for isolation of DNA from environmental samples, which involves vortexing the DNA extract with glass beads (to physically disrupt the sturdy survival structures of pathogenic fungi); addition of skim milk to the extraction buffer (to increase the efficiency of the extraction); purification of the extracted DNA (to remove PCR inhibitors); and addition of bovine serum albumin to the PCR mixture (to relieve interference by PCR inhibitors). Li et al. (2011) also used a method (with a modified DNA purification step) to detect *Phytophthora cactorum* and *P. nicotianae* by multiplex PCR in 89 fields having different soil textures and soil groups.

Loop-mediated isothermal amplification of DNA

The Loop-mediated isothermal Amplification of DNA (LAMP) is a simple, rapid, specific and cost-effective nucleic acid amplification method developed by Notomi et al. (2000). It is characterized by the use of four different primers specifically designed to recognize six distinct regions on the target gene and the reaction process proceeds at a constant temperature using strand displacement reaction. In LAMP assay, amplification and detection of target gene could be completed in a single step by incubating the mixture of samples, primers, DNA polymerase with strand displacement action and substrates at a stable temperature ($\sim$65°C) (Sano and Itano 2010). It provides high amplification efficiency, with DNA being amplified 10^9-10^{10} times in 15-60 min. Owing to the high specificity, the presence of amplified product indicates the presence of target gene.

The technique LAMP has been widely employed to diagnose various plant pathogens and it is also applied as a tool to diagnose the *Pythium* species by various researchers. It is devoid of thermo cycler as it demands the reaction proceeds at a constant temperature (Dai et al. 2012, Fukuta et al. 2013, Gosh et al. 2015, Feng et al. 2015). Successful amplification of *P. aphanidermatum* from hydroponically grown tomato plants using LAMP assay was reported by Fukuta et al. (2013). They also reported the sensitivity of LAMP as more than the actual PCR. Furthermore, the assay allowed the direct detection of *P. aphanidermatum* in infected tomato roots, suggesting that the method could be used for on-site diagnoses in the field conditions (Kageyama 2014). In a nutshell, an array of molecular tools and techniques employed for identification of various *Pythium* species or to differentiate the *Pythium* population are presented in Table 12.4.

Recent advances in *Pythium* genome analysis

In the era of molecular biology, one could witness several recent developments in genome analysis as the ultimate source of information, wherein targeted organism will be precisely deciphered. The genomic analysis is capable of identifying as well as comparing the genomic features of DNA sequence, structural variation, gene expression, regulatory and functional element annotation. Methods for genomic analysis obviously require high-throughput sequencing or microarray hybridization and bioinformatics. In this context, a few *Pythium* species have been subjected for genome analysis and provided vital information. The genome of *Pythium insidiosum* and *P. ultimum* have been analyzed and compared with other Oomycetes. However, there are many important plant-pathogenic *Pythium* spp. that need to be analyzed for precise understanding of the biology of the pathogen, which will help to decide suitable control measures.

The genome of *P. ultimum* is 42.8 Mb, which encodes 15290 genes and possesses extensive sequence similarity with related *Phytophthora* species (e.g. *P. infestans* caused the Irish famine) (Lévesque et al. 2010). The whole transcriptome sequence analysis revealed expression of 86% of genes with measurable differential expression of suites of genes under abiotic stress in the presence of a host. Absence of cutinases in the *P. ultimum* genome suggests a significant difference in virulence mechanisms between *P. ultimum* and other more host-specific Oomycete members. Although there was more degree of similarity between the genome of *Pythium* and *Phytophthora*, a

Table 12.3. Species-specific real time PCR primers developed based on ITS-rDNA for the diagnosis of *Pythium* spp. using RT-PCR technique

Pythium species	Primer	Sequence (5'-3')	Host and source	Reference
P. abappressorium	ABA1bF ABA1bF	GTTGTTGTTGCGTCTGCGGATTTG GTTGTTGTTGCGTCTGCGGATTTG	Wheat	Schroeder et al. (2006)
P. attrantheridium	ATT3F ATT2R	GTTTGTTGGTCATTTTGGCTGCG CGCTACTAACAAAGCAGATCCCAG	Wheat	Schroeder et al. (2006)
*P. dissotocum**	up_504 lo_702 183_KNA FAM	GTTTGGATCGCTTTGCT CCGAAGCTAGAGCGCTT TGACTGGAGTTGTTTTCTGTT	Tomato	Vallance et al. (2009) Le Floch et al. (2007)
P. heterothallicum	HET4F HET2R	GTGAAGTGTCTCGCGCACTTG GTAACCATGCATGCTGCACCA	Wheat	Schroeder et al. (2006)
P. intermedium	Pf002 Pr002b	GAGTTGCTTTGCTCTCCGGC ACACTTCACGTCTGCCACA	-	Li et al. (2010)
P. irregulare (G-1)	IRR3cF IRR3R	GCTGTGGTTGGTGTTTGTTGTTTGC CTGTACAATTGCACACACAAGTATG	Wheat	Schroeder et al. (2006)
P. irregulare (G-I & II)	PiF PiR	GTAGCATGCGTGTTTGCTTA GCAAGCTGTGCATTCATTGC	American ginseng	Kernaghan et al. (2008)
P. irregulare (G-I & II)	PirF1 PirR3	AGTGTGTGTGGCACGTTGTC GATCAACCCGGAGTATACAAAAC	Grapevine	Spies et al. (2011)
P. irregulare (G-IV)	IIV7F ABA1R	GTATCGTCTTGGCGGAGTGG TGCATAAACGAATATACCAACCGC	Wheat	Schroeder et al. (2006)
P. oligandrum	HybriOLI-1 HybriOLI-2	GTCTGCGTCTATTTTGGATG GGATTTGCTGATGTTATTTT	Tomato	Takenaka et al. (2008)
*P. oligandrum**	up_F1 lo_146 142_LNA FAM	TGCTTCGTCGCAAGACT CGTATTCGGAGTATAGTTCAGT AGTCTGCGTCTATTTTGGA	Tomato	Vallance et al. (2009) Le Floch et al. (2007)
P. paroecandrum	PAR2F PAR3R	TGGTTGGCGTTCGTTGTTTG GGATCAACCCGGAGTACACTAATT	Wheat	Schroeder et al. (2006)
P. porphyrae	PP-1 PP-2	TGTGTTCGTGTGCTCCTCTCG CCCAAATTGGTGTTGCCTCC	Seaweed	Park et al. (2001)

(Contd.)

Table 12.3. (*Contd.*)

Pythium species	Primer	Sequence (5'-3')	Host and source	Reference
P. rostratifingens	ROS4F ROS3R	GGTGTAGTCCGGCTTGGAGAAGGA GCGCACCGACAAACGCATACC	Wheat	Schroeder et al. (2006)
P. sylvaticum	SYL1F SYL2R	GTGTCTCGCTGTGGTTGGTATATTTG CTTCTGCCAATTGCACAAGTGC	Wheat	Schroeder et al. (2006)
P. ultimum	AFP276 ITS4	TGTATGGAGACGCTGCATT TCCTCCGCTTATTGATATGC	Tomato	Lievens et al. (2006)
P. ultimum*	92F 166R 116T FAM	TGTTTTCATTTTTGGACACTGGA TCCATCATAACTTGCATTACAACAGA CGGGAGTCAGCAGGACGAAGGTTG	Potato	Cullan et al. (2007)
P. ultimum	PuF PuR	ATGATGGACTAGCTGATGAA TTCCATTACACTTCATAGAA	American ginseng	Kernaghan et al. (2008)
P. ultimum var. ultimum	PultRPB-F PultRPB-R	GCAGATTGTCCCGGATATTAAC CACAATAACCGAGAATCAAAG	Potato	Atallah and Stevenson (2006)
P. ultimum var. ultimum	ULT1F ULT4R	GACACTGGAACGGGAGTCAGC AAAGGACTGACAGATTCTCGATC	Wheat	Schroeder et al. (2006)
P. ultimum var. ultimum*	PulF2 PulR2 PulP2 VIC	GCAGGACGAAGGTTGGTCTG GTCCCCACAGTATAAATCAGTATTTAGGT TGGACTAGCTGATGAACTT	Grapevine	Spies et al. (2011)
P. ultimum var. ultimum	SplendF5 SplendR5	TGTAGGTTTGAGACTTC CTATCGATCAAGGTTGTAT	Peach	Bent et al. (2009)
P. vexans	VexansF2 VexansR2	TATACAACCTTGATCGAC GATGGAAAATTGCAACC	Peach	Bent et al. (2009)
P. vexans*	PvF1 PvR1 VexP1 HEX™	TTTCCGTTTTGTGCTTGATG AGCGAACACACCCAATAAGC CCGTGTCTGCTGGCGGGTC	Grapevine	Spies et al. (2011)
Pythium sp.	Pyth664Fwd Pyth712Rev	GCCCTTTCGGGTGTGTTACTAG CTGAATGGCAGAAGAACATCCTC	Asplanchna (rotifer)	Thomas et al. (2011)

*- Based on TaqMan probe method

Table 12.4. Molecular techniques used for identification, differentiation and classification of *Pythium* species

Molecular technique/ marker	Employed for	Application	Example	Reference
RAPD	In genetic analysis of population of a species and its variation	Identification and differentiation of species to subspecies	*Pythium ultimum* var. *ultimum*	Tojo et al. (1998)
AFLP			*P. helicoides*	Garzon et al. (2005)
SSR			*P. aphanidermatum* and *P. irregulare*	Lee and Moorman 2008
ISSR			*P. aphanidermatum*	Amini et al. (2014)
SSU-rDNA*	For identification of taxon (from kingdom to genera)	Identification and phylogenetic analysis/systematics	Associated with ITS amplification through PCR and further sequencing analysis will help to identify the species	
LSU-rDNA*	For identification of taxon (from class to species)			
ITS	For identification of taxon (from family to species)		*P. rhizosaccharum* and *P. terrestris*	Singh et al. (2003), Paul (2002)
COX-I	Identification	Identification and phylogenetic analysis/systematics		Robideau et al. (2011)
COX-II				Martin (2000), Villa et al. (2006)
SNPs	Differentiate the species	Identification and differentiation of species to subspecies	*P. ramorum*	Kroon et al. (2004)
RT-PCR	Identification of species	Identification and quantification of the amplicon directly	*P. aphanidermatum*	Kageyama et al. (2014)
Multiplex PCR	In a single process, many species could be detected using a species-specific probe designed exclusively for a known species		*P. aphanidermatum, P. arrhenomanes, P. graminicola, P. torulosum* and *P. vanterpoolii*	Asano et al. (2010)
DNA arrays	To detect specific sequences	Identification of known species	*P. aphanidermatum*	Tambong et al. (2006)
LAMP				Fukuta et al. (2013)
Genome analysis	To understand the biology of an organism	Used in all aspects of biology	*P. insidiosum*	Patumcharoenpol et al. (2017)

few novel features are unique to *P. ultimum* including expansion of genes involved in proteolysis. A small gene family of cadherins (type-1 transmembrane proteins) involved in cell adhesion was also reported in *P. ultimum* during its genome analysis. The genome and transcriptome analysis revealed the core pathogenic mechanisms within Oomycetes along with lineage-specific genes associated with alternative virulence and lifestyles found within the Pythiaceae lineages compared to the Perenosporaceae (Lévesque et al. 2010).

Pythium insidiosum (ATCC 200269 strain CDC-B5653), isolated from necrotizing lesions on the mouth and eye of a 2-year-old boy (Memphis, Tennessee), was sequenced using a combination of Illumina MiSeq (300 bp paired-end, 14 millions reads) and PacBio (10 Kb fragment library, 356,001 reads) (DDBJ/EMBL/GenBank accession JRHR00000000.1). The sequencing data assembled using SPAdes version # 3.1.0, yielded a total genome size of 45.6 Mb contained in 8992 contigs, N50 of 13 Kb, 57% G + C content and 17,867 putative protein-coding genes (Ascunce et al. 2016). The draft genome sequence of *Pythium insidiosum,* an important pathogen, causes a life-threatening infectious disease called pythiosis in humans and animals with 53.2 Mb + 14,962 open reading frames (Rujirawat et al. 2015). Patumcharoenpol et al. (2017) reported the draft genome of *P. insidiosum* (causal agent of the pythiosis in humans as well as animals) isolated from the environment. Genome sequences of *P. insidiosum* isolated from humans are accessible in public databases. Along with these, they reported two additional draft genome sequences of the *P. insidiosum* strain CBS 573.85 (35.6 Mb in size; accession number, BCFO00000000.1) isolated from a horse suffering from pythiosis and the strain CR02 (37.7 Mb in size; accession number, BCFR00000000.1) has been isolated from the environment. These genome data will further facilitate to elucidate biology, pathogenicity and evolution-related genomic and genetic studies using *P. insidiosum.*

Comparative genome analysis of various *Pythium* species has been carried out in the recent past. Adhikari et al. (2013) reported sequencing, assembly and annotation of six *Pythium* genomes in comparison with other stramenopiles including photosynthetic diatoms and plant pathogenic Oomycetes like *Hyaloperonospora arabidopsidis, Phytophthora* and *Pythium ultimum* var. *ultimum.* Similarly, mitochondrial genome of *P. insidiosum* was analyzed (454-based nuclear genome sequencing method), which provided useful insights to explore the biology and evolution of *P. insidiosum* and other Oomycetes (Tangphatsornruang et al. 2016). Overall, they identified a complete 54.9 kb mt genome sequence, which contains two large inverted repeats and 65 different genes including those of two ribosomal RNA genes, 25 tRNA genes and 38 genes encoding NADH dehydrogenase, cytochrome b oxidase, cytochrome c oxidase, ATP synthase and ribosomal protein. Nearly 39 among the 65 genes have duplications resulting in a total of 104 genes. Phylogenetic analysis using a set of 30 conserved protein coding genes from the mt genomes of *P insidiosum*, 11 from other Oomycetes and two from diatoms (as an outlier) revealed the presence of two lineages comprising Saprolegnialean and Perenosporalean; thus, *P. insidiosum* was more closely related to *Pythium ultimum* than other Oomycetes (Tangphatsornruang et al. 2016).

Conclusions

The members of the genus *Pythium* are the major constraints to agriculture as they are associated with serious plant diseases. Isolation and identification are very crucial to adopt and mitigate disease management strategies. Soil-borne plant pathogens like *Pythium* species are traditionally not easy to study. Many species of *Pythium* could cause disease in plants, animals and humans. Accurate and robust detection and quantification of fungi are essential for diagnosis and surveillance. Molecular biological techniques enable a deeper understanding of natural microbial communities, particularly when those fungi are difficult or impossible to cultivate. In the last decade, valuable molecular sequencing platforms and probes have been developed and diverse quantitative PCR techniques have revolutionized the research on fungal detection and identification especially *Pythium* species. Numerous detection methods are now accessible, but use of these techniques is dependent on the sensitivity, accuracy, robustness, frequency of testing and cost effectiveness. In spite of many novel techniques being accessible, the challenge still remains to identify the uncultivable fungi especially the cryptic fungi and to characterize their diversity in fungal communities in different environments. The advanced molecular techniques described in this chapter provide up-to-date information. Aforesaid molecular tools and techniques could be employed to illustrate the evolution, taxonomy and ecology of *Pythium* species. However, it is not possible to subject all species for molecular identification. The molecular data play a key role in understanding the biology and its interaction with host as they are controlled by host and/or pathogen genetic machinery.

Acknowledgements

The first author (SM) is grateful to the Council of Scientific and Industrial Research (CSIR), Government of India, New Delhi for the award of Research Associate Fellowship and also acknowledges the support and encouragement by the Department of Studies in Botany, University of Mysore. One of the authors (KRS) is grateful to Mangalore University for the award of Adjunct Professorship in the Department of Biosciences.

References

Abbott, C.L., Gilmore, S.R., Lewis, C.T., Chapados, J.T., Peters, R.D., Platt, H.W., Coffey, M.D. and Lévesque, C.A. 2010. Development of a SNP genetic marker system based on variation in microsatellite flanking regions of *Phytophthora infestans*. Can. J. Pl. Pathol. 32: 440-457.

Adhikari, B.N., Hamilton, J.P., Zerillo, M.M., Tisserat, N., Lévesque, C.A. and Buell, C.R. 2013. Comparative genomics reveals insight into virulence strategies of plant pathogenic Oomycetes. PLoS ONE 8: e75072.

Ahoonmanesh, A., Hajimorad, M.R., Ingham, B.J. and Francki, R.I.B. 1990. Indirect double antibody sandwich ELISA for detecting alfalfa mosaic virus in aphids after short probes on infected plants. J. Virol. Met. 30: 271-282.

Al-Sadi, A.M., Drenth, A., Deadman, M.L., De Cock, A.W.A.M. and Aitken, A.E.B. 2007. Molecular characterization and pathogenicity of *Pythium* species associated with damping-off in greenhouse cucumber (*Cucumis sativus* L.) in Oman. Pl. Pathol. 56: 140-149.

Al-Sadi, A.M., Drenth, A., Deadman, M.L., De Cock, A.W.A.M., Al-Said, F.A. and Aitken, A.E.B. 2008a. Genetic diversity, aggressiveness and metalaxyl sensitivity of *Pythium spinosum* infecting cucumber in Oman. J. Phytopathol. 156: 29-35.

Al-Sadi, A.M., Drenth, A., Deadman, M.L. and Aitken, E.A.B. 2008b. Genetic diversity, aggressiveness and metalaxyl sensitivity of *Pythium aphanidermatum* populations infecting cucumber in Oman. Pl. Pathol. 57: 45-56.

Al-Sadi, A.M., Al-Ghaithi, A.G., Al-Balushi, Z.M. and Al-Jabri, A.H. 2012. Analysis of diversity in *Pythium aphanidermatum* populations from a single greenhouse reveals phenotypic and genotypic changes over 2006 to 2011. Pl. Dis. 96: 852-858.

Ali-Shtayeh, M.S. 1985. Value of oogonial and oospore dimensions in *Pythium* species differentiation. Trans. Br. Mycol. Soc. 85: 761-764.

Ali-Shtayeh, M.S., Len, L.C. and Dick, M.W. 1986. An improved method and medium for quantitative estimates of populations of *Pythium* species from soil. Trans. Br. Mycol. Soc. 86: 39-47.

Amini, M., Safaie, N. and Saify Nabiabad, H. 2014. Assessment of genetic diversity in *Pythium aphanidermatum* isolates using ISSR and rep-CR methods. Iran. J. Gen. Pl. Breed. 3: 41-52.

Asano, T., Senda, M., Suga, H. and Kageyama, K. 2010. Development of multiplex PCR to detect five *Pythium* species related to turfgrass diseases. J. Phytopathol. 158: 609-615.

Ascunce, M.S., Huguet-Tapia, J.C., Braun, E.L., Ortiz-Urquiza, A., Keyhani, N.O. and Goss, E.M. 2016. Whole genome sequence of the emerging oomycete pathogen *Pythium insidiosum* strain CDC-B5653 isolated from an infected human in the USA. Genom. Data 7: 60-61.

Atallah, Z.K. and Stevenson, W.R. 2006. A methodology to detect and quantify five pathogens causing potato tuber decay using real-time quantitative polymerase chain reaction. Phytopathol. 96: 1037-1045.

Avila, F.J., Yuen, G.Y. and Klopfenstein, N.B. 1995. Characterization of a *Pythium ultimum* specific antigen and factors that affect its detection using a monoclonal antibody. Phytopathol. 85: 1378-1387.

Bala, K., Robideau, G.P., Lévesque, C.A., De Cock, A.W.A.M., Abad, Z.G., Lodhi, A.M., Shahzad, S., Ghaffar, A. and Coffey, M.D. 2010. *Phytopythium* Abad et al. gen. nov. and *Phytopythium sindhum* Lodhi et al. sp. nov. Fungal Planet 49. Persoonia 24: 136-137.

Barr, D.J.S., Warwick, S.I. and Desaulniers, N.L. 1997. Isozyme variation in heterothallic species and related asexual isolates of *Pythium*. Can. J. Bot. 75: 1927-1935.

Bent, E., Loffredo, A., Yang, J.I., McKenry, M.V., Becker, J.O. and Brneman, J. 2009. Investigations into peach seedling stunting caused by a replant soil. FEMS Microbiol. Ecol. 68: 192-200.

Biazon, L., Meirelles, P.G., Ono, M.A., Itano, E.N., Taniwaki, M.H., Sugiura, Y., Ueno, Y., Hirooka, E.Y. and Ono, E.Y.S. 2006. Development of polyclonal antibodies against *Fusarium verticillioides* exo-antigens. Food Agric. Immunol. 17: 69-77.

Bilodeau, G.J., Koike, S.T., Uribe, P. and Martin, F.N. 2011. Development of an assay for rapid detection and quantification of *Verticillium dahliae* in soil. Phytopathol. 102: 331-343.

Capote, N., Pastrana, A.M., Aguado, A. and Sánchez-Torres, P. 2012. Molecular tools for detection of plant pathogenic fungi and fungicide resistance. pp. 151-202. *In*: Cumagun, C.J. (ed.). Plant Pathology. ISBN: 978-953-51-0489-6, In Tech. (http: //www.intechopen.com/books/plantpathology/molecular-tools-for-detection-of-plant-pathogenic-fungi-and-fungicide-resistance on 20 September 2018).

Chen, W. 1994. Development of specific primers for identification and detection of *Pythium arrhenomanes*. Phytopathol. 84: 1087.

Clark, M.F. and Adams, A.N. 1977. Characteristics of the microplate method of enzyme linked immuno-sorbent assay for the detection of plant viruses. J. Gen. Virol. 34: 475-483.

Cook, R.J. andVeseth, R.J. 1991. Wheat Health Management. American Phytopathological Society, St. Paul, Minnesota.

Cullen, D.W., Toth, I.K., Boonham, N., Walsh, K., Barker, I. and Lees, A.K. 2007. Development and validation of conventional and quantitative polymerase chain reaction assays for the detection of storage rot potato pathogens, *Phytophthora erythroseptica*, *Pythium ultimum* and *Phoma foveata*. J. Phytopathol. 155: 309-315.

Dai, T., Lu. C., Lu, J., Dong, S., Ye, W., Wang, Y. and Zheng, X. 2012. Development of a loop-mediated isothermal amplification assay for detection of Phytophthorasoae. FEMS Microbiol. Lett. 334: 27-34.

de Cock, A.W., Lodhi, A.M., Rintoul, T.L., M'baya, J., Vear, F., Tourvielle, J., Walser, P. and de Labrouhe, D.T. 2014. *Phytopythium*: Molecular phylogeny and systematics. Persoonia 34: 25-39.

Delmotte, F., Giresse, X., Richard-Cervera, S., M'baya, J., Vear, F., Tourvielle, J., Walser, P. and de Labrouhe, D.T. 2008. Single nucleotide polymorphisms reveal multiple introductions into France of *Plasmopara halstedii*, the plant pathogen causing sunflower downy mildew. Infect. Genet. Evol. 8: 534-540.

Es-selman, E.J., Jianqiang, L., Crawford, D.J., Windus, J.L. and Wolfe, A.D. 1999. Clonal diversity in the rare *Calamagrostis porteri* ssp. *insperata* (Poaceae): Comparative results for allozymes and random amplified polymorphic DNA (RAPD) and intersimple sequence repeat (ISSR) markers. Mol. Ecol. 8: 443-451.

Feng, W., Ishiguro, Y., Hotta, K., Watanabe, H., Suga, H. and Kageyama, K. 2015. Simple detection of *Pythium irregulare* using loop-mediated isothermal amplification assay. FEMS Microbiol. Lett. 362: piifnv 174.

Filion, M., St-Arnaud, M. and Jabaji-Hare, S.H. 2003a. Quantification of *Fusarium solani* f. sp. *phaseoli* in mycorrhizal bean plants and surrounding mycorrhizosphere soil using real-time polymerase chain reaction and direct isolations on selective media. Phytopathol. 93: 229-235.

Filion, M., St-Arnaud, M. and Jabaji-Hare, S.H. 2003b. Direct quantification of fungal DNA from soil substrate using realtime PCR. J. Microbiol. Meth. 53: 67-76.

Fleurat-Lessard, P., Luini, E., Berjeaud, J.M. and Roblin, G. 2010. Diagnosis of grapevine esca disease by immunological detection of *Phaeomoniella chlamydospora*. Aust. J. Gr. Wine Res. 16: 455-463.

Fukuta, S., Takahashi, R., Kuroyanagi, S., Miyake, N., Nagai, H., Suzuki, H., Hashizumae, F., Tsui, T., Taguchi, H., Watanabe, H. and Kageyama, K. 2013. Detection of *Pythium aphanidermatum* in tomato using loop-mediated isothermal amplification (LAMP) with species specific primers. Eur. J. Pl. Pathol. 136: 689-701.

Garzón, C.D., Geiser, D.M. and Moorman, G.W. 2005. Diagnosis and population analysis of *Pythium* species using AFLP finger printing. Pl. Dis. 89: 81-89.

Garzón, C.D., Yánez, J.M. and Moorman, G.W. 2007. *Pythium cryptoirregulare*, a new species within the *P. irregulare* complex. Mycologia 99: 291-301.

Gautam, Y., Cahill, D.M. and Hardham, A.R. 1999. Development of a quantitative immunodipstick assay for Phytophthora nicotianae. Food Agric. Immunol. 11: 229-242.

Ghosh, R., Nagavardhini, A., Sengupta, A. and Sharma, M. 2015. Development of Loop-Mediated Isothermal Amplification (LAMP) assay for rapid detection of *Fusarium oxysporum* f. sp. *ciceris* – wilt pathogen of chickpea. BMC Res. Notes 8: 40, DOI 10.1186/s13104-015-0997-z.

Godfrey, S.A.C., Monds, R.D., Lash, D.T. and Marshall, J.W. 2003. Identification of *Pythium oligandrum* using species-specific ITS rDNA PCR oligonucleotides. Mycol. Res. 107: 790-796.

Gomez-Alpizar, L., Saalu, E., Picado, I., Tambong, J.T. and Saborío, F. 2011. A PCR-RFLP assay for identification and detection of *Pythium myriotylum*, causal agent of the ccoyam root rot disease. Let. Appl. Microbiol. 53: 185-192.

Goud, J.C. andTermorshuizen A.J. 2003. Quality of methods to quantify microsclerotia of *Verticillium dahliae* in soil. Eur. J. Pl. Pathol. 109: 523-534.

Gupta, R., Lee, S.E., Agrawal, G.K., Rakwal, R., Park, S., Wang, Y. and Kim, S.T. 2015. Understanding the plant-pathogen interactions in the context of proteomics-generated apoplastic proteins inventory. Front. Pl. Sci. 6: 1-7.

Harvey, P.R., Butterworth, P.J., Hawke, B.G. and Pankhurst, C.E. 2000. Genetic variation among populations of *Pythium irregulare* in Southern Australia. Pl. Pathol. 49: 619-627.

Harvey, P.R., Butterworth, P.J., Hawke, B.G. and Pankhurst, C.E. 2001. Genetic and pathogenic variation among cereal, medic and sub clover isolates of *Pythium irregulare*. Mycol. Res. 105: 85-93.

Hendrix, F.F. and Campbell, W.A. 1973. Pythiums as plant pathogens. Ann. Rev. Phytopathol. 11: 77-98.

Hendrix, F.F. and Papa, K.E. 1974. Taxonomy and genetics of *Pythium*. Proc. Am. Phytopathol. Soc. 1: 200-207.

Herrero, M.L. and Klemsdal, S.S. 1998. Identification of *Pythium aphanidermatum* using the RAPD-technique. Mycol. Res. 102: 136-140.

Hong, C., Richardson, P.A. and Kong, P. 2002. Comparison of membrane filters as a tool for isolating pythiaceous species from irrigation water. Phytopathol. 92: 610-616.

Hughes, C.R. and Queller, D.C. 1993. Detection of highly polymorphic microsatellite loci in a species with little allozyme polymorphism. Mol. Ecol. 2: 131-137.

Ishiguro, Y., Asano, T., Otsubo, K., Suga, H. and Kageyama, K. 2013. Simultaneous detection by multiplex PCR of the high-temperature-growing *Pythium* species: *P. aphanidermatum, P. helicoides* and *P. myriotylum*. J. Gen. Pl. Pathol. 79: 350-358.

Jeffers, S.N. and Martin, S.B. 1986. Comparison of two media selective for *Phytophthora* and *Pythium* species. Pl. Dis. 70: 1038-1043.

Kageyama, K. and Ui, T. 1982. Survival structure of *Pythium* spp. in the soils of bean fields. Ann. Phytopathol. Soc. Japan 48: 308-313.

Kageyama, K. and Ui, T. 1983. Host range and distribution of *Pythium myriotylum* and unidentified *Pythium* sp. contributed to the monoculture injury of bean and soybean plants. Ann. Phytopathol. Soc. Japan 49: 148-152.

Kageyama, K., Tachi, M., Umetsu, M. and Hyakumachi, M. 1996. Epidemiology of *Pythium sulcatum* associated with brown-blotted root rot of carrots. Ann. Phytopathol. Soc. Japan 62: 130-133.

Kageyama, K., Ohyama, A. and Hyakumachi, M. 1997. Detection of *Pythium ultimum* using polymerase chain reaction with species specific primers. Pl. Dis. 81: 1155-1160.

Kageyama, K., Uchino, H. and Hyakumachi, M. 1998. Characterization of the hyphal swelling group of *Pythium*: DNA polymorphisms and cultural and morphological characteristics. Pl. Dis. 82: 218-222.

Kageyama, K. and Nelson, E.B. 2003. Differential inactivation of seed exudate stimulation of *Pythium ultimum* sporangium germination by *Enterobacter cloacae* influences biological control efficacy on different plant species. Appl. Environ. Microbiol. 69: 1114-1120.

Kageyama, K., Komatsu, T. and Suga, H. 2003a. Refined PCR protocol for detection of plant pathogens in soil. J. Gen. Pl. Pathol. 69: 153-160.

Kageyama, K., Suzuki, M., Priyatmojo, A., Oto, Y., Ishiguro, K., Suga, H., Aoyagi, T. and Fukui, H. 2003b. Characterization and identification of asexual strains of *Pythium* associated with root rot of rose in Japan. J. Phytopathol. 151: 485-491.

Karagyozov, L., Kalcheva, I.D. and Chapman, V. 1993. Construction of random small-insert genomic libraries highly enriched for simple sequence repeats. Nucl. Acids Res. 21: 3911-3912.

Keeratijarut, A., Lohnoo, T., Yingyong, W., Sriwanichrak, K. and Krajaejun, T. 2013. A peptide ELISA to detect antibodies against *Pythium insidiosum* based on predicted antigenic determinants of exo-1,3-beta-glucanase. Southeast Asian J. Trop. Med. Pub. Health. 44: 672-680.

Kernaghan, G., Reeleder, R.D. and Hoke, S.M.T. 2008. Quantification of *Pythium* populations in ginseng soils by culture dependent and real-time PCR methods. Appl. Soil Ecol. 40: 447-455.

Klassen, G.R., Balcerzak, M. and de Cock, A.W.A.M. 1996. 5S-ribosomal RNA gene spacers as species-specific probes for eight species of *Pythium*. Phytopathol. 86: 581-587.

Klemsdal, S.S., Herrero, M.L., Wanner, L.A., Lund, G. and Hermansen, A. 2008. PCR-based identification of *Pythium* spp. causing cavity spot in carrots and sensitive detection in soil samples. Pl. Pathol. 57: 877-886.

Koenig, R. 1978. ELISA in the study of homologous and heterologous reactions of plant viruses. J. Gen. Virol. 40: 309-318.

Kroon, L.P.N.M., Verstappen, E.C.P., Kox, L.F.F., Flier, W.G. and Bonants, P.J. 2004. A rapid diagnostic test to distinguish between American and European populations of *Phytophthora ramorum*. Phytopathol. 94: 613-620.

Le, D.P., Smith, M.K. and Aitken, E.A.B. 2017. Genetic variation in *Pythium myriotylum* based on SNP typing and development of a PCR-RFLP detection of isolates recovered from *Pythium* soft rot of ginger. Let. Appl. Microbiol. 65: 319-326.

Le Floch, G., Tambong, J., Vallance, J., Tirilly, Y., Lévesque, A. and Rey, P. 2007. Rhizosphere persistence of three *Pythium oligandrum* strains in tomato soilless culture assessed by DNA macroarray and real-time PCR. FEMS Microbiol. Ecol. 61: 317-326.

Lee, S. and Moorman, G.W. 2008. Identification and characterization of simple sequence repeat markers for *Pythium aphanidermatum, P. cryptoirregulare* and *P. irregulare* and the potential use in *Pythium* population genetics. Curr. Genet. 53: 81-93.

Lévesque, C.A., Vrain, T.C. and De Boer, S.H. 1994. Development of a species-specific probe for *Pythium ultimum* using amplified ribosomal DNA. Phytopathol. 84: 474-478.

Lévesque, C.A., Harlton, C.E. and de Cock, A.W.A.M. 1998. Identification of some oomycetes by reverse dot blot hybridization. Phytopathol. 88: 213-222.

Lévesque, C.A. and de Cock, A.W.A.M. 2004. Molecular phylogeny and taxonomy of the genus *Pythium*. Mycol. Res. 108: 1363-1383.

Lévesque, C.A., Brouwer, H., Cano, L., Hamilton, J.P., Holt, C., Huitema, E., Raffaele, S., Robideau, G.P., Thines, M., Win, J., Zerillo, M.M., Beakes, G.W., Boore, J.L., Busam, D., Dumas, B., Ferriera, S., Fuerstenberg, S.I., Gachon, C.M., Gaulin, E., Govers, F., Grenville-Briggs, L., Horner, N., Hostetler, J., Jiang, R.H., Johnson, J., Krajaejun, T., Lin, H., Meijer, H.J., Moore, B., Morris, P., Phuntmart, V., Puiu, D., Shetty, J., Stajich, J.E., Tripathy, S., Wawra, S., van West, P., Whitty, B.R., Coutinho, P.M., Henrissat, B., Martin, F., Thomas, P.D., Tyler, B.M., De Vries, R.P., Kamoun, S., Yandell, M., Tisserat, N. and Buell, C.R. 2010. Genome sequence of the necrotrophic plant pathogen *Pythium ultimum* reveals original pathogenicity mechanisms and effect or repertoire. BMC Gen. Biol. 11R73: 1-22.

Li, M., Senda, M., Komatsu, T., Suga, H. and Kageyama, K. 2010. Development of real-time PCR technique for the estimation of population density of *Pythium intermedium* in forest soils. Microbiol. Res. 165: 695-705.

Li, M., Asano, T., Suga, H. and Kageyama, K. 2011. A multiplex PCR for the detection of *Phytophthora nicotianae* and *P. cactorum* and a survey of their occurrence in strawberry production areas of Japan. Pl. Dis. 95: 1270-1278.

Lian, C. and Hogetsu, T. 2002. Development of microsatellite markers in black locust (*Robinia pseudoacacia*) using dual suppression-PCR technique. Mol. Ecol. Notes 2: 211-213.

Lievens, B., Brouwer, M., Vanachter, A.C.R.C., Cammue, B.P.A. and Thomma, B.P.H.J. 2006. Real-time PCR for detection and quantification of fungal and oomycete tomato pathogens in plant and soil samples. Pl. Sci. 171: 155-165.

Litt, M. and Luty, J.A. 1989. A hyper variable microsatellite revealed by *in vitro* amplification of a dinucleotide repeat within the cardiac muscle actin gene. Am. J. Hum. Genet. 44: 397-401.

Liu, Y.G. and Whittier, F. 1995. Thermal asymmetric interlaced PCR: Automatable amplification and sequencing of insert end fragment from P1 and YAC clones for chromosome walking. Genomics 25: 674-681.

MacDonald, J.D., Stites, J. and Kabashima, J. 1990. Comparison of serological and culture plate methods for detecting species of *Phytophthora*, *Pythium* and *Rhizoctonia* in ornamental plants. Pl. Dis. 74: 655-659.

Martin, F. 2000. Phylogenetic relationships among some 20 *Pythium* species inferred from sequence analysis of the mitochondrially encoded cytochrome oxidase II gene. Mycologia 92: 711-727.

Martin, F. and Tooley, P. 2003. Phylogenetic relationships of *Phytophthora ramorum*, *P. nemorosa*, and *P. pseudosyringae*, three species recovered from areas in California with sudden oak death. Mycol. Res. 107: 1379-1391.

Martin, F.N. and Kistler, H.C. 1990. Species-specific banding patterns of restriction endonuclease – digested mitochondrial DNA from the genus *Pythium*. Exp. Mycol. 14: 32-46.

Martin, F.N. 1991. Selection of DNA probes useful for isolate identification of two *Pythium* spp. Phytopathol. 81: 742-746.

Martin, F.N. 1992. *Pythium*. pp. 39-49. *In*: Singleton, L.L., Mihail, J.D. and Rush, C.M. (eds.). Methods for Research on Soil Borne Phytopathogenic Fungi. American Phytopathological Society, St. Paul, Minnesota.

Martin, F.N. 1995. Meiotic instability of *Pythium sylvaticum* as demonstrated by inheritance of nuclear markers and karyotype analysis. Genetics 139: 1233-1246.

Martin, F.N. 2008. Mitochondrial haplotype determination in the oomycete plant pathogen *Phytophthora ramorum*. Curr. Genet. 54: 23-34.

Martin, F.N. 2009. *Pythium* Genetics. pp. 213-239. *In*: Lamour, K. and Kamoun, S. (eds.). Oomycete Genetics and Genomics. John Wiley and Sons, New Jersey.

Matsumoto, C., Kageyama, K., Suga, H. and Hyakumachi, M. 1999. Phylogenetic relationships of *Pythium* species based on ITS and 5,8 s sequences of the ribosomal DNA. Mycoscience 40: 321-331.

Matsumoto, C., Kageyama, K., Suga, H. and Hyakumachi, M. 2000. Intraspecific DNA polymorphisms of *Pythium irregulare*. Mycol. Res. 104: 1333-1341.

Morita, Y. and Tojo, M. 2007. Modifications of PARP medium using fluazinam, miconazole, and nystatin for detection of *Pythium* spp. in soil. Pl. Dis. 91: 1591-1599.

Morrison, T.M., Weis, J.J. and Wittwer, C.T. 1998. Quantification of low-copy transcripts by continuous SYBR green I monitoring during amplification. Biotechniques 24: 954-962.

Nitsche, A., Steuer, N., Schmidt, C.A., Landt, O. and Siegert, W. 1999. Different real-time PCR formats compared for the quantitative detection of human cytomegalovirus DNA. Clin. Chem. 45: 1932-1937.

Notomi, T., Okayama, H., Masubuchi, H., Yonekawa, T., Watanabe, K., Amino, N. and Hase, T. 2000. Loop-mediated isothermal amplification of DNA. Nucl. Acids Res. 28: E63.

Park, C., Kakinuma, M. and Amano, H. 2001. Detection and quantitative analysis of zoospores of *Pythium porphyrae*, causative organism of red rot disease in *Porphyra* by competitive PCR. J. Appl. Phycol. 13: 433-441.

Park, C. 2006. Rapid detection of *Pythium porphyrae* in commercial samples of dried *Porphyra yezoensis* sheets by polymerase chain reaction. J. Appl. Phycol. 18: 203-207.

Patumcharoenpol, P., Rujirawat, T., Lohnoo, T., Yingyong, W., Vanittanakom, N., Kittichotirat, W. and Krajaejun, T. 2017. Draft genome sequences of the oomycete *Pythium insidiosum* strain CBS 573.85 from a horse with pythiosis and strain CR02 from the environment. Data in brief. 16: 47-50.

Paul, B. 2002. *Pythium terrestris*, a new species isolated from France, its ITS region, taxonomy and its comparison with related species. FEMS Microbiol. Lett. 212: 255-260.

Plaats-Niterink, A.J.V. 1981. Monograph of the Genus *Pythium*. Stu. Mycol. # 21. Centraal Bureauvoor Schimmel Cultures, Netherlands.

Ray, M., Dash, S., Achary, K.G., Nayak, S. and Singh, S. 2018. Development and evaluation of polyclonal antibodies for detection of *Pythium aphanidermatum* and *Fusarium oxysporum* in ginger. Food Agric. Immunol. 29: 204-215.

Robideau, G.P., de Cock, A.W.A.M. and Coffey, M.D. 2011. DNA barcoding of oomycetes with cytochrome c oxidase subunit I and internal transcribed spacer. Mol. Ecol. Resour. 11: 1002-1011.

Rujirawat, T., Patumcharoenpol, P., Lohnoo, T., Yingyong, W., Lerksuthirat, T., Tangphatsornruang, S., Suriyaphol, P., Grenville-Briggs, L.J., Garg, G., Kittichotirat, W. and Krajaejun, T. 2015. Draft genome sequence of the pathogenic oomycete *Pythium insidiosum* strain Pi-S, isolated from a patient with pythiosis. Genome Announcement 3: e00574-15.

Sano, A. and Itano, E.N. 2010. Applications of loop-mediated isothermal amplification methods (LAMP) for identification and diagnosis of mycotic diseases: Paracoccidioidomycosis and *Ochroconis gallopava* infection. pp. 417-437. *In*: Gherbawy, Y. and Voigt, K. (eds.). Molecular Identification of Fungi. Springer, Berlin, Heidelberg.

Sawaya, S., Bagshaw, A., Buschiazzo, E., Kumar, P., Chowdhury, S., Black, M.A. and Gemmel, N. 2013. Microsatellite tandem repeats are abundant in human promoters and are associated with regulatory elements. PLoS ONE 8: e54710.

Schnerr, H., Niessen, L. and Vogel, R.F. 2001. Real-time detection of the tri5 gene in *Fusarium* by Light Cycler-PCR using SYBR Green I for continuous fluorescence monitoring. Int. J. Food Microbiol. 71: 53-61.

Schroeder, K.L., Okubara, P.A., Tambong, J.T., Lévesque, C.A. and Paulitz, T.C. 2006. Identification and quantification of pathogenic *Pythium* spp. from soils in eastern Washington using real-time polymerase chain reaction. Phytopathol. 96: 637-647.

Schroeder, K.L., Martin, F.N., de Cock, A.W.A.M., Lévesque, C.A., Spies, C.F.J., Okabura, P.A. and Paulitz, T.C. 2013. Molecular detection and quantification of *Pythium* species: Evolving taxonomy, new tools, and challenges. Pl. Dis. 97: 4-20.

Schröter, J. 1897. Saprolegnineae III Pythiaceae. Engler &Prantl, Nat. Pfl. Fam. 1: 104-105.

Shane, W.W. 1991. Prospects for early detection of *Pythium* blight epidemics on turf grass by antibody-aided monitoring. Pl. Dis. 75: 921-925.

Singh, K.K., Mathew, R., Masih, I.E. and Paul, B. 2013. ITS region of the rDNA of *Pythium rhizosaccharum* sp. nov. isolated from sugarcane roots: Taxonomy and comparison with related species. FEMS Microbiol. Lett. 221: 233-236.

Spies, C.F.J., Mazzola, M. and McLeod, A. 2011. Characterisation and detection of *Pythium* and *Phytophthora* species associated with grapevines in South Africa. Eur. J. Pl. Pathol. 131: 103-119.

Stanghellini, M.E. and Hancock, J.G. 1970. A quantitative method for isolation of *Pythium ultimum* from soil. Phytopathol. 60: 551-552.

Suzuki, H., Tsuji, T., Hashizume, F., Fajita, A. and Kuroda, K. 2013. Population dynamics of high-temperature-growing *Pythium* species on tomato in hydroponic and effects of water temperature on incidence. Jap. J. Phytopathol. 79: 58.

Takenaka, S., Sekiguchi, H., Nakaho, K., Tojo, M., Masunaka, A. and Takahashi, H. 2008. Colonization of *Pythium oligandrum* in the tomato rhizosphere for biological control of bacterial wilt disease analyzed by real-time PCR and confocal laser-scanning microscopy. Phytopathol. 98: 187-195.

Tambong, J.T., de Cock, A.W.A.M., Tinker, N.A. and Lévesque, C.A. 2006. Oligonucleotide array for identification and detection of *Pythium* species. Appl. Environ. Microbiol. 72: 2691-2706.

Tangphatsornruang, S., Ruang-Areerate, P., Sangsrakru, D., Rujirawat, K., Lohnoo, T., Kittichotirat, W., Patumcharoenpol, P., Grenville-Briggs, L.J. and Krajaejun, T. 2016. Comparative mitochondrial genome analysis of *Pythium insidiosum* and related Oomycete species provides new insights into genetic variation and phylogenetic relationships. Genes 575: 34-41.

Tautz, D. 1989. Hyper variability of simple sequences as a general source for polymorphic DNA markers. Nul. Acids Res. 17: 6463-6471.

Thomas, S.H., Housley, J.M., Reynolds, A.N., Penczykowski, R.M., Kenline, K.H., Hardegree, N., Schmidt, S. and Duffy, M.A. 2011. The ecology and phylogeny of oomycete infections in *Asplanchna rotifers*. Freshwat. Biol. 56: 384-394.

Tojo, M., Nakazono, E., Tsushima, S., Morikawa, T. and Matsumoto, N. 1998. Characterization of two morphological groups of isolates of *Pythium ultimum* var. *ultimum* in a vegetable field. Mycoscience 39: 135-144.

Tsao, P.H. 1970. Selective media for isolation of pathogenic fungi. Ann. Rev. Phytopathol. 8: 157-186.

Uzuhashi, S., Tojo, M. and Kakishima, M. 2010. Phylogeny of the genus *Pythium* and description of new genera. Mycosci. 51: 337-365.

Vallance, J., Le Floch, G., Déniel, F., Barbier, G., Lévesque, C.A. and Rey, P. 2009. Influence of *Pythium oligandrum* biocontrol on fungal and oomycete population dynamics in the rhizosphere. Appl. Environ. Microbiol. 75: 4790-4800.

van der Plaats-Niterink, A.J. 1981. Monograph of the Genus *Pythium*. Stud. Mycol. 21: 1-224.

Vasseur, V., Rey, P., Bellanger, E., Brygoo, Y. and Tirilly, Y. 2005. Molecular characterization of *Pythium* group F isolates by ribosomal- and inter microsatellite-DNA regions analysis. Eur. J. Pl. Pathol. 112: 301-310.

Villa, N.O., Kageyama, K. and Asano, T. 2006. Phylogenetic relationships of *Pythium* and *Phytophthora* species based on ITS rDNA, cytochrome oxidase II and beta-tubulin gene sequences. Mycologia 98: 410-422.

Wang, P.H. and Chang, C.W. 2003. Detection of the low-germination-rate resting oospores of *Pythium myriotylum* from soil by PCR. Lett. Appl. Microbiol. 36: 157-161.

Wang, P.H., Chung, C.Y., Lin, Y.S. and Yeh, Y. 2003a. Use of polymerase chain reaction to detect the soft rot pathogen, *Pythium myriotylum*, in infected ginger rhizomes. Lett. Appl. Microbiol. 36: 116-120.

Wang, P.H., Wang, Y.T. and White, J.G. 2003b. Species specific PCR primers for *Pythium* developed from ribosomal ITS1 region. Lett. Appl. Microbiol. 37: 127-132.

Ward, E., Foster, S.J., Fraaije, B.A. and McCartney, A.H. 2004. Plant pathogen diagnostics: Immunological and nucleic acid-based approaches. Ann. Appl. Biol. 145: 1-16.

Watanabe, H., Kageyama, K., Taguchi, Y., Horinouchi, H. and Hyakumachi, M. 2008. Bait method or detect *Pythium* species that grow at high temperatures in hydroponic solutions. J. Gen. Pl. Pathol. 74: 417-424.

Watanabe, T. 1981. Distribution and populations of *Pythium* species in the northern and southern part of Japan. Ann. Phytopathol. Soc. Japan 47: 449-456.

Watanabe, T. 1989. Kinds, distribution, and pathogenicity of *Pythium* species isolated from soils of Kyushu Island in Japan. Ann. Phytopathol. Soc. Japan 55: 32-40.

Waterhouse, G.M. 1967. Key to *Pythium pringsheim*. Mycol. Pap. 109: 1-15.

Weber, J.L. and May, P.E. 1989. Abundant class of human DNA polymorphism which can be typed using the polymerase chain reaction. Am. J. Hum. Genet. 44: 388-396.

White, T., Bruns, T., Lee, S. and Taylor, J. 1990. Amplification and direct sequencing of fungal ribosomal RNA genes for phylogenetics. pp. 315-322. *In*: Innis, M.A., Gelfand, D.H., Sninsky, J.J. and White, T.J. (eds.). PCR Protocols: A Guide to Methods and Applications. Academic Press, New York.

Wittwer, C.T., Herrmann, M.G., Moss, A.A. and Rasmussen, R.P. 1997. Continuous fluorescence monitoring of rapid cycle DNA amplification. Biotechniques 22: 130-138.

Yin-Ling, Zhou, W., Motohashi, K., Suga, H., Fukui, H. and Kageyama, K. 2009. Development of microsatellite markers for *Pythium helicoides*. FEMS Microbiol. Lett. 293: 85-91.

Yuen, G.Y., Craig, M.L. and Avila, F. 1993. Detection of *Pythium ultimum* with a species specific monoclonal antibody. Pl. Dis. 77: 692-698.

Yuen, G.Y., Xia, J.Q. and Sutula, C.L. 1998. A sensitive ELISA for *Pythium ultimum* using polyclonal and species-specific monoclonal antibodies. Pl. Dis. 82: 1029-1032.

Zane, L., Bargelloni, L. and Patarnello, T. 2002. Strategies for microsatellite isolation: A review. Mol. Ecol. 11: 1-16.

Zietkiewicz, E., Rafalski, A. and Labuda, D. 1994. Genome fingerprinting by simple sequence repeat (SSR)-anchored polymerase chain reaction amplification. Genomics 20: 176-183.

PART III
As a Human Pathogen

13

Pythium insidiosum – An Emerging Mammalian Pathogen

Erico S. Loreto[1], Juliana S.M. Tondolo[1,2] and Janio M. Santurio[2*]

[1] Sobresp Faculdade de Ciências da Saúde, Rua Appel n° 520, Santa Maria-RS, Brazil
[2] Laboratório de Pesquisas Micológicas (LAPEMI), Departamento de Microbiologia e Parasitologia, Universidade Federal de Santa Maria (UFSM), Av. Roraima n° 1000, Prédio 20, sala 4139, Santa Maria-RS, Brazil

Introduction

Oomycetes are saprotrophic or opportunistic pathogens that can cause devastating diseases in both animals and plants (van West and Beakes 2014, Kamoun et al. 2015). Although some species of oomycetes, including *Aphanomyces astaci*, *Saprolegnia parasitica*, and *Halioticida noduliformans*, can cause disease in animals (Phillips et al. 2008, Derevnina et al. 2016), *Pythium insidiosum* stands out as a mammalian pathogen and can cause life-threatening disease in apparently healthy hosts, including equines, canines, and humans.

The earliest literary descriptions of pythiosis occurred in the late 19th century in India (Smith 1884, Drouin 1896) and the name "bursattee" (Smith 1884) was used to describe a fungal disease of horses grazing near or within areas containing stagnant water. Although these authors observed the mycelial nature of the disease, their identification was not possible because the microorganism did not sporulate, and it was only in the early 20th century that De Haan and Hoogkamer (1901) made the first extensive description of the disease in affected horses and called it *Hyphomycosis destruens* and subsequently as *Hyphomycosis destruens equi* (Haan and Hoogkamer 1903). In the 1920s, Witkamp expanded clinical, and laboratory data on pythiosis and first described that experimental pythiosis could be reproduced in rabbits (Witkamp 1924, 1925).

The disease has subsequently been described under different names, including expressions commonly used in many countries, such as equine phycomycosis (Bridges and Emmons 1961, Hutchins and Johnston 1972), as well as by regional designations such as *bursattee* or *bursatti* in India (Smith 1884), *swamp cancer* (Austwick and Copland 1974) and phycomycotic granuloma (Johnston and Henderson 1974) in Australia, granular dermatitis in Japan (Ichitani and Amemiya 1980), *Florida horse leeches* in the United States (Bitting 1894), *espundia* in Latin America (Gonzalez-Ch and Ruiz-M 1975), and *ferida da moda* in Brazil (Leal et al. 2001).

From the earliest description of pythiosis and throughout most of the twentieth century, it was believed that the microorganism causing cutaneous granulomas in horses was a fungus since hyphae were observed in the clinical material recovery from infected animals. In the 1960s, Bridges and Emmons (1961) designated the etiological agent of pythiosis as *Hyphomyces destruens*. However, until the 1970s, only vegetative mycelial growth was described in microbiological cultures.

*Corresponding author: janio.santurio@gmail.com

Austwick and Copland (1974), evaluating a clinical isolate from an equine granuloma in New Zealand, considered the similarity of *H. destruens* and *Mortierella* spp. to perform a series of cultures with the aim of inducing the production of sporangia. Among the methods used, the observation of biflagellate zoospores (asexual reproduction) was possible after the transfer of portions of colonies from Sabouraud dextrose agar to a Petri dish containing water and sterilized rotten maize silage. With this observation, the authors suggested that *H. destruens* should be included in the genus *Pythium*. In the 1980s, the descriptions of sexual reproduction initially classified the microorganism isolated from granuloma of equines as *Pythium gracile* Schenk (Ichitani and Amemiya 1980) but the term "pythiosis" was only used in 1987 when the name *Pythium insidiosum* was proposed after a detailed description of the sexual reproduction of the microorganism (De Cock et al. 1987). In the same year, Shipton (1987) analyzed an Australian equine isolate and classified it as a new species: *Pythium destruens*. However, in 1989, serological tests carried out with antigens and rabbit-antisera of *P. insidiosum* as well as *P. destruens* demonstrated the same antigenic profile and established the definitive nomenclature for the etiological agent of pythiosis: *Pythium insidiosum* (Mendoza and Marin 1989).

The genus *Pythium* belongs to the kingdom Straminipila, the class Oomycota, the order Pythiales, and the family Pythiaceae (Rujirawat et al. 2015, McCarthy and Fitzpatrick 2017). These eukaryotes are microscopic, form filamentous cells and belong to SAR supergroup (Straminipila-Alveolata-Rhizaria), which includes animal and plant pathogens (Adl et al. 2012, Beakes and Thines 2017, McCarthy and Fitzpatrick 2017).

Phylogenetic analyses suggest that the genus *Pythium* is polyphyletic, that is, its species descend from two or more independent ancestors. These results indicate that revisions in the taxonomy of the genus *Pythium* are necessary and some authors suggest at least the creation of five new genera (Uzuhashi et al. 2010, Huang et al. 2013). *Pythium* can be divided into 10 clades (A to J), with *P. insidiosum* belonging to clade C. The main morphological difference between these clades refers to the sporangial morphology (filamentous, elongated, ovoid or globose) (Levesque and de Cock 2004, Uzuhashi et al. 2010, McCarthy and Fitzpatrick 2017, McGowan and Fitzpatrick 2017).

Description of the agent

P. insidiosum is a eukaryotic microorganism which belongs to the oomycetes, a diverse group of pathogens as well as free-living saprophytes that morphologically resemble filamentous fungi (Thines 2018). However, it is more phylogenetically related to brown alga, microalgae, autotrophic diatoms, phytoplankton, and *Blastocystis* than to fungi (McCarthy and Fitzpatrick 2017).

Oomycetes and fungi can be distinguished biochemically by the differences observed in the composition of the cell wall. Unlike fungi, which contain chitin as its main structural component, oomycetes have a small amount of this carbohydrate, contains cellulose and several other polysaccharides, particularly β-glucans (Aronson et al. 1967, Judelson and Blanco 2005, Tondolo et al. 2017a). Other differences can be observed in the structure of mitochondria (Powell et al. 1985), actin cytoskeleton (Meijer et al. 2014) and repertoire of proteins (Seidl et al. 2011).

Although oomycetes can incorporate sterols, such as ergosterol, from external sources for the maintenance of their physiological functions (Hendrix 1964, Gaulin et al. 2010), most of these microorganisms, including *P. insidiosum* (Lerksuthirat et al. 2017), have an incomplete sterol biosynthesis pathway. Consequently, they are resistant or exhibit reduced susceptibility to sterol-binding drugs and sterol biosynthesis inhibitors, such as amphotericin B and triazole antifungal (Loreto et al. 2014a). Sterols, in general, are important for the production of reproductive structures *in vitro* but are not necessary for the growth of vegetative hyphae (Grooters 2003).

Microscopically, the development of the vegetative mycelium of *P. insidiosum* is like fungi, particularly resembling agents of entomophthoromycosis and mucormycosis. The induction of the asexual reproduction *in vitro* (a useful laboratory tool for a presumptive diagnosis of animal pythiosis) are only observed in water culture, which consist of colonized grass leaves or small piece of mycelium with agar submerged in sterile distilled water containing positive ions (mainly K^+, Ca^{2+} and Mg^{2+}) or soil extract, followed by incubation at 37°C (De Cock et al. 1987, Mendoza and Prendas 1988, Tondolo et al. 2017b).

Under water-culture (Fig. 13.1), the zoospores develop within large vesicles (20 to 60 µg in diameter) formed terminally on the branches. After the rupture of this vesicle (zoosporangia), the biflagellate zoospores are liberated and, after a few minutes of swimming, they come to rest, lose their flagella and encyst. Under favorable conditions (for example, contact with a mammal tissue), young hyphae (germ tubes) growing from encysted zoospores can actively penetrate the host tissues, exerting a force up to 6.9 µN (pressures of 0.3 µN µm^{-2}) (Ravishankar et al. 2001). Considering that pressure is not sufficient to penetrate the undamaged skin by mechanics alone, it is presumed that there is secretion of proteolytic enzymes that reduce tissue resistance, allowing the infectious process (Ravishankar et al. 2001, MacDonald et al. 2002).

P. insidiosum rarely produces, on routine microbiological culture media, sexual reproductive structures, such as oogonia and antheridia, which are morphological features evaluated for the identification of oomycetes (De Cock et al. 1987, Vilela and Mendoza 2015).

The genome size of oomycetes ranges from 33 to 229 megabases (Mb) and *Albugo* and *Pythium* species harbor smaller genomes (33 to 53 Mb) (Rujirawat et al. 2018). *P. insidiosum* genome size, isolated from a Thai patient with vascular pythiosis, is 53.2 Mb (Rujirawat et al. 2015). A total

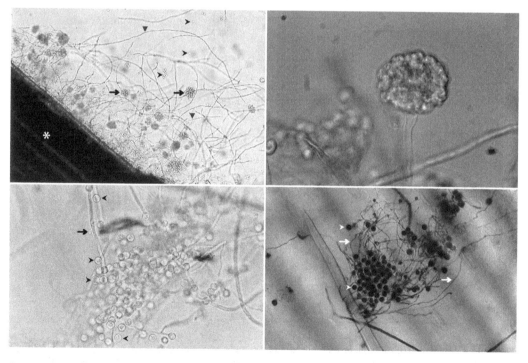

Figure 13.1. Water culture of *P. insidiosum*. Upper left: grass leaves (*Paspalum notatum*, ∗) colonized by *P. insidiosum*. Note that zoosporangia (arrows, →) containing zoospores were formed at the tip of the hyphae (arrowheads, ▼). Encysted zoospores (arrowheads, ▶) can be seen nearby (100x); upper right: a close-up of a lobed zoosporangium of P. insidiosum, strain CBS 101555. This zoosporangium is in the intermediate stage and has six minutes of development (400x); lower left: numerous encysted zoospores (arrowheads, ▶), some forming germ tube (arrow, →) (400x); lower right: in lactophenol blue stain, several encysted zoospores (arrowheads, ▶) initiating new colonization on the leaves of grass. Observe the formation of long filaments (hyphae, arrows, →) (400x). Photos from author's collection

genome size of 45.6 Mb was described for a strain isolated from an infected human in the USA (Ascunce et al. 2016). These genome data are publicly available and will be useful in future studies to explore the genetic and evolutionary characteristics of *P. insidiosum*.

Phylogenetic analysis based on ribosomal DNA internal transcribed spacers (ITS) (Schurko et al. 2003b, Chaiprasert et al. 2009, Kammarnjesadakul et al. 2011, Azevedo et al. 2012), cytochrome oxidase II (COX II) DNA coding sequences (Kammarnjesadakul et al. 2011, Azevedo et al. 2012) and *exo-1,3-β-glucanase* gene (*exo1*) (Ribeiro et al. 2017) demonstrated that *P. insidiosum* isolates from around the world can be phylogeographic subdivided into three distinct clades or groups. Although group nomenclature and the isolates origin studied can differ among these studies, the results have a close concordance. For example, Schurko et al. (2003b) described that Clade I consisted of isolates from the western hemisphere, while clade II included isolates from Asia, Australia and New Zealand and clade III comprised isolates from Americas and Asia (Schurko et al. 2003b).

Azevedo et al. (2012) data supported evolutionary proximity among *P. insidiosum* isolated from countries of the Americas (PI-A group) and a paraphyletic relation between groups of isolates from Thailand (PI-C) in relation to Brazilian isolates (Fig. 13.2). Based on these data, it can be suggested that these *P. insidiosum* groups may form (i) a phylogenetic complex of closely related species or (ii) a complex of different strains belonging to a single species.

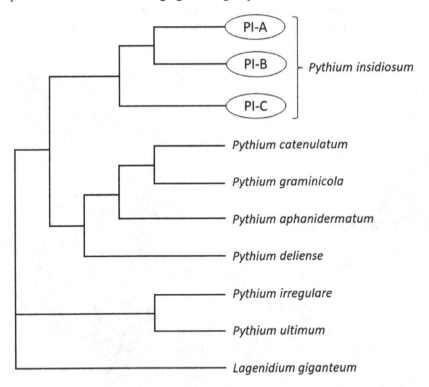

Figure 13.2. Phylogeny of *Pythium insidiosum* based on cytochrome oxidase II (COX II) sequences. The figure illustrates the three major polytomous groups: PI-A represents the America's (Brazil, Ecuador, and the USA) isolates whereas PI-B and PI-C contain isolates from Asia (Thailand). Modified from Azevedo et al. (2012)

Epidemiology and pathogenesis

Infections caused by *P. insidiosum* have, until recently, been considered rare, exotic and restricted to animals in tropical regions and humans in endemic areas, such as Thailand, with few or no occurrences in other geographical areas. However, in the last two decades, the number of cases of pythiosis in animals and humans has increased in all continents (Thianprasit et al. 1996, Krajaejun et al. 2006, Gaastra et al. 2010, Vilela and Mendoza 2015, Hilton et al. 2016, Bagga et al. 2018), suggesting that pythiosis has in the past been misdiagnosed as a fungal infection or that the disease has emerged in the last years.

Pythiosis has been described in tropical, subtropical and temperate areas of the world. In the Americas, the disease is common in the tropical regions of Central, North and South America, and most cases have been described in Brazil, Colombia, Costa Rica, the United States and Venezuela (De Cock et al. 1987, Gaastra et al. 2010, Santos et al. 2011c, d, Marcolongo-Pereira et al. 2012, Santos et al. 2014, Buitrago-Mejía et al. 2017). Studies in the Rio Grande do Sul, a southern Brazilian state, described a seroprevalence for *P. insidiosum* of 11.1% in horses (Weiblen et al. 2016) and the potential of this pathogen to circulate across broad areas in Argentina, Brazil and Uruguay (Machado et al. 2018). Also, the Brazilian Pantanal region is probably the area of the highest incidence and prevalence of equine pythiosis in the world (Mendoza et al. 1996, Leal et al. 2001).

In Asia, pythiosis has been reported in South Korea, the Pacific Islands, India, Indonesia, Japan, and Thailand and also in areas close to this continent, such as Australia, New Guinea and New Zealand (Gaastra et al. 2010, Vilela and Mendoza 2015). Tropical Africa, although presenting ideal environmental conditions for pythiosis, has only one case of canine pythiosis described (Rivierre et al. 2005).

The life cycle of *P. insidiosum* (Fig. 13.3) was first proposed in the 1980s by Miller (1983) and later described by other authors (Mendoza et al. 1993, Fonseca et al. 2014). *P. insidiosum* grows in areas with aquatic vegetation through the colonization or infection of plants (Fig. 13.3, right), in which it develops mycelia and zoosporangia. Biflagellate mobile zoospores, which were released from zoosporangia, can swim through the aquatic environment and attached to a new host. After a stage of encystment, germ-tube formation occurs, and new colonization or infection begins. The environmental conditions are determinant for the development of the organism in its ecosystem. According to Miller and Campbell (1982), the production of zoospores requires temperatures between 30 and 40°C and the accumulation of water in wetlands, ponds, and swamps. Also, the higher incidences of pythiosis cases were observed during or after the rainy season. In the Brazilian Pantanal, the most cases of equine pythiosis are recorded between February and May (summer-autumn), a period that corresponds to the peak of the floods (Leal et al. 2001).

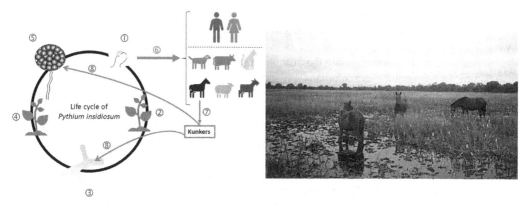

Figure 13.3. The proposed life cycle of *Pythium insidiosum* (**left**): zoospores in water (**1**) can attach to a host plant (**2**) and germinate forming vegetative mycelia (**3**). Under favorable conditions, the mycelia on the natural host plant (**4**) can form zoosporangia (**5**) (asexual reproduction), and the zoospores released into the environment (**1**) may attach to a new host plant. Mammals, when they come in contact with zoospores (**6**), can develop pythiosis. The *kunkers* (**7**), found inside sinus in equines tissues with pythiosis, can be released into the aquatic environment (**7**) and originate new mycelia and zoosporangia (**8**), restarting the biological cycle; **right**: horses in their habitat in the Brazilian Pantanal region. It stands out the massive proliferation of aquatic plants, producing ideal conditions for the life cycle of *P. insidiosum*. Photo from author's collection

In horses with pythiosis, the formation of *kunkers*, which are necrotic masses containing *P. insidiosum* hyphae, can be recovered easily from the lesions. From the *kunkers,* if released in humid environments, zoosporangia can be observed, and based on *in vitro* studies, can "restart" the asexual reproduction cycle of *P. insidiosum* (Fonseca et al. 2014). Furthermore, the most frequent anatomic areas affected by pythiosis are the distal regions of the limbs, ventral portion of the abdomen and face, which coincidentally are regions where the animals have limitations in scaring hematophagous insects, such as some species of flies (Santos et al. 2011c, d, 2014). Although the role of these insects in the biological cycle of *P. insidiosum* remains uncertain, it is possible that flies may act as direct or indirect agents of injuries and abrasions in animals (which could favor the penetration of zoospores in injured skin) or as direct disease (vectors), since *P. insidiosum* was recovered from insect larvae (Schurko et al. 2003a, Vilela et al. 2018).

Zoospores have, as shown by Miller (1983) and Mendoza et al. (1993), a strong tropism for the skin tissue, hair of humans and horses, water lily and grass leaves, as well as a weak attraction for other varieties of leaves. In this context, it is believed that the infection in mammals occurs after the zoospores of *P. insidiosum* come into contact with the skin or the gastrointestinal tract of susceptible hosts. Although *P. insidiosum* has been recovered or identified from environmental sources and areas for agriculture in both aquatic and soil environments (Supabandhu et al. 2008, Vanittanakom et al. 2014, Presser and Goss 2015), laboratory data have determined that hyphae of *P. insidiosum* have no characteristics (force) to invade healthy (intact) epithelial tissue (Ravishankar et al. 2001, MacDonald et al. 2002), suggesting that lesions or micro-injuries in the tissue are necessary for the onset of infection.

Most cases of pythiosis occur in apparently healthy humans and other mammals. Due to the aggressive and destructive nature of infections, it is believed that this pathogen - as well as other oomycetes - has acquired the necessary weaponry to invade mammalian tissue at a more recent evolutionary stage, with evidence that the major genes of virulence were acquired by horizontal transfer from other eukaryotic and prokaryotic species (Jiang and Tyler 2012).

P. insidiosum potential virulence factors include (i) genes involved in the processes of antioxidation, thermal adaptation, immunomodulation, iron metabolism and sterol binding (Krajaejun et al. 2011) and urease (Rujirawat et al. 2018); (ii) cellulose-binding elicitor lectin type proteins (possible involvement in adhesion and hemagglutination of the host cell), elicitins (possibly involved in host cell modulation) (Krajaejun et al. 2011, 2014), glucan 1,3-beta-glucosidase, heat shock protein (HSP) 70 and enolase (Chechi et al. 2018).

P. insidiosum, after entering and adhering to the host, requires mechanisms to avoid its removal by the immune system of the host. Mendoza et al. (2003) proposed that the antigens released by the hyphae of *P. insidiosum* modulate the immune response of the host and may be responsible for maintaining an eosinophilic granulomatous response, blocking the immune system in an immune response with T helper type 2 (Th2) lymphocytes through the continuous stimulatory production of more eosinophils and mast cells, which in turn help to protect the microorganism from cells of the host immune system and leads to a worsening condition, and if not treated properly, can lead to host death. Consequently, the hyphae of *P. insidiosum* surrounded by degranulated eosinophils would be camouflaged within the eosinophilic microabscesses, preventing their complete presentation to the immune system and, thus, ensuring their viable presence in the infected tissues.

This cell recruitment to infection sites is regulated by Toll-like receptors (TLRs) after sensing the pathogen-associated molecular patterns (PAMPs). Information about this pathway in pythiosis is scarce, but it was demonstrated that Toll-deficient *Drosophila melanogaster* becomes susceptible to infection instead of wild-type flies that were resistant (Zanette et al. 2013c). Recently, a study on human corneal epithelial cells and macrophages showed that both zoospore and hypha form of *P. insidiosum* could be recognized by TLR2 promoting inflammatory cytokine induction and contributing to the pathogenesis of pythiosis (Wongprompitak et al. 2018).

These characteristics and the subsequent finding of elevated levels of IgE in humans and horses with the disease validated the concept of a Th2 modulation by *P. insidiosum* during infection (Mendoza et al. 2003, Mendoza and Newton 2005). Elevated levels of interleukin (IL)-4, IL-5 and IL-10 were also detected in human patients with pythiosis, confirming a Th2 immune response (Thitithanyanont et al. 1998, Wanachiwanawin et al. 2004). Additionally, in a murine immunosuppressive model of pythiosis, high levels of IL-10 were observed, and this cytokine is associated to Th1/Th2 balance in fungal infections, with high levels leading to down-regulation of Th1 cytokines (Cyktor and Turner 2011, Tondolo et al. 2017b).

Transmission of *P. insidiosum* between humans, between animals or from animal to human (or vice versa) has not been described (Gaastra et al. 2010, Vilela and Mendoza 2015).

Clinical findings and experimental pythiosis

Pythiosis in humans

Although pythiosis affects apparently healthy individuals (Triscott et al. 1993, Thianprasit et al. 1996, Bagga et al. 2018), in Thai patients the disease has a higher frequency in individuals with thalassemia and other blood disorders (Sathapatayavongs et al. 1989, Thitithanyanont et al. 1998, Wanachiwanawin et al. 2004, Laohapensang et al. 2005, Krajaejun et al. 2006, Sermsathanasawadi et al. 2016). The different forms of pythiosis in humans are:

(i) Superficial infections (keratitis): they resemble those caused by fungi and other microorganisms and usually start after trauma, with subsequent development of conjunctivitis, photophobia, corneal ulcers and hypopyon (Badenoch et al. 2001, 2009, Lekhanont et al. 2009, Tanhehco et al. 2011, Barequet et al. 2013, Hung and Leddin 2014, Lelievre et al. 2015, Bagga et al. 2018, Neufeld et al. 2018, Rathi et al. 2018).

(ii) Cutaneous and subcutaneous forms: granulomatous lesions characterized by plaques and/ or ulcers that remain localized. Papules and itching may occur if the pathogen reaches the subcutaneous tissue (Triscott et al. 1993, Bosco et al. 2005, Salipante et al. 2012, Khunkhet et al. 2015). Also included here is orbital pythiosis, which is a rare subcutaneous presentation of the disease (Mendoza et al. 2004, Kirzhner et al. 2015).

(iii) Vascular: associated with traumatic injury, usually in the lower limbs, followed by dissemination of the pathogen to nearby arteries. Infected skin shows signs of dry gangrene and the formation of painful necrotic ulcers can be observed in some patients. Clinical symptoms include claudication of the affected limb, local ischemia, edema, pain, and absence of the wrist on the affected foot. Ascending arteritis, with the formation of thrombi and aneurysms, is the main characteristic observed as the infection progresses (Sudjaritruk and Sirisanthana 2011, Keoprasom et al. 2012, Hahtapornsawan et al. 2014, Permpalung et al. 2015, Reanpang et al. 2015, Sermsathanasawadi et al. 2016, Chitasombat et al. 2018).

(iv) Systemic infection: vascular pythiosis, when untreated, may spread through the arteries and reach the iliac and renal arteries and the abdominal aorta, causing disseminated pythiosis (Krajaejun et al. 2006, Schloemer et al. 2013), which is usually fatal (Wanachiwanawin et al. 1993, Chanprasert et al. 2015).

Pythiosis in horses

Horses are the animals most affected by pythiosis, with no predisposition of age, race or sex. Subcutaneous lesions are predominant and are located mainly in the distal extremities of the limbs (Fig. 13.4, left), thoracoabdominal (ventral portion) and face (Reis et al. 2003, Santos et al. 2011c, d, Grooters 2013), probably due to the longer contact time with water containing zoospores.

Clinically, the animals present good initial body condition, but rapid weight loss occurs from the second or third week of the evolution of cutaneous lesions (Santos et al. 2011c, d). The lesions are ulcerative, granulomatous and of tumor appearance, increase progressively and can form large tissue masses with more than 50 cm in diameter. The size of the lesions is dependent on the duration and site of the infection and commonly present bloody mucous discharge, and the animals usually mutilate the lesion, probably to relieve the discomfort of the intense pruritus.

The lesions have irregular borders and contain hyphae covered by necrotic cells, which form white-yellow masses like corals, known as *kunkers*. The *kunkers*, found inside sinus, are formed in the granular tissue, measure 2 to 10 mm in diameter, have an irregular shape and sandy aspect, being considered pathognomonic of equine pythiosis. The *kunkers* are usually easily extracted when pressure is exerted on the injuries.

Claudication is common among horses with limb injuries, and most clinical reports describe only one lesion in each animal. However, there is a description of multifocal skin lesions (Santurio and Ferreiro 2008, Santos et al. 2011c, d), dissemination to internal organs (Reis et al. 2003) and atypical cases in the Brazilian Pantanal where the animals with pythiosis present progressive thinning, with deforming lesions covered with darkened and thick skin (Fig. 13.4, right), but without the constant discharge of secretions and maintenance of appearance and size after one year of observation (Leal et al. 2001).

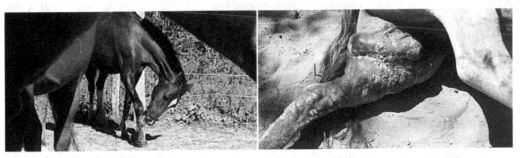

Figure 13.4. Cases of equine pythiosis in equines from the Brazilian Pantanal region. **Left**: Equine with classical pythiosis on the lower limb in self-mutilation process. Horses with pythiosis present clinical signs of intense pruritus at the site of the lesion; **right**: atypical lesion of pythiosis described by Leal et al. (2001). A non-suppurative lesion is observed, covered with dark skin. Under pressure, *kunkers* can be easily recovered from the lesions. Photos from author's collection

Intestinal pythiosis represents the second most common form of the disease in horses and is characterized by episodes of colic due to the presence of tumor masses that cause a decrease or obstruction of the intestinal lumen. Data from surgical excision and necropsy show intestinal ulceration and nodular masses up to 20 cm in diameter located on the wall of the jejunum (Santurio and Ferreiro 2008, Grooters 2013).

In addition to cutaneous and intestinal pythiosis, other organs and tissues may be affected secondary to cutaneous lesions. Bone lesions are usually limited to bones adjacent to chronic subcutaneous limb lesions, which are usually refractory to conventional treatments and characterized by exostoses, areas of osteolysis and signs of osteomyelitis on radiological examination, and identification of hyphae in areas of necrosis at the histopathological examination (Alfaro and Mendoza 1990, Poole and Brashier 2003). Spread of infection through the lymphatic system to the lungs and regional lymph nodes have also been reported (Alfaro and Mendoza 1990, Reis et al. 2003).

P. insidiosum infections in other mammalian and non-mammalian hosts

Canine pythiosis may occur with involvement of the subcutaneous tissues as well as the gastrointestinal tract. In the subcutaneous lesions ulcerations, swelling, sinus tracts and involvement of lymph nodes can be observed (Thomas and Lewis 1998, Dykstra et al. 1999, Krockenberger et al. 2011, Oldenhoff et al. 2014). Intestinal pythiosis is characterized by large, tumor-like masses in the affected intestinal areas. Diarrhea, fever, lethargy, anorexia, weight loss, and death are common in cases of canine pythiosis without medical treatment (Fischer et al. 1994, Patton et al. 1996, Thomas and Lewis 1998, Helman and Oliver 1999, Berryessa et al. 2008). Extension of infection may occur to the pancreas, mesenteric lymph nodes and bile ducts (Thomas and Lewis 1998, Grooters et al. 2003, Berryessa et al. 2008).

The first description of pythiosis in sheep occurred in Brazil in 2004, in the form of epizootic events (Tabosa et al. 2004), in animals that grazed for long periods on the banks of ponds. Ulcerative and swollen lesions in the limbs, abdominal region and pre-scapular of the animals are observed in cutaneous sheep pythiosis (Tabosa et al. 2004, do Carmo et al. 2015). The disease can also be manifested as granulomatous rhinitis, which has been described as an enlargement and facial deformity, dyspnea, sneezing and fetid serous-bloody nasal discharge, bilaterally infarcted submandibular and retropharyngeal lymph nodes, cachexia and progressive thinning (Santurio et al. 2008, Portela et al. 2010, Bernardo et al. 2015, Mustafa et al. 2015). Cases of digestive tract infection, where food regurgitation, lethargy, and anorexia were observed (Pessoa et al. 2012) and pulmonary metastasis, characterized by multifocal nodules, have also been reported (Tabosa et al. 2004).

Pythiosis in cattle affects animals sporadically (Miller et al. 1985, Santurio et al. 1998, Silva et al. 2011) or in epizootic events (Perez et al. 2005, Gabriel et al. 2008, Grecco et al. 2009, Luis-León et al. 2009). In most cases, granulomatous lesions are cutaneous/subcutaneous and localized in the limbs. However, multifocal cutaneous lesions have also been described in the region of the neck, abdomen, sternum, tail and nasal chamfer (Gabriel et al. 2008, Grecco et al. 2009, Cardona Álvarez et al. 2012). The disease can evolve with the spontaneous healing of the lesions or be fatal, especially in cases where the lesions are painful, and the animal cannot stand, which can result in severe dehydration (Gaastra et al. 2010).

Pythiosis in domestic and wild cats is infrequent. In domestic cats, nasal and retrobulbar (Bissonnette et al. 1991), gastrointestinal (Rakich et al. 2005) and sublingual (Fortin et al. 2017) infections were described. In wild cats, one case of fatal pneumonia was described in a 7-month-old Jaguar (*Panthera onca*) (Camus et al. 2004) and an abdominal infection was described in a 6-year-old Bengal tiger (*Panthera tigris tigris*) kept in captivity (Buergelt et al. 2006). Anorexia, vomiting and involuntary loss of feces, as well as abdominal mass, may be observed in gastrointestinal infections (Camus et al. 2004, Rakich et al. 2005, Buergelt et al. 2006).

One case of pythiosis affecting the face and stomata and two reports of granulomatous vulvar lesions were described for camels (*Camelus dromedarius*) kept in captivity (Wellehan et al. 2004, Videla et al. 2012). Only one case of pythiosis in birds has been reported. The infected animal was a nestling white-faced ibis (*Plegadis chihi*) with multifocal cutaneous ulcerations in the wings, neck, head, and limbs. The microscopic characteristics of the disease, including intense and necrotizing eosinophilic and granulomatous inflammation, resembled those previously described in mammals with pythiosis (Pesavento et al. 2008). *P. insidiosum* was also described as an opportunistic pathogen for shrimp (*Litopenaeus vannamei*). Mortality was related to respiratory failure as well as due to the necrosis of the gill (Otta et al. 2018).

Experimental models of pythiosis

Witkamp (1924) was the first researcher to describe the experimental reproduction of pythiosis using rabbits. Since then, this animal was recognized, until recently, as the only option for conducting *in vivo* experimental pythiosis (Fig. 13.5). In the 1980s, Miller and Campbell (1983) used *P. insidiosum* zoospores as infectious units and found that both groups of healthy and immunosuppressed rabbits developed the disease, concluding that this disease could be experimentally induced in immunocompetent rabbits. Since then, this model has been used to evaluate the diagnosis, antimicrobial susceptibility, immunotherapy and immune response of the pythiosis (Santurio et al. 2003, Leal et al. 2005, Pereira et al. 2007, 2008, Bach et al. 2010, Botton et al. 2011, Argenta et al. 2012, Loreto et al. 2012, Bach et al. 2013, Pires et al. 2013, Fonseca et al. 2015a, Jesus et al. 2016, Zanette et al. 2013a, b, 2015).

Figure 13.5. Experimental pythiosis in rabbits. The figure shows the macroscopic aspect of subcutaneous lesion developed 30 days after subcutaneous inoculation of *Pythium insidiosum* zoospores (20,000 cells/mL). The region of the lesion was trichotomized. Photo from author's collection

In recent years, a number of studies dedicated to exploring new experimental models have demonstrated that pythiosis can be induced in *Drosophila melanogaster* Toll-deficient (Zanette et al. 2013c), in immunosuppressed BALB/c (Tondolo et al. 2017b) and Swiss (Loreto et al. 2018a) mice, and in embryonated chicken eggs (Verdi et al. 2018).

Laboratory diagnosis

Direct examination, culture, and zoosporogenesis

The hyphae of *P. insidiosum* are broad (4 to 10 μm in diameter), hyaline, branched at a 90° angle and sparsely septate and can be observed on direct examination with 10% potassium hydroxide (KOH) from clinical specimens, such as macerated tissue or sliced *kunkers* preparations, in the cytological examination. The presence of fragments of short and broad hyphae observed through the staining of Gram and Giemsa were also described in some cases of pythiosis (Neto et al. 2010, Connolly et al. 2012, Vilela and Mendoza 2015, Chatterjee and Agrawal 2018, Rathi et al. 2018).

Isolation of *P. insidiosum* can be performed in several culture media routinely used in the microbiological routine, such as Sabouraud dextrose agar, 5% sheep blood agar, corn meal agar, agar potato and nutrient agar (Fig. 13.6, left). It is recommended to avoid the cooling of the samples and to send the collected material to the laboratory as soon as possible, preferably within 24 hours (Gaastra et al. 2010, Vilela and Mendoza 2015). The growth of *P. insidiosum* mycelia around minocycline disks (30 μg) is inhibited (Fig. 13.6, right) and this test can be used as a presumptive method for the differentiation of *P. insidiosum* from filamentous fungi (Tondolo et al. 2013).

In horses, recovery of the agent is more likely to occur from the *kunkers* than from the infected tissue (Grooters et al. 2002b). Tissue samples and *kunkers* should be sliced into fragments approximately 2 to 10 mm in diameter and physically adhered to the surface (or slightly submerged) of the agar (Vilela and Mendoza 2015). Clinical materials suspected of environmental microbial contamination may be washed with sterile water or with an antibiotic solution (Grooters et al. 2002b). However, it should be considered that antibacterial drugs acting through inhibition of protein synthesis, such as chloramphenicol, may inhibit the *in vitro* growth of *P. insidiosum* (Loreto et al. 2014a) and should be avoided in both wash solutions and selective culture media. *P. insidiosum* grows as radial or irregularly radiate, submerged, white or colorless colonies after incubation at 37°C within 24 to 48 hours.

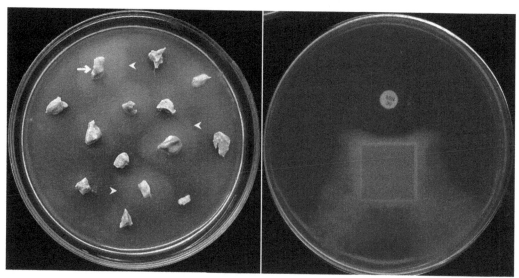

Figure 13.6. The microbiological culture of *Pythium insidiosum* on corn meal agar (CMA). **Left**: mycelia growth of *P. insidiosum* (yellow arrowheads) around sliced tissues (yellow arrow) from rabbits experimentally induced pythiosis after 24 h of incubation at 37°C; **right**: inhibitory effect of minocycline disk (30 μg) on the growth of *P. insidiosum* after 72 h of incubation at 35°C. Photos from author's collection

The induction of zoosporogenesis (asexual formation of zoospores) can be achieved by placing previously autoclaved grass blades on the surface of colonies previously cultured in water agar or corn meal agar for 24 to 48 hours, re-incubation at 37°C for 24 hours, followed by transfer of infected grass blades to an induction medium composed of various mineral salts, which are incubated again for 8 to 24 hours prior to observation of zoosporangia formation and release of zoospores (Mendoza and Prendas 1988, Grooters 2013, Tondolo et al. 2017b) (Fig. 13.1). Although zoospore production is an important feature for the diagnosis of pythiosis, it is not specific for *P. insidiosum* since other oomycetes also produce zoospores. However, it excludes true fungi as agents of disease.

Serological and molecular methods

Detection of anti-*P. insidiosum* antibodies in the serum of infected patients can be used for the diagnosis of pythiosis. Enzyme-Linked Immuno-Sorbent Assay (ELISA) and Western Blot (WB) techniques are both very sensitive and specific for the detection of anti-*P. insidiosum* IgG antibodies (Mendoza et al. 1992a, Grooters et al. 2002a, Krajaejun et al. 2002, Vanittanakom et al. 2004, Keeratijarut et al. 2013) but cross-reactivity with anti-*Lagenidium* antibodies can be observed (Grooters 2003). Detection of anti-*P. insidiosum* can be done through immunochromatographic (ICT) tests and has the advantage of producing fast results and without the need for specialized laboratory machinery (Krajaejun et al. 2009). Comparison of the immunodiffusion (ID), ELISA, ICT, and haemagglutination (HA) tests for pythiosis diagnosis shows that ID is relatively insensitive, and HA is fast but has a sensitivity of 84% and specificity of 82%. ELISA and ICT presented the best results, with sensitivity and specificity of 100% (Chareonsirisuthigul et al. 2013).

The detection of hyphae (or its elements) in clinical samples of pythiosis cases by molecular methods is an important tool for the early diagnosis of the disease and in cases in which the microorganism cannot be recovered in microbiological cultures. Preliminary identification of *P. insidiosum* is possible after extraction of total genomic DNA from clinical samples and subsequent detection of amplicons obtained by PCR using specific primers.

The set of PI-1 and PI-2 primers, which amplify a region of 105-base pairs (bp) from the *P. insidiosum* ITS-1 region, was proposed by Grooters and Gee (2002) and has been used in other studies for the diagnosis and/or confirmation of pythiosis (Jaeger et al. 2002, Vanittanakom et al.

2004, Botton et al. 2011), being totally specific when tested against several filamentous fungi (Vilela and Mendoza 2015). Other primer sets include Cox Pi 5 and Cox Pi 6 (Worasilchai et al. 2018a, b), P24F and P24R (Ribeiro et al. 2017).

P. insidiosum was also identified from clinical specimens by sequencing part of the 18S ribosomal DNA using the NS1, NS2 and ITS universal primers (Badenoch et al. 2001, Vanittanakom et al. 2004) and by dot-blot hybridization, constructing a 530-bp species-specific DNA probe, which specifically binds to the intergenic spacer 1 (IGS1) of this pathogen (Schurko et al. 2004). In addition to the diagnosis of pythiosis, this technique is suggested for the detection of *P. insidiosum* in the environment, since it has not been shown to hybridize with the genomic DNA of 23 other species of *Pythium, Lagenidium gigantum* or several pathogenic fungi, including *Conidiobolus coronatus* and *Basidiobolus ranarum* (Vilela and Mendoza 2015).

The use of molecular techniques exclusively for the diagnosis of pythiosis should be evaluated with caution, since the specificity of the primers is still under investigation, and the amplicons obtained by PCR must be submitted to sequencing to confirm the identity of the microorganism (Schurko et al. 2003b, Vanittanakom et al. 2004, Vilela and Mendoza 2015).

Histology and immunohistochemistry

P. insidiosum appears as hyaline, coenocytic or sparsely hyphae, with a diameter of 6 to 10 μm in histopathological preparations stained with hematoxylin and eosin (H&E). This microorganism triggers, in the host tissue, eosinophilic granulomas with giant cells, mast cells and other inflammatory cells. Hyphae (or their elements) are found in the center of microabscesses with numerous eosinophils that usually degranulate on the hyphae of the microorganism (Splendore-Hoeppli phenomenon). These characteristics are also commonly observed in fungi infections of the order Entomophthorales (*Basidiobolus* spp. and *Conodiobolus* spp.), of which the pythiosis must be differentiated. Hyphae are easily visualized in Grocott-Gomori silver methenamine (GMS) stained but not with periodic acid Schiff (PAS) (Martins et al. 2012, Vilela and Mendoza 2015).

The immunohistochemistry technique using polyclonal anti-*P.insidiosum* antibodies was first described by Brown et al. (1988) and have been used as a confirmatory diagnostic test of pythiosis in several studies (Patton et al. 1996, Reis and Nogueira 2002, Pedroso et al. 2009, Dória et al. 2014). This technique has the advantage of being applicable in paraffin-embedded tissues, in addition to differentiating the hyphae of *P. insidiosum* from other fungi (Martins et al. 2012, Grooters 2013).

Treatment

Surgical removal of all affected tissues is the traditional treatment most used in pythiosis and shows good results in small and superficial lesions. However, removal of the lesion with a safety margin to avoid relapses is often hampered by the anatomical regions normally involved in the pythiosis. Consequently, success rates of approximately 45% are observed with surgical procedures (Mendoza 2009). Nevertheless, the surgical removal of the area affected by the disease represents the last therapeutic resource in many cases of pythiosis, especially in humans with vascular and ocular involvement (Krajaejun et al. 2006, Barequet et al. 2013, Permpalung et al. 2015, Sermsathanasawadi et al. 2016).

Antifungal and antibacterial drugs

Studies from the 1960s reinforced the idea that *P. insidiosum* was probably a fungus, and since then, antifungal drugs have been (and continue to be) used in the treatment of pythiosis. However, these drugs have shown little therapeutic success since the beginning of their use (McMullan et al. 1977, Thianprasit et al. 1996). From the reclassification of this pathogen as an oomycete in the 1970s, it was understood that the unsatisfactory response of antifungal agents, such as amphotericin B and imidazole drugs, was associated with an incomplete set of sterol biosynthetic enzymes described for *P. insidiosum* (Lerksuthirat et al. 2017). Despite this, reports of cure of some cases of canine, equine and human pythiosis with antifungal drugs (Shenep et al. 1998, Hummel et al. 2011, Dória et al. 2015) were described and, additionally, *in vitro* studies demonstrated that *P. insidiosum* presents a varied susceptibility to antifungal drugs (Pereira et al. 2007, Argenta et al. 2008, Cavalheiro et al. 2009, Loreto et al. 2014a). However, monotherapy with antifungal drugs can be considered ineffective for the treatment of pythiosis, especially in humans (Krajaejun et al. 2006, Permpalung et al. 2015, Reanpang et al. 2015, Sermsathanasawadi et al. 2016).

Pythium species differentiate from true fungi because they are quite sensitive to antibacterial drugs that inhibit protein synthesis, such as tetracycline, chloramphenicol, streptomycin, and erythromycin (Marchant and Smith 1968, Rawn and Vanetten 1978, Guo and Ko 1994), and these antimicrobials are likely to inhibit the growth of the microorganism by interfering with cytoplasmic and mitochondrial protein synthesis (Rawn and Vanetten 1978). In this context, several classes of antibacterial drugs were reported for their *in vitro* anti-*P. insidiosum* activity (Fig. 13.7), such as macrolides (azithromycin and clarithromycin), tetracyclines (doxycycline, minocycline) and tigecycline, and oxazolidinones (linezolid and sutezolid) (Loreto et al. 2014a, Bagga et al. 2018, Loreto et al. 2018b).

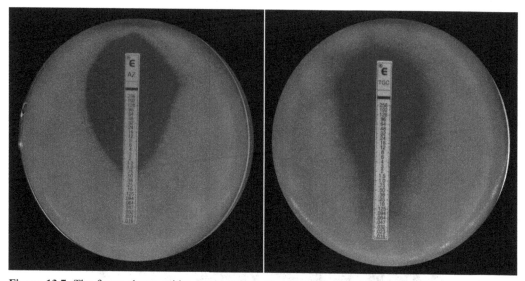

Figure 13.7. The figure shows azithromycin (AZ, left) and tigecycline (TGC, right) Etest strips on agar. For both drugs, ~0.5 µg/mL (observe the intersection of the mycelia growth ellipse with the Etest strip) was considered the minimal inhibitory concentration (MIC) of the antimicrobial substances. Photos from author's collection

The first *in vivo* evaluation of the antimicrobial activity of azithromycin, clarithromycin, minocycline, and tigecycline was performed in an experimental model of pythiosis in rabbits (Jesus et al. 2016). In this model, complete remission or reduction of subcutaneous lesions was observed in all treatments, except for clarithromycin. Regarding human pythiosis treatment, in two cases of presumptive *P. insidiosum* keratitis, the therapy with a combination of azithromycin and linezolid (Ramappa et al. 2017) or voriconazole (Chatterjee and Agrawal 2018) resulted in the successful resolution of the lesions. Additionally, one case of ocular pythiosis was successfully treated with surgery and minocycline (Ros Castellar et al. 2017). Both topical and oral regimen treatments were used in these cases. However, favorable (Bagga et al. 2018) and unfavorable (Agarwal et al. 2018) responses to azithromycin/linezolid therapy in a series of cases of patients with *P. insidiosum* keratitis have been reported.

Immunotherapy

The poor clinical response to the treatment of pythiosis with antifungal drugs stimulated the research of alternative treatments, such as immunotherapy. Miller (1981) and Miller and Campbell (1982) were the first investigators to report the use of *P. insidiosum* antigens with therapeutic potential when injected into horses. Subsequent studies, using different methods for the preparation of *P. insidiosum* antigens (Mendoza and Alfaro 1986, Mendoza et al. 1992b, Monteiro 1999, Mendoza et al. 2003, Santurio et al. 2003), demonstrated that this therapy demonstrated cure rates above 70% and 55% in horses and humans, respectively, but low cure rates in other animals, as reviewed by Loreto et al. (2014b).

Santurio et al. (2003) performed a comparative analysis of the activity of immunotherapeutics prepared from hyphae submitted to sonication, maceration (or liquefaction) or the combination of these two techniques in the experimental pythiosis in rabbits and observed that only the macerated immunotherapeutic was effective, leading to a reduction of 71.8% in lesion size with clinical cure of two rabbits ($n = 5$). Studies of equines from the Brazilian Pantanal using the macerated immunotherapeutic (PitiumVac®, Fig. 13.8) showed cure rates of 83% ($n = 110$) (Monteiro 1999), 75% ($n = 8$) (Santos et al. 2011d), 79% ($n = 34$) (Santos et al. 2014) and 84.6% ($n = 13$) (Santos et al. 2014) to 90% ($n = 11$) (Santos et al. 2011c) when it was associated with surgical excision.

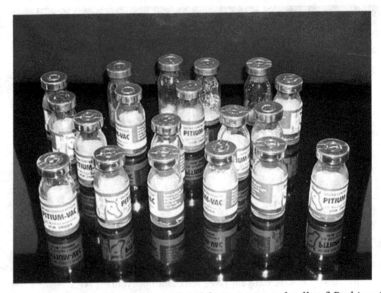

Figure 13.8. Vaccine (immunotherapy) produced from macerated cells of *Pythium insidiosum* (PitiumVac®). Photo from author's collection.

Considering the higher incidence of pythiosis in horses, most of the efficacy data of immunotherapy are described for these species. However, there is also the description of its use in canines (Thitithanyanont et al. 1998, Wanachiwanawin et al. 2004, Laohapensang et al. 2005, Pupaibool et al. 2006, Lekhanont et al. 2009, Sudjaritruk and Sirisanthana 2011, Keoprasom et al. 2012, Salipante et al. 2012, Schloemer et al. 2013, Thanathanee et al. 2013), camels (Wellehan et al. 2004, Videla et al. 2012) and sheep (Carrera et al. 2013), with or without adjuvant therapies and with variable clinical outcomes.

Immunotherapy in human pythiosis has been described both in successful treatments and in therapeutic failures - and in association with surgical procedures and the use of various antimicrobial drugs (Thitithanyanont et al. 1998, Wanachiwanawin et al. 2004, Laohapensang et al. 2005, Pupaibool et al. 2006, Lekhanont et al. 2009, Sudjaritruk and Sirisanthana 2011, Keoprasom et al. 2012, Salipante et al. 2012, Schloemer et al. 2013, Thanathanee et al. 2013).

The mechanism proposed for the therapeutic success of immunotherapy is based on the change in the cellular response. Immunotherapeutic antigens produce the expression of T lymphocytes type 1 (Th1), production of IL-2 and interferon-γ, with the mobilization of T lymphocytes and macrophages that destroy the cells of *P. insidiosum* (Mendoza and Newton 2005, Loreto et al. 2014b). Also, there is an increase in the number of anti-*P. insidiosum* antibodies following immunotherapeutic treatment, which is directly correlated with the aid in curing the disease (Newton and Ross 1993). However, the use of immunotherapy against pythiosis does not induce a protective response against reinfections. Animals with pythiosis that evolved to cure through immunotherapy are susceptible to reinfections if kept in an aquatic environment that contains zoospores of *P. insidiosum* (Santos et al. 2011b).

Efforts have been made to improve the knowledge about how the *P. insidiosum* antigens stimulate the immune system response. In this way, Tondolo et al. (2017a) extracted and identified a highly branched (1,3)(1,6)-β-d-glucan from *P. insidiosum* mycelium that has showed to induce a significant and specific Th17 cellular immune response in pre-immunized BALB/c mice. Other study tested primed dendritic cells (DCs) with this same glucan, heat inactivated zoospores and the PitiumVac® immunotherapic. The results showed that lymphocyte proliferation induced by DCs primed with these antigens promotes the activation of the Th1 and Th17 subset of cytokines (Ledur et al. 2018).

Other antimicrobial compounds and therapeutic proposals

Iodides, as agents proposed in pythiosis control, have been used since the earliest descriptions of the treatment of this disease (Hutchins and Johnston 1972, Murray et al. 1978, Ader 1979). Although iodides are described in monotherapy or associated with other therapies in several studies (Salomão-Nascimento et al. 2010, Videla et al. 2012, Hahtapornsawan et al. 2014, Fujimori et al. 2016, Ros Castellar et al. 2017), it can be considered a compound that presents varied results in pythiosis treatment and, in addition, can cause severe toxic side effects in some treated mammals (Bridges and Emmons 1961, Sterling and Heymann 2000, Costa et al. 2013).

Other adjuvant treatments include (i) the use of neodymium: yttrium-aluminum-pomegranate laser in pythiosis granulomas in equines (Sedrish et al. 1997), which was associated with a decreased risk of recurrence; (ii) photodynamic therapy, which demonstrated *in vitro* inhibition of *P. insidiosum* (Pires et al. 2014) and control of the disease in experimentally infected rabbits (Pires et al. 2013) and (iii) the use of iron chelating therapy (Zanette et al. 2013a) associated or not with immunotherapy. Also, several studies have described anti-*P. insidiosum* activity of compounds of plant origin, such as lignans from *Alyxia schlechteri* (Sriphana et al. 2013a), Clauraila E from *Clausena harmandiana* (Sriphana et al. 2013b), essential oils from *Origanum vulgare*, *Mentha piperita* (Fonseca et al. 2015a) and species of the family Lamiaceae (Fonseca et al. 2015b), carvacrol and thymol (Jesus et al. 2015), oil and nanoemulsion of *Melaleuca alternifolia* (Valente et al. 2016), propolis and geopropolis (Araujo et al. 2016), tannins from *Stryphnodendron adstringens* (Trolezi et al. 2017) and furanocoumarin from *Scaevola taccada* (Suthiwong et al. 2017).

Prevention

Most cases of pythiosis are associated with previous exposure of the host, with lesions or micro-lesions on the skin, in aquatic or to soil and grass in endemic areas, which contain the infective propagules of *P. insidiosum* (zoospores or hyphae). In endemic areas, susceptible hosts should avoid these sites, especially if skin lesions are observed. Transmission between animal, human or animal-human has not yet been described. However, it is recommended that health care professionals and persons in direct contact with pythiosis hosts use personal protective equipment, such as gloves and goggles, to manipulate the lesions. Contact lens wearers should be aware of the correct contact lens and hand sanitizer procedures, even if they do not reside in endemic areas of the disease.

Conclusion and future perspectives

P. insidiosum is a species that appears to be closely related to several other species of *Pythium* but that evolutionarily acquired mechanisms of virulence and specificity to cause infections in mammals. A clear understanding of the pathogenicity mechanisms of *P. insidiosum* is necessary for the development of strategies that will allow a faster and more accurate diagnosis to more effective therapeutic options.

The difficulty in diagnosis is related to the morphological similarities of this species with true fungi and the fact that few physicians and clinical laboratories are familiar with this pathological agent. In this context, new laboratory tools that allow the precise identification of *P. insidiosum* are required. Moreover, standardization of antimicrobial susceptibility testing and definition of clinical breakpoints are required for more effective pharmacological support.

Despite the morphological similarity with filamentous fungi, the antifungal drugs have little or no antimicrobial action against *P. insidiosum*, probably due to the diverse structures of the membrane and cell wall and the particularities in the pathway of synthesis of sterols. Noteworthy, several classes of antibacterial drugs that act by inhibiting protein synthesis of microorganisms have demonstrated anti-*P. insidiosum* activity *in vitro* and *in vivo*. Further clinical studies are needed and should expand the knowledge of the therapeutic potential of these drugs. However, the use of these drugs, even if proven effective, may be limited by the high financial cost for the treatment of animal pythiosis, particularly for large animals and in the epizootic cases of the disease. In these cases, immunotherapy is advantageous.

Immunotherapy has been widely used in the treatment of pythiosis in horses and has favorable results. Current immunotherapeutic formulations use crude *P. insidiosum* antigens. The identification and the *in vitro* and *in vivo* evaluation of immunodominant antigens should stimulate the development of more effective immunotherapies and adjuvants that may result in protective vaccination strategies.

References

Ader, P.L. 1979. Phycomycosis in fifteen dogs and two cats. J. Am. Vet. Med. Assoc. 174(11): 1216-1223.

Adl, S.M., Simpson, A.G., Lane, C.E., Lukes, J., Bass, D., Bowser, S.S., Brown, M.W., Burki, F., Dunthorn, M., Hampl, V., Heiss, A., Hoppenrath, M., Lara, E., Le Gall, L., Lynn, D.H., McManus, H., Mitchell, E.A., Mozley-Stanridge, S.E., Parfrey, L.W., Pawlowski, J., Rueckert, S., Shadwick, L., Schoch, C.L., Smirnov, A. and Spiegel, F.W. 2012. The revised classification of eukaryotes. J. Eukaryot. Microbiol. 59(5): 429-493. doi: 10.1111/j.1550-7408.2012.00644.x.

Agarwal, S., Iyer, G., Srinivasan, B., Benurwar, S., Agarwal, M., Narayanan, N., Lakshmipathy, M., Radhika, N., Rajagopal, R., Krishnakumar, S. and K, L.T. 2018. Clinical profile, risk factors and outcome of medical, surgical and adjunct interventions in patients with *Pythium insidiosum* keratitis. Br. J. Ophthalmol. doi: 10.1136/bjophthalmol-2017-311804.

Alfaro, A.A. and Mendoza, L. 1990. Four cases of equine bone lesions caused by *Pythium insidiosum*. Equine. Vet. J. 22(4): 295-297.

Araujo, M.J., Bosco, S.M. and Sforcin, J.M. 2016. *Pythium insidiosum*: Inhibitory effects of propolis and geopropolis on hyphal growth. Braz. J. Microbiol. 47(4): 863-869. doi: 10.1016/j.bjm.2016.06.008.

Argenta, J.S., Santurio, J.M., Alves, S.H., Pereira, D.I.B., Cavalheiro, A.S., Spanamberg, A. and Ferreiro, L. 2008. *In vitro* activities of voriconazole, itraconazole, and terbinafine alone or in combination against *Pythium insidiosum* isolates from Brazil. Antimicrob Agents Chemother. 52(2): 767-769. doi: 10.1128/Aac.01075-07.

Argenta, J.S., Alves, S.H., Silveira, F., Maboni, G., Zanette, R.A., Cavalheiro, A.S., Pereira, P.L., Pereira, D.I.B., Sallis, E.S.V., Potter, L., Santurio, J.M. and Ferreiro, L. 2012. *In vitro* and *in vivo* susceptibility of two-drug and three-drug combinations of terbinafine, itraconazole, caspofungin, ibuprofen and fluvastatin against *Pythium insidiosum*. Vet. Microbiol. 157(1-2): 137-142. doi: 10.1016/j.vetmic.2011.12.003.

Aronson, J.M., Cooper, B.A. and Fuller, M.S. 1967. Glucans of Oomycete cell walls. Science 155(3760): 332-335. doi: 10.1126/science.155.3760.332.

Ascunce, M.S., Huguet-Tapia, J.C., Braun, E.L., Ortiz-Urquiza, A., Keyhani, N.O. and Goss, E.M. 2016. Whole genome sequence of the emerging oomycete pathogen *Pythium insidiosum* strain CDC-B5653 isolated from an infected human in the USA. Genom Data 7: 60-61. doi: 10.1016/j.gdata.2015.11.019.

Austwick, P.K. and Copland, J.W. 1974. Swamp cancer. Nature 250(5461): 84-84.

Azevedo, M.I., Botton, S.A., Pereira, D.I.B., Robe, L.J., Jesus, F.P.K., Mahl, C.D., Costa, M.M., Alves, S.H. and Santurio, J.M. 2012. Phylogenetic relationships of Brazilian isolates of *Pythium insidiosum* based on ITS rDNA and cytochrome oxidase II gene sequences. Vet. Microbiol. 159(1-2): 141-148. doi: 10.1016/j.vetmic.2012.03.030.

Bach, B.C., Leal, D.B.R., Ruchel, J.B., Souza, V.D.G., Maboni, G., Dal Pozzo, M., Schlemmer, K.B., Alves, S.H. and Santurio, J.M. 2010. Immunotherapy for pythiosis: Effect on NTPDase activity in lymphocytes of an experimental model. Biomed. Pharmacother. 64(10): 718-722. doi: 10.1016/j.biopha.2010.09.016.

Bach, B.C., Leal, D.B., Jaques, J.A., Souza Vdo, C., Ruchel, J.B., Schlemmer, K.B., Zanette, R.A., Hecktheuer, P.A., de Lima Pereira, P., Casali, E.A., Alves, S.H. and Santurio, J.M. 2013. E-ADA activity in lymphocytes of an experimental model of pythiosis treated with immunotherapy. Cell Biochem. Funct. 31(6): 476-481. doi: 10.1002/cbf.2921.

Badenoch, P.R., Coster, D.J., Wetherall, B.L., Brettig, H.T., Rozenbilds, M.A., Drenth, A. and Wagels, G. 2001. *Pythium insidiosum* keratitis confirmed by DNA sequence analysis. Br. J. Ophthalmol. 85(4): 502-503.

Badenoch, P.R., Mills, R.A.D., Chang, J.H., Sadlon, T.A., Klebe, S. and Coster, D.J. 2009. *Pythium insidiosum* keratitis in an Australian child. Clin. Exp. Ophthalmol. 37(8): 806-809. doi: 10.1111/j.1442-9071.2009.02135.x.

Bagga, B., Sharma, S., Madhuri Guda, S.J., Nagpal, R., Joseph, J., Manjulatha, K., Mohamed, A. and Garg, P. 2018. Leap forward in the treatment of *Pythium insidiosum* keratitis. Br. J. Ophthalmol. 102: 1629-1633. doi: 10.1136/bjophthalmol-2017-311360.

Barequet, I.S., Lavinsky, F. and Rosner, M. 2013. Long-term follow-up after successful treatment of *Pythium insidiosum* keratitis in Israel. Semin. Ophthalmol. 28(4): 247-250. doi: 10.3109/08820538.2013.788676.

Beakes, G.W. and Thines, M. 2017. Hyphochytriomycota and oomycota. pp. 435-505. *In*: Archibald, John M., Simpson, Alastair, G.B. and Slamovits, Claudio H. (eds.). Handbook of the Protists. Springer Berlin Heidelberg, New York, NY.

Bernardo, F.D., Conhizak, C., Ambrosini, F., Jesus, F.P.K., Santurio, J.M., Kommers, G.D., Elias, F. and Franciscato, C. 2015. Pythiosis in sheep from Paraná, southern Brazil. Pesq. Vet. Bras. 35(6): 513-517. doi: 10.1590/s0100-736x2015000600004.

Berryessa, N.A., Marks, S.L., Pesavento, P.A., Krasnansky, T., Yoshimoto, S.K., Johnson, E.G. and Grooters, A.M. 2008. Gastrointestinal pythiosis in 10 dogs from California. J. Vet. Intern. Med. 22(4): 1065-1069. doi: 10.1111/j.1939-1676.2008.0123.x.

Bissonnette, K.W., Sharp, N.J.H., Dykstra, M.H., Robertson, I.R., Davis, B., Padhye, A.A. and Kaufman, L. 1991. Nasal and retrobulbar mass in a cat caused by *Pythium insidiosum*. J. Med. Vet. Mycol. 29(1): 39-44.

Bitting, A. 1894. Leeches or leeching. Florida University AES Bulletin. 25: 37-48.

Bosco, S.D.G., Bagagli, E., Araujo, J.P., Candeias, J.M.G., de Franco, M.F., Marques, M.E.A., Mendoza, L., de Camargo, R.P. and Marques, S.A. 2005. Human pythiosis, Brazil. Emerging Infect. Dis. 11(5): 715-718.

Botton, S.A., Pereira, D.I.B., Costa, M.M., Azevedo, M.I., Argenta, J.S., Jesus, F.P.K., Alves, S.H. and Santurio, J.M. 2011. Identification of *Pythium insidiosum* by nested PCR in cutaneous lesions of Brazilian horses and rabbits. Curr. Microbiol. 62(4): 1225-1229. doi: 10.1007/s00284-010-9781-4.

Bridges, C. and Emmons, C. 1961. A phycomycosis of horses caused by *Hyphomyces destruens*. J. Am. Vet. Med. Assoc. 138: 579-589.

Brown, C.C., McClure, J.J., Triche, P. and Crowder, C. 1988. Use of immunohistochemical methods for diagnosis of equine pythiosis. Am. J. Vet. Res. 49(11): 1866-1868.

Buergelt, C., Powe, J. and White, T. 2006. Abdominal pythiosis in a Bengal tiger (*Panthera tigris tigris*). J. Zoo Wildl. Med. 37(2): 186-189. doi: 10.1638/05-003.1.

Buitrago-Mejía, J.A., Díaz-Cueto, M.E. and Cardona-Álvarez, J.A. 2017. Geographic distribution of bovine clinical casuistry in the outpatient service for large animals at the Universidad de Córdoba (Colombia). Rev. Med. Vet. (Bogota). 34(supl. 1): 101-114.

Camus, A.C., Grooters, A.M. and Aquilar, R.E. 2004. Granulomatous pneumonia caused by *Pythium insidiosum* in a central American jaguar, *Panthera onca*. J. Vet. Diagn. Invest. 16(6): 567-571. doi: 10.1177/104063870401600612.

Cardona Álvarez, J.A., Vargas Viloria, M. and Perdomo, S.C. 2012. Frequency of presentation of bovine cutaneous pythiosis (*Pythium insidiosum*) in three cattle farms in Córdoba, Colombia. Ces. Med. Vet. Zootec. 7(2): 47-54.

Carrera, M.V., Peixoto, R.M., Gouveia, G.V., Pessoa, C.R.M., Jesus, F.P.K., Santurio, J.M., Botton, S.A. and Costa, M.M. 2013. Pythiosis in sheep from Pernambuco and Bahia States, Brazil. Pesq. Vet. Bras. 33(4): 476-482. doi: 10.1590/S0100-736X2013000400011.

Cavalheiro, A.S., Maboni, G., de Azevedo, M.I., Argenta, J.S., Pereira, D.I.B., Spader, T.B., Alves, S.H. and Santurio, J.M. 2009. *In vitro* activity of terbinafine combined with caspofungin and azoles against *Pythium insidiosum*. Antimicrob. Agents Chemother. 53(5): 2136-2138. doi: 10.1128/Aac.01506-08.

Chaiprasert, A., Krajaejun, T., Pannanusorn, S., Prariyachatigul, C., Wanachiwanawin, W., Sathapatayavongs, B., Juthayothin, T., Smittipat, N., Vanittanakom, N., Chindamporn, A. 2009. *Pythium insidiosum* Thai isolates: Molecular phylogenetic analysis. Asian Biomed. 3(6): 623-633.

Chanprasert, S., Ungprasert, P., Thongprayoon, C., Cheungpasitporn, W., Srivali, N. and Bruminhent, J. 2015. Pythiosis with renal involvement – Rare but fatal disease: A systematic review. J. Nat. Sci. 1(5): e92.

Chareonsirisuthigul, T., Khositnithikul, R., Intaramat, A., Inkomlue, R., Sriwanichrak, K., Piromsontikorn, S., Kitiwanwanich, S., Lowhnoo, T., Yingyong, W., Chaiprasert, A., Banyong, R., Ratanabanangkoon, K., Brandhorst, T.T. and Krajaejun, T. 2013. Performance comparison of immunodiffusion, enzyme-linked immunosorbent assay, immunochromatography and hemagglutination for serodiagnosis of human pythiosis. Diagn. Microbiol. Infect. Dis. 76(1): 42-45. doi: 10.1016/j.diagmicrobio.2013.02.025.

Chatterjee, S. and Agrawal, D. 2018. Azithromycin in the management of *Pythium insidiosum* keratitis. Cornea. 37(2): e8-e9. doi: 10.1097/ICO.0000000000001419.

Chechi, J.L., Franckin, T., Barbosa, L.N., Alves, F.C.B., Leite, A.L., Buzalaf, M.A.R., Delazari Dos Santos, L. and Bosco, S.M.G. 2018. Inferring putative virulence factors for *Pythium insidiosum* by proteomic approach. Med. Mycol. 57(1): 92-100. doi: 10.1093/mmy/myx166.

Chitasombat, M.N., Larbcharoensub, N., Chindamporn, A. and Krajaejun, T. 2018. Clinicopathological features and outcomes of pythiosis. Int. J. Infect. Dis. 71: 33-41. doi: 10.1016/j.ijid.2018.03.021.

Connolly, S.L., Frank, C., Thompson, C.A., Van Alstine, W.G., Gelb, H., Heng, H.G., Klosterman, E., Kiupel, M. and Grooters, A.M. 2012. Dual infection with *Pythium insidiosum* and *Blastomyces dermatitidis* in a dog. Vet. Clin. Pathol. 41(3): 419-423. doi: 10.1111/j.1939-165X.2012.00447.x.

Costa, R.O., Macedo, P.M., Carvalhal, A. and Bernardes-Engemann, A.R. 2013. Use of potassium iodide in dermatology: Updates on an old drug. An. Bras. Dermatol. 88(3): 396-402. doi: 10.1590/abd1806-4841.20132377.

Cyktor, J.C. and Turner, J. 2011. Interleukin-10 and immunity against prokaryotic and eukaryotic intracellular pathogens. Infect. Immun. 79(8): 2964-2973. doi: 10.1128/IAI.00047-11.

De Cock, A.W.A.M., Mendoza, L., Padhye, A.A., Ajello, L. and Kaufman, L. 1987. *Pythium insidiosum* sp. nov., the etiologic agent of pythiosis. J. Clin. Microbiol. 25(2): 344-349.

De Haan, J. and Hoogkamer, L.J. 1901. Hypho-mycosis destruens. Veeartsenijkundige bladen voor Nederlandsch-Indie. 13: 350-374.

Derevnina, L., Petre, B., Kellner, R., Dagdas, Y.F., Sarowar, M.N., Giannakopoulou, A., De la Concepcion, J.C., Chaparro-Garcia, A., Pennington, H.G., van West, P. and Kamoun, S. 2016. Emerging oomycete threats to plants and animals. Philos. Trans. R. Soc. Lond. B Biol. Sci. 371(1709): 20150459. doi: 10.1098/rstb.2015.0459.

do Carmo, P.M., Portela, R.A., Silva, T.R., Oliveira-Filho, J.C. and Riet-Correa, F. 2015. Cutaneous pythiosis in a goat. J. Comp. Pathol. 152(2-3): 103-105. doi: 10.1016/j.jcpa.2014.11.005.

Dória, R.G.S., Freitas, S.H., Mendonça, F.S., Arruda, L.P., Boabaid, F.M., Martins Filho, A., Colodel, E.M. and Valadão, C.A.A. 2014. Utilização da técnica de imuno-histoquímica para confirmar casos de pitiose cutânea equina diagnosticados por meio de caracterização clínica e avaliação histopatológica. Arq. Bras. Med. Vet. Zootec. 66(1): 27-33. doi: 10.1590/s0102-09352014000100005.

Dória, R.G.S., Carvalho, M.B., Freitas, S.H., Laskoski, L.M., Colodel, E.M., Mendonca, F.S., Silva, M.A., Grigoletto, R. and Fantinato Neto, P. 2015. Evaluation of intravenous regional perfusion with amphotericin B and dimethylsulfoxide to treat horses for pythiosis of a limb. BMC Vet. Res. 11(1): 152. doi: 10.1186/s12917-015-0472-z.

Drouin, V. 1896. Sur une nouvelle mycose du cheval. Rec. Med. Vet. Ec. Alfort. 3: 337-344.

Dykstra, M.J., Sharp, N.J.H., Olivry, T., Hillier, A., Murphy, K.M., Kaufman, L., Kunkle, G.A. and Pucheu-Haston, C. 1999. A description of cutaneous-subcutaneous pythiosis in fifteen dogs. Med. Mycol. 37(6): 427-433.

Fischer, J.R., Pace, L.W., Turk, J.R., Kreeger, J.M., Miller, M.A. and Gossner, H.S. 1994. Gastrintestinal pythiosis in Missouri dogs: Eleven cases. J. Vet. Diagn. Invest. 6(3): 380-382.

Fonseca, A.O., Botton Sde, A., Nogueira, C.E., Correa, B.F., Silveira Jde, S., de Azevedo, M.I., Maroneze, B.P., Santurio, J.M. and Pereira, D.I. 2014. *In vitro* reproduction of the life cycle of *Pythium insidiosum* from *kunkers'* equine and their role in the epidemiology of pythiosis. Mycopathologia. 177(1-2): 123-127. doi: 10.1007/s11046-013-9720-6.

Fonseca, A.O., Pereira, D.I., Botton, S.A., Potter, L., Sallis, E.S., Junior, S.F., Filho, F.S., Zambrano, C.G., Maroneze, B.P., Valente, J.S., Baptista, C.T., Braga, C.Q., Ben, V.D. and Meireles, M.C. 2015a. Treatment of experimental pythiosis with essential oils of *Origanum vulgare* and *Mentha piperita* singly, in association and in combination with immunotherapy. Vet. Microbiol. 178(3-4): 265-269. doi: 10.1016/j.vetmic.2015.05.023.

Fonseca, A.O., Pereira, D.I., Jacob, R.G., Maia Filho, F.S., Oliveira, D.H., Maroneze, B.P., Valente, J.S., Osorio, L.G., Botton, S.A. and Meireles, M.C. 2015b. *In vitro* susceptibility of Brazilian *Pythium insidiosum* isolates to essential oils of some Lamiaceae family species. Mycopathologia. 179(3-4): 253-258. doi: 10.1007/s11046-014-9841-6.

Fortin, J.S., Calcutt, M.J. and Kim, D.Y. 2017. Sublingual pythiosis in a cat. Acta. Vet. Scand. 59(1): 63. doi: 10.1186/s13028-017-0330-z.

Fujimori, M., Lopes, E.R., Lima, S.R., Paula, D.A.J.d., Almeida, A.d.B.P.F.d., Colodel, E.M., Pescador, C.A., Néspoli, P.E.B., Nakazato, L., Dutra, V., Souza, R.L.d. and Sousa, V.R.F. 2016. *Pythium insidiosum* colitis in a dog: Treatment and clinical outcome. Ciênc Rural. 46(3): 526-529. doi: 10.1590/0103-8478cr20150081.

Gaastra, W., Lipman, L.J.A., de Cock, A.W.A.M., Exel, T.K., Pegge, R.B.G., Scheurwater, J., Vilela, R. and Mendoza, L. 2010. *Pythium insidiosum*: An overview. Vet. Microbiol. 146(1-2): 1-16. doi: 10.1016/j.vetmic.2010.07.019.

Gabriel, A.L., Kommers, G.D., Trost, M.E., Barros, C.S.L., Pereira, D.B., Schwendler, S.E. and Santurio, J.M. 2008. Outbreak of cutaneous pythiosis in cattle. Pesq. Vet. Bras. 28(12): 583-587. doi: 10.1590/S0100-736X2008001200003.

Gaulin, E., Bottin, A. and Dumas, B. 2010. Sterol biosynthesis in oomycete pathogens. Plant Signal. Behav. 5(3): 258-260. doi: 10.1111/j.14698137.2009.02895.x.

Gonzalez-Ch, H. and Ruiz-M, A. 1975. Espundia equina: Etiologia y patogenesis de una ficomicosis. Revista ICA (Colombia). 10: 175-185.

Grecco, F.B., Schild, A.L., Quevedo, P., Assis-Brasil, N.D., Kommers, G.D., Marcolongo-Pereira, C. and Soares, M.P. 2009. Cutaneous pythiosis in cattle in the Southern region of Rio Grande do Sul, Brazil. Pesq. Vet. Bras. 29(11): 938-942.

Grooters, A.M. and Gee, M.K. 2002. Development of a nested polymerase chain reaction assay for the detection and identification of *Pythium insidiosum*. J. Vet. Intern. Med. 16(2): 147-152.

Grooters, A.M., Leise, B.S., Lopez, M.K., Gee, M.K. and O'Reilly, K.L. 2002a. Development and evaluation of an enzyme-linked immunosorbent assay for the serodiagnosis of pythiosis in dogs. J. Vet. Intern. Med. 16(2): 142-146.

Grooters, A.M., Whittington, A., Lopez, M.K., Boroughs, M.N. and Roy, A.F. 2002b. Evaluation of microbial culture techniques for the isolation of *Pythium insidiosum* from equine tissues. J. Vet. Diagn. Invest. 14(4): 288-294. doi: 10.1177/104063870201400403.

Grooters, A.M. 2003. Pythiosis, lagenidiosis, and zygomycosis in small animals. Vet. Clin. North Am. Small Anim. Pract. 33(4): 695-720. doi: 10.1016/S0195-5616(03)00034-2.

Grooters, A.M., Hodgin, E.C., Bauer, R.W., Detrisac, C.J., Znajda, N.R. and Thomas, R.C. 2003. Clinicopathologic findings associated with *Lagenidium* sp. infection in 6 dogs: Initial description of an emerging oomycosis. J. Vet. Intern. Med. 17(5): 637-646.

Grooters, A.M. 2013. Pythiosis and zygomycosis. pp. 415-421. *In*: Sellon, Debra C. and Long, Maureen T. (eds.). Equine Infectious Diseases. Saunders/Elsevier, St. Louis, Missouri.

Guo, L.Y. and Ko, W.H. 1994. Growth rate and antibiotic sensitivities of conidium and selfed-oospore progenies of heterothallic *Pythium splendens*. Can. J. Bot. 72(11): 1709-1712.

Haan, J. and Hoogkamer, L. 1903. Hyphomycosis destruens equine. Archiv Wiss Prak Tierheilk. 29: 395-410.

Hahtapornsawan, S., Wongwanit, C., Chinsakchai, K., Hongku, K., Sermsathanasawadi, N., Ruangsetakit, C. and Mutirangura, P. 2014. Suprainguinal vascular pythiosis: Effective long-term outcome of aggressive surgical eradication. Ann. Vasc. Surg. 28(7): 1797 e1791-1796. doi: 10.1016/j.avsg.2014.04.020.

Helman, R.G. and Oliver, J. 1999. Pythiosis of the digestive tract in dogs from Oklahoma. J. Am. Anim. Hosp. Assoc. 35(2): 111-114.

Hendrix, J.W. 1964. Sterol induction of reproduction and stimulation of growth of *Pythium* and *Phytophthora*. Science. 144(3621): 1028-1029.

Hilton, R.E., Tepedino, K., Glenn, C.J. and Merkel, K.L. 2016. Swamp cancer: A case of human pythiosis and review of the literature. Br. J. Dermatol. 175(2): 394-397. doi: 10.1111/bjd.14520.

Huang, J.-H., Chen, C.-Y., Lin, Y.-S., Ann, P.-J., Huang, H.-C. and Chung, W.-H. 2013. Six new species of *Pythiogeton* in Taiwan, with an account of the molecular phylogeny of this genus. Mycoscience 54(2): 130-147. doi: 10.1016/j.myc.2012.09.007.

Hummel, J., Grooters, A., Davidson, G., Jennings, S., Nicklas, J. and Birkenheuer, A. 2011. Successful management of gastrointestinal pythiosis in a dog using itraconazole, terbinafine, and mefenoxam. Med. Mycol. 49(5): 539-542. doi: 10.3109/13693786.2010.543705.

Hung, C. and Leddin, D. 2014. Keratitis caused by *Pythium insidiosum* in an immunosuppressed patient with Crohn's disease. Clin. Gastroenterol. Hepatol. 12(10): A21-22. doi: 10.1016/j.cgh.2014.04.023.

Hutchins, D.R. and Johnston, K.G. 1972. Phycomycosis in the horse. Aust. Vet. J. 48(5): 269-278.

Ichitani, T. and Amemiya, J. 1980. *Pythium gracile* isolated from the foci of granular dermatitis in the horse (*Equus caballus*). T. Mycol. Soc. Jpn. 21(2): 263-265.

Jaeger, G.H., Rotstein, D.S. and Law, J.M. 2002. Prostatic pythiosis in a dog. J. Vet. Intern. Med. 16(5): 598-602.

Jesus, F.P., Ferreiro, L., Bizzi, K.S., Loreto, E.S., Pilotto, M.B., Ludwig, A., Alves, S.H., Zanette, R.A. and Santurio, J.M. 2015. *In vitro* activity of carvacrol and thymol combined with antifungals or antibacterials against *Pythium insidiosum*. J. Mycol. Med. 25(2): e89-93. doi: 10.1016/j.mycmed.2014.10.023.

Jesus, F.P., Loreto, E.S., Ferreiro, L., Alves, S.H., Driemeier, D., Souza, S.O., Franca, R.T., Lopes, S.T., Pilotto, M.B., Ludwig, A., Azevedo, M.I., Ribeiro, T.C., Tondolo, J.S. and Santurio, J.M. 2016. *In vitro* and *in vivo* antimicrobial activities of minocycline in combination with azithromycin, clarithromycin, or tigecycline against *Pythium insidiosum*. Antimicrob Agents Chemother. 60(1): 87-91. doi: 10.1128/AAC.01480-15.

Jiang, R.H. and Tyler, B.M. 2012. Mechanisms and evolution of virulence in oomycetes. Annu. Rev. Phytopathol. 50(1): 295-318. doi: 10.1146/annurev-phyto-081211-172912.

Johnston, K.G. and Henderson, A.W. 1974. Phycomycotic granuloma in horses in the Northern Territory. Aust. Vet. J. 50(3): 105-107.

Judelson, H.S. and Blanco, F.A. 2005. The spores of *Phytophthora*: Weapons of the plant destroyer. Nat. Rev. Microbiol. 3(1): 47-58. doi: 10.1038/nrmicro1064.

Kammarnjesadakul, P., Palaga, T., Sritunyalucksana, K., Mendoza, L., Krajaejun, T., Vanittanakom, N., Tongchusak, S., Denduangboripant, J. and Chindamporn, A. 2011. Phylogenetic analysis of *Pythium insidiosum* Thai strains using cytochrome oxidase II (COX II) DNA coding sequences and internal transcribed spacer regions (ITS). Med. Mycol. 49(3): 289-295. doi: 10.3109/13693786.2010.511282.

Kamoun, S., Furzer, O., Jones, J.D., Judelson, H.S., Ali, G.S., Dalio, R.J., Roy, S.G., Schena, L., Zambounis, A., Panabieres, F., Cahill, D., Ruocco, M., Figueiredo, A., Chen, X.R., Hulvey, J., Stam, R., Lamour, K., Gijzen, M., Tyler, B.M., Grunwald, N.J., Mukhtar, M.S., Tome, D.F., Tor, M., Van Den Ackerveken, G., McDowell, J., Daayf, F., Fry, W.E., Lindqvist-Kreuze, H., Meijer, H.J., Petre, B., Ristaino, J., Yoshida, K., Birch, P.R. and Govers, F. 2015. The Top 10 oomycete pathogens in molecular plant pathology. Mol. Plant Pathol. 16(4): 413-434. doi: 10.1111/mpp.12190.

Keeratijarut, A., Lohnoo, T., Yingyong, W., Sriwanichrak, K. and Krajaejun, T. 2013. A peptide ELISA to detect antibodies against *Pythium insidiosum* based on predicted antigenic determinants of exo-1,3-beta-glucanase. Southeast Asian J. Trop. Med. Public Health. 44(4): 672-680.

Keoprasom, N., Chularojanamontri, L., Chayakulkeeree, M., Chaiprasert, A., Wanachiwanawin, W. and Ruangsetakit, C. 2012. Vascular pythiosis in a thalassemic patient presenting as bilateral leg ulcers. Med. Mycol. Case Rep. 2: 25-28. doi: 10.1016/j.mmcr.2012.12.002.

Khunkhet, S., Rattanakaemakorn, P. and Rajatanavin, N. 2015. Pythiosis presenting with digital gangrene and subcutaneous nodules mimicking medium vessel vasculitis. JAAD Case Rep. 1(6): 399-402. doi: 10.1016/j.jdcr.2015.09.005.

Kirzhner, M., Arnold, S.R., Lyle, C., Mendoza, L.L. and Fleming, J.C. 2015. *Pythium insidiosum*: A rare necrotizing orbital and facial infection. J. Pediatric Infect. Dis. Soc. 4(1): e10-13. doi: 10.1093/jpids/piu015.

Krajaejun, T., Kunakorn, M., Niemhom, S., Chongtrakool, P. and Pracharktam, R. 2002. Development and evaluation of an in-house enzyme-linked immunosorbent assay for early diagnosis and monitoring of human pythiosis. Clin. Diagn. Lab. Immunol. 9(2): 378-382. doi: 10.1128/Cdli.9.2.378-382.2002.

Krajaejun, T., Sathapatayavongs, B., Pracharktam, R., Nitiyanant, P., Leelachaikul, P., Wanachiwanawin, W., Chaiprasert, A., Assanasen, P., Saipetch, M., Mootsikapun, P., Chetchotisakd, P., Lekhakula, A., Mitarnun, W., Kalnauwakul, S., Supparatpinyo, K., Chaiwarith, R., Chiewchanvit, S., Tananuvat, N., Srisiri, S., Suankratay, C., Kulwichit, W., Wongsaisuwan, M. and Somkaew, S. 2006. Clinical and epidemiological analyses of human pythiosis in Thailand. Clin. Infect. Dis. 43(5): 569-576. doi: 10.1086/506353.

Krajaejun, T., Imkhieo, S., Intaramat, A. and Ratanabanangkoon, K. 2009. Development of an immunochromatographic test for rapid serodiagnosis of human pythiosis. Clin. Vaccine Immunol. 16(4): 506-509. doi: 10.1128/Cvi.00276-08.

Krajaejun, T., Khositnithikul, R., Lerksuthirat, T., Lowhnoo, T., Rujirawat, T., Petchthong, T., Yingyong, W., Suriyaphol, P., Smittipat, N., Juthayothin, T., Phuntumart, V. and Sullivan, T.D. 2011. Expressed sequence tags reveal genetic diversity and putative virulence factors of the pathogenic oomycete *Pythium insidiosum*. Fungal Biol. 115(7): 683-696. doi: 10.1016/j.funbio.2011.05.001.

Krajaejun, T., Lerksuthirat, T., Garg, G., Lowhnoo, T., Yingyong, W., Khositnithikul, R., Tangphatsornruang, S., Suriyaphol, P., Ranganathan, S. and Sullivan, T.D. 2014. Transcriptome analysis reveals pathogenicity and evolutionary history of the pathogenic oomycete *Pythium insidiosum*. Fungal Biol. Rev. 118(7): 640-653. doi: 10.1016/j.funbio.2014.01.009.

Krockenberger, M.B., Swinney, G., Martin, P., Rothwell, T.R.L. and Malik, R. 2011. Sequential opportunistic infections in two German Shepherd dogs. Aust. Vet. J. 89(1-2): 9-14. doi: 10.1111/j.1751-0813.2010.00666.x.

Laohapensang, K., Rerkasem, K., Supabandhu, J. and Vanittanakom, N. 2005. Necrotizing arteritis due to emerging *Pythium insidiosum* infection in patients with thalassemia: Rapid diagnosis with PCR and serological tests – Case reports. Int. J. Angiol. 14(3): 123-128. doi: 10.1007/s00547-005-2012-3.

Leal, A.B.M., Leal, A.T., Santurio, J.M., Kommers, G.D. and Catto, J.B. 2001. Equine pythiosis in the Brazilian Pantanal region: Clinical and pathological findings of typical and atypical cases. Pesq. Vet. Bras. 21(4): 151-156.

Leal, A.T., Santurio, J.M., Leal, A.B.M., Catto, J.B., Flores, E.F., Lubeck, I. and Alves, S.H. 2005. Characterization of the specificity of the humoral response to *Pythium insidiosum* antigens. J. Mycol. Med. 15(2): 63-68. doi: 10.1016/j.mycmed.2005.01.004.

Ledur, P.C., Tondolo, J.S.M., Jesus, F.P.K., Verdi, C.M., Loreto, E.S., Alves, S.H. and Santurio, J.M. 2018. Dendritic cells pulsed with *Pythium insidiosum* (1,3)(1,6)-beta-glucan, Heat-inactivated zoospores and immunotherapy prime naive T cells to Th1 differentiation *in vitro*. Immunobiology. 223(3): 294-299. doi: 10.1016/j.imbio.2017.10.033.

Lekhanont, K., Chuckpaiwong, V., Chongtrakool, P., Aroonroch, R. and Vongthongsri, A. 2009. *Pythium insidiosum* keratitis in contact lens wear: A case report. Cornea. 28(10): 1173-1177. doi: 10.1097/ICO.0b013e318199fa41.

Lelievre, L., Borderie, V., Garcia-Hermoso, D., Brignier, A.C., Sterkers, M., Chaumeil, C., Lortholary, O. and Lanternier, F. 2015. Imported *Pythium insidiosum* keratitis after a swim in Thailand by a contact lens-wearing traveler. Am. J. Trop. Med. Hyg. 92(2): 270-273. doi: 10.4269/ajtmh.14-0380.

Lerksuthirat, T., Sangcakul, A., Lohnoo, T., Yingyong, W., Rujirawat, T. and Krajaejun, T. 2017. Evolution of the sterol biosynthetic pathway of *Pythium insidiosum* and related oomycetes contributes to antifungal drug resistance. Antimicrob. Agents Chemother. 61(4): e02352-02316. doi: 10.1128/AAC.02352-16.

Levesque, C.A. and de Cock, A.W. 2004. Molecular phylogeny and taxonomy of the genus *Pythium*. Mycol. Res. 108(Pt 12): 1363-1383. doi: 10.1017/S0953756204001431.

Loreto, E.S., Alves, S.H., Santurio, J.M., Nogueira, C.W. and Zeni, G. 2012. Diphenyl diselenide *in vitro* and *in vivo* activity against the oomycete *Pythium insidiosum*. Vet. Microbiol. 156(1-2): 222-226. doi: 10.1016/j.vetmic.2011.10.008.

Loreto, E.S., Tondolo, J.S., Pilotto, M.B., Alves, S.H. and Santurio, J.M. 2014a. New insights into the *in vitro* susceptibility of *Pythium insidiosum*. Antimicrob. Agents Chemother. 58(12): 7534-7537. doi: 10.1128/AAC.02680-13.

Loreto, E.S., Tondolo, J.S.M., Zanette, R.A., Alves, S.H. and Santurio, J.M. 2014b. Update on pythiosis immunobiology and immunotherapy. World J. Immunol. 4(2): 88-97. doi: 10.5411/wji.v4.i2.88.

Loreto, E.S., Tondolo, J.S.M., de Jesus, F.P.K., Verdi, C.M., Weiblen, C., de Azevedo, M.I., Kommers, G.D., Santurio, J.M., Zanette, R.A. and Alves, S.H. 2018a. Efficacy of azithromycin and miltefosine in experimental systemic pythiosis in immunosuppressed mice. Antimicrob. Agents Chemother. doi: 10.1128/AAC.01385-18.

Loreto, E.S., Tondolo, J.S.M., Oliveira, D.C., Santurio, J.M. and Alves, S.H. 2018b. *In vitro* activities of miltefosine and antibacterial agents from the macrolide, oxazolidinone, and pleuromutilin classes against *Pythium insidiosum* and *Pythium aphanidermatum*. Antimicrob. Agents Chemother. 62(3): e01678-01617. doi: 10.1128/AAC.01678-17.

Luis-León, J.J., Pérez, R.C., Vivas, J.L., Mendoza, L. and Alonso, F.T. 2009. Confirmation of *Pythium insidiosum* as an etiologic agent of enzootic bovine granulomatosis by sequence analysis. Salus online. 12(Sup 1): 205-215.

MacDonald, E., Millward, L., Ravishankar, J.P. and Money, N.P. 2002. Biomechanical interaction between hyphae of two *Pythium* species (Oomycota) and host tissues. Fungal Genet. Biol. 37(3): 245-249. doi: 10.1016/S1087-1845(02)00514-5.

Machado, G., Weiblen, C. and Escobar, L.E. 2018. Potential distribution of *Pythium insidiosum* in Rio Grande do Sul, Brazil, and projections to neighbour countries. Transbound Emerg. Dis. doi: 10.1111/tbed.12925.

Marchant, R. and Smith, D.G. 1968. The effect of chloramphenicol on growth and mitochondrial structure of *Pythium ultimum*. J. Gen. Microbiol. 50(3): 391-397. doi: 10.1099/00221287-50-3-391.

Marcolongo-Pereira, C., Sallis, E.S.V., Raffi, M.B., Pereira, D.I.B., Hinnah, F.L., Coelho, A.C.B. and Schild, A.L. 2012. Epidemiology of equine pythiosis in southern of Rio Grande do Sul State, Brazil. Pesq. Vet. Bras. 32(9): 865-868.

Martins, T.B., Kommers, G.D., Trost, M.E., Inkelmann, M.A., Fighera, R.A. and Schild, A.L. 2012. A comparative study of the histopathology and immunohistochemistry of pythiosis in horses, dogs and cattle. J. Comp. Pathol. 146(2-3): 122-131. doi: 10.1016/j.jcpa.2011.06.006.

McCarthy, C.G.P. and Fitzpatrick, D.A. 2017. Phylogenomic reconstruction of the oomycete phylogeny derived from 37 genomes. mSphere. 2(2): e00095-00017. doi: 10.1128/mSphere.00095-17.

McGowan, J. and Fitzpatrick, D.A. 2017. Genomic, network, and phylogenetic analysis of the oomycete effector arsenal. mSphere. 2(6). doi: 10.1128/mSphere.00408-17.

McMullan, W.C., Joyce, J.R., Hanselka, D.V. and Heitmann, J.M. 1977. Amphotericin B for the treatment of localized subcutaneous phycomycosis in the horse. J. Am. Vet. Med. Assoc. 170(11): 1293-1298.

Meijer, H.J., Hua, C., Kots, K., Ketelaar, T. and Govers, F. 2014. Actin dynamics in *Phytophthora infestans*; rapidly reorganizing cables and immobile, long-lived plaques. Cell Microbiol. 16(6): 948-961. doi: 10.1111/cmi.12254.

Mendoza, L. and Alfaro, A.A. 1986. Equine pythiosis in Costa Rica – Report of 39 cases. Mycopathologia 94(2): 123-129.

Mendoza, L. and Prendas, J. 1988. A method to obtain rapid zoosporogenesis of *Pythium insidiosum*. Mycopathologia 104(1): 59-62.

Mendoza, L. and Marin, G. 1989. Antigenic relationship between *Pythium insidiosum* de Cock et al. 1987 and its synonym *Pythium destruens* Shipton 1987. Mycoses. 32(2): 73-77.

Mendoza, L., Nicholson, V. and Prescott, J.F. 1992a. Immunoblot analysis of the humoral immune response to *Pythium insidiosum* in horses with pythiosis. J. Clin. Microbiol. 30(11): 2980-2983.

Mendoza, L., Villalobos, J., Calleja, C.E. and Solis, A. 1992b. Evaluation of two vaccines for the treatment of pythiosis insidiosi in horses. Mycopathologia 119(2): 89-95.

Mendoza, L., Hernandez, F. and Ajello, L. 1993. Life cycle of the human and animal oomycete pathogen *Pythium insidiosum*. J. Clin. Microbiol. 31(11): 2967-2973.

Mendoza, L., Ajello, L. and McGinnis, M.R. 1996. Infections caused by the oomycetous pathogen *Pythium insidiosum*. J. Mycol. Med. 6(4): 151-164.

Mendoza, L., Mandy, W. and Glass, R. 2003. An improved *Pythium insidiosum*-vaccine formulation with enhanced immunotherapeutic properties in horses and dogs with pythiosis. Vaccine. 21(21-22): 2797-2804. doi: 10.1016/S0264-410x(03)00225-1.

Mendoza, L., Prasla, S.H. and Ajello, L. 2004. Orbital pythiosis: A non-fungal disease mimicking orbital mycotic infections, with a retrospective review of the literature. Mycoses. 47(1-2): 14-23. doi: 10.1046/j.1439-0507.2003.00950.x.

Mendoza, L. and Newton, J.C. 2005. Immunology and immunotherapy of the infections caused by *Pythium insidiosum*. Med. Mycol. 43(6): 477-486. doi: 10.1080/13693780500279882.

Mendoza, L. 2009. *Pythium insidiosum* and mammalian hosts. pp. 387-405. *In*: Lamour, K. and Kamoun, S. (eds.). Oomycete Genetics and Genomics: Diversity, Interactions and Research Tools. John Wiley & Sons, Hoboken, N.J.

Miller, R.I. 1981. Treatment of equine phycomycosis by immunotherapy and surgery. Aust. Vet. J. 57(8): 377-382.

Miller, R.I. and Campbell, R.S.F. 1982. Clinical observations on equine phycomycosis. Aust. Vet. J. 58(6): 221-226.

Miller, R.I. 1983. Investigation into the biology of three 'phycomycotic' agents pathogenic for horses in Australia. Mycopathologia 81(1): 23-28.

Miller, R.I. and Campbell, R.S.F. 1983. Experimental pythiosis in rabbits. Sabouraudia. 21(4): 331-341.

Miller, R.I., Olcott, B.M. and Archer, M. 1985. Cutaneous pythiosis in beef calves. J. Am. Vet. Med. Assoc. 186(9): 984-986.

Monteiro, A.B. 1999. Immunotherapy of equine pythiosis: Testing the efficacy of a biological and evaluation of the leukocytic response to the treatment in horses naturally infected with *Pythium insidiosum*. Master in Veterinary Medicine, Federal University of Santa Maria.

Murray, D.R., Ladds, P.W., Johnson, R.H. and Pott, B.W. 1978. Metastatic phycomycosis in a horse. J. Am. Vet. Med. Assoc. 172(7): 834-836.

Mustafa, V.S., Guedes, K.M.R., Lima, E.M.M., Borges, J.R.J. and Castro, M.B. 2015. Nasal cavity diseases of small ruminants in Federal District and Goias State, Brazil. Pesq. Vet. Bras. 35(7): 627-636. doi: 10.1590/s0100-736x2015000700005.

Neto, R.T., de, M.G.B.S., Amorim, R.L., Brandao, C.V., Fabris, V.E., Estanislau, C. and Bagagli, E. 2010. Cutaneous pythiosis in a dog from Brazil. Vet. Dermatol. 21(2): 202-204. doi: 10.1111/j.1365-3164.2009.00779.x.

Neufeld, A., Seamone, C., Maleki, B. and Heathcote, J.G. 2018. *Pythium insidiosum* keratitis: A pictorial essay of natural history. Can. J. Ophthalmol. 53(2): e48-e50. doi: 10.1016/j.jcjo.2017.07.002.

Newton, J.C. and Ross, P.S. 1993. Equine pythiosis – An overview of immunotherapy. Comp. Cont. Educ. Pract. Vet. 15(3): 491-493.

Oldenhoff, W., Grooters, A., Pinkerton, M.E., Knorr, J. and Trepanier, L. 2014. Cutaneous pythiosis in two dogs from Wisconsin, USA. Vet. Dermatol. 25(1): 52-e21. doi: 10.1111/vde.12101.

Otta, S.K., Praveena, P.E., Raj, R.A., Saravanan, P., Priya, M.S., Amarnath, C.B., Bhuvaneswari, T., Panigrahi, A. and Ravichandran, P. 2018. *Pythium insidiosum* as a new opportunistic fungal pathogen for Pacific white shrimp, *Litopenaeus vannamei*. Indian J. Mar. Sci. 47(5): 1036-1041.

Patton, C.S., Hake, R., Newton, J. and Toal, R.L. 1996. Esophagitis due to *Pythium insidiosum* infection in two dogs. J. Vet. Intern. Med. 10(3): 139-142.

Pedroso, P.M.O., Bezerra, P.S., Pescador, C.A., Dalto, A.G.C., da Costa, G.R., Pereira, D.I.B., Santurio, J.M. and Driemeier, D. 2009. Immunohistochemical diagnostic of cutaneous pythiosis in horses. Acta. Sci. Vet. 37(1): 49-52.

Pereira, D.I.B., Santurio, J.M., Alves, S.H., Argenta, J.S., Potter, L., Spanamberg, A. and Ferreiro, L. 2007. Caspofungin *in vitro* and *in vivo* activity against Brazilian *Pythium insidiosum* strains isolated from animals. J. Antimicrob. Chemother. 60(5): 1168-1171. doi: 10.1093/Jac/Dkm332.

Pereira, D.I.B., Santurio, J.M., Alves, S.H., de Azevedo, M.I., Silveira, F., da Costa, F.F., Sallis, E.S.V., Potter, L. and Ferreiro, L. 2008. Comparison between immunotherapy and caspofungin as agents to treat experimental pythiosis in rabbits. J. Mycol. Med. 18(3): 129-133. doi: 10.1016/j.mycmed.2008.05.001.

Perez, R.C., Luis-Leon, J.J., Vivas, J.L. and Mendoza, L. 2005. Epizootic cutaneous pythiosis in beef calves. Vet. Microbiol. 109(1-2): 121-128. doi: 10.1016/j.vetmic.2005.04.020.

Permpalung, N., Worasilchai, N., Plongla, R., Upala, S., Sanguankeo, A., Paitoonpong, L., Mendoza, L. and Chindamporn, A. 2015. Treatment outcomes of surgery, antifungal therapy and immunotherapy in ocular and vascular human pythiosis: A retrospective study of 18 patients. J. Antimicrob. Chemother. 70(6): 1885-1892. doi: 10.1093/jac/dkv008.

Pesavento, P.A., Barr, B., Riggs, S.M., Eigenheer, A.L., Pamma, R. and Walker, R.L. 2008. Cutaneous pythiosis in a nestling white-faced ibis. Vet. Pathol. 45(4): 538-541. doi: 10.1354/vp.45-4-538.

Pessoa, C.R.M., Riet-Correa, F., Pimentel, L.A., Garino, F., Dantas, A.F.M., Kommers, G.D., Tabosa, I.M. and Reis, J.L. 2012. Pythiosis of the digestive tract in sheep. J. Vet. Diagn. Invest. 24(6): 1133-1136. doi: 10.1177/1040638712462026.

Phillips, A.J., Anderson, V.L., Robertson, E.J., Secombes, C.J. and van West, P. 2008. New insights into animal pathogenic oomycetes. Trends Microbiol. 16(1): 13-19. doi: 10.1016/j.tim.2007.10.013.

Pires, L., Bosco Sde, M., da Silva, N.F. Jr. and Kurachi, C. 2013. Photodynamic therapy for pythiosis. Vet. Dermatol. 24(1): 130-136 e130. doi: 10.1111/j.1365-3164.2012.01112.x.

Pires, L., Bosco Sde, M., Baptista, M.S. and Kurachi, C. 2014. Photodynamic therapy in *Pythium insidiosum* – An *in vitro* study of the correlation of sensitizer localization and cell death. PLoS One. 9(1): e85431. doi: 10.1371/journal.pone.0085431.

Poole, H.M. and Brashier, M.K. 2003. Equine cutaneous pythiosis. Comp. Cont. Educ. Pract. Vet. 25(3): 229-235.

Portela, R.D., Riet-Correa, F., Garino, F., Dantas, A.F.M., Simoes, S.V.D. and Silva, S.M.S. 2010. Diseases of the nasal cavity of ruminants in Brazil. Pesq. Vet. Bras. 30(10): 844-854. doi: 10.1590/S0100-736x2010001000007.

Powell, M.J., Lehnen, L.P. and Bortnick, R.N. 1985. Microbody-like organelles as taxonomic markers among oomycetes. BioSyst. 18(3-4): 321-334. doi: 10.1016/0303-2647(85)90032-2.

Presser, J.W. and Goss, E.M. 2015. Environmental sampling reveals that *Pythium insidiosum* is ubiquitous and genetically diverse in North Central Florida. Med. Mycol. 53(7): 674-683. doi: 10.1093/mmy/myv054.

Pupaibool, J., Chindamporn, A., Patrakul, K., Suankratay, C., Sindhuphak, W. and Kulwichit, W. 2006. Human pythiosis. Emerg. Infect. Dis. 12(3): 517-518. doi: 10.3201/eid1205.051044.

Rakich, P.M., Grooters, A.M. and Tang, K.N. 2005. Gastrointestinal pythiosis in two cats. J. Vet. Diagn. Invest. 17(3): 262-269. doi: 10.1177/104063870501700310.

Ramappa, M., Nagpal, R., Sharma, S. and Chaurasia, S. 2017. Successful medical management of presumptive *Pythium insidiosum* keratitis. Cornea 36(4): 511-514. doi: 10.1097/ICO.0000000000001162.

Rathi, A., Chakrabarti, A., Agarwal, T., Pushker, N., Patil, M., Kamble, H., Titiyal, J.S., Mohan, R., Kashyap, S., Sharma, S., Sen, S., Satpathy, G. and Sharma, N. 2018. *Pythium* keratitis leading to fatal cavernous sinus thrombophlebitis. Cornea 37(4): 519-522. doi: 10.1097/ICO.0000000000001504.

Ravishankar, J.P., Davis, C.M., Davis, D.J., MacDonald, E., Makselan, S.D., Millward, L. and Money, N.P. 2001. Mechanics of solid tissue invasion by the mammalian pathogen *Pythium insidiosum*. Fungal Genet. Biol. 34(3): 167-175. doi: 10.1006/fgbi.2001.1304.

Rawn, C.D. and Vanetten, J.L. 1978. Mechanism of antibacterial antibiotic sensitivity in *Pythium ultimum*. J. Gen. Microbiol. 108(1): 133-139.

Reanpang, T., Orrapin, S., Orrapin, S., Arworn, S., Kattipatanapong, T., Srisuwan, T., Vanittanakom, N., Lekawanvijit, S.P. and Rerkasem, K. 2015. Vascular pythiosis of the lower extremity in northern Thailand: Ten years' experience. Int. J. Low. Extrem. Wounds. 14(3): 245-250. doi: 10.1177/1534734615599652.

Reis, J.L. and Nogueira, R.H.G. 2002. Anatomopathological and immunohistochemical study of pythiosis in naturally infected horse. Arq. Bras. Med. Vet. Zootec. 54(4): 358-365.

Reis, J.L., de Carvalho, E.C.Q., Nogueira, R.H.G., Lemos, L.S. and Mendoza, L. 2003. Disseminated pythiosis in three horses. Vet. Microbiol. 96(3): 289-295. doi: 10.1016/j.vetmic.2003.07.005.

Ribeiro, T.C., Weiblen, C., de Azevedo, M.I., de Avila Botton, S., Robe, L.J., Pereira, D.I., Monteiro, D.U., Lorensetti, D.M. and Santurio, J.M. 2017. Microevolutionary analyses of *Pythium insidiosum* isolates of Brazil and Thailand based on exo-1,3-beta-glucanase gene. Infect. Genet. Evol. 48: 58-63. doi: 10.1016/j.meegid.2016.11.020.

Rivierre, C., Laprie, C., Guiard-Marigny, O., Bergeaud, P., Berthelemy, M. and Guillot, J. 2005. Pythiosis in Africa. Emerg. Infect. Dis. 11(3): 479-481. doi: 10.3201/eid1103.040697.

Ros Castellar, F., Sobrino Jimenez, C., del Hierro Zarzuelo, A., Herrero Ambrosio, A. and Boto de Los Bueis, A. 2017. Intraocular minocycline for the treatment of ocular pythiosis. Am. J. Health Syst. Pharm. 74(11): 821-825. doi: 10.2146/ajhp160248.

Rujirawat, T., Patumcharoenpol, P., Lohnoo, T., Yingyong, W., Lerksuthirat, T., Tangphatsornruang, S., Suriyaphol, P., Grenville-Briggs, L.J., Garg, G., Kittichotirat, W. and Krajaejun, T. 2015. Draft genome sequence of the pathogenic oomycete *Pythium insidiosum* strain Pi-S, isolated from a patient with pythiosis. Genome Announc. 3(3): e00574-00515. doi: 10.1128/genomeA.00574-15.

Rujirawat, T., Patumcharoenpol, P., Lohnoo, T., Yingyong, W., Kumsang, Y., Payattikul, P., Tangphatsornruang, S., Suriyaphol, P., Reamtong, O., Garg, G., Kittichotirat, W. and Krajaejun, T. 2018. Probing the phylogenomics and putative pathogenicity genes of *Pythium insidiosum* by Oomycete genome analyses. Sci. Rep. 8(1): 4135. doi: 10.1038/s41598-018-22540-1.

Salipante, S.J., Hoogestraat, D.R., Sengupta, D.J., Murphey, D., Panayides, K., Hamilton, E., Castaneda-Sanchez, I., Kennedy, J., Monsaas, P.W., Mendoza, L., Stephens, K., Dunn, J.J. and Cookson, B.T. 2012. Molecular diagnosis of subcutaneous *Pythium insidiosum* infection using PCR screening and DNA sequencing. J. Clin. Microbiol. 50(4): 1480-1483. doi: 10.1128/JCM.06126-11.

Salomão-Nascimento, R.B., Frazão-Teixeira, E. and Oliveira, F.C.R.d. 2010. Hepatic and renal analysis in horses with pythiosis treated with potassium iodate, through the detection of serum proteins, nitrogenated substances and enzymes. Rev. Bras. Med. Vet. 32(2): 105-110.

Santos, C.E.P., Juliano, R.S., Santurio, J.M. and Marques, L.C. 2011a. Efficacy of immunotherapy in the treatment of facial horse pythiosis. Acta Sci. Vet. 39(1): 955.

Santos, C.E.P., Marques, L.C., Zanette, R.A., Jesus, F.P.K. and Santurio, J.M. 2011b. Does immunotherapy protect equines from reinfection by the oomycete *Pythium insidiosum*? Clin. Vaccine Immunol. 18(8): 1397-1399. doi: 10.1128/Cvi.05150-11.

Santos, C.E.P., Santurio, J.M., Colodel, E.M., Juliano, R.S., Silva, J.A. and Marques, L.C. 2011c. Contribution to the study of cutaneous pythiosis in equidae from northern Pantanal, Brazil. Ars Veterinaria. 27(3): 134-140.

Santos, C.E.P., Santurio, J.M. and Marques, L.C. 2011d. Pythiosis of livestock in the Pantanal, Mato Grosso, Brazil. Pesq. Vet. Bras. 31(12): 1083-1089. doi: 10.1590/S0100-736X2011001200008.

Santos, C.E.P., Ubiali, D.G., Pescador, C.A., Zanette, R.A., Santurio, J.M. and Marques, L.C. 2014. Epidemiological survey of equine pythiosis in the Brazilian Pantanal and nearby areas: Results of 76 cases. J. Equine. Vet. Sci. 34(2): 270-274. doi: 10.1016/j.jevs.2013.06.003.

Santurio, J.M., Monteiro, A.B., Leal, A.T., Kommers, G.D., de Sousa, R.S. and Catto, J.B. 1998. Cutaneous Pythiosis insidiosi in calves from the Pantanal region of Brazil. Mycopathologia 141(3): 123-125.

Santurio, J.M., Leal, A.T., Leal, A.B.M., Festugatto, R., Lubeck, I., Sallis, E.S.V., Copetti, M.V., Alves, S.A. and Ferreiro, L. 2003. Three types of immunotherapics against pythiosis insidiosi developed and evaluated. Vaccine 21(19-20): 2535-2540. doi: 10.1016/S0264-410x(03)00035-5.

Santurio, J.M., Argenta, J.S., Schwendler, S.E., Cavalheiro, A.S., Pereira, D.I.B., Zanette, R.A., Alves, S.H., Dutra, V., Silva, M.C., Arruda, L.P., Nakazato, L. and Colodel, E.M. 2008. Granulomatous rhinitis associated with *Pythium insidiosum* infection in sheep. Vet. Rec. 163(9): 276-277.

Santurio, J.M. and Ferreiro, L. 2008. Pitiose: Uma abordagem micológica e terapêutica. 1 ed. ed. Porto Alegre: Editora da UFRGS.

Sathapatayavongs, B., Leelachaikul, P., Prachaktam, R., Atichartakarn, V., Sriphojanart, S., Trairatvorakul, P., Jirasiritham, S., Nontasut, S., Eurvilaichit, C. and Flegel, T. 1989. Human pythiosis associated with thalassemia hemoglobinopathy syndrome. J. Infect. Dis. 159(2): 274-280.

Schloemer, N.J., Lincoln, A.H., Mikhailov, T.A., Collins, C.L., Di Rocco, J.R., Kehl, S.C. and Chusid, M.J. 2013. Fatal disseminated *Pythium insidiosum* infection in a child with Diamond-Blackfan anemia. Infect. Dis. Clin. Pract. 21(4): e24-e26. doi: 10.1097/IPC.0b013e318278f3b5.

Schurko, A.M., Mendoza, L., de Cock, A.W.A.M. and Klassen, G.R. 2003a. Evidence for geographic clusters: Molecular genetic differences among strains of *Pythium insidiosum* from Asia, Australia and the America are explored. Mycologia 95(2): 200-208. doi: 10.1080/15572536.2004.11833105.

Schurko, A.M., Mendoza, L., Levesque, C.A., Desaulniers, N.L., de Cock, A.W.A.M. and Klassen, G.R. 2003b. A molecular phylogeny of *Pythium insidiosum*. Mycol. Res. 107(5): 537-544. doi: 10.1017/S0953756203007718.

Schurko, A.M., Mendoza, L., de Cock, A.W.A.M., Bedard, J.E.J. and Klassen, G.R. 2004. Development of a species-specific probe for *Pythium insidiosum* and the diagnosis of pythiosis. J. Clin. Microbiol. 42(6): 2411-2418.

Sedrish, S.A., Moore, R.M., Valdes Vasquez, M.A., Haynes, P.F. and Vicek, T. 1997. Adjunctive use of a neodymium: Yttrium-aluminum-garnet laser for treatment of pythiosis granulomas in two horses. J. Am. Vet. Med. Assoc. 211(4): 464-465.

Seidl, M.F., Van den Ackerveken, G., Govers, F. and Snel, B. 2011. A domain-centric analysis of oomycete plant pathogen genomes reveals unique protein organization. Plant Physiol. 155(2): 628-644. doi: 10.1104/pp.110.167841.

Sermsathanasawadi, N., Praditsuktavorn, B., Hongku, K., Wongwanit, C., Chinsakchai, K., Ruangsetakit, C., Hahtapornsawan, S. and Mutirangura, P. 2016. Outcomes and factors influencing prognosis in patients with vascular pythiosis. J. Vasc. Surg. 64(2): 411-417. doi: 10.1016/j.jvs.2015.12.024.

Shenep, J.L., English, B.K., Kaufman, L., Pearson, T.A., Thompson, J.W., Kaufman, R.A., Frisch, G. and Rinaldi, M.G. 1998. Successful medical therapy for deeply invasive facial infection due to *Pythium insidiosum* in a child. Clin. Infect. Dis. 27(6): 1388-1393.

Shipton, W.A. 1987. *Pythium destruens* sp. nov, an agent of equine pythiosis. J. Med. Vet. Mycol. 25(3): 137-151.

Silva, T.R.d., Neto, E.G.d.M., Medeiros, J.M., Melo, D.B.d. and Dantas, A.F.M. 2011. Cutaneous pythiosis in ruminants. Vet. Zootec. 18(4-Supl. 3): 871-874.

Smith, F. 1884. The pathology of bursattee. Vet. J. 19(7): 16-17.

Sriphana, U., Thongsri, Y., Ardwichai, P., Poopasit, K., Prariyachatigul, C., Simasathiansophon, S. and Yenjai, C. 2013a. New lignan esters from *Alyxia schlechteri* and antifungal activity against *Pythium insidiosum*. Fitoterapia. 91: 39-43. doi: 10.1016/j.fitote.2013.08.005.

Sriphana, U., Thongsri, Y., Prariyachatigul, C., Pakawatchai, C. and Yenjai, C. 2013b. Clauraila E from the roots of *Clausena harmandiana* and antifungal activity against *Pythium insidiosum*. Arch. Pharm. Res. 36(9): 1078-1083. doi: 10.1007/s12272-013-0115-5.

Sterling, J.B. and Heymann, W.R. 2000. Potassium iodide in dermatology: A 19th century drug for the 21st century-uses, pharmacology, adverse effects, and contraindications. J. Am. Acad. Dermatol. 43(4): 691-697. doi: 10.1067/mjd.2000.107247.

Sudjaritruk, T. and Sirisanthana, V. 2011. Successful treatment of a child with vascular pythiosis. BMC Infect. Dis. 11(1): 33. doi: 10.1186/1471-2334-11-33.

Supabandhu, J., Fisher, M.C., Mendoza, L. and Vanittanakom, N. 2008. Isolation and identification of the human pathogen *Pythium insidiosum* from environmental samples collected in Thai agricultural areas. Med. Mycol. 46(1): 41-52. doi: 10.1080/13693780701513840.

Suthiwong, J., Thongsri, Y. and Yenjai, C. 2017. A new furanocoumarin from the fruits of *Scaevola taccada* and antifungal activity against *Pythium insidiosum*. Nat. Prod. Res. 31(4): 453-459. doi: 10.1080/14786419.2016.1188100.

Tabosa, I.M., Riet-Correa, E., Nobre, V.M.T., Azevedo, E.O., Reis-Junior, J.L. and Medeiros, R.M.T. 2004. Outbreaks of pythiosis in two flocks of sheep in Northeastern Brazil. Vet. Pathol. 41(4): 412-415. doi: 10.1354/vp.41-4-412.

Tanhehco, T.Y., Stacy, R.C., Mendoza, L., Durand, M.L., Jakobiec, F.A. and Colby, K.A. 2011. *Pythium insidiosum* keratitis in Israel. Eye & Contact Lens. 37(2): 96-98. doi: 10.1097/ICL.0b013e3182043114.

Thanathanee, O., Enkvetchakul, O., Rangsin, R., Waraasawapati, S., Samerpitak, K. and Suwan-apichon, O. 2013. Outbreak of *Pythium* keratitis during rainy season: A case series. Cornea 32(2): 199-204. doi: 10.1097/Ico.0b013e3182535841.

Thianprasit, M., Chaiprasert, A. and Imwidthaya, P. 1996. Human pythiosis. Curr. Top. Med. Mycol. 7(1): 43-54.

Thines, M. 2018. Oomycetes. Curr. Biol. 28(15): R812-R813. doi: 10.1016/j.cub.2018.05.062.

Thitithanyanont, A., Mendoza, L., Chuansumrit, A., Pracharktam, R., Laothamatas, J., Sathapatayavongs, B., Lolekha, S. and Ajello, L. 1998. Use of an immunotherapeutic vaccine to treat a life-threatening human arteritic infection caused by *Pythium insidiosum*. Clin. Infect. Dis. 27(6): 1394-1400.

Thomas, R.C. and Lewis, D.T. 1998. Pythiosis in dogs and cats. Comp. Cont. Educ. Pract. Vet. 20(1): 63-75.

Tondolo, J.S., Loreto, E.S., Denardi, L.B., Mario, D.A., Alves, S.H. and Santurio, J.M. 2013. A simple, rapid and inexpensive screening method for the identification of *Pythium insidiosum*. J. Microbiol. Methods 93(1): 52-54. doi: 10.1016/j.mimet.2013.02.002.

Tondolo, J.S.M., Ledur, P.C., Loreto, E.S., Verdi, C.M., Bitencourt, P.E.R., de Jesus, F.P.K., Rocha, J.P., Alves, S.H., Sassaki, G.L. and Santurio, J.M. 2017a. Extraction, characterization and biological activity of a (1,3)(1,6)-beta-d-glucan from the pathogenic oomycete *Pythium insidiosum*. Carbohydr. Polym. 157: 719-727. doi: 10.1016/j.carbpol.2016.10.053.

Tondolo, J.S.M., Loreto, E.S., Ledur, P.C., Jesus, F.P.K., Silva, T.M., Kommers, G.D., Alves, S.H. and Santurio, J.M. 2017b. Chemically induced disseminated pythiosis in BALB/c mice: A new experimental model for *Pythium insidiosum* infection. PLoS One. 12(5): e0177868. doi: 10.1371/journal.pone.0177868.

Triscott, J.A., Weedon, D. and Cabana, E. 1993. Human subcutaneous pythiosis. J. Cutan. Pathol. 20(3): 267-271.

Trolezi, R., Azanha, J.M., Paschoal, N.R., Chechi, J.L., Dias Silva, M.J., Fabris, V.E., Vilegas, W., Kaneno, R., Fernandes Junior, A. and Bosco, S.M. 2017. *Stryphnodendron adstringens* and purified tannin on *Pythium insidiosum*: *in vitro* and *in vivo* studies. Ann. Clin. Microbiol. Antimicrob. 16(1): 7. doi: 10.1186/s12941-017-0183-3.

Uzuhashi, S., Kakishima, M. and Tojo, M. 2010. Phylogeny of the genus *Pythium* and description of new genera. Mycoscience 51(5): 337-365. doi: 10.1007/s10267-010-0046-7.

Valente, J.S.S., Fonseca, A.O.S., Brasil, C.L., Sagave, L., Flores, F.C., de Bona da Silva, C., Sangioni, L.A., Potter, L., Santurio, J.M., de Avila Botton, S. and Pereira, D.I. 2016. *In vitro* activity of *Melaleuca alternifolia* (Tea Tree) in its free oil and nanoemulsion formulations against *Pythium insidiosum*. Mycopathologia 181(11-12): 865-869. doi: 10.1007/s11046-016-0051-2.

van West, P. and Beakes, G.W. 2014. Animal pathogenic Oomycetes. Fungal Biol. 118(7): 525-526. doi: 10.1016/j.funbio.2014.05.004.

Vanittanakom, N., Supabandhu, J., Khamwan, C., Praparattanapan, J., Thirach, S., Prasertwitayakij, N., Louthrenoo, W., Chiewchanvit, S. and Tananuvat, N. 2004. Identification of emerging human-pathogenic *Pythium insidiosum* by serological and molecular assay-based methods. J. Clin. Microbiol. 42(9): 3970-3974. doi: 10.1128/Jcm.42.9.3970-3974.2004.

Vanittanakom, N., Szekely, J., Khanthawong, S., Sawutdeechaikul, P., Vanittanakom, P. and Fisher, M.C. 2014. Molecular detection of *Pythium insidiosum* from soil in Thai agricultural areas. Int. J. Med. Microbiol. 304(3-4): 321-326. doi: 10.1016/j.ijmm.2013.11.016.

Verdi, C.M., Jesus, F.P.K., Kommers, G., Ledur, P.C., Azevedo, M.I., Loreto, E.S., Tondolo, J.S.M., Andrade, E.N.C., Schlemmer, K.B., Alves, S.H. and Santurio, J.M. 2018. Embryonated chicken eggs: An experimental model for *Pythium insidiosum* infection. Mycoses. 61(2): 104-110. doi: 10.1111/myc.12710.

Videla, R., van Amstel, S., O'Neill, S.H., Frank, L.A., Newman, S.J., Vilela, R. and Mendoza, L. 2012. Vulvar pythiosis in two captive camels (*Camelus dromedarius*). Med. Mycol. 50(2): 219-224. doi: 10.3109/13693786.2011.588970.

Vilela, R. and Mendoza, L. 2015. Lacazia, lagenidium, pythium, and rhinosporidium. pp. 2196-2208. *In*: Jorgensen, James H., Pfaller, Michael A. and Carroll, Karen C. (eds.). Manual of Clinical Microbiology. American Society for Microbiology, ASM Press, Washington, DC.

Vilela, R., Montalva, C., Luz, C., Humber, R.A. and Mendoza, L. 2018. *Pythium insidiosum* isolated from infected mosquito larvae in central Brazil. Acta Trop. 185: 344-348. doi: 10.1016/j. actatropica.2018.06.014.

Wanachiwanawin, W., Thianprasit, M., Fucharoen, S., Chaiprasert, A., Ayudhya, N.S.N., Sirithanaratkul, N. and Piankijagum, A. 1993. Fatal arteritis due to *Pythium insidiosum* infection in patients with thalassemia. Trans. R. Soc. Trop. Med. Hyg. 87(3): 296-298.

Wanachiwanawin, W., Mendoza, L., Visuthisakchai, S., Mutsikapan, P., Sathapatayavongs, B., Chaiprasert, A., Suwanagool, P., Manuskiatti, W., Ruangsetakit, C. and Ajello, L. 2004. Efficacy of immunotherapy using antigens of *Pythium insidiosum* in the treatment of vascular pythiosis in humans. Vaccine. 22(27-28): 3613-3621. doi: 10.1016/j.vaccine.2004.03.031.

Weiblen, C., Machado, G., Jesus, F.P.K.d., Santurio, J.M., Zanette, R.A., Pereira, D.S.B., Diehl, G.N., Santos, L.C.d., Corbellini, L.G. and Botton, S.d.A. 2016. Seroprevalence of *Pythium insidiosum* infection in equine in Rio Grande do Sul, Brazil. Ciênc Rural. 46(1): 126-131. doi: 10.1590/0103-8478cr20150056.

Wellehan, J.F., Farina, L.L., Keoughan, C.G., Lafortune, M., Grooters, A.M., Mendoza, L., Brown, M., Terrell, S.P., Jacobson, E.R. and Heard, D.J. 2004. Pythiosis in a dromedary camel (*Camelus dromedarius*). J. Zoo Wildl. Med. 35(4): 564-568. doi: 10.1638/03-098.

Witkamp, J. 1924. Bijdrage tot de kennis van de Hyphomycosis destruens. Ned Ind Blad voor Diergeneeskd en Dierenteelt. 36: 229-245.

Witkamp, J. 1925. Het voorkomen van metastasen in de regionaire lymphlieren by Hyphomycosis destruens. Ned Ind Blad voor Diergeneeskd en Dierenteelt. 37: 79-102.

Wongprompitak, P., Pleewan, N., Tantibhedhyangkul, W., Chaiprasert, A., Prabhasawat, P., Inthasin, N. and Ekpo, P. 2018. Involvement of toll-like receptor 2 on human corneal epithelium during an infection of *Pythium insidiosum*. Asian Pac. J. Allergy Immunol. doi: 10.12932/AP-110518-0311.

Worasilchai, N., Chaumpluk, P., Chakrabarti, A. and Chindamporn, A. 2018a. Differential diagnosis for pythiosis using thermophilic helicase DNA amplification and restriction fragment length polymorphism (tHDA-RFLP). Med. Mycol. 56(2): 216-224. doi: 10.1093/mmy/myx033.

Worasilchai, N., Permpalung, N. and Chindamporn, A. 2018b. High-resolution melting analysis: A novel approach for clade differentiation in *Pythium insidiosum* and pythiosis. Med. Mycol. 56(7): 868-876. doi: 10.1093/mmy/myx123.

Zanette, R.A., Alves, S.H., Pilotto, M.B., Weiblen, C., Fighera, R.A., Wolkmer, P., Flores, M.M. and Santurio, J.M. 2013a. Iron chelation therapy as a treatment for *Pythium insidiosum* in an animal model. J. Antimicrob. Chemother. 68(5): 1144-1147. doi: 10.1093/Jac/Dks534.

Zanette, R.A., Bitencourt, P.E., Alves, S.H., Fighera, R.A., Flores, M.M., Wolkmer, P., Hecktheuer, P.A., Thomas, L.R., Pereira, P.L., Loreto, E.S. and Santurio, J.M. 2013b. Insights into the pathophysiology of iron metabolism in *Pythium insidiosum* infections. Vet. Microbiol. 162(2-4): 826-830. doi: 10.1016/j.vetmic.2012.10.036.

Zanette, R.A., Santurio, J.M., Loreto, E.S., Alves, S.H. and Kontoyiannis, D.P. 2013c. Toll-deficient *Drosophila* is susceptible to *Pythium insidiosum* infection. Microbiol. Immunol. 57(10): 732-735. doi: 10.1111/1348-0421.12082.

Zanette, R.A., Bitencourt, P.E., Kontoyiannis, D.P., Fighera, R.A., Flores, M.M., Kommers, G.D., Silva, P.S., Ludwig, A., Moretto, M.B., Alves, S.H. and Santurio, J.M. 2015. Complex interaction of deferasirox and *Pythium insidiosum*: Iron-dependent attenuation of growth *in vitro* and immunotherapy-like enhancement of immune responses *in vivo*. PLoS One. 10(3): e0118932. doi: 10.1371/journal.pone.0118932.

PART IV
Management of Diseases Caused by *Pythium*

Damping-off Caused by *Pythium* Species: Disease Profile and Management

Mohammad Imad Khrieba

Plant Pathology and Biological Control, National Commission of Biotechnology (NCBT), Damascus, Syria
E-mail: imadkhrieba@gmail.com

Introduction

Damping-off is caused due to the attack of seedlings by soil-inhabiting fungi usually just at the soil level on to hypocotyls or upper taproot. This causes partial or complete rot and the seedlings suddenly topple over in a characteristic manner (Moorman 2011). The damping-off has been known for a long time in both Europe and America (Alexopoulos et al. 1996, Tomioka et al. 2013). By the beginning of the twentieth century in the United States, as early as 1901 to 1905, the damping-off disease was found in forest nursery seedbeds and began to receive attention (Smith et al. 2014). This disease is encountered in seedbeds where plants are propagated to be transplanted later in the fields. However, greenhouses and nursery seedbeds are ideal places for outbreak of damping-off, if moisture and temperature conditions are favorable (Bahramisharif et al. 2013). Damping-off generally appears as a pre-emergence loss involving failure to emerge and as post-emergence loss involving rotting and collapsing of seedlings at their base (Alexopoulos et al. 1996, Tomioka et al. 2013). Damping-off has been found to affect up to 80% (in some cases even 100%) of seedlings, resulting in poor seedling stands in nurseries of various crops (Agrios 2005). Almost any kind of plant may be attacked by damping-off fungi while in the young, tender, succulent stage of development. The incidence of disease is dependent more upon the conditions under which the seedlings are grown than the particular species of plant concerned (Bahramisharif et al. 2014). The most common root-rotting fungi of greenhouse plants and garden flowers are *Rhizoctonia solani*, species of *Pythium* and *Phytophthora*, and *Thielaviopsis basicola*. Species of *Pythium* and *Phytophthora* are most likely to cause damping-off and root-rot in cool/cold, wet, poorly aerated soils, whereas the majority of the other fungi usually cause disease under warmer and drier conditions (Alcala et al. 2016). The disease may affect 5% to 80% of the seedlings, thereby inducing heavy economic consequences for farmers. Much research efforts have been made in recent years to develop bio-control solutions for damping-off, which have interesting future perspectives. Damping-off management requires integrated pest management (IPM), which is an approach combining both preventive and curative tactics and strategies (Hill and Waller 1990). This chapter aims to highlight the major features of damping-off diseases, especially those caused by *Pythium* spp., currently used disease management strategies, knowledge gaps, and suggest key challenges and future priorities for a sustainable production of crop plants.

Importance

Damping-off can be a devastating disease and cause huge economic loss. Damping-off mainly affects plants prior to seed germination and throughout the seedling stage (Agrios 1997). Crop losses due to damping-off disease vary from place to place. Moderate to severe incidences, leading to crop loss of more than 50% to 80%, have been reported because of this disease (Punja and Yip 2003, Smith et al. 2014, Lamichhane et al. 2017). The amount of crop loss depends on the stage of crop growth at which the infection starts. If it occurs early, total crop loss of the affected clump will result into huge loss, whereas the loss is partial if the plant is affected at a later stage (Agrios 2005). *Pythium* spp. also cause catastrophic loss in storage. Crop losses inflicted by soil-borne pathogens continue to increase and become a limiting factor in stabilizing and maximizing crop yields on a worldwide basis (Kageyama 2015, Kennelly 2016).

Conditions of disease development

The risk of *Pythium* blight is high during humid weather when day temperatures are 30°C to 35°C and a consistent decrease overnight to 20°C at least (Kennell 2016). The disease is most common when soils are saturated with water, due to excessive rainfall or irrigation. Also, long dew periods, high relative humidity, and lush and dense turfgrass growth favors the disease development. Areas of low-altitude, poor airflow and poor drainage are particularly vulnerable (Mundel et al. 1995).The soil pH influences some phases of life-cycle of the *Pythium* species, such as the formation of oospores and sporangia. *Pythium* species prefer to grow in alkaline soils (above pH 7) (Lumsden et al. 1987). The soil pH between 5.2 and 5.7 (moderately acidic) prevents damping-off problems. Aluminum sulfate drenches, sulfur (200 to 500 pounds/acre), and acid peat applications can be used to maintain the soil pH. As an example, when aluminum sulfate is used, beds should be kept moist to prevent burning of the roots. Abundant moisture and a warm temperature are conducive to the invasion of the seedlings. The soil temperature of 20°C to 28°C are ideal for the host infection by direct penetration. Excessive moisture in the surface layer of the soil due to overwatering, thick sowing in the seedbed causing heavy stands, lack of aeration, and too much shade, combine to encourage disease initiation and development (Lumsden et al. 1987, Russell 1990, Kennell 2016). *Pythium* species differ in soil temperature requirements. Some *Pythium* species prefer low soil temperature (5 to 10°C), such as *P. ultimum* and *P. irregulare*, while others thrive in warmer temperatures (25 to 30°C), such as *P. aphanidermatum*. Spore germination and germ tube growth are influenced by soil temperature, with greater zoospore discharge from sporangia at lower soil temperature, while oospore germination and germ tube growth occurs at relatively higher temperatures (Mundel et al. 1995, Agrios 2005).

Symptoms of damping-off

Based on the time of appearance, the disease symptoms can be classified into two phases. The first is pre-emergence, in which sprouting seeds decay in soil and young seedlings rot before emergence. The second is post-emergence phase, which is characterized by the infection of the juvenile tissues of the collar at the ground level (Bacharis et al. 2010). Damping-off symptoms are most commonly seen as poor stand establishment in seeded areas, often giving them a 'patchy' look, since the plants are highly vulnerable after germination when a little damage can kill the plant (Phillips and Burdekin 1982, Tomioka et al. 2013). Typical symptoms of damping-off are rotting stems at or near the soil, and line and root decay (post-emerging damping off). Affected areas in the seedbed are usually a 30 cm or more in diameter with shriveled brown, collapsed or stunted seedlings (Olson et al. 2016). *Pythium* attacks often at root tips (Moorman 2011), while moldy fungal growth may be observed on affected plants at the soil line. Also, germinating seeds can be attacked by these fungi before they emerge from the soil (pre-emerging damping off), resulting in poor stands. Usually, affected plants occur in patches in nursery beds or in the fields at low elevations (Olson et al. 2016).

Seedlings that survive from infected nurseries often show the so-called "wire stem" or "stem girdling" syndrome (i.e. a blackish sunken lesion girdles the base of the stem, where the stem becomes thin and hard, especially at soil level, and may bend or twist) resulting in reduced vigor (Lamichhane et al. 2017). If the infection occurs before the emergence, the germinating plants become soft, brown, and decompose, while if the infection happens after emergence, water-soaked lesions form about one cm above or below the soil line (Alcala et al. 2016). The stem will be softened and cannot support the seedling, which collapses and die. Stunting of plants due to root-rot may also occur. In this case, the root system rots becomes brown and develops few secondary roots (Wick 1998, Kennelly 2016). In flat fields, affected plants are generally found in scattered parts (Horst 2013). Infection by the pathogen may occur much later after emergence; in this case, the infection usually is not lethal but plant growth and yield may be reduced (Moorman 2011). Severe infection in the root region may cause wilting in warm or windy weather. Also, extensive root rotting induces symptoms of nutritional deficiencies in the plant, since nutrients cannot move sufficiently from the soil up through the plant. However, pathogen isolation and identification is needed to confirm diagnosis because several pathogens can cause similar symptoms (Horst 2013, Olson et al. 2016).

The fungus attacks the seedlings at hypocotyls or taproot and when the tissue is susceptible and contributes to the initiation of a thin cell wall during rapid growth (Scheuerell et al. 2005). The disease symptoms can be observed from seeding until four to six weeks post-sowing (Horst 2013, Olson et al. 2016). The only evidence of pre-emergence damping-off in nursery beds is that the germinating seedlings are sparse and patchy (Agrios 2005). This phase is difficult to detect, but sometimes may be diagnosed by digging up seeds that have not emerged and checking to see whether seeds or germinated seeds are decayed or withered. Post-emergence damping-off, which occurs at the cotyledon stage, causes wither and collapse of the seedlings. When the succulent root-collar tissue or the roots is penetrated by the pathogen, the disease is defined as soil-infection damping-off, whereas when the fungal invasion occurs at a higher level on the stem or cotyledons, it is called top infection damping-off (Moorman 2011).

There are multiple cases that have similar symptoms to damping-off, where sometimes the seed may not germinate if it is old or if conditions have not been favorable for germination. Also, if tender young seedlings are over fertilized, roots appear shriveled and desiccated, and the plants die due to the high salts. In addition, hot water, heat stress, lack of water and chemical sprays can also prompt death of the tender young seedlings (Daughtery 1995).

Etiology of damping-off disease caused by *Pythium*

Damping-off is caused by a number of *Pythium* species, including *P. aphanidermatum* (Edson) Fitzp, *P. ultimum* Trow, and *P. irregulare* (Punja et al. 2003). *Pythium ultimum* (Genus: *Pythium*; Family: Pythiaceae; Order: Pythiales; Class: Oomycetes; Phylum: Oomycota; Kingdom: Chromista and Domain: Eukaryota) (Uzuhashi et al. 2010). It is a necrotrophic, aseptate and coenocytic plant pathogen with a wide host range, which can reproduce sexually through oospores and asexually through sporangia and zoospores (Okubara et al. 2014). The genus *Pythium* generally causes rotting of fruit, roots and stems, as well as pre- or post-emergence damping-off of seedlings (Uzuhashi et al. 2010). *Pythium* is a microscopic fungus and the largest genus of family Pythiaceae, which in most of the classifications has been included in class Oomycetes. Oomycetes differ from the true fungi (Eumycota) in having coenocytic hyphae containing β-1–3 glucans and cellulose in their cell walls (Cheung et al. 2008). *Pythium* is a common problem in nurseries, so much so that some plant pathologists have called it the 'common cold' of plants. *Pythium* spp. are common worldwide, which are associated with plant roots and other parts in contact with the proximity of soil (Le et al. 2014, Kamoun et al. 2015). *Pythium* constitutes one of the most destructive groups of plant pathogens, causing many diseases in a wide range of hosts, from algae to mammals (Park et al. 2016).

Description

Temperature optimum for growth of *P. aphanidermatum* is 34°C, with a minimum at 10°C and a maximum above 43°C. On artificial media, cottony white aerial mycelium is produced with aseptate hyphae 7 to 10 µm in diameter. In *P. aphanidermatum,* zoosporangia mostly terminal, sometimes intercalary with inflated structures, oogonia terminal, globose, smooth (21–25 µm) in diameter, antheridia intercalary, sometimes terminal (10–15 µm) long and (8–11 µm) wide, one per oogonium, oospores aplerotic (20-22 µm) in diameter, oospore wall (1–2 µm) thick. A site of formation of sporangia, oogonia and antheridia as well as their shape and dimensions slightly vary for the other *Pythium* species (Russell 1990, Howell 2002). *Pythium* is similar to *Phytophthora* in their capability of moving through water as zoospores via flagella, and are commonly referred to as water molds. More than 300 species are reported in the soil-borne neurotropic oomycete genus *Pythium*.

Host range and distribution

Pythium has been found in Australia, Canada, Brazil, China, Korea, Japan, South Africa and many other countries in the world including most states within the United States of America. It causes *Pythium* blight of turfgrass, inflicting very serious damage to golf courses (Allen et al. 2004). It can also infect crops like cabbage, broccoli, cucumber, carrot, melon, cotton and wheat (El-Mohamedy 2012). Surveys of the diversity of *Pythium* species in agricultural soils have demonstrated that it can be common to isolate more than one species in a field, including both pathogenic and non-pathogenic species. However, several pathogenic *Pythium* species can be associated with damping-off in a particular field, making management of the disease more complicated (Broders et al. 2007, 2009).

Survival and reproduction

Ordinarily, *Pythium* species survive saprophytically in soil because of their inability to compete with other plant pathogens (Agrios 2005). However, if oversaturated soil containing potential hosts is available (i.e. seeds or seedlings), infection can occur. The fungus is able to spread from the infected seedling due to the separation of the host cells from the breakdown of the middle lamella. The breakdown of these cell components is likely the result of pectinolytic enzymes secreted from the hyphal tips of *Pythium* (Jadhav and Ambadkar 2007).

The reproductive phase sets on before the death of the host in the case of parasitic species. However, asexual mode is the most common while sexual reproduction occurs mostly towards the end of the growing season (Paulitz et al. 2003). The life-cycle of *Pythium* is haploidic, which means that, except the zygote (oospore), all the stages represent the haploid (n) phase. According to this reality, the somatic mycelium, which is haploid in nature, later on towards the end of growing season, bears haploid gametangia (sex organs). The contents of the male and female gametangia transform into male and female gametes, respectively. The gametes fuse to form diploid (2n) oospore representing the diplophase in the life cycle. All other structures, which include zoospores, mycelium, gametangia, gametes and asexual reproductive bodies, represent haplophase in the life cycle (Fig. 14.1) (van West et al. 2003, Moorman 2011, Kamoun et al. 2015).

Sporangia and oospores of *Pythium* spp. germinate directly by producing a germ tube or indirectly by producing zoospores. They are attracted primarily to the region of root elongation, root hair emergence or the root cap where they encyst and penetrate the host. Under non-favorable conditions, zoospores may encyst in the soil, remaining viable as long as soil moisture content and temperature are suitable (Bradley 2008, Moorman 2011). The sporangia get detached from the hyphae bearing them and are blown by the wind or dispersed by water. On falling or reaching a suitable host, the sporangium germinates like a conidium if the temperature is above the optimum. It directly puts out a germ tube (G) which finds its way into the host either through a stomata or directly through an epidermal cell by secreting an enzyme at its tip. Inside the host, it grows and branches to form the mycelium (H). In this case, the mycelium develops without the intervention of zoospores. Some mycologists call such sporangia, which germinate directly and thus behave like conidia, conidiosporangia. Others hold that these sporangia-like structures are in fact conidia. The production of conidiosporangia or conidia permits spread of the disease under relatively dry conditions (Bacharis et al. 2010, Olson et al. 2016).

The sexual oospores of *Pythium* species can survive either in plant debris or in soil for years. Germination occurs under the right conditions to produce the mobile zoospores or to directly infect vulnerable plant tissue within four hours of exposure to plant exudates. The response of zoospores to root exudate components differ among *Pythium* species. Host tissue is penetrated either with or without appresorium formation. *Pythium* spp. commonly reproduce within diseased tissues, but are also good saprophytes colonizing organic material in the soil. The mycelium remains in cortical cells (host plant) during the early attack. The host plant collapses and falls to the ground, which is called damping off disease (Agrios 2005). Later hyphae spread to other parts of the plant through vascular bundle and absorb the food of plant. Then the plant dies and *Pythium* becomes saprophytic, releases its enzymes and decomposes the plant body (Moorman 2011, Kamoun et al. 2015). In soil, resistant survival structure is more important than saprophytic persistence for *Pythium* spp. For short and intermediate periods, survival is in the form of zoospores and sporangia, while long-term survival depends primarily on oospores (Trigiano et al. 2004). Survival of sporangia of *Pythium* spp. has been shown to extend up to 11 months in soil and can germinate 1.5 hours after nutrient addition, reactivating their pathogenic capacity (Agrios 1997).

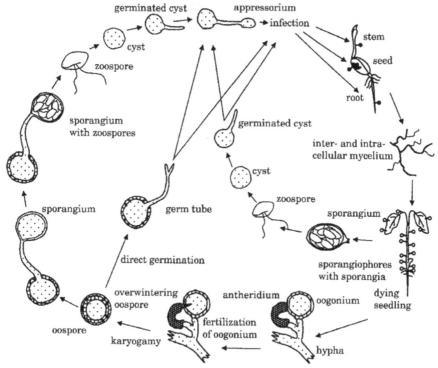

Figure 14.1. Life cycle of a typical root infecting *Pythium* species. This consists of two cycles that are usually stimulated by differing environmental conditions. The asexual cycle is characterized by the production of sporangia. Sporangia may germinate either directly in liquid or on a surface to produce a germ tube (direct germination) or may differentiate by a process of cytoplasmic cleavage to form uninucleate, biflagellate zoospores (indirect germination). In root pathogenic oomycete, the zoospores are often retained within a discharge vesicle. Reproduced from "Physiological and Molecular Plant Pathology Journal" (van West et al. 2003)

Epidemiology

Pythium species have been isolated from the soil of arable land, pastures, forests, nurseries, marshes, swamps, and water (Agrios 1997, Trigiano et al. 2004). They occur most abundantly in cultivated soils near the root region in superficial soil layers, and less commonly in uncultivated or acid soils where fungistasis keeps them suppressed (Agrios 1997, Trigiano et al. 2004 provide recent references). Moist soil is particularly important for *Pythium* species, for which water is required for the movement of spores, and its low oxygen level encourages the release of exudates from seeds (Nzungize et al. 2011). Compost is a very common source of contamination in seedbeds. Sometimes, these materials are obtained from an infected source and of course, the organism is introduced along with the soil or compost (Bacharis et al. 2010). Infected seeds, soils and seedlings serve as a primary inoculum source for epidemics in nurseries and production fields of various crops (Sharma and Razdan 2012).

Integrated disease management

Utilization of a range of control strategies such as biological, cultural, and chemical is needed to keep the disease below the economic thresholds of a sustainable agro-ecosystem. Integrated disease management has been reported to be quite effective for control of soil-borne plant pathogens (Castillo 2004). Preventing the disease is considered one of the most effective control strategies, as there is no cure for damping-off once it occurs (Milijasevic et al. 2017). Even though some fungicides can control this pathogen, nowadays researchers are more interested in biological control agents and their antifungal metabolites due to the development of resistance in the pathogen. Biological control is an alternative approach to the chemical fungicides and it may be a safe, effective and ecofriendly method for plant disease management (Rajendraprasad et al. 2017).

Prevention and cultural management

Economical control of plant disease is rarely achieved by a single procedure. It must be supported by the use of planting material that is free of the pathogen, tillage methods that are unfavorable to the pathogen but favorable to the host, the use of antagonists, sanitation, use of fungicides, and especially by host plant resistance. Such integrated control is based on the fact that different methods work best at different times and places or under different conditions, each compensating to some extent for the deficiencies of the others (Kumar et al. 2012). Overall, public awareness of the environmental impact of chemical pesticides in agriculture has been shifting the focus away from high-energy input agro-ecosystems towards more ecologically sustainable systems. In these systems, synthetic chemicals are at least minimized, and biologically based processes are used for both nutrient supply and pathogen control (Tilman et al. 2002).

In general, control of soil-borne disease is difficult. Several management methods like physical, chemical and biological have proved to be effective against soil-borne pathogens. Chemical control of these soil-borne pathogens is uneconomical and not advisable owing to the risk of groundwater pollution, death of non-target beneficial flora and evolution of fungicide resistant variants of pathogen. Hence, it is recommended to adopt integrated disease management. Accordingly, we mention several general methods used to prevent damping-off:

- Proper soil preparation and management to provide good soil drainage, structure, aeration, water-holding capacity and plant nutrition by including proper amounts of fertilizer and lime according to the soil test report (Bradley 2008).
- Proper soil treatment with heat or chemicals to reduce the level of fungi that cause damping-off (Milijasevic et al. 2017).
- Proper seeding rates to avoid thick plant stands, poor air movement and low light intensity (Horst 2013).

- Proper planting depth and soil temperature to assure rapid seeding emergence and growth (Moorman 2011).
- The growth medium should be free of chemicals and other toxic residues. Use plant containers with one or more openings in the bottom to provide proper drainage. Avoid excess watering. Place the plants at the best site possible for maximum growth (Russell 1990, Moorman 2011).
- Use a sterile potting mix, rather than soil from your farm. Outdoor soil can harbor fungus spores.
- Avoid watering before taking cuttings or otherwise handling plants. The chances of transmitting pathogens are reduced when only dry plants are handled. Cuttings should be made as far from the soil surface as practical to minimize contamination (Russell 1990).
- Give your seedlings plenty of heat and light, so they germinate and grow quickly. Damping-off only affects seedlings. If you can get them past the seedling stage, they're safe (Bradley 2008).
- Avoid spreading soil from infested areas. Carefully dig up diseased plants, remove them from growing areas, and place them in a covered trash container (Milijasevic et al. 2017).

If the above preventive control measures fail, several fungicides are available that might be effective if applied as a drench or heavy spray as soon as the first symptoms of damping-off are observed. Growing conditions should be improved and flats or areas of bed with damping-off should be carefully removed from the growing area (Scheuerell et al. 2005, Kleczewski and Egel 2011).

Chemical control

Once pathogens become established on crops, especially perennials, and reach damaging levels, chemical control is needed. Pests and pathogens may be killed by exposing them to toxic substances. Toxic chemicals lethal to all forms are used in agriculture only as sterilants for the eradication of pests and pathogens from soil. Chemical control has to be followed with each outbreak. It is, however, very quick in action and still the method by which best results were obtained for mortality of disease outbreaks (Hill and Waller 1990).

Fungicide overuse threatens the human health and causes ecological concerns. This practice has led to the emergence of pesticide-resistant microorganisms in the environment. Thus, there are increasing concerns to develop sustainability and durable damping-off management strategies that are less reliant on conventional pesticides. Achieving such a goal requires a better knowledge of pathogen biology and disease epidemiology in order to facilitate the decision-making process (Abawi et al. 2006). It also demands using all available non-chemical tools that can be adapted to regional and specific production situations (Ringer et al. 1997). For control, the fungicides are best used as a spot treatment or as a preventative approach; however, application will rarely cure an infection alone and the pathogen may reappear (Kleczewski and Egel 2011). The management of *Pythium* damping-off relies on the use of fungicides. However, fungicide application has resulted in the accumulation of residual toxicity in soil and environmental pollution, and has altered the biological balance in the soil by decimating non-target and beneficial microorganisms. Development of fungicide resistance in the pathogen has also been reported (Bharathi et al. 2004). Little is known about the relative efficacy of fungicides for *Pythium* root-rot control. Fungicides have continued to be an important element for the control of *Pythium* disease (Bradley 2008).

There are many specific pesticides such as benomyl, captafol, captan, carboxin, metalaxyl, propamocarb hydrochloride and etridiazole, which have already proven to be effective in controlling *Pythium* root-rot diseases (Abawi et al. 2006). The compounds are general biocides and they can control not only fungi in the soil but also bacteria, nematodes, insects, mites, germinating weed seeds, and most other forms of life in the area, including the roots of trees and shrubs. Since methyl bromide, chloropicrin, and mixtures of the two are restricted pesticides, they should be applied by a certified pesticide applicator trained in soil-fumigation techniques. Methyl bromide and chloropicrin require the use of a gas-proof cover (such as polyethylene or vinyl) over the treated area for 48 to 72 hours or more (Bradley 2008). Several applications of the fungicide may be necessary.

If the specific fungus causing damping-off is not known, one broad-spectrum fungicide (captan or ferbam), two specific fungicides (benomyl plus, etridiazole or metalaxyl) or a combination of fungicides (etridiazole + thiophanate methyl) should be used (Bradley 2008). Fungicides and sterilants can reduce pathogenic fungi on the seed coat and improve germination. However, the same treatments can have phytotoxic effects depending on the species of seed, condition of the seed coat, and application method (Milijasevic et al. 2017).

A number of fungicides currently in use by nurseries are very specific concerning the damping-off pathogens control. For *Pythium* and *Phytophthora*, a pre-plant application of metalaxyl, fosetyl aluminum, or etridiazole can be used to prevent damping-off. Drench applications of thiophanate-methyl may reduce damping-off by *Fusarium*, *Rhizoctonia*, *Botrytis*, and *Cylindrocladium*. Iprodione is a pre-plant drench for control of *Rhizoctonia* and a foliar spray for control of *Botrytis* (James 1988, Russell 1990). The first post-plant fungicide application should be made when most seedlings have emerged and the seeds begin to drop from cotyledon leaves. A good all-purpose preventive treatment for damping-off is a 50-50 mixture of captan and benomyl applied as a drench at rates recommended on the label. If frequent applications are planned, alternation of the captan-benomyl mix with other fungicides is advised (Boyce 1961).

If the soil in flowerbeds becomes severely infested with one or more pathogens, fumigate the soil (Omar and Rahall 1993, Kleczewski and Egel 2011). The most common chemical means of soil disinfection is by the use of a formaldehyde drench. This method is very effective but necessitates a thorough wetting of the soil and a delay of 10 to 14 days before planting to allow the gas to escape from the soil. Further, it cannot be used in greenhouses with growing plants. The ideal fungicide used for soil disinfection should be a dust that can be mixed with the soil and that causes no injury when seeds are sown immediately (O'Neill et al. 2005).

Seed treatment is good insurance for better stands of more vigorous seedlings, especially if the soil is cold and wet after planting, but in general, treatment of seeds with fungicides is not recommended. In previous years seed treatment was customary, but fungicides applied to the seed coat offer little or no protection to the emerging seedling. In addition, some seed treatments are phytotoxic. Although the effectiveness of fungicides to control damping-off is highly variable, many growers use them. Several fungicides are registered for use in forest nurseries to control soil-borne diseases. Certain fungicides or combinations of fungicides seem to work better in one nursery than another (Russell 1990, Hill and Waller 1990).

Biological control

In an attempt to reduce the use of pesticides, there is an increasing interest in introducing biological agents and putting to use plant compounds as natural commercial products for managing soil-borne pathogens (Cook 1993). Biological control of soil-borne diseases is particularly complex because the pathogens occur in a dynamic environment at the rhizosphere interface. The rhizosphere is typified by intense microbial activity involving, firstly, a high population of microorganisms, and secondly, a rapid change in pH, in salt concentrations, and in osmotic and water potential (Handelsman et al. 1996, Larsen et al. 2003). Microorganisms can protect the plant from fungal attacks through the production of antifungal metabolites, competition with the pathogen for nutrients, niche exclusion, parasitism or lysis of the pathogen, or through induction of plant resistance mechanisms (Whipps 2001). The use of biological control agents such as *Trichoderma* can provide some protection when pathogen pressures are low; however, in case of over-watering or using excess fertilizers, their efficacy will be reduced significantly (Kumar and Hooda 2007). *Trichoderma* spp. are well documented as effective biological control agents of plant diseases caused by soil-borne fungi (Abdel-Kader 1997, Larsen et al. 2003).

Biological control using beneficial microorganisms has been considered a viable alternative strategy for the replacement of chemical methods. The most common antagonistic agents that have been used to control damping-off include bacteria such as *Pseudomonas*, *Bacillus* and *Burkholderia* isolates (Ramarathnam et al. 2011, Sallam et al. 2013), whose presence in composted materials is very significant. Their involvement both in the compost suppressive abilities and plant growth promotion is widely known (Haruta et al. 2005). The biocontrol efficacy of these bacteria is most likely related to their numerous antagonistic mechanisms of action. These mechanisms include the production of antibiotics (Prasanna et al. 2009), siderophores, cell wall degrading enzymes or hydrogen cyanide (Voisard et al. 1989). Organic soil amendment may induce chemical and physical changes that also affect soil microflora. An increasing number of research projects are focusing on using plant residues and organic matter to increase the population of microorganisms antagonistic to plant pathogens (Gamliel et al. 2000).

In addition, some organic matter can stimulate the production of lytic enzymes involved in the degradation of plant pathogens. Degradation of *P. ultimum* cell walls was accomplished by b-glucanase and cellulose enzymes (Inglis and Kawchuk 2002). Production of these enzymes by antagonistic fungi was induced by the presence of *P. ultimum* cell walls (Inglis and Kawchuk 2002). The presence of organic substances rich in cellulose was also shown to stimulate the growth of organisms with cellulose activity, and the production of this enzyme was proposed as the mode of action of antagonistic microorganisms against *Pythium*. Microbial degradation of plant residues can also produce secondary products with antifungal activity (Boehm et al. 1993).

Organic production systems also need additional control methods for soil-borne pathogens as these production systems rely largely on host resistance and other biologically based methods for control. Biologically based control methods, such as microbial biological control agents and organic amendments, suffer from inconsistent performance (Roberts and Kobayashi 2011) and in the case of organic matter amendments, can take several years before they provide benefit (Bonanomi et al. 2007). The efficacy of certain biological control agents can be enhanced by incorporating them into seed priming processes, a variety of procedures that enhance seed germination and vigor through the addition of moisture. Even in the absence of specific microbial inoculants, primed seeds, compared with seeds that are not primed prior to planting, emerge in greater frequencies from soils infested with *Pythium* spp. (Haruta et al. 2005). Seed treatment with antagonist is the cheapest method for delivery of fungal antagonists to the rhizosphere of crop plants to be protected from the soil-borne disease for success beyond the suppression of seed rots and damping-off of young seedlings. The antagonist applied to seeds must be able to multiply in the soil (or) rhizosphere to inhibit a given pathogen by competition, mycoparasitism or antibiosis. The ideal seed inoculum must be an effective antagonist and be an aggressive spermosphere and rhizosphere colonizer (Papavizas 1985).

Seed treatment with biological agents is considered as a means to substantially increase the value of the seed and to improve plant growth and productivity. The bio-agents proliferate on the seed coat of germinating seed and colonize the additional plant parts such as the roots and the collar region. The treatment is quite inexpensive and eco-friendly as compared to other methods of disease control and is successfully exploited for the control of a wide range of seed as well as soil-borne diseases (Mukhopadhyay 1995).

Soil sterilization

Soil solarization is a non-chemical approach for soil disinfestation and one of the most promising methods to control soil-borne pathogens. It utilizes solar radiation and a thin film of transparent mulch, usually of polyethylene, to heat the soil to a range of 38 to 50°C to a depth of about 10 to 20 cm after sufficient irrigation (Gamliel and Katan 2012) for soil pasteurization. It is also a method for controlling,mesophilic organisms, which include most plant pathogens (such as soil-borne fungi) and other pests, without destroying the beneficial bacteria and mycorrhizal fungi (Minuto et al. 1995). Soil-borne pathogens, pests and weeds become inactivated by high temperature and excessive moisture during the hot season in July and August (Al-Karaghouli and Al-Kayssi 2001). Under suitable climatic conditions, solarization can successfully control a wide range of soil-borne diseases and pests. Since soil solarization is a climate-dependent measure, it should therefore be adapted to specific regions and seasons only (Katan 1999).

There is debate on the subject of sterilization of growing medium. It is often a recommended technique for preventing fungal spores from spoilage of seedlings, such as baking your soil at 140 degrees for 30 minutes. Other schools of thought believe that these methods kill off beneficial microorganisms that would otherwise keep pathogens in check (Mihajlovic et al. 2017). It is probably best to use new potting soil for seeds every time, perhaps pasteurized or treated soils can also be considered (Abbasi and Lazarovits 2005).

Conclusions

To sum up, *Pythium* is one of the most important pathogen causing damping-off and root-rot of many crops. The evidence is now available on the occurrence of these soil-borne pathogens as serious root-rot pathogens that could be important source of inoculum. The information obtained will serve as basic knowledge regarding pathogenicity of *Pythium* species and will facilitate the development of pathogen-specific disease control practices. Various disease control options are available, including chemical control, biological control, genetic resistance methods and cropping practices. However, the crucial issue is the non-availability of resistant seed varieties to users.

Thus, in order to improve the sustainability of the resistance-based control method, it is essential to use diversified sources of resistance and to integrate other control methods, mainly appropriate for cropping practices. This will reduce the risk and encourage the rapid build-up of *Pythium* inocula. Initial results obtained by combining different methods for the control of soil-borne diseases imply a necessity to continue research in this area in order to ensure long-lasting sustainability of crop protection.

Given the complex nature of damping-off and the numerous factors involved in its occurrence, we recommend further research on critical niches of complexity, such as seeds, seedbed, associated microbes and their interfaces, using novel and robust experimental and modeling approaches based on the research priorities described in this chapter. Many research efforts have been made in recent years to develop biocontrol solutions for damping-off and there are interesting future perspectives. As with most nursery pest and disease issues, an integrated pest and disease approach is the preferred option for *Pythium* management.

Acknowledgement

I would like to thank Dr. Yosef Shahin Al Shoffe, a research associate at Cornell University in the United States who helped me review and verify the text in terms of formulation.

References

Abawi, G.S., Ludwig, J.W. and Gugino, B.K. 2006. Bean root rot evaluation protocols currently used in New York. Annu. Rep. Bean Improv. Cooperative 49: 83-84.

Abbasi, P.A. and Lazarovits, G. 2005. Effects of AG3 phosphonate formulations on incidence and severity of *Pythium* damping-off of cucumber seedlings under growth room, microplot, and field conditions. Can. J. Plant Pathol. 27: 420-429.

Abdel-Kader, M.M. 1997. Field application of *Trichoderma harzianum* as biocide for control bean root rot disease. Egypt. J. Phytopath. 25: 19-25.

Agrios, G.N. 1997. Plant Pathology, fourth ed. Academic Press, New York.

Agrios, G.N. 2005. Plant Pathology. Fifth edition. Elsevier Academic Press, London, U.K.

Alcala, A.V.C., Paulitz, T.C. and Schroeder, K.L. 2016. *Pythium* species associated with damping-off of pea in certified organic fields in the Columbia basin of Central Washington. Plant Dis. 100: 916-925.

Alexopoulos, C.J., Mims, C.W. and Blackwell, M. 1996. Introductory Mycology, 4th ed. John Wiley and Sons, Inc., Nova York, 868 pp.

Al-Karaghouli, A. and Al-Kayssi, A.W. 2001. Influence of soil moisture content on soil solarization efficiency. Renew. Energy 24(1): 131-144.

Allen, T.W., Martinez, A. and Burpee, L.L. 2004. *Pythium* blight of turfgrass. [Online] Available: http://www.apsnet.org/edcenter/intropp/lessons/fungi/Oomycetes/Pages/PythiumBlight.aspx [2012, June 16].

Bacharis, C., Gouziotis, A. and Kalogeropoulou, P. 2010. Characterization of *Rhizoctonia* spp. isolates associated with damping-off disease in cotton and tobacco seedlings in Greece. Plant Dis. 94: 1314-1322.

Bahramisharif, A., Lamprecht, S.C., Calitz, F. and McLoed, A. 2013. Suppression of *Pythium* and *Phytophthora* damping-off of rooibos by compost and a combination of compost and non-pathogenic *Pythium* taxa. Plant Dis. 97: 1605-1610.

Bahramisharif, A., Lamprecht, S.C., Spies, C.F.J., Botha, W.J., Calitz, F.J. and McLoed, A. 2014. *Pythium* spp. associated with rooibos seedlings, and their pathogenicity toward rooibos, lupin, and oat. Plant Dis. 98: 223-232.

Barnett, J.P. and Pesacreta, T.C. 1993. Handling long leaf pine seeds for optimal nursery performance. South. J. Applied Forestry 17(4): 180-186.

Ben-Yephet, Y. and Nelson, E.B. 1999. Differential suppression of damping-off caused by *Pythium aphanidermatum*, *P. irregulare*, and *P. myriotylum* in composts at different temperatures. Plant Dis. 83: 356-360.

Bharathi, R., Vivekananthan, R., Harish, S., Ramanathan, A. and Samiyappan, R. 2004. Rhizobacteria-based bio-formulations for the management of fruit rot infection in chillies. Crop Prot. 23: 835-843.

Boehm, M.J., Madden, L.V. and Hoitink, H.A.J. 1993. Effect of organic matter decomposition level on bacterial species diversity and composition in relationship to *Pythium* damping-off severity. Appl. Environ. Microbiol. 59: 4171-4179.

Bonanomi, G., Antignani, V., Pane, C. and Scala, F. 2007. Suppression of soilborne fungal diseases with organic amendments. J. Plant Pathol. 89: 311-324.

Boyce, J.S. 1961. Forest Pathology. 3d ed. New York: McGraw-Hill. 572 p.

Bradley, C.A. 2008. Effect of fungicide seed treatments on stand establishment, seedling disease and yield of soybean in North Dakota. Plant Dis. 92: 120-125.

Broders, K.D., Lipps, P.E., Paul, P.A. and Dorrance, A.E. 2007. Characterization of *Pythium* spp. associated with corn and soybean seed and seedling disease in Ohio. Plant Dis. 91: 727-735.

Broders, K.D., Wallhead, M.W., Austin, G.D., Lipps, P.E., Paul, P.A., Mullen, R.W. and Dorrance, A.E. 2009. Association of soil chemical and physical properties with *Pythium* species diversity, community composition, and disease incidence. Phytopathology 99: 957-967.

Castillo, J.V. 2004. Inoculating composted pine bark with beneficial organisms to make a disease suppressive compost for container production in Mexican forest nurseries. N.P.J. 5: 181-185.

Chang, L. 2007. Soil borne plant pathogen. Class Project 728 pp.

Cheung, F., Win, J. and Lang, J.M. 2008. Analysis of the *Pythium ultimum* Transcriptome using Sanger and Pyrosequencing approaches. BMC Genomics 9: 542-552.

Cook, R.J. 1993. Making greater use of introduced microorganisms for biological control of plant pathogens. Annu. Rev. Phytopathol. 31(1): 53-80.

Daughtery, M.L., Wick, R.L. and Peterson, J.L. 1995. Compendium of Flowering Potted Plant Diseases. APS Press, St. Paul, MN.

El-Mohamedy, R.S. 2012. Biological control of *Pythium* root rot of broccoli plants under greenhouse conditions. I.J.A.T. 8: 1017-1028.

Gamliel, A., Austerweil, M. and Kritzman, G. 2000. Non-chemical approach to soilborne pest management-organic amendments. Crop Prot. 19: 847-853.

Gamliel, A. and Katan, J. 2012. Soil Solarization: "Theory and Practice". American Phytopathological Society, St. Paul.

Handelsman, J. and Stabb, E.V. 1996. Biocontrol of soilborne plant pathogens. Plant Cell 8: 1855-1869.

Haruta, S., Nakayama, T., Nakamura, K., Hemmi, H., Ishii, M., Igarashi, Y. and Nishino, T . 2005. Microbial diversity in biodegradation and reutilization processes of garbage. J. Biosci. Bioeng. 99(1): 1-11.

Hill, D.S. and Waller, J.M. 1990. Pests and Diseases of Tropical Crops, Volume I: Principles and Methods of Control. pp. 175. Longman, Group Ltd. Hong Kong.

Hoitink, H.A.J. and Fahy, P.C. 1986. Basis for the control of soilborne plant pathogens with composts. Annu. Rev. Phytopathol. 24: 93-114.

Hoitink, H.A.J., Inbar, Y. and Boehm, M.J. 1991. Status of compost-amended potting mixes naturally suppressive to soilborne diseases of floricultural crops. Plant Dis. 75: 869-873.

Horst, R.K. 2013. Damping-off. Westcott plant disease handbook. Springer Netherlands, Ordrecht. 177 pp.

Howell, C.R. 2002. Cotton seedling pre emergence damping-off incited by *Rhizopus oryzae* and *Pythium* spp. and its biological control with *Trichoderma* spp. Phytopathology 92: 177-180.

Inglis, G.D. and Kawchuk, L.M. 2002. Comparative degradation of oomycete, ascomycete, and basidiomycete cell walls by mycoparasitic and biocontrol fungi. Can. J. Microbiol. 48: 60-70.

Jadhav, V.T. and Ambadkar, C.V. 2007. Effect of *Trichoderma* spp. on seedling emergence and seedling mortality of tomato, chilli and brinjal. J. Plant Dis. Sci. 2: 190-192.

James, R.L. 1988. Principles of fungicide usage in container tree seedling nurseries. Tree Planters' Notes 39: 22-25.

Kageyama, K. 2015. Studies on the taxonomy and ecology of oomycete pathogens. J. G. P. P. 81: 461-465.

Kamoun, S., Furzer, O., Jones, J.D.G., Judelson, H.S., Ali, G.S., Dalio, R.J.D., Roy, S., Schena, L., Zambounis, A., Panabieres, F., Cahill, D., Ruocco, M., Figueiredo, A., Chan, X.-R., Hulvey, J., Stam, R., Lamour, K., Gijzen, M., Tyler, B.M., Grünwald, N.J., Mukhtar, M.S., Tome, D.F.A., Tor, M., Van Den Ackerveken, G., McDowell, J., Daayf, F., Fry, W.E., Lindqvist-Kreuze, H., Meijer, H.J.G., Petre, B., Ristaino, J., Yoshida, K., Birch, P.R.J. and Govers, F. 2015. The top 10 oomycete pathogens in molecular plant pathology. Mol. Plant Pathol. 16: 413-434.

Katan, J. 1999. The methyl bromide issue: Problems and potential solutions. J. G. P. P. 81: 153-159.

Kennelly, M. 2016. *Pythium* blight of turfgrass. Plant Pathology EP-159.

Kleczewski, N.M. and Egel, D.S. 2011. Sanitation for Disease and Pest Management. Purdue Extension HO-250.

Kumar, M.R. and Hooda, I. 2007. Evaluation of antagonistic properties of *Trichoderma* species against *Pythium aphanidermatum* causing damping-off of tomato. J. Mycol. Pl. Pathol. 37(2): 240-243.

Kumar, V., Gupta P. and Dwivedi, S. 2012. Bio-efficacy of fly-ash based *Trichoderma* formulations against damping-off and root-rot diseases in tomato (*Lycopersicum esculentum*). Indian Phytopath. 65(4): 404-405.

Lamichhane, J.R., Dürr, C., Schwanck, A.A., Robin, M.H., Sarthou, J.P., Cellier, V., Messéan, A. and Aubertot, J.N. 2017. Integrated management of damping-off diseases: A review. Agron. Sustain. Dev. 37: 10.

Larsen, J., Ravnskov, S. and Jakobsen, I. 2003. Combined effect of an Arbuscular mycorrhizal fungus and a biocontrol bacterium against *Pythium ultimum* in soil. Folia Geobotanica 38: 145-154.

Lumsden, R.D., Garcia-E, R., Lewis, J.A. and Frias-T, G.A. 1987. Suppression of damping-off caused by *Pythium* spp. in soil from the indigenous Mexican chinampa agricultural system. Soil Biol. Biochem. 19: 501-508.

Milijasevic, S., Rekanovic, E., Hrustic, J., Grahovac, M. and Tanovic, B. 2017. Methods for management of soilborne plant pathogens. Pestic. Phytomed. 32(1): 9-24.

Minuto, A., Migheli, Q. and Garibaldi, A. 1995. Integrated control of soil-borne plant pathogens by solar heating and antagonistic microorganisms. Act Hort. (ISHS) 238: 138-144.

Moorman, G.W. 2011. Damping off. Pennsylvania State University Cooperative Extension. Plant Disease Fact Sheet.

Mukhopadhyay, A.N. 1995 Exploitation of *Gliocladiumvirens* and *Trichoderma harzianum* for biological seed treatment against soil borne diseases. Indian J. of Mycol. Pl. Pathol. 25(1 and 2): 124.

Mundel, H.H., Huang, H.C., Kozub, G.C. and Barr, D.J.S. 1995. Effect of soil moisture and temperature on seedling emergence and incidence of *Pythium* damping-off in safflower. *Carthamus tinctorius* L. Can. J. Plant Sci. 75: 505-509.

Nzungize, J., Gepts, P., Buruchara, R., Buah, S., Ragama, P., Busogoro, J.P. and Baudoin, J.P. 2011. Pathogenic and molecular characterization of *Pythium* species inducing root rot symptoms of common bean in Rwanda. Afr. J. Microbiol. 5: 1169-1181.

Okubara, P.A., Dickman, M.B. and Blechl, A.E. 2014. Molecular and genetic aspects of controlling the soilborne necrotrophic pathogens *Rhizoctonia* and *Pythium*, Plant Sci. 228: 61-70.

Olson, J.D., Damicone, J.P. and Kahn, B.A. 2016. Identification and characterization of isolate of *Pythium* and *Phytophthora* spp. from snap beans with cottony leak. Plant Dis. 100: 1446-1453.

Omar, S.A.M. and Rahhal, M.M.H. 1993. Influence of fungicides on damping-off disease and seed yield of soybean. Egy. J. Agric. Res. 71: 65-74.

O'Neill, T.M., Green, K.R. and Ratcliffe, T. 2005. Evaluation of soil steaming and a formaldehyde drench for control of *Fusarium* wilt in column stock. Act. Horticul. 689(16): 129-134.

Papavizas, G.C. 1985. Trichoderma and Gliocladium: Biology, ecology and potential for bio-control. Annu. Rev. Phytopathol 23: 23-54.

Park, M.J., Han, K.S., Kim, J.W., Park, J.H. and Shin, H.D. 2016. *Pythium aphanidermatum* causing *Pythium* Rot on *Lampranthus spectabilis* in Korea. J. Phytopathol. 164: 567-570.

Phillips, D.H. and Burdekin, D.A. 1982. Diseases of Forest and Ornamental Trees. London: McMillan. 435 p.

Prasanna, K.R.B., Rao, K.S. and Reddy, K.R.N. 2009. Sheath blight disease of *Oryza sativa* and its management by biocontrol and chemical control *in vitro*. Electron. J. Environ. Agric. Food Chem. 8(8): 639-646.

Punja, Z.K. and Yip, R. 2003. Biological control of damping-off and root rot caused by *Pythium aphanidermatum* on greenhouse cucumbers. Can. J. Plant Pathol. 25: 411-417.

Rajendraprasad, M., Vidyasagar M.B., Uma Devi, G. and Koteswar, S.R. 2017. Biological control of tomato damping off caused by *Pythium debaryanum*. Int. J. Chem. Stud. 5(5): 447-452.

Ramarathnam, R., Fernando, W.D. and de Kievit, T. 2011. The role of antibiosis and induced systemic resistance, mediated by strains of *Pseudomonas chlororaphis*, *Bacillus cereus* and *B. amyloliquefaciens*, in controlling blackleg disease of canola. Biol. Control 56: 225-235.

Ringer, C.E., Millner, P.D., Teerlinck, L.M. and Lyman, B.W. 1997. Suppression of seedling damping-off disease in potting mix containing animal manure composts. Compost. Sci. Util. 5: 6-14.

Roberts, D.P. and Kobayashi, D.Y. 2011. Impact of spatial heterogeneity within spermosphere and rhizosphere environments on performance of bacterial biological control agents. pp. 111-130. *In*: Maheshwari, D.K. (ed.). Bacteria in Agrobiology: Crop Ecosystems. Springer Verlag, New York, NY.

Russell, K. 1990. Damping-off. pp. 2-5. *In*: Hamm, P.B., Campbell, S.J. and Hansen, E.M. (eds.). Growing Healthy Seedlings: Identification and Management of Pests in Northwest Forest Nurseries. Corvallis (OR): Forest Research Laboratory, Oregon State University. Special Publication.

Sallam, N.A., Riad, S.N., Mohamed, M.S. and El-Eslam, A.S. 2013. Formulations of Bacillus spp. and Pseudomonas fluorescens for biocontrol of cantaloupe root caused by *Fusarium solani*. J. Plant Prot. Res. 53(3): 295-300.

Scheuerell, S.J., Sullivan, D.M. and Mahaffee, W.F. 2005. Suppression of seedling damping-off caused by *Pythium ultimum*, *P. irregulare*, and *Rhizoctonia solani* in container media amended with a diverse range of Pacific Northwest compost sources. Phytopathology 95: 306-315.

Sharma, N. and Razdan, V.K. 2012. Perpetuation of *Phomopsis vexans* on infected seeds, leaf and fruit parts of *Solanum melongena* L. (Brinjal). VEGETOS: Int. J. Plant Res. 25: 196-202.

Smith, M., Le, D.P., Hudler, G.W. and Aitken, E. 2014. *Pythium* soft rot of ginger: Detection and identification of the causal pathogens, and their control. Crop Protect. 65: 153-167.

Tilman, D., Cassman, K.G., Matson, P.A., Naylor, R. and Polasky, S. 2002. Agricultural sustainability and intensive production practices. Nature 418: 671-677.

Tomioka, K., Takehara, T., Osaki, H., Sekiguchi, H., Nomiyama, K. and Kageyama, K. 2013. Damping-off of soybean caused by *Pythium myriotylum* in Japan. J. G. P. P. 79: 162-164.

Trigiano, R.N., Windham, M.T. and Windham, A.S. 2004. Pythiaceae. pp. 174-176. *In*: Plant Pathology Concepts and Laboratory Exercises. CRC Press. Danvers, MA.

Uzuhashi, S., Tojo, M. and Kakishima, M. 2010. Phylogeny of the genus *Pythium* and description of new genera. Mycoscience 51: 337-365.

Van West, P., Appiah, A.A. and Gow, N.A.R. 2003. Advances in research on oomycete rootpathogens. Physiol. Mol. Plant P. 62(2): 99-113.

Voisard, C., Keel, C., Haas, D. and Défago, G. 1989. Cyanide production by Pseudomonas fluorescens helps suppress black root rot of tobacco under gnotobiotic conditions. EMBO J. 8: 351-358.

Whipps, J.M. 2001. Microbial interactions and biocontrol in the rhizosphere. J. Exp. Bot. 52: 487-511.

Wick, R.L. 1998. Damping-off of Bedding Plants and Vegetables. University of Massachusetts.

15

The Genus *Pythium*: Genomics and Breeding for Resistance

Ramadan A. Arafa*, Said M. Kamel and Kamel A. Abd-Elsalam

Plant Pathology Research Institute, Agricultural Research Center, Giza, 12619, Egypt

Introduction

Pythium causes damping-off disease of seedlings, and also affects seeds and roots. However, the highest damage is done to the seedling roots during germination emergence. The maximum losses are achieved when high soil moisture and low temperature exist (Rey et al. 1998). Extremely large seedlings in seed beds are fully broken by the damping-off or they get infected soon after they are transplanted. In many cases, the evil germination of seeds or poor emergence of seedlings is the result of damping-off infections in the pre-emergence phase (Martin 1992). Older plants are killed when infected with the damping-off pathogen, but when they develop root and stem lesions and root rots, their growth may be retarded considerably, and their yields may be reduced. Some species of oomycetes attack the plants, which rot in the field or in storage.

Of these, damping-off incited by *Pythium* species, caused more than 60% death of seedlings in both nurseries and the main fields (Manoranjitham and Prakasam 2000, Jadhav and Ambadkar 2007). Among these, *P. aphanidermatum* (Edson) Fitzp is one of the most pathogenic species with a host range and causes severe damage to many economically important crops.

It is known to cause infections on a wide range of plant species, belonging to different families viz., *Amaranthaceae, Amaryllidaceae, Araceae, Basellaceae, Bromeliaceae, Cactaceae, Chenopodiaceae, Compositae, Coniferae, Convolvulaceae, Cruciferae, Cucurbitaceae, Euphorbiaceae, Gramineae, Leguminosae, Linaceae, Malvaceae, Moraceae, Passifloraceae, Rosaceae, Solanaceae, Umbelliferae, Violaceae, Vitaceae* and *Zingiberaceae* (Abdelzaher et al. 2004). Among the oomycete, classes of fungi like the genus *Pythium* is one of the largest genera. It consists of more than 130 accepted species which are isolated from different crops and regions of the world (Robideau et al. 2011). Almost all the *Pythium* species are known to affect crop plants and ultimately cause severe damage (Kucharek and Mitchell 2000). In vegetable pre- and post-emergence, damping-off caused by *Pythium* spp. is economically very important around the world (Whipps and Lumsden 1991). It becomes very difficult to control *Pythium*, when the sporangia are exposed to the seeds and roots exudate and volatile compounds (Whipps and Lumsden 1991). Damping-off disease in vegetables and other field crops is considered an important limiting factor in the successful cultivation of crop plants worldwide. Yield loss due to *Pythium* species in different crops has been estimated to be approximately multibillion-dollar worldwide (Van West et al. 2003).

Inheritance of *Pythium* disease resistance can be conferred by quantitative and qualitative resistance which is controlled by a multi and single gene, respectively (Mahuku et al. 2005).

*Corresponding author: arafa.r.a.85@gmail.com

Characterization of molecular pathogenesis mechanisms and population dynamics are essential for the enhancement of *Pythium* disease's resistance on vegetable crops. Therefore, implementation of NGS strategies is required for more genomics and genetics research of serious *Pythium* species along with economic host species. Such technologies play a crucial role in multidisciplinary "omics" approaches including phenomics, genomics, transcriptomics, proteomics, and metabolomics. NGS-based breeding for *Pythium* species resistance provides an innovative answer for the current agricultural problems especially in vegetable crops through investigating the agro-genomics of wild relative species and *Pythium* spp. populations. In the current chapter, we focus on the impact of *Pythium* species on vegetable crop productivity and management of this serious oomycete pathogen. NGS-based genotyping technology and its advantages in accelerating vegetable crop breeding programs and identifying quantitative trait loci to gain durable resistance against *Pythium* diseases have also been discussed.

Symptoms

When seeds of susceptible plants are planted in infested soils and are attacked by the *Pythium* sp., they fail to germinate, become soft and turn brown, wilt and lastly shatter. The infected tissue becomes soft and water soaked, the collar region rots and the seedlings in the end collapse and die. The guaranteed supply of disease-free seedlings in the required quantities is a major pre-requisite for stabilized production of vegetable and cereal crops. While raising seedlings in nursery beds, farmers face a major problem of damping off incited by *Pythium* spp. Damping-off is a common nursery disease in a number of vegetable crop plants and it causes reduced germination percentage, vigor, quality and yield of crops. It also causes seedling rot, root rot, cottony-leak, cottony blight, stalk rot, etc. The damping-off in the solanaceous crop is caused by *Pythium* spp. including *P. aphanidermatum, P. irregulare, P. debaryanum,* and *P. ultimum,* which can cause pre-emergence damping-off and result in seed decay before the plants come out of the soil. The post-emergence damping-off is designated by infection of the young tissue at the collar of the stem at aboveground level (Ramamoorthy et al. 2002).

Classification of *Pythium* species

The *Pythium* species are fungal-like organisms. Commonly referred to as moisture-loving fungi (Domain Eukaryota; Kingdom Chromista; phylum Oomycota; class Oomycetes; order Pythiales and family Pythiaceae), they are present all over the world and are found in wide variety of habitats ranging from terrestrial or aquatic environments, in cultivated or fallow soils, in plants or animals, in saline or fresh water.

Plant disease caused by *Pythium*

Several species of *Pythium* cause damping-off. A white, rapidly growing mycelium is produced by *Pythium*. The mycelium gives rise to sporangia, which germinate through producing one to several germ tubes or by producing a short hypha at the end of which forms a balloon-parallel secondary sporangium called a vesicle. In the vesicle, 100 or more zoospores are produced, which, when released, flock about for a few minutes, round off to form a sac and then germinate by producing a germ tube Pythium blight, root rot and soft-rot diseases caused by *Pythium* species and then reduce the germination rate and dramatically decrease vegetable crop productivity. However, resistance level to soft-rot may increase in older plants, but root-rot delays or prevents plant growth and causes wilt for yellow plants as well as stunting in case of intense infection occurs. Infected seeds appear to rot and soil sticks to them. Infected seedlings have water-soaked lesions on the hypocotyl and cotyledons that develop into a brown soft rot. Diseased plants are easily pulled from the soil because of rotted roots.

Impact of *Pythium* species on crop production

The agricultural sector is the backbone for the national, regional and global economy, especially in the developing countries. Population in many African countries lives in provincial regions where high levels of poverty, diseases, privation, and other socio-economic problems are very common. For instance, Ghana, Kenya, and Zambia have wide achievements in the exportation of some crops such as cocoa, tea, cut flowers and cotton which reflect the positive effect of agriculture in the recovery of the national economy (Diao et al. 2007). On the other hand, the Green Revolution in Asia through 1970s led to a dramatic increase in national income of Asian countries. Transforming traditional agriculture into a contemporary domain revealed the essential contribution of agriculture as a starting point for economic development and enhancement (Adelman 2001). The agriculture sector plays a pivotal role in saving global food security, especially post the fast progress in world population growth in the 1960s (Khush 2001). Crop improvement strategies and durable resistance to pests is still the main way to solve the dilemma of 'hidden hunger' (Welch and Graham 2004). Without compromising, economic crops suffer from different types of losses due to biotic (animals, insects, weeds and pathogens) and abiotic (drought, salinity, weather conditions, soils types) stresses that lead to reduction and negative influence on crop production (Kaur and Singla 2016). Plant pathogens are one of the most biotic stresses which cause a dramatic decrease in both quality and quantity of crop productivity worldwide. Annually, plant diseases can cause losses in global agricultural productivity from 20 to 30% (Vipinadas and Thamizharasi 2015). The genus of *Pythium* is one of the most important genera among oomycetes members. The genus *Pythium* includes over 280 reported species (Lévesque and de Cock 2004, HO 2009); most of these species are plant pathogens that are responsible for seed rot, damping-off, root, stem and fruit rot, foliar blight, and postharvest diseases. Root rot disease is a serious problem in many vegetable crops, resulting in around 100% losses in crop species productivity, particularly in susceptible genotypes (Mukankusi 2007).

Interestingly, there is a certain species of *Pythium* such as *Pythium insidiosum* which can cause humans and animal diseases in different regions worldwide (Mendoza 2009). New species are considered causal agents of new diseases worldwide. The number of characterized species is dramatically rising upon reconnaissance of new behavior of these species in nature. The environmental expansion played an essential role in the genetic variation within *Pythium* species along with releasing aggressive physiological races. Interestingly, the habitats of *Pythium* species are still incompletely understood in ecosystems. Although there are many soil-borne pathogens, *Pythium* genus was represented in the maximum of soil samples tested in an oak forest in Germany (Jung et al. 1996). *Pythium* genus does not include only soil-borne species, but also foliar diseases species. Lévesque (2011) mentioned that environmental conditions played pivotal roles in the repression of plant diseases that are caused by oomycete pathogens and decrease the rate of plant infection in the ecosystem. Although oomycete plant pathogens can cause many biotic stresses for different crops, they have miscellaneous roles in natural systems where Packer and Clay (2000) reported that *Pythium* spp. can attack the saplings of *Prunus serotine* and decrease the competition among trees. The absence of this organism in some European countries led to a high degree of growth and infection of *P. serotine* (Reinhart et al. 2010). Moreover, Allain-Boulé et al. (2004) mentioned the same situation with *Pythium attrantheridium*, where this species colonizes in the root of adult appeal trees and prevents the growth of seedlings and young trees. In this ecosystem, it is useful to utilize weak races which are avirulent to mature plants and possess aggressive genes in young trees to reduce the level of competition among plant species. Management and control strategies for plant diseases relied on different fungicides with a wide range of chemical groups. However, many people who are responsible for making the decision worldwide appeal to decreasing the pesticides used in all crops without affecting the productivity of these crops.

Annually, millions of kg of pesticides are used to protect crops from plant diseases with extreme cost reaching to billions of dollars. Moreover, the utilization of pesticides with overdoses many times will emerge and release new aggressive pathotypes resistant to these pesticides (Saville et al. 2015). The use of several hazardous agrochemicals can destroy human, animal, fish and plant life and can be injurious for natural enemies. Also, agricultural soils, water, and air can be contaminated by chemical compounds. The delicate diagnosis of plant diseases is a key tool for the achievement of integrated disease management (IDM) strategy. Recognition and characterization of serious plant diseases manually is time-consuming, exert painstaking effort and low level of accuracy. Therefore, advanced and promising tools for rapid and precise detection of plant pathogens, particularly wide host range-oomycetes like *Pythium* species, have become an imperative request. Thus, great efforts to develop new effective and environmentally safe and friendly approaches for management of plant pathogens are indispensable. Identification of resistance genes (R-genes) in economic crops are alternative ways instead of hazard fungicides. Utilization of molecular markers in the resistance mechanism and follow up of gene expression can enable many researchers to characterize and identify effective genes (Shehzadi et al. 2017).

Molecular markers and phytopathogenic oomycetes

Although the plant pathogenic oomycetes include a lot of notorious genus, here we will discuss the *Pythium* genus that threatens a plenty of ornamental, horticulture, field crops and fruits and forest trees. Molecular markers played a crucial role in contemporary classification of plant pathogenic oomycetes and classified them in chromista (stramenopiles) kingdom (Baldauf et al. 2000, Cooke et al. 2000, Margulis and Schwartz 2000) instead of kingdom of fungi, which is very close to diatoms and brown algae (Kamoun and Smart 2005). So, oomycete pathogens are not true fungi, but they are fungi-like microorganisms. Molecular genetic markers are also considered the cornerstone in the detection and characterization of plant pathogenic oomycetes, particularly the *Pythium* species in the genomic era. The convergence of *Pythium*, convenient environmental conditions and susceptible hosts will draw gruesome conceptualization about the future of agriculture in all countries. Therefore, in this case, the utilization of novel molecular approaches to skip this sophisticated problem is the elixir of life for all farmers and plantations.

DNA sequencing and single nucleotide polymorphisms

Unambiguously, the strategies of DNA sequencing played a key role in the deep understanding of the organism's biology and diagnosis, characterizing and identifying new species. It is important to develop novel methods for DNA sequencing to face increasing biological challenges. DNA sequencing is the most popular and widely used method with single nucleotide polymorphism (SNPs) to detect differences in DNA genome sequences with high throughput genotyping. Whole genome shotgun (WGS) technology for dozens of economic crops such as tomatoes, potatoes, rice, soybeans, and other plants has been achieved and sequenced. Furthermore, some *Pythium* species like *Pythium insidiosum* strain CDC-B5653, *P. ultimum, P. vexans, P. iwayamai, P. arrhenomanes, P. irregulare* and *P. aphanidermatum* (Lévesque et al. 2010, Ascunce et al. 2016, Rujirawat et al. 2018) have been sequenced (Fig. 15.1). Strikingly, comparative genomic analysis of *Pythium* species using advanced bioinformatics tools leads to future prospects regarding virulence and resistance mechanisms.

SNP means the differences between two individuals or DNA sequencing or some species under the same genus at the level of a single nucleotide base. Since 2000, the scientific community has shifted to SNPs because very high density of markers for diseases and agronomy studies are needed. Insertion, deletion (indels) or substitutions of single nucleotides are very common in humans, animals, plants, and microorganisms. It is not available in many cases to discriminate between sequencing and polymorphism data especially when two peaks are located on the same site in heterozygotes. So, based on the obtained data from sequencing, many approaches can be used to discover SNP markers of the genome. The SNP discovery is a good tool to detect the differences

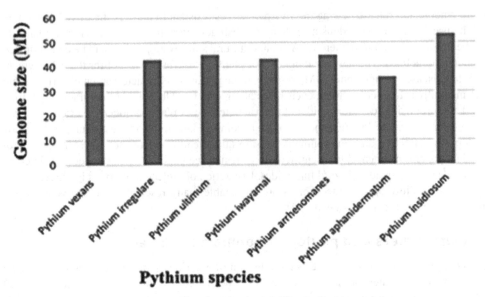

Pythium species

Figure 15.1. The genome size of different *Pythium* species

between mapping populations and individuals, and cascades of markers may be developed. Edwards et al. (2007) and Xu (2010) mentioned that the SNP can repeat the whole genome in every 100-300 bp in plants, but in the human genome one SNP is present in every 1000 bp (Sachidanandam et al. 2001). Some researchers in many labs used SNPs as an alternative tool to microsatellite and minisatellite in genetic biodiversity projects in animals, plants, and microorganisms. Moreover, it is available to detect high-quality SNP markers, with high efficiency. So, it is projected that the number of SNP markers will be dramatically increased and used for various purposes with many organisms, especially after the great evolution in sequencing technology. One of the most beneficial SNP markers is a high throughput analysis with moderate costs (Chen and Sullivan 2003). The generation of whole genome sequences is a suitable strategy to understand the way of SNP identification and gene annotation in the genome of the species (Weber and Myers 1997). However, SNPs assay has some limitations such as marker development and the requirement of high quality of DNA. Also, well-qualified staff in genetic laboratories and molecular breeding programs are needed.

Molecular breeding for increased resistance in host plants

Conventional plant breeding has been initiated in numerous crops since many years ago worldwide. Classical breeding was used to produce a string of varieties with desirable agronomical traits. Unfortunately, this type of breeding has some drawbacks where the time-consuming process is more than ten years, it is difficult to apply in some crops like potatoes due to its polyploidy ($2n = 4x = 48$) and needs a lot of effort. Therefore, molecular marker-assisted breeding (MAB) contributes to create a new field in breeding science in order to develop a novel cultivar which is consistent with consumer requests. With the beginning of the twenty-first century, a great revolution of molecular markers and genetic tools have provided in various benefits of plant science such as genotype evaluation, lines selection, characterization of agronomic traits, genetic biodiversity and construction of genetic linkage maps. We have projected molecular markers, genomic selection, next-generation sequencing, whole-genome re-sequencing, and genome-wide association mapping broadening our conception regarding breeding program pathways in the future. So, modern and contemporary breeding with markers assisted selection (MAS) has some advantages better than conventional breeding where selection for target traits can be achieved at seedling stage and undesirable characteristics using MAS can be discarded, thus decreasing the period needed for phenotypic data at adult stage (Jiang

2013). Also, MAB is not influenced by environmental conditions, thus QTLs for tested traits (disease and insect resistance, drought and salinity) can be accomplished under any conditions. In addition, genotypic tools are the taster and more precise than phenotypic selection, thus MAB saves not only considerable time and effort but via molecular markers and genotype data, we can also predict individual's phenotype. Microsatellite and SNP are co-dominating markers, which play a vital and essential role in modern breeding which displays the role of recessive alleles in heterozygous status.

Globally, breeding for plant disease resistance has a wide range of acceptance in both private and public sectors. This branch of science creates a good media and a strong relationship among different fields of plant sciences which are plant breeding, plant pathology, and plant genetics. Therefore, working in this career brews the spirit of collaboration between different institutes in many countries. Furthermore, plant disease resistance strategy using resistant cultivars and established breeding programs is one of the most successful alternative ways to protect the crops and alleviate yield impairment and losses due to pathogens. Intriguingly, researchers exert painstaking efforts to pinpoint the MAB, particularly in cases of intractable and stubborn diseases. Two major categories of plant disease resistance have been identified: (i) qualitative resistance or race-specific resistance (vertical resistance) controlled by a single resistance gene (major gene), (ii) quantitative resistance or field resistance (horizontal resistance) conferred by multiple genes (minor genes) (Poland et al. 2009). Based on historical records of breeding programs, quantitative resistance has been used to release and develop varieties with desirable and economic traits without information about resistance locus (Parlevliet and van Ommeren 1988). However, in molecular markers and genomic era, MAS was applied in various breeding programs in order to crop improvement and determination of suitable markers for targeted traits (Moloney et al. 2009). MAS also supports the plant breeder's intake decisions with a high level of precision and accuracy at low cost and high throughput. Root rot disease of beans caused by various *Pythium* species is one of the serious problems hindering common bean productivity where the elite of cultivars have a fat chance to escape from such pathogens. Actually, there are many ways to manage the *Pythium* spp., for instance agricultural practices, seed and soil treatment, and various fungicides being the most common. Unfortunately, the aforementioned approaches are uneconomical, and several hazardous chemicals can destroy both human and animal life. Therefore, great efforts to develop new, effective and environmentally safe approaches for management of *Pythium* diseases are needed to improve the production of vegetable crops and reduce loss and fungicide applications. Globally, different breeding programs in various vegetable and oil crops for *Pythium* spp. resistance have been conducted. Interestingly, genotypes of common bean have been evaluated against *P. ultimum* where colored seeds conferred a high resistance level to this oomycete plant pathogen more than white seed varieties (Lucas and Griffiths 2004). Additionally, inheritance of quantitative and qualitative resistance linked to damping-off and seed rot have been reported (Mahuku et al. 2005). Campa et al. (2010) mapped a major gene (*Py-1*) and two quantitative traits loci (ER3XC and SV6XC) (Table 15.1) associated with *P. ultimum* resistance in the recombinant inbred lines (RIL) of common beans. Moreover, Namayanja et al. (2014) investigated the genetic analysis of common bean resistance to *P. ultimum* through crossing between the resistant and susceptible parental lines. *Pythium* root rot resistance was conferred by a single gene where the genotype RWR 719 was utilized as a genetic resource to control *P. ultimum* severity.

Using the entire genome re-sequencing approach, two QTLs related to *P. ultimum* var. *ultimum* resistance in the soybean were detected. These QTLs showed phenotypic variation of 7.5–13.5% for the first one on chromosome 6, whereas the second QTL was identified on Chr.8 and revealed 7.5–13.5% phenotypic variation. Additionally, high-quality SNP markers such as a useful marker-assisted selection for damping-off disease were discovered (Klepadlo et al. 2018). The whole genome shotgun strategy revealed that the genome size of *P. ultimum* was 42.8 Mb with 15,290 genes. The effector proteins (RXLR) were not detected in the genome; however, a few Crinkler genes were detected compared to other phytopathogenic oomycetes (Lévesque et al. 2010). The

Table 15.1. Quantitative trait loci and resistance genes associated with *Pythium* species resistance in different crops

No.	Gene/QTL name	Trait	Chromosome	Host	Pathogen	Reference
1	ER3[xc]	Emergence rate	LG3	Common Bean (*Phaseolus vulgaris* L.)	*P. ultimum*	Campa et al. 2010
2	SV6[xc]	Seedling vigor	LG6	Common Bean (*Phaseolus vulgaris* L.)	*P. ultimum*	Campa et al. 2010
3	qRRW20	Ratio of root weight	20 (56.7 cM)	Soybean (*Glycine max* L.)	*P. irregulare*	Lin et al. 2018
4	qRRW11	Ratio of root weight	11 (126.4 cM)	Soybean (*Glycine max* L.)	*P. irregulare*	Lin et al. 2018
5	RpiQ1319-1	Resistance to stalk rot		Maize (*Zea mays*)	*P. inflatum*	Song et al. 2015
6	RpiQ1219-2	Resistance to stalk rot		Maize (*Zea mays*)	*P. inflatum*	Song et al. 2015
7	NA	Resistance to root rot	14	Soybean (*Glycine max* L.)	*P. irregulare*	Stasko et al. 2016
8	NA	Resistance to root rot	19-2	Soybean (*Glycine max* L.)	*P. irregulare*	Stasko et al. 2016
9	*Rpa1*	*Pythium* damping-off resistance	MLG F (10.6 and 26.6 cM)	Soybean (*Glycine max* L.)	*P. aphanidermatum*	Rosso et al. 2008
10	NA	Root rot severity	3	Soybean (*Glycine max* L.)	*P. oopapillum*	Lerch et al. 2017
11	NA	Root rot severity	4	Soybean (*Glycine max* L.)	*P. oopapillum*	Lerch et al. 2017
12	NA	Pre- and post-emergence damping-off of seedlings and root rot resistance	6	Soybean (*Glycine max* L.)	*P. ultimum* var. *ultimum*	Klepadlo et al. 2018
13	NA	Pre- and post-emergence damping-off of seedlings and root rot resistance	8	Soybean (*Glycine max* L.)	*P. ultimum* var. *ultimum*	Klepadlo et al. 2018

QTL = Quantitative trait loci; LG = Linkage group; NA= No data available breeding program for *P. irregulare* in soybean was established to identify quantitative trait loci using RILs where 1032 SNP markers were mapped. As a result, two QTLs associated with *P. irregulare* resistance were identified with phenotypic variation ranging from 2 to 13.6% (Stasko et al. 2016).

On the other hand, molecular marker's assays introduce a major paradigm shift in contemporary breeding which allows gene pyramiding (combining) technology in crops. Gene pyramiding strategy relies on transferring and stacking more than one gene of interest into a single cultivar which leads to promising varieties and durable resistance versus plenty of pests (pathogens and insects). For instance, the common bean genotypes G2333 and PI207262 (as a genetic resource for anthracnose resistance) and RWR719 (*Pythium* resistance parent) were crossed with four susceptible cultivars, K132, NABE 4, NABE 13 and NABE 14, to pyramid different genes. This combination displayed satisfactory results in the dimension of tackling the disease compared with the effect of each gene separately (Kiryowa et al. 2015).

A significant stride in the identification of QTLs tightly related to *Pythium* disease resistance by molecular markers has been carried (Table 15.1). The candidate QTLs have been validated using different types of markers which are used as an MAS in vegetable breeding programs for *Pythium* resistance. In the same trend, SSR and AFLP markers attached to quantitative resistance were implemented and showed benefit and speed results in population selection as MAB. Moreover, molecular markers were used in economical crops where more than 35,000 microsatellite were introduced and mapped onto 20 linkage groups in soybean (Song et al. 2010).

However, we believe that modern breeding strategy with molecular markers and MAS has more advantages than classical breeding. However, molecular marker-assisted breeding is still not perfect and suffers from some flaws. Jiang (2013) mentioned that some molecular markers are not powerful for plant breeders and should be converted to other markers like convert RAPD to SCAR markers. In addition, MAB methods have to be applicable and acceptable to plant breeders in different breeding programs. Technical equipment, high cost, and experience are still considered stumbling blocks for implementation of MAB with a broad spectrum, particularly in developing countries (Ribaut et al. 2010).

So, compatibility and integration between MAB and classical breeding is a promising strategy for the future of plant breeding science in different plant species. It is projected that this combination seeks to progressively overcome the disadvantages of both modern and conventional breeding and allows too wide an application of MAB in a lot of breeding programs as well as in different countries. Particka and Hancock (2008) conducted a breeding program on strawberry to black root rot (BRR) that was caused by soil-borne plant pathogens including *Pythium*, where nine genotypes ranged from high tolerance to low tolerance against BRR disease.

NGS-based strategies have closely mirrored significant insights into microbe genomes and agro-genomic studies while revealing new consultations for their contribution to plant breeding programs. Combinations between the NGS approaches have facilitated crop improvement through the generation of thousands of SNPs that are associated with traits of interest, particularly plant disease resistance (Vlk and Řepková 2017). Modern and smart agriculture needs combinations between multidisciplinary in the fields of science to achieve our target effectively (Fig. 15.2). Additionally, genetic linkage map construction and fine mapping of targeted genomic regions among breeding populations can lead to rapid identification of resistant genes tightly linked to *Pythium* species resistance. Therefore, genotyping of plant species and pathogens can gain substantial benefit through reduction of the impact of epidemic losses that are generated by a wide host range pathogen like *Pythium* spp. as well as a selection of resistant lines to biotic stress among mapping breeding populations. Finally, the implementation of NGS-based strategies in the agricultural domain generates a major paradigm shift in crop improvement and plant disease control.

Basically, the genetic resources (lines, landraces, wild species, etc.) is the first step to accomplish durable resistance through gaining new resistance genes (R-genes). Then there are genetic and genomic-based breeding approaches including marker-assisted selection, marker-assisted breeding,

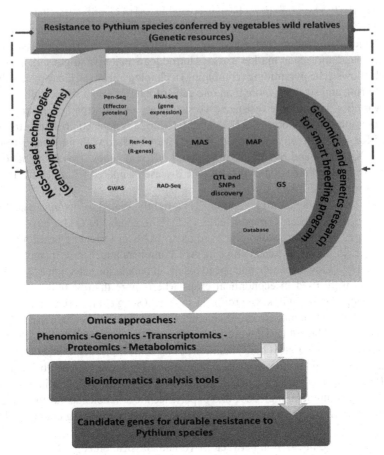

Figure 15.2. Genomic trend towards achieving smart breeding programs required for sustainable vegetable crop improvement and durable resistance to *Pythium* species

identification of quantitative trait resistance (QTLs) and SNP markers and genomic selection (GS) that lead to smart breeding. Next-generation sequencing (NGS) strategies offer high-throughput platforms and genotyping technologies including restriction site-associated DNA sequencing (RAD-seq), genotyping-by-sequencing (GBS), resistance gene enrichment sequencing (Ren-Seq), genome-wide association studies (GWAS) and pathogen enrichment sequencing (Pen-Seq). Furthermore, different bioinformatics analysis tools can be conducted on omics approaches towards detection of candidate genes tightly related to Pythium disease resistance.

Next-generation sequencing strategies and resistance genes

Next-generation sequencing can be used for detection and analysis of a range of markers, including SNPs. Actually, application of NGS approaches can be implemented in the multidisciplinary genome (DNA-Seq), transcriptome (RNA-Seq) and epigenome (ChIP-Seq). NGS-based genotyping methods include genotyping-by-sequencing (GBS), complexity reduction of polymorphic sequences (CRoPS), reduced-representation libraries (RRLs), and restriction site-associated DNA sequencing (RAD-seq). Currently, NGS is one of the most significant and vastly used technologies for genetic marker validation and estimation in molecular genetics and genomics research (Davey et al. 2011). Actually, there are a variety of platforms like Illumina (Bentley et al. 2008), Roche 454 (Rothberg and Leamon 2008) and AB SOLiD (Pandey et al. 2008) that can be now used to generate gigabases of DNA reads at a critical time and effective cost where whole-genome resequencing can be easily conducted. Classical genetic markers' (Vos et al. 1995, Jarne and Lagoda 1996) development suffers from some constraints, for example they are time-consuming and costly; however, NGS technologies overcame these serious problems and developed marker-assisted selection in breeding programs. Furthermore, NGS can rapidly produce a huge amount of genetics and genomics sequencing data with high precision and accuracy, high-throughput, and cost-effectively. Platforms of NGS strategies can generate a huge dataset that can be analyzed using advanced bioinformatics tools. Dataset analysis and mining can be conducted based on following and tracking several sequence steps (Fig. 15.3). NGS technology is strongly inserted into molecular plant breeding programs to alleviate the drawbacks, disadvantages, and circumstances of the classical plant breeding methods.

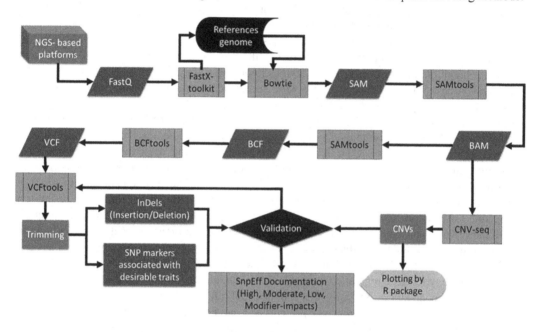

Figure 15.3. The general workflow of bioinformatics tools to NGS dataset analysis

Color version at the end of the book

The possible schematic is to be conducted for analysis and mining of high-throughput and high-quality data starting from an individual's genotyping and reference genome to the identification of candidate SNPs and genomic regions associated with the trait of interest. Furthermore, NGS strategies were employed in plant pathogens domain to preclude outbreaks of epidemic plant disease. Plenty of plant species genomes have been sequenced which exceeded 100 species with the onset of NGS technology in 2005 (Michael and Jackson 2013, Michael and VanBuren 2015). Transcriptomes can be completely sequenced in a vast number of populations and species using next-generation cDNA sequencing (RNA-seq) to discover millions of SNPs and assess gene expression (Barbazuk and Schnable 2011, Ozsolak and Milos 2011). Because sequencing data is ongoing for duplication every five months and novel genomes are sequenced continuously (Stein 2010), advanced tools to comprehend and manipulate sequencing throughput become an imperative request. Subsequently, NGS-based genotyping platforms have different approaches compared to WGS, where a small genomic region is sequenced for the discovery of SNPs. The use of this method has been started in the human genome project (Altshuler et al. 2000) to develop new SNP candidates, then become convenient with NGS for implementation in other different organisms, i.e. cattle (van Tassell et al. 2008), swine (Ramos et al. 2009), maize (Gore et al. 2009), rainbow trout (Sánchez et al. 2009), soybean (Hyten et al. 2010), tomato (Arafa et al. 2017), and *Phytophthora infestans* (Arafa et al. 2019). CRoPS technology is also one of the reduced-representation methods that mitigates a large amount of sequencing data for successful identification of polymorphism (van Orsouw et al. 2007). The former strategy has been efficiently used in the development of SNPs to fix duplication and repetition trouble in the genome of maize (van Orsouw et al. 2007, Mammadov et al. 2010). RAD-Seq method procedure mainly depends on the use of restricted enzymes to shear the genome in the specific-sites to a fragment from 300-700bp (Miller et al. 2007, Davey and Blaxter 2010). This approach has also been applied to the NGS for genotyping and the discovery of targeted SNPs (Baird et al. 2008). Moreover, data generated from RAD-Seq have been widely used in population genomics study for genetic linkage maps construction and quantitative trait loci (QTLs) detection in different types of populations (Chutimanitsakun et al. 2011, Shirasawa et al. 2016). On the other hand, genetic analysis and comparatives between genomes of non-model organisms without prior genome sequencing was effectively achieved via RAD-Seq technique (Baxter et al. 2011).

GBS is a new approach to NGS technologies for genotyping and the development of novel molecular markers into crop populations (Sonah et al. 2013, He et al. 2014). Libraries' preparation of GBS is based on the use of restricted enzymes to decrease complications in crop genomes like maize and barely with affordable, straightforward, rapid and precise techniques (Elshire et al. 2011, Romay et al. 2013). In contrast, the GBS tool can be conducted using a set of PCR markers, in which it is designed to target genomic regions for the identification of SNPs. The GBS procedure throughput has been strongly applied in divergent ways in crop breeding programs such as SNP discovery, genetic variability, QTL identification, candidate gene detection and genome-wide association study (Fu et al. 2014). Additionally, the genome-wide association study (GWAS) is an innovative statistical analysis method to pave the way for novel insights and conceptualizations in crop improvement. Furthermore, GWAS is a powerful approach for SNPs' discovery of interest throughout association mapping between phenotype and genotype data in a vast number of individuals (Korte and Farlow 2013). GWAS strategy possesses high flexibility where it can be used to anchor the position of SNPs in a different species with plenty of NGS technologies such as RAD-Seq, GBS and CRoPS. For around a decade, the NGS technology has been combined with a promising strategy for gene regulation, namely Chromatin immunoprecipitation sequencing (ChIP-seq) (Weinmann et al. 2002). Basically, this genomic tool relies on the use of specific antibodies for immunoprecipitation of DNA-binding site proteins and sequencing for specific genomic loci (Horak and Snyder 2002). ChIP-seq facilitates the investigations of the relationship between amino acid and nucleic acid on the large-scale of genome research. This strategy also plays an effective role in the identification of transcription factors, which are responsible for the expression of a string of genes under different conditions (Landt et al. 2012). Intriguingly, diverse diseases and some biological processes in the cell may be detected based on the application of ChIP-seq technique as an NGS technology (Mundade et al 2014).

Plant pathogens can generate a highly versatile race with fitness and adaptation under undesirable environmental conditions. Therefore, it is prudent utilization of advanced technologies for genotyping, characterization and identification of these strains. NGS-based approaches were utilized in a broad-spectrum for identification of new molecular markers associated with resistance genes to reduce crop losses (Devran et al. 2015). NGS provides a novel prospect into understanding the evolution and epidemiology of serious plant pathogens through generating whole-genome re-sequencing data. NGS created a great revolution in the field of plant-pathogen interactions where substantial information has to be provided. Resistance genes are a key factor in significantly decreasing crop yield losses and continuing sustainable agriculture in commercially planted species. Resistance gene enrichment sequencing (RenSeq) is one of the most important NGS technologies in the field of resistance gene annotation for plant pathogens (Jupe et al. 2013). The RenSeq strategy was widely implemented in the identification of nucleotide binding-site leucine-rich repeat (NB-LRR) gene family and detection of new molecular markers matched with resistance loci in potato (Jupe et al. 2013). Recently, this promising approach was applied in cloning resistance genes *Sr22* and *Sr45* in wheat to accomplish durable resistance against stem rust disease (Steuernagel et al. 2016).

The plant pathogens with a high population dynamic have sophisticated genomic structure, gene flow, and high rate of mutations, which contribute to breaking down resistance genes in economic crops (McDonald and Linde 2002). These pathogens can generate unprecedented genotypes with high fitness and aggressiveness to morbidity in croplands worldwide. As a result, painstaking efforts to pinpoint plant-pathogen interaction, identification of clonal lineage, and novel strains need to be conducted. Interestingly, NGS has been tightly contributed to confer quantitative resistance in some crops against epidemic disease, for instance in peanut (*Arachis hypogaea* L.) (Leal-Bertioli et al. 2015), in Upland cotton (*Gossypium hirsutum* L.) (Wang et al. 2015) and in *Lolium perenne* (Pfender et al. 2011). In recent years, NGS strategies have played a crucial role in filling the gaps in plant pathogens, revolution mechanisms, understanding and characterization of plant disease resistance pathways. We conclude that RenSeq provides an opportunity to quickly map functional NB-LRR-type R genes to control important diseases in crop plants, and a method for improvement of existing plant genome annotations.

Conclusion and future perspectives

Currently, the agro-genomic approach is one of the main domains that save global food security and sustainable development. But track and trace plant pathogenic oomycetes, particularly *Pythium* species, have become an inevitable consequence of durable resistance to serious economic crops in the genomic era. In the future, identification of genetic resources (wild types, GeneBank germplasm) associated with *Pythium* species resistance would be a highly efficient step towards the successful breeding program in vegetable crops. NGS-based technologies like pathogen enrichment sequencing (PenSeq) facilitate the detection of effector proteins and pathogenicity factors as well as population dynamics of *Pythium* species. Thus, the characterization of *Pythium* species and the identification of SNP markers tightly linked to *Pythium* resistance is an unambiguously effective way towards integrated disease management. The development of marker-assisted selection, whereby breeders select for molecular markers related to R-genes, enables pyramiding of more than one effective R gene in the individual lines. Marker-assisted selection and gene pyramiding play a serious role as a powerful and useful technology in modern plant breeding to accelerate the release of elite promising lines of resistance to *Pythium* diseases. Markers linked to the gene(s) of interest help to select plants genotypically that are genetically similar to the recurrent parent possessing the desired traits. Efforts to identify and introgression of new QTLs conferring horizontal resistance in crops are imperative. Moreover, a clear description of *Pythium* spp. lifestyle can be conducted via contemporary bioinformatics analysis approaches on various omics datasets. Therefore, in light of technological advancement, smart plant breeding using DNA-based molecular markers is considered the cornerstone for crop improvement and elimination of malnutrition and absolute poverty, particularly in developing countries.

References

Abdelzaher, H.M.A., Imamm, M.M., Shoulkamy, M.A. and Gherbawy, Y.M.A. 2004. Biological control of Pythium damping-off of bushokrausing rhizosphere strains of *Pseudomonas fluorescens*. Mycobiology 32: 139-147.

Adelman, I. 2001. Fallaciesin development theory and their implications for policy. pp. 103-135. *In*: Meier, G.M. and Stiglitz, J.E. (eds.). Frontiers of Development Economics: The Future in Perspective. World Bank, NewYork.

Allain-Boulé, N., Tweddell, R., Mazzola, M., Bélanger, R. and Lévesque, C.A. 2004. *Pythium attrantheridium* sp. nov.-taxonomy and comparison with related species. Mycol. Res. 108: 795-805.

Altshuler, D., Pollara, V.J., Cowles, C.R., Van Etten, W.J., Baldwin, J., Linton, L. and Lander, E.S. 2000. An SNP map of the human genome generated by reduced representations hotgun sequencing. Nature 407: 513-516.

Arafa, R.A., Rakha, M.T., Soliman, N.E.K., Moussa, O.M., Kamel, S.M. and Shirasawa, K. 2017. Rapid identification of candidate genes for resistance to tomato late blight disease using next-generation sequencing technologies. PLoSOne 12(12): e0189951.

Arafa, R.A., Kamel, S.M., Rakha, M.T., Soliman, N.E.K., Moussa, O.M. and Shirasawa, K. 2019. Analysis of the lineage of *Phytophthora infestans* isolates using mating type assay, traditional markers, and next generation sequencing technologies. Plos ONE (BioRxiv).

Ascunce, M.S., Huguet-Tapia, J.C., Braunc, E.L., Ortiz-Urquiza, A., Keyhani, N.O. and Goss, E.M. 2016. Whole genome sequence of the emerging oomycete pathogen *Pythium insidiosum* strain CDC-B5653 isolated from an infected human in the USA. Genomics Data 7: 60-61.

Baird, N.A., Etter, P.D., Atwood, T.S., Currey, M.C., Shiver, A.L., Lewis, Z.A., Selker, E.U., Cresko, W.A. and Johnson, E.A. 2008. Rapid SNP discovery and genetic mapping using sequenced RAD markers. PloSOne. 3(10): e3376.

Baldauf, S.L., Roger, A.J., Wenk-Siefert, I. and Doolittle, W.F. 2000. A kingdom-level phylogeny of eukaryotes based on combined protein data. Science 290: 972-977.

Barbazuk, W.B. and Schnable, P.S. 2011. SNP discovery by transcriptome pyrosequencing. Methods Mol. Biol. 729: 225-246.

Baxter, S.W., Davey, J.W., Johnston, J.S., Shelton, A.M., Heckel, D.G., Jiggins, C.D. and Blaxter, M.L. 2011. Linkage mapping and comparative genomics using next-generation RAD sequencing of a non-model organism. PLoSOne 6(4): e19315.

Bentley, D.R., Balasubramanian, S., Swerdlow, H.P., Smith, G.P., Milton, J., Brown, C.G., Hall, K.P., Evers, D.J., Barnes, C.L., Bignell, H.R. and Boutell, J.M. 2008. Accurate whole human genome sequencing using reversible terminator chemistry. Nature 456: 53-59.

Campa, A., Pérez-Vega, E., Pascual, A. and Ferreira, J.J. 2010. Genetic analysis and molecular mapping of quantitative trait loci in common bean against *Pythium ultimum*. Phytopathology 100: 1315-1320.

Chen, X. and Sullivan, P.F. 2003. Single nucleotide polymorphism genotyping: Biochemistry, protocol, cost and throughput. Pharmacogenomics J. 3: 77-96.

Chutimanitsakun, Y., Nipper, R.W., Cuesta-Marcos, A., Cistué, L., Corey, A., Filichkina, T., Johnson, E.A. and Hayes, P.M. 2011. Construction and application for QTL analysis of a restriction site associated DNA (RAD) linkage emapin barley. BMC Genomics 12: 4.

Cooke, D.E.L., Drenth, A., Duncan, J.M., Wagels, G. and Brasier, C.M. 2000. A molecular phylogeny of *Phytophthora* and related oomycetes. Fungal Genet. Biol. 30(1): 17-32.

Davey, J.W. and Blaxter, M.L. 2010. RAD Seq: Next-generation population genetics. Brief Funct. Genomics. 9: 416-423.

Davey, J.W., Hohenlohe, P.A., Etter, P.D., Boone, J.Q., Catchen, J.M. and Blaxter, M.L. 2011. Genome-wide genetic marker discovery and genotyping using next-generation sequencing. Nat. Rev. 12: 499-510.

Diao, X., Hazell, P., Resnick, D. and Thurlow, J. 2007. The role of agriculture in development: Implications for sub-Saharan Africa. Resrep. 153. Washington, DC: IFPRI.

Edwards, D., Forster, J.W., Chagne, D. and Batley, J. 2007. What is SNPs? pp. 41-52. *In*: Oraguzie, N.C., Rikkerink, E.H.A., Gardiner, S.E. and de Silva, H.N. (eds.). Association Mapping in Plants. Springer, Berlin.

Fu, Y.B., Cheng, B. and Peterson, G.W. 2014. Genetic diversity analysis of yellow mustard (*Sinapis alba* L.) germ plasm based on genotyping by sequencing. Genet. Resour. Crop Evol. 61: 579-594.

Gore, M.A., Chia, J.M., Elshire, R.J., Sun, Q., Ersoz, E.S., Hurwitz, B.L., Peiffer, J.A., McMullen, M.D., Grills, G.S., Ross-Ibarra, J. and Ware, D.H. 2009. A first-generation haplo type map of maize. Science 326: 1115-1117.

He, J., Zhao, X., Laroche, A., Lu, Z.-X., Liu, H. and Li, Z. 2014. Genotyping-by-sequencing (GBS), an ultimate marker-assisted selection (MAS) tool to accelerate plant breeding. Front. Plant Sci. 5: 484.

Ho, H.-H. 2009. The genus Pythium in Taiwan, China (1) – A synoptic review. Front. Biol. China. 4(1): 15-28.

Hyten, D.L., Cannon, S.B., Song, Q., Weeks, N., Fickus, E.W., Shoemaker, R.C., Specht, J.E., Farmer, A.D., May, G.D. and Cregan, P.B. 2010. High throughput SNP discovery through deeper sequencing of a reduced representation library to anchor and orient scaffolds in the soybean whole genome sequence. BMC Genomics 11: 38.

Jadhav, V.T. and Ambadkar, C.V. 2007. Effect of *Trichoderma* spp. on seedling emergence and seedling mortality of tomato, chilli and brinjal. J. Plant Dis. Sci. 2: 190-192.

Jarne, P. and Lagoda, P.J. 1996. Microsatellites, from molecules to populations and back. Trends Ecol. Evolut. 11(10): 424-429.

Jiang, G.L. 2013. Molecular markers and marker-assisted breeding in plants. pp. 45-83. *In*: Anderson, S.B. (ed.). Plant Breeding from Laboratories to Fields. InTech, Croatia.

Jiang, G.L. 2013. Plant marker-assisted breeding and conventional breeding: Challenges and perspectives. Adv. Crop Sci. Tech. 1: e105.

Jung, T., Blaschke, H. and Neumann, P. 1996. Isolation, identification and pathogenicity of *Phytophthora* species from declining oak stands. Eur. J. Pathol. 26: 253-272.

Jupe, F., Witek, K., Verweij, W., Sliwka, J., Pritchard, L., Etherington, G.J., Maclean, D., Cock, P.J., Leggett, R.M., Bryan, G.J. and Cardle, L. 2013. Resistance gene enrichment sequencing (Ren Seq) enables reannotation of the NB-LRR gene family from sequenced plant genomes and rapid mapping of resistance loci in segregating populations. Plant J. 76: 530-544.

Kamoun, S. and Smart, C.D. 2005. Late blight of potato and tomato in the genomics era. Plant Dis. 89: 692-699.

Kaur, V. and Singla, S. 2016. Are view on the plant leaf disease detection techniques. Int. J. Innov. Res. Sci. Eng. Technol. 7: 539-543.

Khush, G.S. 2001. Green revolution: The way forward. Nat. Rev. Genet. 2: 815-822.

Kiryowa, M., Nkalubo, S.T., Mukankusi, C., Talwana, H., Gibson, P. and Ukamuhabwa, P. 2015. Effect of marker aided pyramiding of Anthracnose and *Pythium* root rot resistance genes on plant agronomic characters among advanced common bean benotypes. J. Agr. Sci. 7(3): 98.

Klepadlo, M., Balk, C.S., Vuong, T.D., Dorrance, A.E. and Nguyen, H.T. 2018. Molecular characterization of genomic regions for resistance to *Pythium ultimum* var. *ultimum* in the soybean cultivar Magellan. Theo. Appl. Genet. 1-13.

Korte, A. and Farlow, A. 2013. The advantages and limitations of trait analysis with GWAS: A review. Plant Methods 9(1): 29.

Kucharek, T. and Mitchell, D. 2000. Diseases of agronomic and vegetable crops caused by *Pythium*. Plant pathology factsheet. University of Florida, p. 53.

Landt, S.G., Marinov, G.K., Kundaje, A., Kheradpour, P., Pauli, F., Batzoglou, S., Bernstein, B.E., Bickel, P., Brown, J.B., Cayting, P. and Chen, Y. 2012. ChIP-seq guidelines and practices of the ENCODE and mod ENCODE consortia. Genome Res. 22(9): 1813-1831.

Leal-Bertioli, S.C., Cavalcante, U., Gouveia, E.G., Ballén-Taborda, C., Shirasawa, K., Guimarães, P.M., Jackson, S.A., Bertioli, D.J. and Moretzsohn, M.C. 2015. Identification of QTLs for rust resistance in the peanut wild species *Arachismagna* and the development of KASP markers for marker assisted selection. G3(Bethesda). 5(7): 1403-1413.

Lerch, E.R. 2017. Soybean-*Pythium* pathosystem: Search for host resistance and interaction with Fusarium species. PhD Thesis. Iowa State University Capstones, pp. 134.

Lévesque, A., Brouwer, H., Cano, L., Hamilton, J.P., Holt, C., Huitema, E., Raffaele, S., Robideau, G.P., Thines, M., Win, J. and Zerillo, M.M. 2010. Genome sequence of the necrotrophic plant pathogen *Pythium ultimum* reveals original pathogenicity mechanisms and effect or repertoire. Genome Biol. 11: R73.

Lévesque, C.A. and deCock, A.W.A.M. 2004. Molecular phylogeny and taxonomy of the genus Pythium. Mycol. Res. 108: 1363-1383.

Lévesque, C.A. 2011. Fifty years of oomycetes from consolidation to evolutionary and genomic exploration. Fungal Divers. 50: 35-46.

Lin, F., Wani, S.H., Collins, P.J., Wen, Z., Gu, C., Chilvers, M.I. and Wang, D. 2018. Mapping Quantitative Trait Loci for Tolerance to *Pythium irregular* in Soybean (*Glycinemax* L.). G3. Genes, Genomes, Genet. 8: 3155-3161.

Lucas, B. and Griffiths, P.D. 2004. Evaluation of common bean accessions for resistance to *Pythium ultimum*. Hort. Science 39(6): 1193-1195.

Mahuku, G.S., Buruchara, R.A. and Navia, M. 2005. A gene that confers resistance to *Pythium* root rot in common bean: Genetic characterization and development of molecular markers. Phytopathology 95: S64.

Mammadov, J.A., Chen, W., Ren, R., Pai, R., Marchione, W., Yalçin, F., Witsenboer, H., Greene, T.W., Thompson, S.A. and Kumpatla, S.P. 2010. Development of highly polymorphic SNP markers from the complexity reduced portion of maize (*Zeamays* L.) genome for use in marker-assisted breeding. Theo. Appl. Genet. 121: 577-588.

Manoranjitham, S.K. and Prakasam, V. 2000. Management of chilli damping of fusing biocontrol agents. Capsicum Eggplant News Lett. 19: 101-104.

Margulis, L. and Schwartz, K.V. 2000. Five Kingdoms: An illustrated Guide to the Phyla of Life on Earth. W.H. Freeman and Co., New York, N.Y.

Martin, F.N. 1992. Pythium. pp. 39-49. *In*: Singleton, L.L., Mihail, J.D. and Rush, C.M. (eds.). Methods for Research on Soilborne Phytopathogenic Fungi. APS Press, St. Paul, MN.

McDonald, B.A. and Linde, C. 2002. Pathogen population genetics, evolutionary potential, and durable resistance. Annu. Rev. Phytopathol. 40(1): 349-379.

Mendoza, L. 2009. *Pythium insidiosum* and mammalian hosts. pp. 387-405. *In*: Lamour, K. and Kamoun, S. (eds.). Oomycete Genetics and Genomics: Diversity, Interactions, and Research Tools. John Wiley & Sons, Inc., Hoboken, NJ.

Michael, T.P. and Jackson, S. 2013. The first 50 plant genomes. Plant Genome 6: 1-7.

Michael, T.P. and Van Buren, R. 2015. Progress, challenges and the future of crop genomes, Curr. Opin. Plant Biol. 24: 71-81.

Miller, M.R., Dunham, J.P., Amores, A., Cresko, W.A. and Johnson, E.A. 2007. Rapid and cost-effective polymorphism identification and genotyping using restriction site associated DNA(RAD) markers. Genome Res. 17: 240-248.

Moloney, C., Griffin, D., Jones, P.W., Bryan, G.J., McLean, K., Bradshaw, J.E. and Milbourne, D. 2009. Development of diagnostic markers for use in breeding potatoes resistant to Globoderapallida pathotype Pa2/3 using germ plasm derived from *Solanum tuberosum* ssp. *andigena* CPC2802. Theor. Appl. Genet. 120(3): 679-689.

Mukankusi, C. 2007. Improving Résistance to Fusarium root rot [*Fusarium solani* (Mart.) Sacc. f. sp. *Phaseoli* (Burkholder) W.C. Synder, H.N. Hans] in Common bean (*Phaseolus vulgaris* L.). PhD Thesis, University of KwaZulu-Natal, South Africa.

Namayanja, A., Msolla, S.N., Buruchara, R. and Namusoke, A. 2014. Genetic analysis of resistance to *Pythium* root rot disease in common bean (*Phaseolus vulgaris* L.) genotypes. J. Crop. Improv. 28: 184-202.

Ozsolak, F. and Milos, P.M. 2011. RNA sequencing: Advances, challenges and opportunities. Nature Rev. Genet. 12: 87-98.

Packer, A. and Clay, K. 2000. Soil pathogens and spatial patterns of seedling mortality in a temperate tree. Nature 404: 278-281.

Pandey, V., Nutter, R.C. and Prediger, E. 2008. Applied biosystems SOLID system: Ligation-based sequencing. pp. 431-444. *In*: Janitz, M. (ed.). Next generation genome sequencing: Towards personalized medicine. Germany: Wiley-VCH, Weinheim.

Parlevliet, J.E. and Van Ommeren, A. 1988. Accumulation of partial resistance in barley to barley leaf rust and powdery mildew through recurrent selection against susceptibility, Euphytica. 37: 261-274.

Particka, C.A. and Hancock, J.F. 2008. Breeding for increased tolerance to blackroot rot in strawberry. Hortscience 43(6): 1698-1702.

Pfender, W.F., Saha, M.C., Johnson, E.A. and Slabaugh, M.B. 2011. Mapping with RAD (restriction-site associated DNA) markers to rapidly identify QTL for stem rust resistance in *Lolium perenne*. Theor. Appl. Genet. 122: 1467-1480.

Poland, J.A., Balint-Kurti, P.J., Wisser, R.J., Pratt, R.C. and Nelson, R.J. 2009. Shades of gray: The world of quantitative disease resistance. Trends Plant Sci. 14: 21-29.

Ramamoorthy, V., Raguchander, T. and Samiyappan, R. 2002. Enhancing resistance of tomato and hot pepper to Pythium diseases by seed treatment with fluorescent pseudomonads. Eur. J. Plant Pathol. 108: 429-441.

Ramos, A.M., Crooijmans, R.P., Affara, N.A., Amaral, A.J., Archibald, A.L., Beever, J.E., Bendixen, C., Churcher, C., Clark, R., Dehais, P. and Hansen, M.S. 2009. Design of a high density SNP genotyping assay in the pig using SNPs identified and characterized by next generation sequencing technology. PLoS One 4: e6524.

Reinhart, K.O., Tytgat, T., Vander Putten, W.H. and Clay, K. 2010. Virulence of soil-borne pathogens and invasion by *Prunus serotina*. New Phytol. 186(2): 484-495.

Rey, P., Benhamou, N. and Tirilly, Y. 1998. Ultrastructural and cytochemical investigation of a symptomatic infection by Pythium spp. Phytopathology 88: 234-244.

Ribaut, J.M., de Vicente, M.C. and Delannay, X. 2010. Molecular breeding in developing countries: Challenges and perspectives. Curr. Opin. Plant Biol. 13: 1-6.

Robideau, G.P., de Cock, A.W., Coffey, M.D., Voglmayr, H., Brouwer, H., Bala, K., Chitty, D.W., Désaulniers, N., Eggertson, Q.A., Gachon, C.M. and HU, C.H. 2011. DNA bar coding of oomycetes with cytochrome coxidase subunit I and internal transcribed spacer. Mol. Ecol. Resour. 11(6): 1002-1011.

Romay, M.C., Millard, M.J., Glaubitz, J.C., Peiffer, J.A., Swarts, K.L., Casstevens, T.M., Elshire, R.J., Acharya, C.B., Mitchell, S.E., Flint-Garcia, S.A. and McMullen, M.D. 2013. Comprehensive genotyping of the USA national maize in bred seedbank. Genome Biol. 14, R55.

Rosso, M.L., Rupe, J.C., Chen, P. and Mozzoni, L.A. 2008. Inheritance and genetic mapping of resistance to *Pythium* damping-off caused by Pythium aphanidermatum in 'Archer' soybean. Crop Sci. 48: 2215-2222.

Rothberg, J. and Leamon, J. 2008. The development and impact of 454 sequencing. Nat. Biotechnol. 26: 1117-1124.

Rujirawat, T., Patumcharoenpol, P., Lohnoo, T., Yingyong, W., Kumsang, Y., Payattikul, P., Tangphatsornruang, S., Suriyaphol, P., Reamtong, O., Garg, G. and Kittichotirat, W. 2018. Probing the phylogenomics and putative pathogenicity genes of *Pythium insidiosum* by oomycete genome analyses. Sci. Rep. 8: 4135.

Sachidanandam, R., Weissman, D., Schmidt, S.C., Kakol, J.M., Stein, L.D., Marth, G., Sherry, S., Mullikin, J.C., Mortimore, B.J., Willey, D.L., Hunt, S.E., Cole, C.G., Coggill, P.C., Rice, C.M., Ning, Zemin., Rogers, Jane., Bentley, David R., Kwok, P., Mardis, E.R., Yeh, R.T., Schultz, B., Cook, L., Davenport, R., Dante, M., Fulton, L., Hillier, L., Waterston, R.H., McPherson, J.D., Gilman, B., Schaffner, S., Van Etten, W.J., Reich, D., Higgins, J., Daly, M.J., Blumenstiel, B., Baldwin, J., Stange-Thomann, N., Zody, M.C., Linton, L., Lander, E.S. and Altshuler, D. 2001. A map of human genome sequence variation containing 1.42 million single nucleotide polymorphisms. Nature 409: 928-933.

Sánchez, C.C., Smith, T.P., Wiedmann, R.T., Vallejo, R.L., Salem, M., Yao, J. and Rexroad, C.E. 2009. Single nucleotide polymorphism discovery in rainbow trout by deep sequencing of a reduced representation library. BMC Genomics 10: 559.

Saville, A., Graham, K., Grünwald, N.J., Myersm, K., Fry, W.E. and Ristaino, J.B. 2015. Fungicide sensitivity of US genotypes of *Phytophthora infestans* to six oomycete-targeted compounds. Plant Dis. 99: 659-666.

Shehzadi, A., Abass, H.M.K., Ahmed, Z. and Saleem, S. 2017. Effect plant disease resistance genes: Recent applications and future perspectives. J. Innov. Bio-Res. 1(1): 86-103.

Shirasawa, K. Hirakawa, H. and Isobe, S. 2016. Analytical work flow of double-digest restriction site-associated DNA sequencing based on empirical and in silico optimization in tomato. DNA Res. 23(2): 145-153.

Sonah, H., Bastien, M., Iquira, E., Tardivel, A., Légaré, G., Boyle, B., Normandeau, É., Laroche, J., Larose, S., Jean, M. and Belzile, F. 2013. An improved genotyping by sequencing (GBS) approach offering increased versatility and efficiency of SNP discovery and genotyping. PloS One. 8: e54603.

Song, F.-J., Xiao, M.-G., Duan, C.-X., Li, H.-J., Zhu, Z.-D., Liu, B.-T., Sun, S.L., Wu, X.F. and Wang, X.M. 2015. Two genes conferring resistance to *Pythium* stalk rot in maize in bredline Qi319. Mol. Genet. Genomics. 290: 1543-1549.

Song, Q., Jia, G., Zhu, Y., Grant, D., Nelson, R.T., Hwang, E-Y., Hyten, D. L., and Cregan P.B. 2010. Abundance of SSR motifs and development of candidate polymorphic SSR markers (BARCSOYSSR_1.0) in soybean. Crop Sci. 50: 1950-1960.

Stasko, A.K., Wickramasinghe, D., Nauth, B.J., Acharya, B., Ellis, M.L., Taylor, C.G., McHale, L.K. and Dorrance, A.E. 2016. High-density mapping of resistance QTL toward *Phytophthora sojae*, *Pythium irregulare*, and *Fusarium graminearum* in the same soybean population. Sci. 56: 2476-2492.

Stein, L.D. 2010. The case for cloud computing in genome informatics. Genome Biol. 11: 207.

Steuernagel, B., Periyannan, S.K., Hernández-Pinzón, I., Witek, K., Rouse, M.N., Yu, G., Hatta, A., Ayliffe, M., Bariana, H., Jones, J.D. and Lagudah, E.S. 2016. Rapid cloning of disease-resistance genes in plants using mutagenesis and sequence capture. Nature Biotechnol. 34: 652-655.

Van Orsouw, N.J., Hogers, R.C., Janssen, A., Yalcin, F., Snoeijers, S., Verstege, E., Schneiders, H., van der Poel, H., Van Oeveren, J., Verstegen, H. and Van Eijk, M.J. 2007. Complexity reduction of polymorphic sequences (CRoPS): A novel approach for large-scale polymorphism discovery in complex genomes. PLoS One 2: e1172.

Van Tassell, C.P., Smith, T.P., Matukumalli, L.K., Taylor, J.F., Schnabel, R.D., Lawley, C.T., Haudenschild, C.D., Moore, S.S., Warren, W.C. and Sonstegard, T.S. 2008. SNP discovery and allele frequency estimation by deep sequencing of reduced representation libraries. Nature Methods 5: 247-252.

Van West, P., Appiah, A.A. and Gow, N.A.R. 2003. Advances in research on oomycete root pathogens. Physiol. Mol. Plant Pathol. 62: 99-113.

Vipinadas, M.J. and Thamizharasi, A.A. 2015. Survey on plant disease identification. Int. J. Comput. Trends Technol. 3: 129-135.

Vlk, D. and Řepková, J. 2017. Application of next-generation sequencing in plant breeding. Czech J. Genet. Plant Breed. 53(3): 89-96.

Vos, P., Hogers, R., Reijans, M., van de Lee, T., Hornes, M., Friters, A., Pot, J., Paleman, J., Kuiper, M. and Zabeau, M. 1995. AFLP a new technique for DNA fingerprinting. Nucl. Acids Res. 23: 4407-4414.

Wang, H., Huang, C., Guo, H., Li, X., Zhao, W., Dai, B., Yan, Z. and Lin, Z. 2015. QTL mapping for fiber and yield traits in upland cotton under multiple environments. PLoS One. 10(6): e0130742.

Waterhouse, G.M. and Waterston, J.M. 1964. *Pythium aphanidermatum*. CMI descriptions of pathogenic fungi and bacteria, No. 36. Commonwealth Mycological Institute: Surrey. Waterhouse.

Waterhouse, G.M. 1968. The genus Pythium. Mycol. Pap. 110: 1-71.

Weber, J.L. and Myers, E.W. 1997. Human whole-genomes hot gun sequencing. Genome Res. 7: 401-409.

Weinmann, A.S., Yan, P.S., Oberley, M.J., Huang, T.H.M. and Farnham, P.J. 2002. Isolating human transcription factor targets by coupling chromatin immunoprecipitation and CpGisl and microarray analysis. Genes Dev. 16(2): 235-244.

Welch, R.M. and Graham, R.D. 2004. Breeding for micro nutrients in staple food crops from a human nutrition perspective. J. Exp. Bot. 55: 353-364.

Whipps, J.M. and Umsden, D.R. 1991. Biological control of *Pythium* species. Biocontrol. Sci. Tech. 1: 75-90.

Xu, Y. 2010. Molecular Plant Breeding. CAB International.

CHAPTER 16

Role of Phytochemicals in Plant Diseases Caused by *Pythium*

Rajendra M. Gade[1*], Mahendra Rai[2], Ranjit S. Lad[3] and Amol V. Shitole[3]

[1] Vasantrao Naik College of Agricultural Biotechnology, Dr. PDKV, Yavatmal (M.S.), 445001, India
[2] Department of Biotechnology, S.G.B. Amravati University, Amravati, Maharashtra 444602, India
[3] Department of Plant Pathology, Dr. PDKV, Akola (M.S.), 444104, India

Introduction

The *Pythium* species are fungus-like organisms, commonly referred to as water molds (kingdom Straminopila; phylum Oomycota; class Oomycetes; subclass Peronosporomycetidae; order Pythiales and family Pythiaceae). They are worldwide in distribution and associated with a wide variety of habitats ranging from terrestrial or aquatic environments, in cultivated or fallow soils, in plants or animals, in saline or fresh water. The genus *Pythium* is one of the largest oomycete genus and consists of more than 130 recognized species which are isolated from different regions of the world. Generally, it is recognized that species of *Pythium* are not host specific and most of the species are known to parasitize and cause infections in the crop plants and ultimately damage them (Van Der 1981). Pre- and post-emergence damping-off disease caused by *Pythium* spp. in vegetable crops is economically very important worldwide.

The indiscriminate use of most of the synthetic fungicides has created different types of environmental and toxicological problems. Recently, in different parts of the world, attention has been paid towards exploitation of higher plant products as novel chemotherapeutants in plant protection. The popularity of botanical pesticides is once again increasing and some plant products are being used globally as green pesticides (Gurjar et al. 2012). Sometimes the fungicide has adverse effects on the non-target organisms. Hence, it is necessary to search new antifungal compounds as a safe, eco-friendly, cheap alternative to synthetic fungicides, since they produce different secondary metabolites which perform defensive role in plants and protect the plants from their invaders.

Plants serve as a rich storehouse of biochemicals, which have the potential to play an important role in the defense against several diseases. These phytochemicals have been exploited as natural pesticides, flavoring agents, fragrances, medicinal compounds, fibers, beverages and food metabolites and were identified by nuclear magnetic resonance (NMR), mass spectrometry (MS) and X-ray analysis techniques (Yazdani et al. 2011). Plants have the ability to synthesize aromatic secondary metabolites, like phenols, phenolic acids, quinones, flavones, flavonoids, flavonols, tannins and coumarins (Cowan 1999). Several previous reports have described antifungal, anti-proliferative and antimicrobial activities of *Lantana camara*. *L. camara* contains secondary metabolites such as alkaloids, terpenoids and phenolics, which could be held partially responsible for some of the biological activities (Ganjewala et al. 2009). *L. camara* also contains lantadenes, the pentacyclic

*Corresponding author: gadermg@gmail.com

triterpenes that possess a number of useful biological activities such as antifungal (Kumar et al. 2006), anti-proliferative (Nagao et al. 2002) and antimicrobial activities (Juliani et al. 2002). *Datura stramonium* contains saponins, tannins, carbohydrates, proteins, steroids, flavonoids, alkaloids, phenol and glycosides and is used in medicine due to its analgesic and anti-asthmatic activities (Soni et al. 2012). Leaves extract of the plant contains different types of secondary metabolites such as glycosides, phenols, lignins, saponins, sterols and tannins (Nain et al. 2013). *Parthenium hysterophorus,* member of the Asteraceae family, is a noxious weed in America, Asia, Africa and Australia. The plant contains a large number of important bioactive compounds, mainly sesquiterpene lactones, flavonoid glycosides and pinenes. The plant possesses potent inhibitory activity against four species of fungi *viz. Trichophyton rubrum, Trichophyton mentagrophytes, Epidermophyton floccosum* and *Microsporum gypseum. Psoralea corylifolia* Linn. (Bavanchi/ babchi) is a rare and endangered herbaceous medicinal plant, distributed throughout the tropical and subtropical regions of the world. It is a source of valuable alkaloids having pharmaceutical importance. *P. corylifolia* contained coumarins, flavonoids, merotepenes etc. such as psoralen, isopsoralen, neobavaislfoavone, bavachin, bavaislfoavone, bavachromene, psoralidin, corylifolinin, bavachinin and bavachalcone. Some of the above phytoextracts are found effective against *Pythium* in earlier studies.

The components with phenolic structures, like carvacrol, eugenol and thymol, were highly active against the pathogen. These groups of compounds show antimicrobial effect and serve as plant defense mechanisms against pathogenic microorganisms (Das et al. 2010) (Fig. 16.3). The most commonly used solvents for preliminary investigations of antimicrobial activity in plants are methanol, ethanol and water. The other solvents used by researchers are dichloro-methane, acetone and hexane. Some authors use a combination of these solvents to obtain the best solvent systems for extraction (Eloff 1998). Phytochemical screening of methanol leaf extracts of *Aegle marmelos* was carried out by More et al. (2016). They also conducted studies on development of phytochemical parameters and investigated the antifungal substances present in methanolic extract obtained from leaves of *A. marmelos.* Preliminary phytochemical screening of the extracts revealed the presence of flavonoids, tannins, alkaloids, saponins and phenolic compounds. Petroleum ether:Ethyle acetate (2:1) solvent system gives maximum numbers of compounds from methanolic extract of *A. marmelos* and these compounds were separated on TLC plates. TLC resulted in identification of three spots found in the methanolic extract. The Rf values of methanolic extract of *A. marmelos* run under Petroleum ether:Ethyl acetate (02:01) solvent system were 0.04, 0.07, 0.14, 0.18, 0.25, 0.35, 0.45, 0.61, 0.70 and 0.84. The study provided referential information for the correct identification of the crude plant extract of *A. marmelos.* Methanol was the best solvent for extraction of antifungal constituents from *A. marmelos* leaves.

Plant products such as marigold water extract are also compatible with bacterial bio-control agents like *Pseudomonas fluorescens* and *Bacillus subtilis* under *in vitro* conditions (Wavare et al. 2015). An experiment was conducted by Pente et al. (2015) in which different combinations of botanicals *viz.,* neem, tulsi, onion, garlic, chrysanthemum and bio-agents were used under *in vitro* and greenhouse conditions against oomycetes fungus *Phytophthora* in citrus. Similarly, Wavare et al. (2017) tested antifungal effects of methanol, acetone, dichloromethane and aqueous extracts of four flowers: marigold (*Tagetes erecta*), *Gaillardia* sp. (*G. aristata*), *Chrysanthemum* sp. (*C. indicum*) and *Calotropis* sp. (*C. gigantea*) were evaluated *in vitro* and *in vivo* against *Sclerotium rolfsii*, the causal agent of collar rot of chickpea alone and in combination with bio-agents. Birari et al. (2018) tested effectiveness of extracts of babchi seeds (*Psoralea corylifolia*), datura leaves (*Datura* sp.) and ghaneri leaves (*Lantana camara*) as biocontrol agents against *Colletotrichum capsici*. The three botanicals *viz. Psoralea corylifolia, Datura* sp. and *Lantana camara* at different concentrations 250 μl, 500 μl, 750 μl and 1000 μl were tested by poisoned food technique. *P. fluorescens* (culture filtrate @ 0.5%) + *T. viride* (culture filtrate @ 0.5%) + methanolic extract of *P. corylifolia* @ 2% proved highly effective in reducing disease intensity (80%) in chili under detached fruit bioassay.

Several researchers conducted studies for the management of *Pythium* spp. with the help of phytochemicals and in this chapter we have mainly focused on the management of *Pythium* using different phytochemicals.

Antifungal activity of phytochemicals against *Pythium* species

Many reports are available on evaluation of antifungal properties of plant extracts against *Pythium* species. Suleiman and Emua (2009) reported that Potato dextrose agar (PDA) medium poisoned with neem leaf extract showed about 55% growth inhibition of *Pythium aphanidermatum*. While ginger rhizome extract reduced 70% damping-off infection on cowpea *in vivo*, ginger and aloe could completely inhibit the mycelial growth of *P. aphanidermatum* causing damping-off under *in vitro* condition. In another study, antifungal efficacy of about 66 medicinal plants belonging to 41 families were evaluated against *P. aphanidermatum*, the causal agent of chilli damping-off (Muthukumar et al. 2010). The Zimmu (*Allium sativum* L. × *Allium cepa* L.) leaf extract at 10% concentration had the highest zone of inhibition (13.7 mm) against mycelial growth of *P. aphanidermatum*. Alhussaen et al. (2011) conducted *in vitro* tests of three concentrations of garlic extract (5, 10 and 100%) against *Pythium ultimum* isolated from tomato seedling roots. The results obtained showed that 100% garlic extract inhibited total mycelial growth of pathogen compared to mean growth area of control (56.72 cm^2). Moreover, diluted garlic extract (10%) reduced the fungal growth to 15.5% (8.84 cm^2) and 5% garlic extract showed growth area of 41% (23.25 cm^2).

Pattnaik et al. (2012) demonstrated bio-efficacy of some plant extracts against fruit spot disease in tomato. The plant leaf extracts of *Pongamia pinnata*, *A. marmelos*, *Azadirachta indica*, *Brassica campestris*, *Piper nigrum*, *Euphorbia tirucalli*, *Vitex negundu*, *Ageratum conyzoides*, *Tagetes patula* and *Zigiphus jujube* were used for testing antifungal activity against pathogens of tomato (*Clavibacter michigenesis*, *Alternaria solani*, and *Septoria lycopersici*, *Pythium debaryanum* and *Phytopthora capsici*). The results obtained showed the significant reduction in disease symptoms (spots on fruit). Among the extract tested, extract of *A. marmelos* showed maximum 61.29% reduction in fruit spots, followed by the extract of *B. campestris* (58.06% reduction); however, all other plant extracts showed the reduction in fruit spots in between 27.41-57.04%. Similarly, *in vitro* studies were undertaken by Gholve et al. (2014) who evaluated bio-efficacy of ten different plants extracts (each at 10, 15 and 20%) against *P. ultimum*. The results recorded revealed that all the tested plant extracts demonstrated fungistatic properties against the test pathogen and significantly inhibited its growth. Among all these plant extracts evaluated, garlic was found most effective and recorded significantly higher growth inhibition (94.83%) followed by adulsa (75.53%) and datura (60.65%). However, other tested extracts, i.e. mehandi, ginger, tulsi, parthenium, neem, turmeric and satawari showed growth inhibition in the range of 20.82 to 56.83%. Similarly, Parveen and Sharma (2014) demonstrated the antifungal potential of 20 different plant extract and found that the extract of *Jacaranda mimosifolia* and *Moringa oleifera*, exhibited 27.7% mycelial growth inhibition of *P. aphanidermatum,* whereas the extracts of *Polyalthia longifolia* and *Terminalia arjuna* showed 22.2% growth inhibition. Besides these, *Lawsonia inermis*, *Aegle marmelos*, *Nigella sativa*, and *Azadirachta indica* extracts also exhibited significant inhibitory activity against *P. aphanidermatum*.

Some botanicals like *Boerhaavia diffusa*, *Lawsonia inermis*, *Phyllanthus niruri* etc. extracted by using different solvents were found very effective against *Pythium* spp. Mohamed and El-Hadidy (2008) reported that cucumber seedlings when soaked in 2000 ppm of butanolic and ethereal extracts of *Verbascum eremobium* before planting induced resistance to root rot disease caused by *Pythium ultimum*. Pandey et al. (2010) reported that seed treatment (10%) and spraying of *Boerhaavia diffusa* root extract reduced the disease incidence of rhizome rot of ginger caused by *Pythium aphanidermatum*. Maximum (62.92%) reduction in disease incidence was recorded in plots where rhizome treatment was supplemented with three foliar sprays of *Boerhaavia diffusa* root extract @ 10% followed by rhizome treatment and three foliar sprays with *B. diffusa* root extract @ 5% (58.00%), rhizome treatment with *B. diffusa* root extract @ 10% (54.85%), three foliar sprays with *B. diffusa* root extract @ 10% (46.70%) and three foliar sprays with *B. diffusa* root extract @ 5% (43.93%). Ambikapathy et al. (2011) tested n-butanol, methanol and aqueous extracts of five medicinal plants against *Pythium debaryanum* by agar well diffusion method. Among these, methanol extract of *Lawsonia inermis* and *Phyllanthus niruri* exhibited maximum inhibition zone of 25 and 20 mm, respectively. Jeyasakthy et al. (2013) studied antifungal activity of sequentially extracted different cold organic solvents extracts like methanol extract and ethyl acetate of *Lantana camara* and *Cassia alata* against *Pythium* spp. and three other phytopathogenic fungi by agar well diffusion method. Interestingly, all solvent extracts of *Cassia alata* exhibited antifungal activity against *Pythium* spp. from 48 hrs to 96 hrs of incubation.

In another study, antifungal activities of acetone, ethanol, methanol and chloroform leaf and fruit extracts of *Aegle marmelos*, *Syzygium cumini* and *Pongamia pinnata* were evaluated against *Pythium debaryanum*. Interaction effects of solvents and plants at different concentrations were tested on pathogen by using following solvents, plants and concentrations – Solvents (S): S1-acetone, S2-ethanol, S3-methanol, S4-chloroform. Plant leaves (P): P1-*A. marmelos* leaf extract, P2-*S. cumini* leaf extract, P3-*P. pinnata* leaf extract. Concentrations (C): C1- 250 µl, C1-250 µl, C2-500 µl, C3-750 µl, C4-1000 µl. The maximum activity was found with methanol extract of *A. marmelos* leaves and fruits. At 1 ml quantity of concentration C4 (1000 µl), complete inhibition of mycelial growth of *P. debaryanum* was observed with extract S3P1 (methanolic extract of *A. marmelos*) (100%), followed by S3P3 (methanolic extract of *P. pinnata*) (97.63%) and S2P1 (95.21%). The lowest inhibition was observed with S4P2 (*S. cumini* extract from chloroform) (75.40%). Pot culture study revealed that tomato seed treatment with *Pseudomonas fluorescens* (10 g/kg) + *Trichoderma viride* (4 g/kg) + methanol extract of *A. marmelos* (4%) was effective to control pre- and post-emergence damping-off caused by *P. debaryanum*. The methanol extract revealed strongest antifungal activity against *P. debaryanum*, followed by ethanol extract and lowest antifungal activity was found in chloroform extract (Table 16.1 and Fig. 16.1) (More et al. 2017).

Similarly, in another *in vitro* and pot experiment about tomato damping off caused by *Pythium debaryanum* performed at Plant Pathology Dept., Dr. PDKV, Akola (M.S.) during 2013-2017, it was reported that among different weed solvent extracts, *P. corylifolia* seeds extract was found effective for reducing dry mycelial weight of the test pathogen at all tested concentrations (125, 250, 375 and 500 µl). Pot experiment showed that seed treatment with metalaxyl (0.2%), thiram (0.2%) and *P. fluorescens* (10 g/kg) + *T. harzianum* (4 g/kg) + *P. corylifolia* seeds methanol extract 4% was most effective for reducing damping-off incidence in tomato (74.73, 68.95 and 71.71%, respectively) (Tables 16.2 and 16.3; Fig. 16.2) (Shitole 2018). Earlier, Wavare et al. (2016) reported that among *in vitro* study on different floral solvent extracts, the marigold distilled water extract was found effective for reducing dry mycelial weight of the *P. debaryanum* at all tested concentrations, i.e. 250, 500, 750 and 1000 µl.

Essential oils, volatile oils or simply the 'oil' extracted from plant, such as "oil of lemongrass", are hydrophobic liquids containing volatile aroma compounds extracted from vegetal materials using steam or hydro distillation techniques. Most of these volatile natural products belong to monoterpenoids compounds (Hanson 2003). The essential oils are important because of their antibacterial, antifungal, antioxidant and anti-carcinogenic properties (Tzortzakis and Economakis 2007). Similarly, Pandey and Dubey (1994) extracted essential oil from the leaves of 30 angiospermic plants and tested their antifungal activity against *P. aphanidermatum* and *P. debaryanum*. Of these, essential oil from *Hyptis suaveolens*, *Murrayaa koenigii* and *Ocimum sanctum* showed strong antifungal activity against both test fungi causing damping-off disease of tomato. Nafiseh et al. (2012) reported the antimicrobial activity of essential oil from *eucalyptus* (*Eucalyptus camaldulensis* Dehnn.) against three post-harvest pathogenic fungi (*Penicillium digitatum*, *Aspergillus flavus* and *Colletotrichum gloeosporioides*) and three soil-borne pathogenic fungi (*P. ultimum*, *Rhizoctonia solani* and *Bipolaris sorokiniana*) under *in vitro* condition. The result

Table 16.1. Effect of interaction means of solvents×plants×concenrations (s×p×c)

(Solvent×plant×concentration)	% Inhibition over control			
	C1 (250 µl)	C2 (500 µl)	C3 (750 µl)	C4 (1000 µl)
S1P1	13.6 (21.68)*	43.04 (41.00)	79.69 (63.21)	86.57 (68.30)
S1P2	10.96 (19.33)	29.98 (33.19)	50.98 (45.56)	81.05 (64.19)
S1P3	14.45 (22.34)	41.68 (40.21)	68.78 (56.03)	88.86 (70.50)
S2P1	15.86 (23.47)	57.63 (49.39)	86.41 (68.37)	95.21 (77.36)
S2P2	12.16(20.41)	33.33 (35.26)	57.76 (49.46)	85.43 (67.56)
S2P3	16.64 (24.07)	45.50 (42.42)	74.45 (59.63)	91.04 (72.59)
S3P1	16.99 (24.34)	86.35 (68.31)	90.91 (72.45)	100.00 (89.76)
S3P2	14.40 (22.30)	37.67 (37.86)	63.22 (52.66)	88.73 (70.39)
S3P3	18.89 (25.76)	48.84 (44.33)	80.01 (63.44)	97.63 (81.15)
S4P1	10.96 (33.11)	29.85 (33.11)	74.11 (59.41)	77.64 (61.78)
S4P2	8.83 (28.88)	23.33 (28.88)	43.01 (40.98)	75.40 (60.26)
S4P3	13.31 (38.57)	38.88 (38.57)	62.18 (52.02)	77.73 (61.84)
Control	0.00 (0.00)	0.00 (0.00)	0.00 (0.00)	0.00 (0.00)
Source		S.E (M)±		**C.D. at (P=0.01)**
Solvent (S)		0.04		0.18
Plants (P)		0.04		0.18
Concentrations (C)		0.04		0.18
Solvent×plants (S×P)		0.08		0.31
Solvent×concentrations (S×C)		0.09		0.36
Plants×concentrations (P×C)		0.08		0.31
Solvent×plants×concentrations (S×P×C)		0.14		0.62

*Numbers in parenthesis are arc sin transformed values. Average of three replications. Solvents (S): S1-acetone, S2-ethanol, S3-methanol, S4-chloroform. Plant leaves (P): P1-*A. marmelos* leaf extract, P2-*S. cumini* leaf extract, P3-*P. pinnata* leaf extract. Concentrations (C): C1- 250 µl, C1-250 µl, C2-500 µl, C3-750 µl, C4-1000 µl.

showed that the oil is very effective for complete mycelial growth inhibition in case of *P. ultimum* and *R. solani* in all concentrations tested; however, it is not found to be effective against post-harvest pathogens. Amini et al. (2012) studied the effectiveness of essential oils from three medicinal plants *viz. Zataria multiflora, Thymus vulgaris* and *Thymus kotschyanus* on the mycelial growth of three pathogenic fungi, i.e. *Pythium aphanidermatum, Fusarium graminearum* and *Sclerotinia*

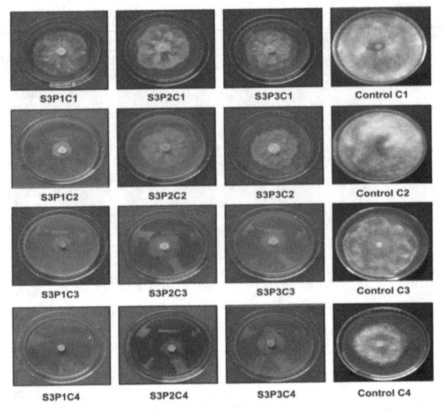

Figure 16.1. Effect of methanol extracts of plants on mycelial growth of *P. debaryanum*
Solvents (S): S1-acetone, S2-ethanol, S3-methanol, S4-chloroform. Plant leaves (P): P1-*A. marmelos* leaf
extract, P2-*S. cumini* leaf extract, P3-*P. pinnata* leaf extract. Concentrations (C): C1-250 µl, C2-500 µl,
C3-750 µl, C4-1000 µl

Color version at the end of the book

sclerotiorum. Complete mycelial growth inhibition of all the tested pathogens was recorded by all the plant extracts at 200 µl/liter concentration. Similarly, Fonseca et al. (2015) demonstrated antifungal activity of essential oil extracted from *Origanum vulgare, O. majorana, Mentha piperita* and *R. officinalis* against *P. insidiosum.* The results revealed that the essential oil obtained from *O. vulgare* showed highest efficacy on *P. insidiosum.*

Separation and identification of active compounds from phytoextracts

Thin layer chromatography (TLC) is used for separation of different compounds present in crude water extract, bioautography for isolation of active compound(s) was also performed. The metabolites which show promising results will be selected for analyzing structural elucidation using spectroscopic techniques *viz.* Gas Chromatography-Mass Spectroscopy (GC-MS), Fourier Transform Infrared Spectroscopy (FT-IR) and Nuclear Magnetic Resonance (NMR) (^{1}H and ^{13}C) which are useful for detection of mass of the compound, functional groups and carbon and hydrogen position within the compound, respectively. Shitole (2018) showed presence of two compounds *viz.* Babchi Seed Compound Red-1 (BSCR1) and Babchi Seed Compound Yellow 2 (BSCY2) in methanol extract of *Psoralea corylifolia* that have antifungal activity on TLC Bioautography plate against *Pythium debaryanum, Aspergillus niger* and *Fusarium oxysporum* f. sp. *ciceri.* It is revealed

Table 16.2. Effect of *P. corylifolia* seeds methanol extract and bio-agents alone and in combination on seedling growth parameters of tomato by paper towel method

Treatment no.	Treatment	% Germination	Mean shoot length (cm)	Mean root length (cm)	SVI
T1	*P. fluorescens* alone (10 g/kg)	73.20	3.90	4.90	644.16
T2	*T. harzianum* alone (4 g/kg)	77.80	3.90	5.40	723.54
T3	*P. corylifolia* seeds methanol extract alone 2%	62.53	2.30	2.50	300.14
T4	*P. corylifolia* seeds methanol extract alone 3%	64.87	3.20	2.70	382.73
T5	*P. corylifolia* seeds methanol extract alone 4%	67.87	3.50	2.90	434.37
T6	*P. fluorescens* (10 g/kg) + *T. harzianum* (4 g/kg)	78.53	5.60	6.90	981.63
T7	*P. fluorescens* (10 g/kg) + *P. corylifolia* seeds methanol extract 2%	75.53	3.90	5.00	672.22
T8	*P. fluorescens* (10 g/kg) + *P. corylifolia* seeds methanol extract 3%	76.53	3.80	5.10	681.12
T9	*P. fluorescens* (10 g/kg) + *P. corylifolia* seeds methanol extract 4%	77.87	4.20	5.60	763.13
T10	*T. harzianum* (4 g/kg) + *P. corylifolia* seeds methanol extract 2%	79.53	3.90	5.60	755.54
T11	*T. harzianum* (4 g/kg) + *P. corylifolia* seeds methanol extract 3%	80.53	4.00	5.60	773.09
T12	*T. harzianum* (4 g/kg) + *P. corylifolia* seeds methanol extract 4%	82.20	4.20	6.60	887.76
T13	*P. fluorescens* (10 g/kg) + *T. harzianum* (4 g/kg) + *P. corylifolia* seeds methanol extract 4%	84.53	6.90	7.70	1234.14
T14	Metalaxyl 0.2%	76.10	5.40	4.20	730.56
T15	Thiram 0.2%	75.20	3.90	6.60	789.60
T16	Carbendazim 0.1%	72.20	7.40	3.70	801.42
T17	Control (pathogen inoculated)	63.20	2.10	1.70	240.16

* Figures in parenthesis are arc sin transformed values. Average of five replications (SVI = Seedling vigor index)

from analytical data and mass library search that partial or most probable structure of compound one isolated from *Psoralea corylifolia* seeds was Phenol, 4-(3,7-dimethyl-3-ethenylocta-1,6-dienyl) and 7H-Furo [3,2-g][1] Benzopyran 7-One as compound two isolated from *Psoralea corylifolia* seeds at Rf values 0.87 and 0.47, respectively.

Similarly, attempt of identification of potential compounds from *P. corylifolia* that was done by Kim et al. (2009) identified the structures of the isolated compounds as psoralen (1),

Table 16.3. Effect of *P. corylifolia* seeds methanol extract alone and in combination with bio-agents on damping-off of tomato caused by *P. debaryanum* in pot culture treatment

Tr. no.	Treatment	Seed emergence (%)	Damping off (%)		Total damping off (%)	Total reduction over control (%)
			Pre	Post		
T1	*P. fluorescens* alone (10 g/kg)	72.67 (58.48)*	13.03 (21.16)	30.89 (33.77)	43.92 (41.51)	49.71 (44.84)
T2	*T. harzianum* alone (4 g/kg)	76.22 (60.82)	8.77 (17.23)	30.04 (33.23)	38.81 (38.53)	55.57 (48.20)
T3	*P. corylifolia* seeds methanol extract alone 2%	62.89 (52.47)	24.73 (29.82)	38.51 (38.36)	63.25 (52.68)	27.59 (31.69)
T4	*P. corylifolia* seeds methanol extract alone 3%	66.44 (54.60)	20.48 (26.90)	35.13 (36.35)	55.60 (48.22)	36.34 (37.07)
T5	*P. corylifolia* seeds methanol extract alone 4%	71.33 (57.63)	14.63 (22.49)	30.53 (33.54)	45.15 (42.22)	48.30 (44.03)
T6	*P. fluorescens* (10 g/kg) + *T. harzianum* (4 g/kg)	78.44 (62.34)	6.12 (14.32)	26.63 (31.07)	32.75 (34.91)	62.51 (52.24)
T7	*P. fluorescens* (10 g/kg) + *P. corylifolia* seeds methanol extract 2%	72.67 (58.48)	13.03 (21.16)	27.51 (31.64)	40.54 (39.55)	53.58 (47.06)
T8	*P. fluorescens* (10 g/kg) + *P. corylifolia* seeds methanol extract 3%	72.89 (58.62)	12.76 (20.93)	26.52 (31.00)	39.29 (38.82)	55.02 (47.88)
T9	*P. fluorescens* (10 g/kg) + *P. corylifolia* seeds methanol extract 4%	76.00 (60.67)	9.04 (17.50)	23.98 (29.32)	33.02 (35.08)	62.19 (52.06)
T10	*T. harzianum* (4 g/kg) + *P. corylifolia* seeds methanol extract 2%	76.44 (60.97)	8.51 (16.96)	26.44 (30.95)	34.95 (36.24)	59.99 (50.76)
T11	*T. harzianum* (4 g/kg) + *P. corylifolia* seeds methanol extract 3%	77.11 (61.42)	7.71 (16.12)	25.66 (30.44)	33.37 (35.29)	61.79 (51.82)
T12	*T. harzianum* (4 g/kg) + *P. corylifolia* seeds methanol extract 4%	78.00 (62.03)	6.65 (14.94)	20.80 (27.13)	27.45 (31.59)	68.58 (55.91)
T13	*P. fluorescens* (10 g/kg) + *T. harzianum* (4 g/kg) + *P. corylifolia* seeds methanol extract 4%	80.44 (63.76)	3.72 (11.12)	20.99 (27.27)	24.71 (29.81)	71.71 (57.87)
T14	Metalaxyl 0.2%	77.56 (61.72)	7.18 (15.54)	14.89 (22.70)	22.07 (28.02)	74.73 (59.82)
T15	Thiram 0.2%	76.44 (60.97)	8.51 (16.96)	18.61 (25.55)	27.12 (31.38)	68.95 (56.14)
T16	Carbendazim 0.1%	65.11 (53.83)	22.08 (28.02)	26.48 (30.97)	48.56 (44.17)	44.41 (41.79)
T17	Control (pathogen inoculated)	54.89 (47.81)	34.31 (35.86)	53.04 (46.74)	87.35 (69.16)	0.00 (0.00)*
F test	-	Sig	Sig	Sig	Sig	Sig
SE (m)±	-	0.40	0.58	0.61	0.58	0.65
CD at (0.01)	-	1.48	2.16	2.24	2.15	2.41

* Figures in parenthesis are arc sin transformed values. Average of three replications

Figure 16.2. Effect of *P. corylifolia* seeds methanol extract, bio-agents and fungicides on damping-off of tomato in pot culture treatment
T18 - Only soil. T17 - Control (pathogen inoculated). T14 - Metalaxyl 0.2%. T15 - Thiram 0.2%
T13- *P. fluorescens* (10 g/kg) + *T. harzianum* (4 g/kg) + *P corylifolia* seeds methanol extract 4%

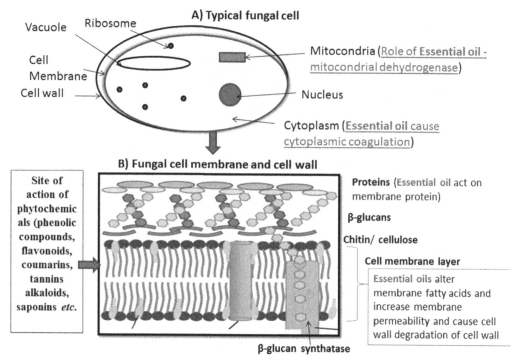

Figure 16.3. Diagrammatic representation of site and mechanism of action of antifungal phytochemicals on fungal cell

Color version at the end of the book

isopsoralen (2), bakuchiol (3), isobavachromene (4), 6-prenylnaringenin (5), corylin (6), bavachinin (7), and Δ^3,2-hydroxybakuchiol (8). Rajurkar (2011) found that *P. corylifolia* extract contains a number of bioactive compounds that are the molecular basis of its action, including flavonoids (neobavaisoflavone, isobavachalcone, bavachalcone, bavachinin, bavachin, corylin, corylifol, corylifolin and 6-prenylnaringenin, coumarins (psoralidin, psoralen, isopsoralen and angelicin) and meroterpenes (bakuchiol and 3-hydroxybakuchiol). Uikey et al. (2011) reported that *P. corylifolia* which includes bavachalcone, bavachromene, neoba-vaislfloavone, borachin, bavaislfavooz, psoralidin, corylifolinin, psoralenoside and isopsoralenoside have antifungal and antibacterial properties. Akhtar et al. (2012) carried out quantitative method using precoated silica gel-60 Lichrosphere high performance thin-layer chromatography (HPTLC) plates, automated band wise sample application and *n*-hexane:acetone:formic acid (2:1:0.025 *v/v/v*) as mobile phase, which has been developed and validated for the analysis of psoralen in marketed formulations and novel solid lipid nanoparticles (SLNs). Densitometric analysis was performed at 250 nm in absorbance mode. Compact bands of psoralen were obtained at Rf 0.32 ± 0.02. Lee et al. (2012) reported isolation of seven flavonoids from the methanol extracts of *P. corylifolia*. The structures of bakuchiol, bavachinin, neobavaisoflavone, corylifol A, corylin, isobavachalcon and bavachin were determined.

In another *in vitro* experiment of More et al. (2016), the methanolic extracts of *A. marmelos* were found effective against *P. debaryanum*. Preliminary phytochemical screening of the extracts revealed the presence of flavonoids, tannins, alkaloids, saponins and phenolic compounds. Petroleum ether: Ethyle acetate (2:1) solvent system gives maximum numbers of compounds from methanolic extract of *A. marmelos* and these compounds were separated on TLC plates. TLC resulted in identification of three spots found in the methanolic extract. The Rf values of methanolic extract of *A. marmelos* run under Petroleum ether:Ethyle acetate (02:01) solvent system were 0.04, 0.07, 0.14, 0.18, 0.25, 0.35, 0.45, 0.61, 0.70 and 0.84.

Conclusion

The genus *Pythium* is one of the largest oomycete genus and diseases caused by *Pythium* spp. in different crops are economically very important worldwide. Plants contain thousands of constituents and are valuable sources of new and biologically active molecules possessing antimicrobial property. At present, scientists are investigating for plant products of antimicrobial properties. It would be advantageous to standardize methods of extraction and *in vitro* antimicrobial efficacy testing so that the search for new biologically active plant products could be more systematic and interpretation of results would be facilitated. Above mentioned studies will help in producing natural pesticides that reduces the loss in yield caused by *Pythium* spp. in different crops. Moreover, these natural pesticides will help in reducing environmental and human health damages caused by chemical pesticides. Further extensive studies are needed to assess the effect of these phytochemicals under field conditions.

References

Akhtar, N., Faiyazuddin, Md., Mustafa, G., Sultana, Y., Baboota, S. and Ali, J. 2012. High-performance thin-layer chromatographic analysis of psoralen in marketed formulations and manufactured Solid Lipid Nanoparticles (SLNs): Validation of the method. Acta Chromatographica 24(4): 603-613.

Alhussaen, K., Hussein, E.I., Al-Batayneh, K.M., Al-Khatib, M., Al-Khateeb, W., Jacob, J.H., Shatnawi, M.A., Khashroum, A. and Hegazy, M.I. 2011. Identification and controlling *Pythium* sp. infecting tomato seedlings cultivated in Jordan Valley using garlic extract. Asian J. Pl. Path. 5(2): 84-92.

Ambikapathy, V., Gomathi, S. and Panneerselvam, A. 2011. Effect of antifungal activity of some medicinal plants against *Pythium debaryanum* (Hesse). Asian J. Pl. Sci. and Res. 1(3): 131-134.

Amini, M., Safaie, N., Salmani, M.J. and Shams-Bakhsh, M. 2012. Antifungal activity of three medicinal plant essential oils against some phytopathogenic fungi. Trakia J. Sci. 10(1): 1-8.

Birari, B.P., Gade, R.M. and Chuodhari, R. 2018. Antifungal efficacy of plant extracts, biocontrol agents against *Colletotrichum capsici* causing anthracnose of chilli. J. Pharma. Phytochem. 7(5): 1368-1373.

Cowan, M.M. 1999. Plant products as antimicrobial agents. Clin. Microbiol. Rev. 12: 564-582.

Das, K., Tiwari, R.K.S. and Shrivastava, D.K. 2010. Techniques for evaluation of medicinal plant products as antimicrobial agent: Current methods and future trends. J. Med. Plants. Res. 4: 104-111.

Eloff, J.N. 1998. A sensitive and quick micro plate method to determine the minimum inhibitory concentration of plant extracts for bacteria. Planta Medica 64: 711-713. doi: 10.1055/s-2006-957563

Fonesca, A.O., Pereiro, D.J., Jacob, R.G., Maia Filho, F.S., Oliverra, D.H., Maroneze, B.P., Valents, J.S., Osorio, L.G., Botton, S.A. and Meireles, M.C. 2015. *In vitro* susceptibility of Brazilian *Pythium insidiosum* isolates to essential oils of some Lamiaceae family species. Mycopathologia 179: 253-258.

Ganjewala, D., Sam, S. and Khan, K.H. 2009. Biochemical composition and antibacterial activities of *Lantana camara* plants with yellow, lavender, red and white flowers. Eur. Asia. J. Bio. Sci. 3: 69-77.

Gholve, V.M., Tatikundalwar, V.R., Suryawanshi, A.P. and Dey, U. 2014. Effect of fungicides, plant extracts/botanicals and bio-agents against damping off in brinjal. Afr. J. Microbiol. Res. 8(30): 2835-2848.

Gurjar, M., Ali, S., Akhtar, M. and Singh, K. 2012. Efficacy of plant extracts in plant disease management. Agricultural Sciences 3(3): 425-433.

Hanson, J.R. 2003. Natural products: The secondary metabolites. Royal Society of Chemistry, UK, pp. 1-33.

Jeyasakthy, S.J., Jeyadevan, P., Thavaranjit, A.C., Manoranjan, T., Srikaran, R. and Krishnapillai, N. 2013. Antifungal activity and qualitative phytochemical analysis of extracts obtained by sequential extraction of some medicinal plants in Jaffna peninsula. Archiv. of Appli. Sci. Res. 5(6): 214-221.

Juliani, H.R., Biurrum, F. and Koroch, A.R. 2002. Chemical constituents and antimicrobial activity of essential oil of *Lantana xenica*. Planta Medica. 68: 762-764.

Kim, Y.J., Lee, H., Park, E. and Shim, S.H. 2009. Inhibition of human 20S Proteasome by compounds from seeds of *Psoralea corylifolia*. Bull. Korean Chem. Soc. 30(80): 1867-1869.

Kumar, V.P., Neelam, S.C. and Harish, P. 2006. Search for antibacterial and antifungal agents from selected Indian medicinal plants. J. of Ethopharmacalogy 107: 182-188.

Lee, S.W., Yun, B.R., Kim, M.H., Park, C.S., Lee, W.S., Oh, W.S. and Rho, M.C. 2012. Phenolic compounds isolated from *Psoralea corylifolia* inhibit IL-6-induced STAT3 activation. Planta Med. 78: 903-906.

Mohamed, N.H. and El-Hadidy, A.M. 2008. Studies on biologically active constituents of *Verbascum eremobium* Murb. and its inducing resistance against some diseases of cucumber. Egypt. J. Phytopathol. 36: 133-150.

More, Y., Gade, R.M., Wavare, S.H. and Shitole, A.V. 2016. Phytochemical investigation and thin layer chromatography of *Aegle marmelos* leaves Methanolic extract. Adv. Life Sci. Res. 5(15): 5685-5690.

More, Y.D., Gade, R.M. and Shitole, A.V. 2017. Evaluation of antifungal activities of extracts of *Aegle marmelos, Syzygium cumini* and *Pongamia pinnata* against *Pythium debaryanum.* Indian J. Pharm. Sci. 79(3): 377-384.

Muthukumar, A., Eswaran, A., Nakkeeran, S. and Sangeetha, G. 2010. Efficacy of plant extracts and biocontrol agents against *Pythium aphanidermatum* inciting chilli damping-off. Crop Protect. 29: 1483-1488.

Nafiseh, N., Mehrded, L. and Ali, G. 2012. Accumulation of chromium and its effect on growth of *Allium cepa* cv. Hybrid. Eur. J. Exp. Biol. 2: 969-979.

Nagao, T., Abe, F. and Kinjo, J. 2002. Antiproliferative constituents in plants: Flavones from the leaves of *Lantana montevidensis* Briq and consideration of structural relationship. Biochem. Pharm. Bull. 25: 875-879.

Nain, J., Bhatt, S., Dhyani, S. and Joshi, N. 2013. Phytochemical screening of secondary metabolites of *Datura stramonium*. Int. J. Curr. Pharm. Res. 5(2): 151-153.

Pandey, A.K., Awasthi, L.P., Srivastva, J.P. and Sharma, N.K. 2010. Management of rhizome rot disease of ginger (*Zingiber officinale* rose L.). J. Phytol. 2(9): 18-20.

Pandey, V.N. and Dubey, N.K. 1994. Antifungal potential of leaves and essential oils from higher plants against soil phytopathogens. Soil Biol. Biochem. 26: 1417-1421.

Parveen, T. and Sharma, K. 2014. Management of "Soft Rot" of ginger by botanicals. Int. J. of Pharm. Life Sci. 5(4): 3478-3484.

Pattnaik, M.M., Kar, M. and Sahu, R.K. 2012. Bio-efficacy of some plant extracts on growth parameters and control of diseases in *Lycopersicum esculentum.* Asian. J. Plant Sci. Res. 2(2): 129-142.

Pente, R., Gade, R.M., Shitole, A.V. and Belkar, Y.K. 2015. Testing of efficacy of bio-agents and botanicals against *Phytophthora* root rot in Nagpur mandarin. *In*: Proceedings of National Conference on Harmony with Nature in Context of Bioresources and Environmental Health, The Ecosca, pp. 332-336.

Rajurkar, B.M. 2011. A comparative study on antimicrobial activity of *Clerodendrum infortunatum, Simarouba glauca* and *Psoralea corylifolia*. Int. J. of Res. and Reviews in Pharm. and Appli. Sci. 1(4): 278-282.

Shitole, A.V. 2018. Detection and antifungal potency of weed extract against phytopathogenic fungi. PhD. (Agri.) Thesis (unpub.), Dr. PDKV Akola (M.S) 444104, India.

Soni, P., Siddiqui, A.A., Dwivedi, J. and Soni, V. 2012. Pharmacological properties of *Datura stramonium* L. as a potential medicinal tree. Asian Pac. J. Trop. Biomed. 2(12): 1002-1008.

Suleiman, M.N. and Emua, S.A. 2009. Efficacy of four plant extract in the control of root rot disease of cow pea (*Vigna unguiculata* Linn. Walp). Afr. J. Biotechnol. 8: 3806-3808.

Tzortzakis, N.G. and Economakis, C.D. 2007. Antifungal activity of lemongrass (*Cympopogon citratus* L.) essential oil against key postharvest pathogens. Innov. Food Sci. Emerging Tech. 8: 253-258.

Uikey, S.K., Yadav, A.S., Sharma, A.K., Rai, A.K., Raghuwanshi, D.K. and Badkhane, Y. 2011. The botany, chemistry, pharmacological and therapeutic application of *Psoralea corylifolia* L. – A review. Int. J. Phytomedicine. 2: 100-107.

Van Der Plaats-Niterink, A.J. 1981. Monograph of the Genus *Pythium*. Studies in Mycology No. 21. Baarn, Netherlands: Central Bureau Voor Schimmelcultures

Wavare, S.H., Gade, R.M., Belkar, Y.K., Vyavhare, G.F. and Gawande, A.D. 2015. Compatibility of Marigold Water Extract with Biocontrol Agents. Trends Biosci. 8(15): 4001-4006.

Wavare, S.H., Gade, R.M., Shitole, A.V. and Ingole, M.N. 2016. Evaluation of floral extracts, biocontrol agents and fungicides for management of damping off of tomato caused *Pythium debaryanum*. The Ecoscan, Special issue, 9: 789-795.

Wavare, S.H., Gade, R.M. and Shitole, A.V. 2017. Effect of plant extracts, bio-agents and fungicides against *Sclerotium rolfsii* causing collar rot in chickpea. Indian J. Pharm. Sci. 79(4): 513-520.

Yazdani, D., Tan, Y.H., Abidin, M.A. and Jaganath, I.B. 2011. A review on bioactive compounds isolated from plants against plant pathogenic fungi. J. Med. Plants Res. 5: 6584-6589.

17

Pythium aphanidermatum and Its Control Measures

Tahira Parveen*, **Mukesh Meena, Tripta Jain, Kavita Rathore, Surbhi Mehta** and **Kanika Sharma**
Department of Botany, Mohanlal Sukhadia University, Udaipur - 313001, Rajasthan, India

Introduction

The species of *Pythium* that resemble with fungi are generally known as water molds. They belong to Straminopila kingdom, Oomycota as phylum, Oomycetes the class and Peronosporomycetidae subclass, Pythiales and Pythiaceae are order and family, respectively. They are worldwide in distribution. *Pythium* is one of the largest genus of Oomycetes, and its species are isolated from various places of the world (Dick 1990, de Cock and Lévesque 2004, Paul et al. 2006, Bala et al. 2010, Robideau et al. 2011). *Pythium* is a group of unicellular protists which are filamentous and physically resemble fungi. However, *Pythium* is different from real fungi in having cellulosic cell wall instead of chitin and aseptate mycelia. In *Pythium*, glycogen, the intracellular storage polysaccharide is replaced by $(1,3)$-β-glucans, which is a major difference compared with true fungi. These reserve carbohydrate are known as 'mycolaminarin' a soluble form of $(1,3)$-β-glucan and 'cellulin' an insoluble granular inclusion (Hospenthal et al. 2011). In its vegetative/asexual stage, they have diploid nuclei unlike haploid nuclei in most fungi.

Pythium can survive, grow and germinate in very broad habitats ranging from aquatic or terrestrial environments and in different types of cultivated or fallow soils, in almost every kind of plant (aquatic/terrestrial), even associated with animals as well as humans. Different types of diseases caused by *Pythium* like rots, pre-emergence and post-emergence damping-off in vegetable crops are economically very important worldwide (Whipps and Lumsden 1991). *Pythium aphanidermatum* also causes root rot on common ice plant worldwide where no other *Pythium* spp. have been recorded previously from this plant (You et al. 2015). *P. aphanidermatum* is also reported to cause infections in human (Hospenthal et al. 2011). It is generally recognized that species of *Pythium* are not host specific; it can parasitize and cause diseases in the crop plants and finally damage them (Hendrix and Campbell 1973, Krober 1985, Kucharek and Mitchell 2000).

The diseases which result from *Pythium* infection act as an important limiting factor for cultivation of vegetables and crop plants because it totally deteriorates the plants, resulting in no yield; in absence of host plant, it acts as saprophyte on decaying plant debris and remains dormant in soil as saprophyte and also by the formation of oospores (Elnaghy et al. 2009, Al- Sheikh and Abdelzaher 2010). They germinate again when there is favorable condition for them such as humidity, temperature and presence of host. The other peculiar and aggressive nature of *Pythium* can be understood by the 'biological threshold of the pathogen' – it is the amount of inoculum that must be present in the growing system which results in subsequent disease development. Biological

*Corresponding author: parveentahira06@gmail.com

threshold of *P. aphanidermatum* is present in different forms, like zoospores, conidia, mycelium, resting spores etc. In case of *P. aphanidermatum,* the concentrations of biological threshold is smaller than one zoospore per milliliter of irrigation water which results in disease in crop plants (Postma et al. 2001). For other pathogens, higher concentrations are needed even for symptom development, which clearly shows the aggressive nature of *P. aphanidermatum.*

Rapid germination of sporangia of *Pythium* after exposure to exudates or volatiles from seeds or roots (Omidbeygi et al. 2007) followed by quick infection has made management and control of *Pythium* very difficult (Bala et al. 2010). *Pythium* species are liable for losses of multibillion dollar worldwide (Van West et al. 2003).

Effective chemical control of this pathogen can be achieved through preventative fungicide applications but as it is not a true fungus, hence only few fungicides are active against it. Moreover, it does not contain ergosterol in its cytoplasmic membrane and as a result does not show inhibition to most of the antifungal agents which shows activity against ergosterol (Hospenthal et al. 2011). On the other hand, the repeated use of selective fungicides which are used to manage this non host-specific pathogen with a single mode of action pathway may result in selection of fungicide resistant strains and hence reduction or loss of efficacy occurs.

It has been reported that mefenoxam and promocarb are commonly used fungicides against *P. aphanidermatum*, and hence the fungus has developed resistance (Goldberg and Stanghellini 1990). Fungicide rotation may be a basic recommendation for prevention of the build-up of resistance in strains. Hence, due to different chemical constituents of this non-host specific pathogen, divergent from real fungi, the most effective and general method to control *P. aphanidermatum* is also at the boundary in order to control the diseases originated by it.

The aim of the chapter is to provide information about the nonspecific host range, virulent nature and various control measures of *P. aphanidermatum*. The described control measures are used by various workers and some are used experimentally in *in vivo* studies by researchers. The aim also includes spreading of information to the persons of various fields so that they can explore the findings and use it further.

Ecology

P. aphanidermatum remains present in the plant throughout the production season and can be found on cuttings and mature plants as well as on debris of plant remaining (Moorman et al. 2002, Hong and Moorman 2005). It favors warm weather and moist conditions, temp - 30-40°C is the optimum range (Littrell and McCarter 1970). Under the favorable conditions, oospores of *P. aphanidermatum* germinate and give rise to sporangia, which develop a discharge tube through which cytoplasm flows and forms a membrane-bound vesicle at the hyphal tip. Inside this vesicle cytoplasm differentiates into zoospores, which get free after maturity by disruption of the membrane (Schroeder et al. 2013). Zoospores are attracted chemotactically to ooze out of root and move by flagella through soil and surface water. Zoospores get encysted when they come in touch/association with root tissues and then they form cell walls and start infection. The pathogen secreting cell-wall-degrading enzymes enter root epidermis and kill the tissues (Schroeder et al. 2013). After colonization, *P. aphanidermatum* produces sporangia, oogonia and antheridia. Oospores serve as the supreme survival structure which are produced through fertilization of oogonia and antheridia.

Inoculum of *P. aphanidermatum* enters in the production system through contaminated water, potting media, infected seeds/propagules as well as infected cuttings (Sanogo and Moorman 1993, Moorman et al. 2002, Hong and Moorman 2005).

Isolation of *Pythium aphanidermatum*

P. aphanidermatum can be isolated by the following methods.

Isolation from infected soil by using baits: In this method, propagules of fungi can be isolated from infected soil using pea seeds as baits at 25±1°C, according to the procedure suggested by Sinclair and Dhingra (1995).

Isolation from infected soil can also be made by Serial Dilution Method. In this method, infected soil (10 gm) is suspended in 100 ml sterile distilled water; after shaking thoroughly, the soil solution is used to isolate *P. aphanidermatum* according to the process suggested by Amoo et al. (2007).

Isolation from infected samples can be made by putting the pieces of infected sample (after surface sterilization) on to different selective media by following the method suggested by Eckert and Tsao (1962), and Jeffers and Martin (1986).

In all the above three processes, the normal as well as selective media is used and according to Dick (1990), *Pythium* species grow much faster than other fungi associated with it; hence actively growing hyphal tips from periphery can be transferred to new Petri plates to get pure culture.

Morphology

As *Pythium* grows faster, it has 30-30.8 mm growth rate/day in a laboratory (favorable) condition. *Pythium* species are differentiated based on morphology as well as on the basis of its heterothallic or homothallic nature. Traditionally, *P. aphanidermatum* is identified on the morphological as well as physiological basis where all the aspects like daily growth parameter and colony morphology in different general as well as specific medium, asexual and sexual structures are taken into contemplation. *P. aphanidermatum* is homothallic in nature and hence on this basis the sexual structures oogonia and antheridia can be studied easily at the same time from a single sample (Fig. 17.1)

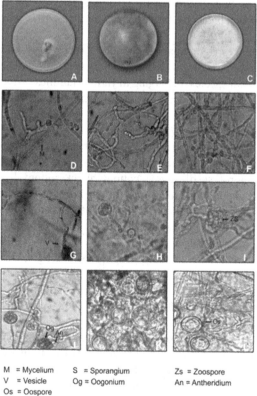

M = Mycelium   S = Sporangium   Zs = Zoospore
V = Vesicle   Og = Oogonium   An = Antheridium
Os = Oospore

Figure 17.1. Microscopic images of *P. aphanidermatum*

Color version at the end of the book

Sexual structures: As *P. aphanidermatum* is homothallic, producing both male and female gametangia, oogonia are globose, smooth, terminal, and size in range of 20 to 25 µm in diameter. Antheridia are mostly intercalary in position, sometimes they may be terminal, broadly sac-shaped, 10 to 14 µm × 10 to 14 µm. Oospores are produced after fertilization of 1 to 2 monoclinous antheridia per oogonium. Oospores produced after fertilization is aplerotic in nature, 18 to 22 µm (average 20.5 µm) in diameter, with a 1 to 2 µm thick wall (Plaats-Niterink, 1981). Sporangia give rise to biflagellate zoospores asexually, with tinsel and whiplash flagella and can swim to access the host for infection. Encysted zoospores measure approximately 12 µm in diameter; zoospores are formed at 25-30°C. *P. aphanidermatum* can also withstand at high temperature at 30°C and can also flourish well at 40°C.

Colony characteristics

Colonies produced by *P. aphanidermatum* on general and specific media do not show any special characteristics and in all media, almost same appearance with white cottony aerial mycelium without a special pattern is reported. CMA (Corn Meal Agar) and WA (Water Agar) produced no heavy aerial mycelium. On PDA (Potato Dextrose Agar) heavy aerial mycelia are generally formed, while on PCA (Potato Carrot Agar) loose aerial mycelia are produced.

Mode of spread, host range and types of disorders caused by P. aphanidermatum

P. aphanidermatum spreads very easily at any stage of crop whether it is seed/seedling/plantlet/mature plant/harvested crop etc. It can even infect plant during transport as well as during storage. The propagules for infection includes mycelium, oospores and sporangium present in organic debris etc. As already mentioned above, potting media contaminated water as well as a small amount of infected plant propagative materials such as seeds also act as infecting materials. *P. aphanidermatum* can also be spread by fungus gnats and shoreflies (Goldberg and Stanghellini 1990, Hyder et al. 2009).

P. aphanidermatum reduces vigor, quality and yield of crops, often killing a large percentage of plants affected. It is an aggressive pathogen at high temperature (Gold and Stanghellini 1985), known to infect a wide variety of plant species, belonging to various families viz. *Araceae, Amaranthaceae, Amaryllidaceae, Bromeliaceae, Cactaceae, Chenopodiaceae, Basellaceae, Compositae, Coniferae, Convolvulaceae, Cruciferae, Cucurbitaceae, Gramineae, Leguminosae, Euphorbiaceae, Moraceae, Linaceae, Malvaceae, Solanaceae, Passifloraceae, Rosaceae, Violaceae, Umbelliferae, Vitaceae* and *Zingiberaceae* (Waterhouse and Waterston 1964). The diseases caused by *P. aphanidermatum* differs with the different host plants. It is the causal agent of many diseases pre-emergence and post emergence of seedlings, root rot, seedling rots, cottony blight, cottony-leak, stalk rot etc. Root rot and damping-off caused by the *P. aphanidermatum* are thought to be the most destructive diseases of the crops grown in greenhouses. A notable number of studies on *P. aphanidermatum* have been done in disparate parts of the world (Plaats-Niterink 1981). It infects nearly all crops grown in every part of the world (Ben Yephet and Nelson 1999).

The great losses in more or less all types of crops belonging to different families has been reported from different parts of the world like Australia, Japan, China, Argentina, India, Oman, Spain, Honduras, USA etc. (Bellgard and Ham 2004, Vanker and Patel 2004, Kavitha and Thomas 2007, Aoki et al. 2007, Deadman et al. 2007, Rosso et al. 2008, Cara et al. 2008, Juan et al. 2009, Uma et al. 2012).

The disorders caused by *P. aphanidermatum* has a long list which includes rots like foot rot, root rot, stem and leaf rot, crown rot of melon, rhizome rot etc. It also causes damping-off (pre-emergence and post-emergence), damping-off of cabbage (Kubota 2010), coriander (Awasthi et al. 2017) and alfalfa seed rot seedling rot besides damping-off, cottony leak and crown necrosis of adult bean plants (Kawarazaki et al. 2008), vascular wilt (groundnuts) (Bellgard and Ham 2004), soft rot of rhizomes of ginger in India (Sagar et al. 2008), and watermelon "sudden death" (Guirado et al. 2009).

It is also associated with crown rot of industrial hemp (Schoener et al. 2017). *P. aphanidermatum* was also found to be associated with the rot of *Cannabis sativa* grown hydroponically (Punja and Rodriguez 2018).

Control measures

There are various agricultural and horticultural practices which are followed by workers, farmers and researchers to control diseases caused by this aggressive pathogen, such as use of resistant cultivars, crop rotation and planting disease-free seeds but the pathogen is non-specific in host range and can live well in every condition in the form of oospores and sporangia for a very long period of time; so these implementations are found to be less significant. Following practices are used to control and prevent the diseases caused by *P. aphanidermatum.*

Cultural practices

Proper sanitation and use of sterilized pots, media and plant material can greatly reduce and help to prevent the risk of infection by *P. aphanidermatum*. Proper distance between plants is necessary so as to circulate the air between plants properly, as proper air circulation can also make the condition unfavorable for disease development. Stagnant water and poor developed drainage system also contributes in dispersal of propagules; on the other hand, availability of water favors the zoospores to move to the host plant for infection.

Good drainage system and optimum irrigation are two important factors to control disease development, as a well drained soil will limit the dispersal of propagules, particularly zoospores. Besides this, growing plants at optimum temperature and pH and use of disease-free plants are also helpful in decreasing the disease and infection caused by *P. aphanidermatum*.

For management of diseases caused by *P. aphanidermatum,* chemical fungicide is most commonly used by the growers. In lawns and gardens, the fungicides like Fore, Subdue Maxx and Alliette are effective in controlling *P. aphanidermatum*. Some other fungicides like Cymoxanil and Metalaxyl are also used against it (Zagade et al. 2012). *Pythium* rot includes the use of mefenoxam, promocarb, cyazofamid, and etridiazole. For *Pythium* blight metalaxyl, fosetyl-Al, Pyraclostrobin and Azoxystrobin etc. are used. Mancozeb, Carbendazim, Ridomil and Topsin etc. have been recommended against *Pythium* (Sheldon 2003, Anomynous 2005, ANSAB 2011, Poudyal et al. 2011). Such synthetic/chemical fungicides bring about the reticence of pathogens either by destroying their cell membrane or changing the permeability or by decreasing metabolic processes of the pathogens and hence are extremely effective (Osburn et al. 1989). But in case of diseases caused by *P. aphanidermatum,* if the disease once enters into the crop field or greenhouse, it is practically impossible to control it properly by using only chemical fungicides. Ridomil Gold and Subdue are also registered fungicides against *P. aphanidermatum*. Lower concentration of fungicides are used for preventive control and higher rates are used for curative purpose.

The use of chemicals to defend the disease is also censured throughout the world due to its deleterious effects on environment. They enter through the food chain and cause several detrimental effects not only to organisms but also to non-living components of ecosystem and also contribute to significant decrease in populations of useful soil microorganisms, soil acidification and compaction, thatch accumulation, and diminished resistance to diseases (Shiva et al. 2004). Moreover, inappropriate, fanatical and indiscriminate use of most of the synthetic fungicides not only has adverse effects on ecosystems, but it also creates a possible carcinogenic risk and toxicological problems higher than that of insecticides and herbicides put together (Cameron et al.1984, Osburn et al.1989, Manoranjitham et al. 2000).

The range of effective registered chemical fungicides currently available to growers for managing *Pythium* diseases is narrow. Several of these fungicides are quite effective against *Pythium* diseases, which involve single mode of action and hence single mode of action can be defeated quickly by the built-out resistance in *Pythium* population. The sleek list of presently available fungicides along with the potential for fast development of fungicide resistance and public concern for the sake of human and environmental health has made the disease management essential for growers.

Moreover, resistance by pathogens to fungicides has rendered certain fungicides ineffective (Ma and Michailides 2005). Development of fungicide resistance by *P. aphanidermatum* further discourages its use for disease control (Whipps and Lumsden 1991). Some populations of *Pythium* are reported to have resistance to metalaxyl, mefenoxam and/or propamocarb. Certain fungicides, usually systemic fungicides, are said to be 'at risk' to the development of resistance if they are used repeatedly (Bardin et al. 2004). It is advocated that chemicals at high risk be used sparsely and in rotation or mixed with chemicals with some other modes of actions.

Besides all these, Folman and his coworkers (2004) reported that chemical fight against this aggressive pathogen is not significant. Hence, it is essential to search for an ecologically safe and economically viable strategy for the control of diseases and to minimize the dependence on the chemical/synthetic fungicides. Many researches have conducted *in vitro* and *in vivo* studies to control *Pythium* diseases and several chemicals have been used to control it but no concrete solution has been found and not even a single chemical has been identified yet to control *P. aphanidermatum* (Poudyal et al. 2011). In spite of devastating effect of chemical fungicides, chemical fight (Folmen et al. 2004) and physical control (Benhamou et al. 1997) of this fungal pathogen are very difficult. Besides these practices, an important physical method, i.e. soil solarization has also been used to control the diseases caused by *P. aphanidermatum.*

Physical control

Soil solarization is also called plasticulture; it is an eco-friendly method for soil disinfestations, especially to control plant pathogens which are soil-borne like *P. aphanidermatum.* It is a process in which the soil is covered by plastic covering/film and hence heated by exposure to sunlight during the warm months. It is a type of hydrothermal process that occurs in moist soil. Due to high temperature the chemical, physical, and biological properties of the soil gets changed and helps to improve the soil health. It is an alternate to soil fumigant agricultural chemicals that have high environmental risk, a negative impact on useful soil microorganisms, and that are not user friendly (Pokharel 2010).

The positive impact of soil solarization against *Pythium* has been reported by many workers. The researchers have also reported that soil solarization is helpful in reducing the diseases caused by *Pythium* and gave significant results against it (Usman et al.1996, Katan 2000). Mathur also used the soil solarization method against *Pythium* to control rot of ginger in Rajasthan college of Agriculture, Udaipur. They used thin transparent sheets/polythene film to cover the soil and found significant reduction of *Pythium* (Mathur et al. 2002). Some of the researchers also reported complete elimination of *P. aphanidermatum* in carrot, watermelon and potato after field soil solarization. Christensen and Thinggaard (1999) reported decrease in *Pythium* root rot in greenhouse by summer soil solarization even in Denmark in a temperate climate. Reddy et al. (2001) reported 91.2% inhibition of *P. aphanidermatum* after soil solarization. Pandey and Pandey (2002) reported 32.4 to 95.6% damping-off reduction in chili and tomato.

As *P. aphanidermatum* is soil borne in nature, it is found to be strongly subdued by the solarizing treatment. Several studies documented an almost complete eradication of *Pythium* spp., and the diseases caused by it in many vegetable and fruit crops. The soil solarization has a positive impact on soil fertility and microflora, hence it helps the agricultural and farming community to manage the disease effectively and also enhance the yield level (Jacobson et al. 1980). Deadman et al. (2006) reported that soil solarization is helpful in decreasing the *Pythium* infection and also enhances the vegetative growth of crop. Soil solarization is a perfect solution for all situations where the use of chemical fertilizers is restricted. Advantages of solarization also include economic convenience, adaptability to many cropping systems as well as ease of use by growers. It also shows full integration with other control tools. Solarization has been reported to improve soil structure and enhance soil content of soluble nutrients such as inorganic nitrogen forms, dissolved organic matter and cations present in the soil.

Organic measure

Organic materials are also used widely to control and manage diseases caused by *Pythium* by many researchers (Shuler et al. 1989, Theodore and Toribio 1995). Organic materials suppress the diseases and increases the quantity of associated useful microorganisms (Ringer et al.1997, Erhart and Burian 1997). Many researchers also reported that poultry slurry, municipal bio-solids and sugarcane residues also gave good suppressing activity against *P. aphanidermatum*. Organic materials actuate a chain of chemical and microbial degradation; as a result, toxicity of soil pathogen ceases (Mandelbaum and Hadar 1990, Teodore and Toribio 1995). On the other hand, the addition of organic manure to the soil enhances the nutrients and fill up the deficiencies of other nutritive materials which hence increases the crop yield (Aharonson and Katan 1991, Gamliel et al. 2000).

As described above, soil solarization is a good alternative to chemical fungicides against *P. aphanidermatum* but if soil solarization, along with other organic materials, is applied against this aggressive pathogen, the results are found to be significantly enhanced, as compared to individual treatment of either soil solarization or use of organic materials. Satya (Satya et al. 2005) reported that combination of both of the control measures over only one gave more significant results. They used manure of composted chicken alone and in combination with heat, and found significant reduction of *Pythium* spp. and almost complete eradication of *Pythium* population from the soil, respectively. More significant results of combined method are seen because the organic soil amendments protect the biomass of microorganisms and enzymatic activities of these biomass from the detrimental effect of heating caused by soil solarization method (Pokharel 2010).

Use of biocontrol agents

The other significant process which is used against the *P. aphanidermatum* besides soil solarization and organic amendment is the use of biocontrol agents. Among the mechanisms that biocontrol agents use to manage *Pythium* diseases are production of antagonistic metabolites, contention for space and nutrients, mycoparasitism, hyphal interactions, enzymes secretion and actually feeding on *Pythium* propagules. Elad et al. (1982) reported that *Trichoderma harzianum* was a significant antagonist against *Pythium* species. Lumsden and Locke (1989) claimed that *Gliocladium virens* can significantly control the damping-off of zinnia, cotton, and cabbage caused by *Pythium*. The studies revealed that the *T. harzianum* was effective in reducing disease incidence and increasing crop germination (Shanmughan et al. 1999). Manasmohandas and Sivaprakasam (1994) and Manorajitham and Prakasam (2000) reported the antagonistic activity of *T. hamatum, T. harzianum, T. viridae, Psudomonas fluorescens* and *T. reesai,* against *P. aphanidermatum*. Bardin et al. (2004) reported that *Pseudomonas fluorescens* was significant in controlling *Pythium* damping-off of sugar beet. Shweta et al. (2014) also reported that *T. harzianum, T. viride* and *Pseudomonas fluorescence* were found to be antagonistic against *Pythium*. Talc based formulation of antagonist in reducing pre- and post-emergence damping-off of chili caused by *P. aphanidermatum* was reported by Haritha et al. (2010). Kamala and Indira (2011) also reported that 32% out of 110 *Trichoderma* isolates were found to be strongly antagonistic against *Pythium* species. *T. hamatum* and *T. harzianum* were reported to show excellent inhibition against *P. aphanidermatum* in chili. Biocontrol agents work best when pathogen pressure is low to moderate because their activities against pathogens are biological by nature. It is possible that they will not be much significant when overwhelmed by high levels of pathogen. In addition, biocontrol agents are generally not effective if once the plants have been infected by *P. aphanidermatum* and thus should not be considered as curative control treatment (Bardin et al. 2004).

Recent trends indulge the use of optional/alternative substances derived from plant extracts. Natural plant products are significant resources of new agrochemicals for the control of plant diseases (Matsuzaki et al. 1998, Kagale et al. 2004). Their use in controlling diseases is considered as an important alternative to chemical fungicides due to their negative impact on the environment, as they do not leave any toxic residues and therefore can effectively replace synthetic fungicides (Cao and Forrer 2001). These natural products of plant origin can act either as leads for chemical synthesis of novel agrochemicals, or as commercial products in their own right, or as a source of innovation to biochemists for the build-out of new bioassays capable of detecting other, structurally simpler compounds with a similar mode of action. The use of plants may offer a new source of antimicrobial agents with significant activity (Munoz-Mingaro et al. 2003, Coelho de Souza et al. 2004).

In modern times, focus on plant research has increased all over the world and a large body of corroboration has been collected to show great potential of plants used in various traditional systems. More than 13,000 plants have been studied during the last five year period. Dahanukar and his coworkers (2000) have reviewed the research on plant based antifungal compounds as a scientific approach and innovative scientific tool from 1994-1998. Antifungal activity of plant extracts against a broad range of fungi has also been reported by many workers (Kurita et al. 1981, Grane and Ahmed 1988, Wilson et al. 1997, Cowan 1999, Abd-Alla et al. 2001). Researchers have suggested that antifungal components of the plant extracts cross the cell membrane interacting with the enzymes and proteins of the membrane, hence producing a flux of protons towards the cell exterior which induces change in the cell and ultimately their death (Noelting and Sandoval 2007, Pane et al. 2011). Other researchers reported the inhibitory effect of these plant extracts to their hydrophobicity characters and their components. This enables them to partition in the lipids of the fungal cell wall membrane and mitochondria disturbing their structure and rendering them more permeable, hence leaking of ions and other cell contents can then occur causing cell death (Burt 2004).

The aqueous extract of *Zygophyllum fabago*, and ethanolic extracts of *Allium sativum*, *Azadirachta indica* and *Curcuma longa* were shown to inhibit mycelial growth of *P. aphanidermatum*, in *in vitro* condition. Ramanathan et al. (2004) reported inhibitory effect of spider lily (*Crinum asiaticum*) against *P. aphinidermatum*. Vanker and Patel (2004) demonstrated the efficacy of leaf extracts of *Lawsonia inermis* and *Emblica officinalis* against damping-off of *Pythium* species. Parveen and Sharma (2014) also reported the antifungal activity of crude aqueous, 50% hydroalcoholic and alcoholic leaf extracts of *Azadiracta indica*, *Aegle marmelos*, *Cassia fistula*, *Clitoria ternatae*, *Delonix regia*, *Eucalyptus globules*, *Jacarandas mimosifolia*, *Justicia gendarussa*, *Moringa oleifera*, *Murraya koenigii*, *Nigella sativa*, *Pongamia pinnata*, *Polyalthia longifolia*, *Tecomella undulatae* and *Terminalia arjuna* against the *P. aphanidermatum*. Among the above screened plants, all of them were showing inhibitory activity except *Pithecelobium dulce* and *Pongamia pinnata*. The maximum inhibition of fungal growth was recorded with 50% hydro-alcohol extracts of *Jacaranda mimosifolia* and *Moringa oleifera*.

Goel et al. (2011) reported the antifungal activity of hexane, ethyl acetate and methanol extracts of *Parmelia reticulata* against *P. aphanidermatum*. Haouala et al. (2008) and Suleiman and Emua (2006) reported that an aqueous extract of Fenugreek and Aloe could inhibit mycelial growth of *P. aphanidermatum*. Sagar et al. (2007) and Dileep et al. (2006) showed the fungitoxic efficacy of *Ferula asafeotida* and *Azadirachta indica* plant extracts and ripe and unripe pericarp extracts of *Polyalthia longifolia* against *P. aphanidematum*. Al-Rahmah et al. (2013) reported that methanolic extracts of *Thymus vulgaris*, *Zingiber officinale*, *Salvadora persica* and *Lantana camara* showed significant antifungal activity against *P. aphanidermatum*. Madduleti and Moses (1995) also reported inhibition of *P. aphanidermatum* damping-off in brinjal, chili and tomato, by using neem cake, neem leaf and eucalyptus leaf amended with soil.

Plants have generated the interest of man for therapeutic values chiefly because of the presence of various secondary metabolites. Plants produce a wide variety of bioactive secondary metabolites which serve as plants' defense mechanisms against pests. Some secondary metabolites give plants their odors (terpenoides), some are liable for plant pigments (quinines and tannins) and others (some of terpenoids) are accountable for plant flavor. The antimicrobial properties of plant extracts are also because of presence of secondary metabolites such as alkaloids, phenols, flavonoids, terpenoids, essential oils etc. (Haritha et al. 2010).

In another screening, antifungal effects of 66 medicinal plants belonging to 41 families were evaluated against *P. aphanidermatum*, the causal agent of chili damping-off (Muthukumar et al. 2010). Wang et al. (2005) reported a chitinase with antifungal activity isolated from *Phaseolus mungo* seeds. Chitinase is also reported to exert antifungal activity against *P. aphanidermatum*, besides other phytopathogenic fungi.

Conclusions

The present scenario of research suggests that there are alternatives to replace the chemical/synthetic fungicides for management of this notorious soil- as well as seed-borne pathogen, *P. aphanidermatum*, which causes diminution of millions of dollars. However, farmers use the common synthetic fungicides which leads to ill effects as well as many of the commonly used synthetic fungicides are unable to control *Pythium* species due to development of resistance against these synthetic fungicides. Hence, there is a need to replace the chemical fungicides by bio-fungicides, prepared from plant extracts and antagonistic microorganisms. Bio-fungicides will also be economical to the farmers. Besides this, the use of bio-fungicides will not leave any harmful effect in water, soil as well as in the environment. It is possible that by combining these approaches (use of plant extracts, antagonistic micro-organisms, organic manure), an economically viable alternative for crop production system can be developed. The use of bio-fungicides has been proved to be economical alternative that can be implemented at the farm level. For the effective production of crops, formulation protocols as well as its using methods should be provided to the farmers. Formulation must have adequate shelf life, stability, and titer. Before any formulated product is marketed, it must first be thoroughly tested by growers, whose comments, critiques, and suggestions for improvement must be considered. If the bioformulations and antagonistic microorganisms are used simply by farmers and growers, then the risk of toxicity and biomagnifications of the toxins produced by these fertilizers could be managed. On the other hand, these bioformulations and chemical free methods will be able to actually enhance not only the yield but also the quality of crop.

References

Abd-Alla, M.S., Atalia, K.M. and El-Sawi, M.A.M. 2001. Effect of some plant waste extracts on growth and aflatoxin production by *Aspergillus flavus*. Annals Agri. Sci. 46: 579-592.

Abdel-Sayed, P., Kaeppeli, A., Siriwardena, T., Darbre, T., Perron, K., Jafari, P., Reymond, J.L., Pioletti, D.P. and Applegate, L.A. 2016. Antimicrobial dendrimers against multidrug-resistant *P. aeruginosa* enhance the angiogenic effect of biological burn-wound bandages. Sci. Rep. 6: 22020.DOI: 10.1038/srep22020

Aharonson, N. and Katan, J. 1991. Pesticide behavior in solarized and disinfected soils. pp. 131-138. In: Katan, J. and De Vay, J.E. (eds.). Soil Solarization. CRC Press, Boca Raton, FL.

Al-Rahmah, A.N., Mostafa, A.A., Abdel-Megeed, A., Yakout, S.M. and Hussein, S.A. 2013. Fungicidal activities of certain methanolic plant extracts against tomato phytopathogenic fungi. Afr. J. Microbiol. Res. 7(6): 517-524.

Amoo, B.G., Balogun, O.S., Oyedunmade, E.E.A. and Fawole, O.B. 2007. Survey of Rhizospheric Mycoflora associated with Sorghum in Kwara State. Proceedings of 5th International Conference Nigerian Society for Experimental Biology (NISEB), 11-12, 2007.

Anonymous. 2005. Experiences in Collaboration – Ginger Pests and Diseases. Indo-Swiss Project Sikkim. Intercooperation India Programme Series 1, Intercooperation Delegation, Hyderabad, India, 57 pp.

ANSAB (Asia Network for Sustainable Agriculture and Bioresources). 2011. A report on value chain analysis of Ginger in Nepal. Report Submitted to The Netherland Development Organization (SNV), Nepal. (Unpublished material)

Aoki, K., Tojo, M., Watanabe, K., Uzuhashi, S. and Kakishima, M. 2007. Cottony leak of scarlet runner bean caused by *Pythium aphanidermatum*. J. General Plant Pathology 73(6): 408-410.

Ashwathi, S., Ushamalini, C., Parthasarathy, S. and Nakkeera, S. 2017. Morphological, pathogenic and molecular characterisation of *Pythium aphanidermatum*: A causal pathogen of coriander damping-off in India. Pharma. Innovation J. 6(11): 44-48.

Bala, K., Robideau, G., de Cock, A.W.A.M., Abad, Z.G., Lodhi, A.M., Shahzad, S., Ghaffar, A., Coffey, M.D. and Lévesque, C.A. 2010. *Phytopythium* gen. nov. Persoonia 24: 136-137.

Bardin, S.D., Huang, H.C. and Moyer, J.R. 2004. Control of *Pythium* damping-off of sugar beet by seed treatment with crop straw powders and a biocontrol agent. Biol. Control 29(3): 453-460.

Bellgard, S. and Ham, C. 2004. Common diseases of peanuts in the top end of the NT *Arachis hypogaea* L. Agdex No: 141/633. Agnote Northern Territory of Australia (162): 7.

Ben Yephet, Y. and Nelson, E.B. 1999. Differential suppression of damping-off caused by *Pythium aphanidermatum, P. irregulare* and *P. myriotylum* in composts at different temperatures. Plant Dis. 83: 356-360.

Benhamou, N., Rey, P., Cherif, M., Hockenhull, J. and Tirilly, Y. 1997. Treatment with the mycoparasite *Pythium oligandrum* tiggers induction of defence-related reactions in tomato roots when challenged with *Fusarium oxysporum* f.sp. radicis-lycopersici. Phytopathol 87: 108-121.

Boehm, M.J. and Hoitink, H.A.J. 1992. Sustenance of microbial activity in potting mixes and its impact on severity of *Pythium* root rot of poinsettia. Phytopathol. 82: 259-264.

Burt, S. 2004. Essential oils: Their antimicrobial properties and potential applications in foods – A review. Int. J. Food Microbiol. 94: 223-253.

Cameron, H.J. and Julian, G.R. 1984. The effects of four commonly used fungicides on the growth of Cyanobacteria. Plant Soil 78: 409-415.

Cao, K.Q. and Forrer, H.R. 2001. Current status and prosperity on biological control of potato late blight. J. Agric. Univ. Hebei 24: 51-58.

Cara, M. de, López, V., Santos, M. and Tello Marquina, J.C. 2008. Association of *Pythium aphanidermatum* with root and crown rot of melons in Honduras. J. Plant Dis. 92(3): 483.

Christensen, L.K. and Thinggaard, K. 1999. Solarization of green house soil for prevention of *Pythium* root rot in organically grown cucumber. J. Plant Pathol. 81(2): 137-144.

Coelho de Souza, G., Has, A.P.S., Von Poser, G.L., Schapoval, E.E.S. and Elisabetski, E. 2004. Ethnopharmacological studies of antimicrobial remedies in the South of Brazil. J. Ethnopharmacol. 90(1): 135-143.

Cowan, M.M. 1999. Plant products as antimicrobial agents. Cli. Microbiol. Rev. 12(4): 564-582.

Dahanukar, S.A., Kulkarni, R.A. and Rege, N.N. 2000. Pharmacology of Medicinal Plants and Natural Products. Indian J. Pharmacol. 32(4): 81-118.

Dana, E.D., Delomas, J.G. and Sanchez, J. 2010. Effects of the aqueous extracts of *Zygophyllum fabago* on the growth of *Fusarium oxyosporumf.* sp. *Melonis* and *Pythium aphanidermatum*, Weed Biol. Manag. 10: 170-175.

de Cock, A.W.A.M. and Lévesque, C.A. 2004. New species of *Pythium* and *Phytophthora*. Stud. Mycol. 50: 481-487.

Deadman, M., Al Hasani, H. and Al Sa'di, A. 2006. Solarization and biofumigation reduce *Pythium aphanidermatum* induced damping-off and enhance vegetative growth of greenhouse cucumber in Oman. J. Plant Pathol. 88(3): 335-337.

Deadman, M., Perret, J., Al-Jabri, S., Al-Maqbali, Y., Al-Sa'di, A., Al-Kiyoomi, K. and Al-Hasani, H. 2007. Epidemiology of damping-off disease in greenhouse cucumber crops in the Sultanate of Oman. Acta Horticulturae 731: 319-326.

Dick, M.W. 1990. *Keys to Pythium*. Department of Botany, School of Plant Sciences, University of Reading, Reading, U.K.

Dileep, N., Junaid, S., Rakesh, K.N., Kekuda, P.T.R. and Nawaz, A.S.N. 2013. Antifungal activity of leaf and pericarp extract of *Polyalthia longifolia* against pathogens causing rhizome rot of ginger. Sci. Technol. Arts Res. J. 2(1): 56-59.

Eckert, J.W. and Tsao, P.H.1962. A selective medium for isolation of *Phytophthora* and *Pythium* from plant roots. Phytopathol. 52: 771-777.

Elad, Y., Kalfon, A. and Chet, I. 1982. Control of *Rhizoctonia splani* in cotton by seed-coating with *Trichoderma* spp. spores. Plant Soil 66: 279-281.

Erhart, E. and Burian, K. 1997. Evaluating quality and supressiveness of Austrian biowaste composts. Composts Sci. Util. 5: 15-24.

Folman, L.B., De Klein, M.J.E.M., Postma, J. and Van Veen, J.A. 2004. Production of antifungal compounds by Lycobasteren zymogenes isolate 3.1T8 under different conditions in relation to its efficacy as a biocontrol agent of *Pythium aphanidermatum* in cucumber. Biol. Control 31: 145-154.

Gamliel, A., Austerweil, M. and Kritzman, G. 2000. Non-chemical approach to soil-borne pest management: Organic amendments. Crop Protect. 19: 847-853.

Goel, Mayurika, Dureja, Prem, Rani, Archana, Uniyal, Prem and Laatsch Hartmurt, L. 2011. Isolation, characterization and antifungal activity of major constituents of the Himalayan lichen *Parmelia reticulata Tayl.* J. Agric. Food Chem. 59(6): 2299-2307.

Gold, S.E. and Stanghellini, M.E. 1985. Effects of temperature on *Pythium* root rot of spinach *Spinacia oleracea* grown under hydroponic conditions. Phytopathol. 75: 33-37.

Goldberg, N.P. and Stanghellini, M.E. 1990. Ingestion-egestion and aerial transmission of *Pythium aphanidermatum* by shore flies (Ephydrinae: *Scatellastagnalis*). Phytopathol. 80: 1244-1246.

Grane, M. and Ahmed, S. 1988. Handbook of Plants with Pest Control Properties. John Wiley and Sons, New York, p. 431.

Grinstein, A., Orion, D., Greenberger, A. and Katan, J. 1979. Solar heating of the soil for the control of *Verticillum dahliae* and *Pratylenchus thornei* in potatoes. pp. 431-438. *In*: Schhippers, B. and Gams, W. (eds.). Soil Borne Plant Pathogens. Academic Press, New York, NY, London.

Guirado, M.L., Sáez, Y., Serrano, E. and Gómez, J. 2009. Aetiology watermelon "sudden death" of greenhouses in the southeast of Spain. Boletín de Sanidad Vegetal, Plagas 35(4): 617-628.

Gujar, J. and Talwankar, D. 2012. Antifungal potential of crude plant extracts on some pathogenic fungi. World J. Sci. Technol. 2(6): 58-62.

Haouala, R., Hawala, S., El-ayeb, A., Khanfir, R. and Boughanmi, N. 2008. Aqueous and organic extracts of *Trigonella foenum-graecum* L. inhibit the mycelia growth of fungi. J. Environ. Sci. 20(12): 1453-1457.

Haritha, V., Gopal, K.,Madhusudhan, P., Vishwanath, K. and Rao, S.V.R.K. 2010. Integrated management of damping-off disease incited by *Pythium aphanidermathum* (Edson) pitzpin tobacco nursery. J. Pl. Dis. Sci. 5(1): 41-47.

Hendrix, F.F. and Campbell, W.A. 1973. *Pythiums* as plant pathogens. Ann. Rev. Phytopathol. 11: 77-98.

Hong, C.X. and Moorman, G.W. 2005. Plant pathogens in irrigation water: Challenges and opportunities. Crit. Rev. Plant Sci. 24: 189-208.

Hospenthal, Duane R., Tatjana, P. Calvano, Peter, J. Blatz, Todd, J. Vento, Brian, L. Wickes, Deanna, A. Sutton, Elizabeth, H. Thompson, Christopher, E. White and Evan, M. Renz. 2011. *Pythium aphanidermathum* infection following combat trauma. J. Clin. Microbiol. Oct 49(10): 3710-3713.

Hyder, N., Coffey, M.D. and Stanghellini, M.E. 2009. Viability of oomycete propagules following ingestion and excretion by fungus gnats, shore flies, and snails. Plant Dis. 93: 720-726.

Jacobson, R., Greenberger, A., Katan, J., Levi, M. and Alon, H. 1980. Control of Egyptian Broomrape (*Orobancheae gyptiaca*) and other weeds by means of solar heating of the soil by polyethylene mulching. Weed Science 28: 312-316.

Jeffers, S.N. and Martin, S.B. 1986. Comparison of two media selective for *Phytophthora* and *Pythium* species. Plant Dis. 70: 1038-1043.

Juan, Z., Xue, Q.H. and Tang, M. 2009. Screening of antagonistic action mycetes against 'Jiashi' *Cucumismelo* L. damping-off. J. Northwest A & F University – Natural Science Edition, 37(5): 144-148.

Kagale, S., Marimuthu, T., Thayumanavan, B., Nandakumar, R. and Samiyappan, R. 2004. Antimicrobial activity and induction of systemic resistance in rice by leaf extract of *Datura metel* against *Rhizoctonia solani* and *Xanthomonas oryzae* pv. *oryzae*. Physiol. Mol. Plant Pathol. 65(2): 91-100.

Kamala, Th. and Indira, S. 2011. Evaluation of indigenous *Trichoderma* isolates from Manipur as biocontrol agent against *Pythium aphanidermatum* on common beans 3. Biotech. Dec 2011, 1(4): 217-225.

Kao, C.W. and Ko, W.H. 1986. Suppression of *Pythium splendens* in Hawaiian soil by calcium and microorganisms. Phytopatology 76: 215-220.

Katan, J. 2000. Physical and culture methods for the management of soil borne pathogens. Crop Protect. 19: 725-731.

Kavitha, P.G. and Thomas, G. 2007. Evaluation of Zingiberaceae for resistance to ginger soft rot caused by *Pythium aphanidermatum* (Edson) Fitzp. J. Plant Genet. Resour. Newsl. 152: 54-57.

Kawarazaki, H., Nara,Y., Kijima, T. and Goto, M. 2008. *Pythium* rot of figmarigold (*Lampranthus spectabile*) caused by *P. aphanidermatum*. J. Gen. Pl. Pathol. 74(1): 94-95.

Krober, H. 1985. *Erfahrungenmit Phytophthora de Bary*and *Pythium Pringsheim*. Biologische Bundesanstaltfur Landund Forstwietschaft, Institut fur Mikrobiologie, Berlin-Dahlem, 225 pp.

Kubota, M. 2010. Diseases of cabbage plug seedlings in Japan and control of the diseases. Bull. Natl. Inst. Veg. Tea Sci. 9: 57-112.

Kucharek, T. and Mitchell, D. 2000. Diseases of agronomic and vegetable crops caused by *Pythium*. Plant Pathology Fact Sheet, University of Florida, 53, http: //128.227.207.2453.

Kurita, N., Makoto, M., Kurane, R. and Takahara, Y. 1981. Antifungal activity of components of essential oils. Agric. Biol. Chem. 45: 945-952.

Kusum Mathur, Ram, D., Poonia, J. and Lodha, B.C. 2002. Integration of soil solarization and pesticides for management of rhizome rot of ginger. Indian Phytopathol. 55: 345-347.

Lin, Y.S., Gung, Y.H. and Huang, J.H. 2002. Control of *Pythium* root rot of vegetable pea seedlings in soilless cultural system. ISHS 578: 221-229.

Littrell, R.H. and McCarter, S.M. 1970. Effect of soil temperature on virulence of *Pythium aphanidermatum* and *Pythium myriotylum* to rye and tomato. Phytopathol. 60: 704-707.

Lumsden, R.D. and Locke, J.C. 1989. Biological control of damping-off caused by *Pythium Ultimum* and *Rhizoctonia solani* with *Gliocladium virens* in soilless mix. Phytopathol. 79: 361-366.

Ma, Z. and Michailides, T.J. 2005. Advances in understanding molecular mechanisms of fungicide resistance and molecular detection of resistant genotypes in phytopathogenic fungi. Crop Prot. 24(10): 853-863.

Madduleti, A. and Moses, G.J. 1995 Plant products and damping-off disease management in vegetables. Indian J. of Mycol. and Pl. Pathol. 25(1&2): 86.

Mandelbaum, R., Hadar, Y. and Chen,Y. 1988. Composting of agricultural wastes for their use as container media: Effect of heat treatments on suppression of *Pythium aphanidermatum* and microbial activities in substrates containing compost. Biol. Wastes 26: 261-274.

Mandelbaum, R. and Hadar, Y. 1990. Effects of available carbon source on microbial activity and suppression of *Pythium aphanidermatum* in compost and peat container media. Phytopathol. 80: 794-804.

Manomohandas, T.P. and Sivaprakasam, K. 1994. Biological control of damping off disease in chili nursery. Crop Disease Innovative Techniques and Management. Kalyani Publi. New Delhi, pp. 199-203.

Manoranjitham, S.K. and Prakasam, V. 2000. Management of chili damping off using biocontrol agents. Capsicum Eggplant News Lett. 19: 101-104.

Masuduzzaman, S., Meah, M.B. and Rashid, M.M. 2008. Determination of inhibitory action of *Allamanda* leaf extracts against some important plant pathogens. J. Agric. Rural Dev. 6(1-2): 107-112.

Matsuzaki, M., Hamaguchi, H. and Shimonasako, H. 1998. The effect of manure application and soil fumigation on the field crops cultivated continuously. Res. Bull. Hokkaido Natl. Agric. Exp. Stn. Hokkaido 66: 1-65.

Maya, C. and Thippanna, M. 2013. *In vitro* evaluation of ethno-botanically important plant extracts against early blight disease (*Alternaria solani*) of tomato. Global J. BioSci. Biotechnol. (GJBB) 2: 248-252.

Mehra, R. 2005. Seed spices: Diseases and their management. Indian J. of Areca Nut, Spices & Medicinal Plants 7(4): 134-143.

Moorman, G.W., Kang, S., Geiser, D.M. and Kim, S.H. 2002. Identification and characterization of *Pythium* species associated with greenhouse floral crops in Pennsylvania. Plant Dis. 86: 1227-1231.

Munoz-Mingaro, D., Acero, N., Llinares, F., Fozuelo, J.M., Galan de Mera, A. and Vicenten, J.A. 2003. Biological activity of extracts from *Catalpa bignonioides* walt (Bignoniaceae). J. Ethnopharmacol. 87: 163-167.

Muthukumar, A., Eswaran, A., Nakkeeran, S. and Sangeetha, G. 2010. Efficacy of plant extracts and biocontrol agents against *Pythium aphanidermatum* inciting chili damping-off. Crop Protect. 29: 1483-1488.

Nelson, E.B. 1987. Rapid germination of sporangia of *Pythium* species in response to volatiles from germination seeds. Phytopathol. 77: 1108-1112.

Noelting, M.C.I. and Sandoval, M.C. 2007. First report of stem canker affecting *Amaranthus Caudatus* in Argentina. Australas. Plant Dis. Notes 2(1): 5.

Omidbeygi, M., Barzegar, M., Hamidi, Z. and Naghdibadi, H. 2007. Antifungal activity of thyme, summer savory and clove essential oils against *Aspergillus flavus* in liquid medium and tomato paste. Food Control 18: 1518-1523.

Osburn, R.M., Schroth, M.N., Hancock, J.G. and Hendson, M. 1989. Dynamics of sugarbeet colonization by *Pythium ultimum* and *Pseudomonas* species: Effects on seed rot and damping-off. Phytopathol. 79: 709-716.

Osman, K.A. and Al-Rehiayam, S. 2003. Risk assessment of pesticide to human and the environment. Saudi J. Biol. Sci. 10: 81-106.

Owen-Going, T.N., Beninger, C.W., Sutton, J.C. and Hall, J.C. 2008. Accumulation of phenolic compounds in plants and nutrient solution of hydroponically-grown preppers inoculated with *Pythium aphanidermatum*. Can. J. Plant Pathol. 30: 214-225.

Palmucci, H.E. and Grijalba, P.E. 2007. Root and stem rot caused by *Pythium aphanidermatum* on poinsettia in a soilless culture system in Buenos Aires Province, Argentina. Australas. Plant Dis. Notes 2(1): 139-140.

Pandey, K.K. and Pandey, P.K. 2002. Study on soil solarization for nursery management of vegetable crops. J. Mycol. Pl. Pathol. 32(3): 423.

Pane, C., Spaccini, R., Piccolo, A., Scala, F. and Bonanomi, G. 2011. Compost amendments enhance peat suppressiveness to *Pythium ultimum*, *Rhizoctonia solani* and *Sclerotinia minor*. Biol. Control 56: 115-124.

Parveen, T. and Sharma, K. 2014. Phytochemical profiling of leaves and stem bark of *Terminalia arjuna* and *Tecomella undulata*. Int. J. Pharm. Bioscience 1(5): 1-7.

Paul, B., Bala, K., Belbahri, L., Calmin, G., Sanchez-Hernandez, E. and Lefort, F. 2006. A new species of *Pythium* with ornamented oogonia: Morphology, taxonomy, ITS region of its rDNA, and its comparison with related species. FEMS Microbiology Letters 254: 317-323.

Paulitz, T.C. and Baker, R. 1987. Biological control of *Pythium* damping-off of cucumbers of *Pythium nunn*: Influence of soil environment and organic amendments. Phytopathol. 77: 341-346.

Plaats-Niterink, A.J. Van Der. 1981. Monograph of the Genus *Pythium*. Studies in Mycology 21: 1-224.

Pokharel, Ramesh. 2010. Western Colorado Research Center. Colorado State University. Soil Solarization, a potential Solution to Replant Diseases.

Poudyal, B.K., Du, G., Zhang, Y., Liu, J. and Shi, Q. 2011. The control of soft rot of ginger by Jeevatu based organic liquid manure. Front. Agric. China 5(1): 45-50.

Punja, Z.K. and Rodriguez, G. 2018. *Fusarium* and *Pythium* species infecting roots of hydroponically grown marijuana (*Cannabis sativa* L.) plants. Can. J. Plant Pathol. 40(4): 498-513.

Radhakrishnan, N. and Balasubramanian, R. 2009. Salicylic acid induced defence responses in *Curcuma longa* (L.) against *P. aphanidermatum* infection. J. Crop Prot. 28(11): 974-979.

Ramanathan, A., Marimuthu, T. and Raguchander, T. 2004. Effect of plant extracts on growth in *Pythium aphanideramtum*. J. Mycol. Plant Pathol. 34: 315-317.

Reddy, G.S., Rao, V.K., Sitaramaiah, K. and Chalam, T.V. 2001. Soil solarization for control of soil borne pathogen complex due to *Meloidogyne incognita* and *Pythium aphanidermatum*. Indian J. Nematol. 31(2): 136-138.

Ringer, C.E., Millner, P.D., Teerlinck, L.M. and Lyman, B.W. 1997. Suppression of seedling damping off disease in potting mix containing animal manure composts. Compost. Sci. Util. 5: 6-14.

Robertson Alison, E. 2018. Plant Disease. First report of alfalfa (Medicago sativa & L.) Seed Rot, Seedling Root Rot, and Damping off caused by Pythium spp. in Sudanese soil. 102(5): 1043.

Robideau, G.P., de Cock, A.W., Coffey, M.D., Voglmayr, H., Brouwer, H., Bala, K., Chitty, D.W., Désaulniers, N., Eggertson, Q.A., Gachon, C.M., Hu, C.H., Küpper, F.C., Rintoul, T.L., Sarhan, E., Verstappen, E.C., Zhang, Y., Bonants, P.J., Ristaino, J.B. and Lévesque, C.A. 2011. DNA barcoding of oomycetes with cytochrome c oxidase subunit I and internal transcribed spacer. Mol. Ecol. Resour. 11(6): 1002-1011.

Rongyi, Z., Zhiqiong, T. and Shanying, C. 2003. First report of leaf rot caused by Fusarium oxysporum and Pythium aphanidermatum on Aechmea fasciata in Hainan Province. China Plant Disease 87(5): 599.

Rosso, M.L., Rupe, J.C., Chen, P.Y. and Mozzoni, L.A. 2008. Inheritance and genetic mapping of resistance to Pythium damping-off caused by Pythium aphanidermatum in 'Archer' soybean. J. of Crop Sci. 48(6): 2215-2222.

Sagar, S.D., Kulkarni, S. and Hegde, Y.R. 2007. Management of rhizome rot of ginger by botanicals. Int. J. Plant Sci. 2(2): 155-158.

Sagar, S.D., Kulkarni, S. and Hegde, Y.R. 2008. Survey, surveillance and etiology of rhizome rot of ginger in Karnataka. J. Plant Dis. Sci. 3(1): 37-39.

Saha, S., Naskar, I., Nayak, D.K. and Sarkar, M.K. 2008. Efficacy of different seed dressing agents in the control of damping off disease of chili caused by Pythium aphanidermatum. J. of Mycopath. Res. 46(1): 121-123.

Sanogo, S. and Moorman, G.W. 1993. Transmission and control of Pythium aphanidermatum in an ebb- and flow subirrigation system. Plant Dis. 77: 287-290.

Satya, V.K., Radhajeyalakshmi, R., Kavitha, K., Paranidharan, V., Bhaskaran, R., Scheuerell, S.J., Sullivan, D.M. and Mahaffee, W.F. 2005. Suppression of seedling damping-off caused by Pythium ultimum, P. irregulare and Rhizoctonia solani in container media amended with a diverse range of pacific northwest compost sources. Phytopathol. 95(3): 306-315.

Schoener, J., Wilhelm, R. and Wang, S. 2017. Pythium aphanidermatum crown rot of industrial hemp. Nevada Department of Agriculture Plant Pathology Laboratory NPDN News 12(9): 10-11.

Serrano, Y., Guirado, M.L., Carmona, M.P., Gómez, J. and Melero-Vara, J.M. 2008. First report of root and crown necrosis of bean caused by Pythium aphanidermatum in Spain. J. of Plant Dis. 92(1): 174.

Shanmugam, V., Varma, A.S. and Surendran, M. 1999. Management of rhizome rot of ginger by antagonistic microorganisms. Madras Agric. J. 86: 339-341.

Shweta, Sharma, Mohinder, Kaur and Durga, Prashad. 2014. Isolation of fluorescent pseudomonas strain from temperate zone of Himachal Pradesh and their evaluation as plant growth promoting rhizobacteria (PGPR). The Bioscan 9(1): 323-328.

Sheldon, M. Elliott. 2003. Rhizome Rot Disease of Ginger. Ministry of Agriculture Research and Development Division. St. Catherine, Jamaica.

Shiva, V., Pande, P. and Singh, J. 2004. Principles of Organic Farming, Renewing the Earth's Harvest, Published by Navdanya, New Delhi.

Shuler, C., Biala, J., Bruns, C., Gottschall, R., Ahlers, S. and Vogtmann, H. 1989. Suppression of root rot on peas, beans and beetroots caused by Pythium ultimum and Rhizoctonia solani through the amendment of growing media with composted organic household waste. J. Phytopathol. 127: 227- 238.

Singh, A.K., Singh, V.K. and Shukla, D.N. 2010. Effect of plant extracts against Pythium Aphanidermatum – the incitant of fruit rot of muskmelon (Cucumis melo). Indian J. Agr. Sci. 80(1): 51-53.

Siva, N., Ganesan, S., Banumathy, N. and Muthuchelian. 2008. Antifungal effect of leaf extract of some medicinal plants against Fusarium oxysporum causing wilt disease of Solanum melogena L. Ethnobot. Leafl. 12: 156-163.

Stephens, C.T., Herr, L.J., Schmitthenner, A.F. and Powell, C.C. 1983. Sources of Rhizoctonia solani and Pythium spp. in the bedding plant greenhouse. Plant Dis. 67: 272-275.

Subhashini, D.V. and Padmaja, K. 2009. Exploitation of Pseudomonas fluorescens for the management of damping-off disease of tobacco in seedbeds. Indian J. Plant Prot. 37: 147-150.

Suleman, M.N. and Emua, S.A. 2009. Efficacy of four plant extracts in the control of root rot disease of cowpea. Afr. J. Biotechnol. 8(16): 3806-3808.

Tabin, T., Arunachalam, A., Shrivastava, K. and Arunachalam, K. 2009. Effect of arbuscular mycorrhizal fungi on damping-off disease in *Aquilaria agallocha* Roxb. Seedlings. J. Trop. Ecol. 50(2): 243-248.

Taechowisan, T., Wanbanjob, A., Tuntiwachwuttikul, P., Shen, Y. and Lumyong, S. 2008. Synergistic activities of 4-arylcoumarins against phytopathogenic fungi. Res. J. Microbiol. 3: 237-245.

Talibi, I., Askarne, L., Boubaker, H., Boudyach, E., Msanda, F., Saadi, B. and Ait Ben Aoumar, A. 2012. Antifungal activity of some Moroccan plants against *Geotrichum candidum*, the causal agent of postharvest citrus sour rot. Crop Prot. 35: 41-46.

Tanina, K., Tojo, M., Date, H., Nasu, H. and Kasuyama, S. 2004. *Pythium* rot of chingensai (*Brassica campestris* L. chinensis group) caused by *Pythium ultimum* var. *ultimum* and *Pythium aphanidermatum*. J. Gen. Plant Pathol. 70(3): 188-191.

Theodore, M. and Toribio, J.A.. 1995. Suppression of *Pythium aphanidermatum* in composts prepared from sugarcane factory residues. Plant Soil 177: 219-233.

Tomioka, K., Sato, T. and Nakanishi, T. 2002. Foot rot of ulluco caused by *Pythium aphanidermatum*. J. Gen. Plant Pathol. 68(2): 189-190.

Uma, T., Mannam, S., Lahoti, J., Devi, K., Kale, R.D. and Bagyaraj, D.J. 2012. Biocidal activity of seed extracts of fruits against soil borne bacterial and fungal plant pathogens. J. Biopest 5(1): 103-105.

Usman, H. and Osuji, J.C. 2007. Phytochemical and *in vitro* antimicrobial assay of the leaf extract of *New bouldia* leaves. Afr. J. Trad. CAM 4(4): 476-480.

Usman, M.N., Balakrishnan, P. and Sarma, Y.R. 1996. Biocontrol of rhizome rot of ginger. J. Plant. Crops 24(Supply): 184-191.

Van West, P., Appiah, A.A. and Gow, N.A.R. 2003. Advances in research on oomycete root pathogens. Physiol. Mol. Plant Pathol. 62: 99-113.

Vankar, H.J. and Patel, B.N. 2004. Efficacy of leaf extracts and botanical pesticide in conjunction with fungicide against damping off of bidi tobacco. J. Mycol. Pl. Pathol. 34(1): 198.

Vinayaka, K.S., Prashitha Kekuda, T.R., Noor Nawaz, A.S., Syed Junaid, Dileep, N. and Rakesh, K.N. 2014. Inhibitory activity of Usneapictoides G. Awasthi (*Parmeliaceae)* against *Fusarium oxysporum* F. sp. *Zingiberi* and *Pythium aphanidermatum* isolated from rhizome rot of ginger. Life Sci. Leafl. 49: 17-22.

Wang, S., Wu, J., Rao, P., Ng, T.B. and Ye, X. 2005b. A chitinase with antifungal activity from the mung bean. Protein Exp. Purif. 40(2): 230-236.

Waterhouse, G.M. and Waterston, J.M. 1964. *Pythium aphanidermatum*. CMI descriptions of pathogenic fungi and bacteria, No. 36. Commonwealth Mycological Institute: Kew, Surrey, UK. Waterhouse.

Whipps, J.M. and Lumsden, D.R. 1991. Biological control of *Pythium* species. Biocontrol Sci. Technol. 1: 75-90.

Wilson, C.L., Solar, J.M., Ghaouth, A.E.I. and Wisniewski, M.E. 1997. Rapid evaluation of plant extracts and essential oils for antifungal activity against *Botrytis cinerea*. Plant Dis. 81: 204-210.

You, X.D., Park, J.E., Takase, M., Wada, T. and Tojo, M. 2015. First report of *Pythium aphanidermatum* causing root rot on common ice plant (*Mesembryanthemum crystallinum*). New Disease Reports (2015) 32: 36. [http: //dx.doi.org/10.5197/j.2044-0588.2015.032.036]

Zagade, S.N., Deshpanday, G.D., Gavade, D.B., Anoorkar, A.A. and Pawar, S.V. 2012. Biocontrol agents and fungicides for management of damping off in chili. World J Agric. Sci. 8(6): 590-597.

Management of *Pythium* Diseases

Rehana Naz Syed[1], Abdul Mubeen Lodhi[2*] and Saleem Shahzad[3]

[1] Department of Plant Pathology, Sindh Agriculture University, Tandojam, Pakistan
[2] Department of Plant Protection, Sindh Agriculture University, Tandojam, Pakistan
[3] Department of Agriculture & Agribusiness Management, University of Karachi, Karachi, Pakistan

Introduction

Pythium species belong to order Pythiales of Phylum Oomycota. Phytopathogens of this genus cause seed, seedling and root decays, i.e. disintegration of seed or seedling before it comes out from the soil (pre-emergence damping-off), or roots and crown of very young seedling get infected immediately after plant emergence from the soil that may result in death of the seedling (post-emergence damping-off). If plants survive, the affected roots absorb less water and nutrient leading to wilt and ultimately death of the plant. *Pythium* species also cause rotting of stem, bulb, tubers and fruits. High soil moisture favors the pathogens growth and infection process by favoring the mobility of motile spores and widespread infection that leads to substantial losses. Various methods have been employed for the management of diseases caused by the pathogenic *Pythium* species that include biological, chemical, cultural and host resistance. Integrated use of two or more methods is also widely practiced to gain best possible management of the pathogens. This review includes various reports on the management of diseases caused by *Pythium* species in different parts of the world.

Before 2010, the genus *Pythium* was a large assemblage of morphologically variable species. Based on the phylogenetic study of the genus by Lévesque and de Cock (2004), several species of the genus have been segregated into new genera *Phytopythium* (Bala et al. 2010), *Globisporangium*, *Elongisporangium* and *Pilasporangium* (Uzuhashi et al. 2010), leaving only the species with filamentous sporangia in the genus *Pythium*. Since this chapter includes the work on the management of *Pythium* species before as well after 2010, we have treated the species transferred to other genera as *Pythium* species.

Chemical control

Use of chemical fungicides is the most effective and reliable control measure for diseases affecting the plants (Gawande and Shukla 1976). Against *Pythium* species, chemical fungicides are either used as seed treatment/dressing, soil treatment or foliar spray. The effective and recommended fungicides for controlling various *Pythium* diseases either change with the time, development of new chemicals or establishment of fungicide resistance in targeted pathogens.

During the whole 19th and most part of the 20th century, the copper fungicides like cuprous oxide, copper carbonate, monohydrated copper sulfate, copper sulfate, copper stearate, corona

---

*Corresponding author: mubeenlodhi@gmail.com, amlodhi@sau.edu.pk

copper carbonate and Bordeaux mixture were frequently used against different groups of plant pathogens, including *Pythium* either as seed or soil treatment (Haenseler 1930, Hobsfall 1932, Nolla 1932, Bayus et al. 1943, Bruna 1946, Ark and Middleton 1949).

Different mercury based compounds such as mercury dust, ethylmercuric phosphate, ethylmercury chloride, and methylmercuric dicyanamide were frequently used as seed treatment against diseases caused by *Pythium* species from early to late 20th century, including damping-off of garden peas, alfalfa, cucumber, table beet, spinach, black rot of sugarcane, and seed decay of cucumber (Haenseler 1930, Buchholtz 1936, 1942, Campbell 1939, Bayus et al. 1943, Dezeeuw 1954).

The dithiocarbamate fungicides such as Thiram, Ferbam, Nabam, Zineb and Maneb introduced in 1940s and Mancozeb in 1961, are broad spectrum fungicides effective against a large number of pathogens as well as *Pythium* (Gerhold 1956, Geard 1959, Iremiren 1987, McDonald et al. 2004). Systemic carboxamide (Carboxin), benzimidazole (Benomyl) and triazole (Triadimefon) fungicides invented from 1969-1976 are highly effective against other groups of phytopathogens but not to Oomycetes (Klittich 2008).

Metalaxyl (Phenylamide) and Fosetyl-aluminium are first two systemic fungicides for Oomycetes, both introduced in 1977 (Schwinn and Staub 1987). Metalaxyl and its relatives provide superb control of all Oomycetes including soil-borne species of *Pythium* and *Phytophthora*. Its excellent efficacy resulted in more frequent and tremendous use as curative or protective, applied as seed, soil or foliar treatment that has caused resistance in many Oomycetes species (Morton and Urech 1988).

Soil disinfection with highly toxic chemical Formalin (37% formaldehyde) is also the most commonly used practice against soil-borne pathogens, including many species of *Pythium* (Haenseler 1930, Agati 1931, Bruna 1946, Dezeeuw 1954, Molin 1955, Sharan 1958, Arslan 1979, Singh et al. 1985, Avikainen et al. 1993, Raj et al. 1998). Till banned in 2005, soil fumigation with methyl bromide was considered the most highly effective treatment to eliminate the propagules of *Pythium* and other pathogens; it was mostly practiced in nurseries, seed beds, potting mix or small areas (Beuzenberg 1966, Cheng et al. 1989, Anon 1971). Chloropicrin, 1,3-dichloropropene, metam sodium and metam potassium appeared as the alternative fumigants of methyl bromide against soil-borne pathogens (Anon 1971, Cook et al. 1987, Cebolla et al. 1995, Aloj et al. 2000, De Cal et al. 2005, Boz et al. 2011). The fumigation with 1,3-dichloropropene, chloropicrin and metam-sodium effectively inhibited *P. irregulare* and *P. myriotylum* inoculum associated with different plant species (Nelson et al. 2002, Desaeger et al. 2008).

Repeated reports of declining efficacy of chemical fungicides against a number of *Pythium* species and development of resistant strains to commonly used fungicide have been made. Under field conditions, most of the pathogenic *Pythium* species including *P. sulcatum, P. violae, P. sylvaticum* and *P. macrosporum* causing cavity spot of carrot have been controlled by Metalaxyl. However, *P. sulcatum* showed some resistance to Metalaxyl, but also signs of biodegradation (Hiltunen and White 2002, Titone et al. 2009). The resulting resistance has been overcome by restricting its applications as well as combining it with other fungicides like Mancozeb or Thiram (Schwinn and Staub 1987). Although in most of the cases, satisfactory level of control has been achieved with the use of Metalaxyl, ever growing disease complexity and potential resistance development in pathogens require some alternate and integrated disease control strategy (Breton and Rouxel 1994). Zoxamide, that showed effectiveness and significantly reduced disease development by *P. sulcatum* (Martinez et al. 2005), may be used as an alternative. Table 18.1 briefly summarizes the fungicides used against different *Pythium* species causing diseases in various plants.

Despite the effectiveness of chemical fungicides, the hazards of these toxic chemicals to men, animal and environment have shifted the emphasis of researchers on the use of alternate methods for plant disease management such as biological and cultural method.

Table 18.1. Summary of the chemicals used against various plant diseases caused
by different *Pythium* species

Chemicals	Host diseases/target Pythium species	Reference
1,3-Dichloropropene	Root diseases of tomato and eggplant (*P. aphanidermatum, P. myriotylum*), root diseases of squash (*P. irregulare*)	Nelson et al. 2002, Desaeger et al. 2008
Aluminium containing salt (Aluminium lactate, Aluminium sulphate, Aluminium chloride)	Cavity spot of carrot (*P. sulcatum*)	Kolaei et al. 2012, 2013
Azoxystrobin	Root rot of flue-cured tobacco seedlings (*P. myriotylum*), tobacco damping-off (*P. debaryanum*)	Gutierrez et al. 2012, Tashkoski et al. 2016
Benzothiadiazole	Cocoyam root diseases (*P. myriotylum*)	Oumar et al. 2015
Boscalid+Pyraclostrobin (Signum)	Tobacco damping-off (*P. debaryanum*)	Tashkoski 2015, Tashkoski et al. 2016
Calcium sulfate	Cavity spot of carrot (*P. sulcatum*)	Kolaei et al. 2012, 2013
Captan	Chili damping-off (*P. aphanidermatum*)	Saha et al. 2008
Carbendazim	Root rot of sugarcane (*P. graminicola*), chili damping-off (*P. aphanidermatum*), rhizome rot of cardamom (*P. vexans*), soft rot of turmeric (*P. myriotylum*)	Shaikh and Kareappa 2008, Sivakumar et al. 2012, Attaullah et al. 2015, Deshmukh et al. 2016
Carbendazim+Mancozeb	Brinjal damping-off (*P. ultimum*)	Gholve et al. 2014
Carboxin	*Ruta graveolens* damping-off (*P. debaryanum*), wheat damping-off (*P. ultimum*)	Helmy et al. 2001, Yanashkov et al. 2017
Chloropicrin	Root diseases of tomato and eggplant (*P. aphanidermatum, P. myriotylum*), seedling diseases of conifer (*P. irregulare*), root diseases of squash (*P. irregulare*)	Nelson et al. 2002, Desaeger et al. 2008, Weiland et al. 2011
Copper oxychloride	Root rot of sugarcane (*P. graminicola*), rhizome rot of ginger (*P. myriotylum*), rhizome rot of cardamom (*P. vexans*), cabbage root diseases (*P. debaryanum*), damping-off in bidi tobacco (*P. aphanidermatum*)	Cuc et al. 2007, Dhanapal and Thomas 2008, Sivakumar et al. 2012, Yadav and Joshi 2012, Lalfakawma et al. 2014, Deshmukh et al. 2016
Cyazofamid (Ranman)	Douglas-fir damping-off (*P. irregulare*)	Linderman et al. 2008
Cymoxanil	*Pythium* damping-off of chili (*P. ultimum*)	Zagade et al. 2012
Cymoxanil+Mancozeb	Tobacco damping-off (*P. aphanidermatum, P. myriotylum*)	Shamarao and Jahagirdar 2012
Difenoconazole	Hemp root diseases (*P. debaryanum*)	Trotus and Naie 2008
Etridiazole	*Pythium* root rot geranium (*P. irregulare*), root rot of flue-cured tobacco seedlings (*P. myriotylum*)	Gutierrez et al. 2012, Krasnow and Hausbeck 2017
Etridiazole+Quintozene (Terraclor Superx)	Root rot and wilt of Kangaroo Paw (*P. myriotylum*)	Tsror et al. 2005

(Contd.)

Fenamidon	Root rot and wilt of Kangaroo Paw (*P. myriotylum*)	Tsror et al. 2005
Fenamidone+Mancozeb	Damping-off in bidi tobacco (*P. aphanidermatum*)	Yadav and Joshi 2012
Fludioxonil+Metalaxyl+ Thiamethoxam (Cruiser)	Canola root diseases (*Pythium* F group, *P. irregulare, P. mamillatum, P. ultimum* var. *ultimum*)	De Villiers and Agenbag 2007
Fosetyl-Al	Cabbage root diseases (*P. debaryanum*), blight of turfgrasses (*P. aphanidermatum*), seedling diseases of pepper (*P. aphanidermatum*), damping-off and root rot in forest nurseries (*P. irregulare, P. dissotocum, P. vipa, P. sylvaticum* and *P. ultimum*)	Cuc et al. 2007, Cook et al. 2009, Mihajlović et al. 2013, Weiland et al. 2014
Hymexazol	Cucumber seedling rot (*P. aphanidermatum*)	Al-Balushi et al. 2018
Kresoxim-Methyl	Tobacco damping-off (*P. debaryanum*)	Tashkoski et al. 2016
Mancozeb	Seedling diseases of pepper (*P. aphanidermatum*)	Mihajlović et al. 2013
Mefenoxam+Copper hydroxide	Pythium leak of snap bean (*P. aphanidermatum, P. ultimum*)	Damicone et al. 2012
Metalaxyl	Peanut pod rot (*P. irregulare, P. myriotylum*), potato leak (*P. ultimum*), damping-off and root rot in forest nurseries (*P. irregulare*), damping-off and root rot of cucumber (*P. aphanidermatum*), papaya damping-off (*P. debaryanum*), cavity spot of carrot (*P. sulcatum, P. violae*), citrus root rot (*P. ultimum*), soybean damping-off (*P. ultimum*), brinjal damping-off (*P. ultimum*), chili damping-off (*P. ultimum*), cabbage root diseases (*P. debaryanum*), corn and soybean root diseases (*P. catenulatum, P. irregulare, P. paroecandrum, P. splendens* and *P. torulosum*), soybean seedling diseases (*Pythium* spp.), seedling blight of field pea (*Pythium* spp.)	Hwang et al. 2001, Hiltunen and White 2002, Dorrance et al. 2004, Taylor et al. 2004, Wheeler et al. 2005, Al-Kiyumi 2006, Cuc et al. 2007, Al-Sa'di et al. 2008, El-Tarabily et al. 2009, Augusto et al. 2010, Kean et al. 2010, Dar et al. 2012, Zagade et al. 2012, Gholve et al. 2014, Weiland et al. 2014, Hudge 2015, Grijalba and Ridao 2017
Metalaxyl/Mefenoxam + Oxathiapiprolin	Seedling diseases of soybean (*P. mercuriale*)	Vargas 2018
Metalaxyl+Azoxystrobin (Soygard)	Soybean seedling diseases (*Pythium* spp.)	Bradley 2008
Metalaxyl+Captan	Brinjal damping-off (*P. ultimum*)	Gholve et al. 2014
Metalaxyl+Carbendazim	Rhizome rot of ginger (*P. myriotylum*)	Meena and Mathur 2003
Metalaxyl+Ethaboxam	Soybean damping-off (*P. irregulare, P. ultimum* var. *ultimum* and *P. ultimum* var. *sporangiiferum*)	Scott 2018

(Contd.)

Table 18.1. (*Contd.*)

Chemicals	Host diseases/target Pythium species	Reference
Metalaxyl+Mancozeb	Tobacco damping-off (*P. aphanidermatum*), Pre-emergence damping-off of soybean (*P. helicoides*), chili damping-off (*P. aphanidermatum*), soft rot in ginger (*P. myriotylum*), damping-off in bidi tobacco (*P. aphanidermatum*), root rot of chilli and brinjal (*P. aphanidermatum*)	Karthikeyan et al. 2000, Subhashini and Padmaja 2010, Shamarao and Jahagirdar 2012, Subhashini 2012, Yadav and Joshi 2012, Kato et al. 2013, Dinesh et al. 2015, Ved et al. 2017
Metalaxyl+Prothioconazole	Seed and seedlings diseases of maize (*P. splendens*), maize root diseases (*P. ultimum*)	Venturini et al. 2014
Metalaxyl+Trifloxystrobin (Trilex 2000)	Soybean seedling diseases (*Pythium* spp.)	Urrea et al. 2013
Metalaxyl-M	Blight of turfgrasses (*P. aphanidermatum*), douglas-fir damping-off (*P. irregulare*), root rot of geranium (*P. aphanidermatum*), *P. irregulare* (*P. ultimum*), seedling diseases of pepper (*P. aphanidermatum*), root rot and wilt of Kangaroo Paw (*P. myriotylum*), damping-off and root rot in forest nurseries (*P. dissotocum, P. irregulare, P. vipa, P. sylvaticum* and *P. ultimum*)	Tsror et al. 2005, Linderman et al. 2008, Cook et al. 2009, Mihajlović et al. 2013, Weiland et al. 2014, Múnera and Hausbeck 2015
Metalaxyl-M+ Difenoconazole (Dividend XL)	*Pythium* root rot of wheat (*P. irregulare*)	Cook et al. 2002
Metalaxyl-M+Fludioxonil+ (Hurricane/Warden RTA)	Douglas-fir damping-off (*P. irregulare*), soybean seedling diseases (*Pythium* spp.)	Bradley 2008, Linderman et al. 2008,
Metalaxyl-M+ Fludioxonil+Azoxystrobin (ApronMaxx+Dynasty)	Soybean seedling diseases (*Pythium* spp.)	Urrea et al. 2013
Metalaxyl-M+Azoxystrobin	Root rot of Russell prairie gentian (*P. spinosum*)	Satou and Fukuta 2016
Metam-sodium	Root diseases of squash (*P. irregulare*)	Desaeger et al. 2008
Methyl bromide	Seedling diseases of conifer (*P. irregulare*)	Weiland et al. 2011
Oxathiapiprolin	Seedling diseases of soybean (*P. mercuriale*)	Vargas 2018
Phosphonate (AG3)	Cucumber damping-off (*P. aphanidermatum, P. ultimum, P. irregulare, P. ultimum* and other *Pythium* spp.)	Abbasi and Lazarovits 2006
Potassium phosphite	Blight of turfgrasses (*P. aphanidermatum*)	Cook et al. 2009
Potassium phosphonate	Rhizome rot of small cardamom (*P. vexans*)	Dhanapal and Thomas 2008
Propamocarb hydrochloride (Previcur)	Papaya damping-off (*P. aphanidermatum*), seedling diseases of pepper (*P. aphanidermatum*)	Male and Vawdrey 2010, Mihajlović et al. 2013
Propioconazole	Chili damping-off (*P. aphanidermatum*)	Ved et al. 2017
Prothiocarb	Cabbage root diseases (*P. debaryanum*)	Cuc et al. 2007
Prothioconazole	Seed and seedling diseases of maize (*P. splendens, P. ultimum*)	Venturini et al. 2014

(*Contd.*)

Thiram+Metalaxyl+ Imidacloprid+Iprodione	Canola root diseases (*Pythium* F group, *P. irregulare, P. mamillatum, P. ultimum* var. *ultimum*)	De Villiers and Agenbag 2007
Thiophanate-methyl	*Ruta graveolens* damping-off (*P. debaryanum*)	Helmy et al. 2001
Thiram	Hemp root diseases (*P. debaryanum*), *Ruta graveolens* damping-off (*P. debaryanum*), soft rot of turmeric (*P. myriotylum*), chili damping-off (*P. aphanidermatum*), soft rot of turmeric (*P. myriotylum*), douglas-fir damping-off (*P. irregulare*)	Helmy et al. 2001, Saha et al. 2008, Shaikh and Kareappa 2008, Trotus and Naie 2008, Linderman et al. 2008
Tubuconazole	Chili damping-off (*P. aphanidermatum*)	Ved et al. 2017
Zoxamide	Cavity spot of carrot (*P. sulcatum*)	Martinez et al. 2005

Biological control

A number of diseases caused by different plant pathogens are effectively controlled through the use of fungal biocontrol agents. There is no single biocontrol agent that is effective for all *Pythium* species. Antagonists are highly variable in their interaction with different organisms. Isolates from a host appeared more effective for controlling the pathogens of same host crop. Usually higher population is found in nutrient-rich and rhizospheric soil than in non-rhizosphere soils (Ponmurugan et al. 2013). Selection of antagonist as well as method and time of application also matter for the effective control. Moreover, efficacy of used antagonist depends on inoculum density in the given field.

Use of fungal antagonists

Seed treatment: Seed dressing with antagonistic microorganisms not only reduces considerable seed loss caused by different pathogens, but also improves germination and plant health. For instance, *T. harzianum* and *T. viride* significantly decreased the incidence of tomato damping-off (Krishnamoorthy and Bhaskaran 1990, Gulhane et al. 2005). Most isolates of both species possessed a strong antagonistic ability against *P. debaryanum* and provided effective protection to beet and cotton seedlings (Dumitras and Fratilescu-Sesan 1979, Balicevic et al. 2007, 2009, Paradikovic et al. 2007). Use of *T. viride* for seed treatment @ 4 g/kg was also found effective in controlling the *P. deliense* causing tomato damping-off (Dinakaran and Ramakrishnan 1996). Seed treatment with different formulations of *T. harzianum, T. viride* and *T. longibrachiatum* prepared in different substrates like saw dust, maize cob + saw dust, neem seed kernel extract, neem oil, wheat bran, talc and cellulose increased the efficacy of antagonists effective in controlling the infection of *Pythium* spp. (Cotes et al. 1996, Mev and Meena 2003, Muthukumar et al. 2007, Kapoor 2008). Root rot caused by *P. myriotylum* is effectively controlled by pretreatment of plant/seed with *Beauveria bassiana* (Ownley et al. 2008, Mbarga et al. 2012). Seed dressing with antagonistic microorganisms, especially *Aspergillus tamari*, reduced considerable seed loss due to *P. debaryanum* in tomato (Keshwal 1981). Coating soybean and Egyptian clover seeds and roots with spores and mycelia or soil application of *Aspergillus sulphureus, A. carneus, A. cervinus, Penicillium islandicum, P. funiculosum, P. nigricans, Paecilomyces variotii, P. lilacinus, Chaetomium globosum* and *Phoma pomorum* gave germinating seeds and seedlings a very good protection from diseases caused by *P. spinosum* (Maghazy et al. 2008, Al-Sheikh and Abdelzaher 2010). *Penicillium damascenum* appeared very effective in providing the protection against *P. vexans* infection; its mycelium develops the protective layers of antifungal compound around the host seedling roots (Yamaji et al. 2001, 2005). Among 32 fungi isolated from maize field, *A. terreus, Chaetomium globosum* and

Myrothecium verrucaria showed strong antagonism to *P. deliense*. Maize seed treatment with these fungi efficiently controlled root rot and damping-off diseases (Abdelzaher et al. 2000). Some non-pathogenic strains of soil-borne pathogenic fungi served as biocontrol agents. *Fusarium* sp. strain AF-967 showed strong biocontrol ability and remarkably inhibited the growth of *P. debaryanum* and other phytopathogens (Alzum et al. 1997). Against damping-off of chili and brussels sprouts caused by *P. irregulare*, *R. solani* AG 4 and *Pythium ultimum* var. *sporangiiferum*, the binucleate isolates of *Rhizoctonia* (BNR1 and BNR2) were more effective than *Bacillus amyloliquefaciens* and *Pseudomonas putida* (Harris and Adkins 1999). The *Epicoccum nigrum* and its culture filtrates showed significant antagonism against *P. debaryanum* and *P. ultimum*. Its application on cotton seeds or seedlings proved helpful to manage *Pythium* damping-off and root rot disease (Hashem and Ali 2004). *Pythium oligandrum*, a mycoparasite, effectively controlled *P. splendens* infection in cucumber seedlings (Thinggaard et al. 1988, Thinggaard 1989); it also showed hyperparasitism on *R. solani* and antibiosis towards *P. ultimum in vitro* and reduced seed rot and damping-off tomato caused by *P. ultimum* and *R. solani* (He et al. 1992).

VAM fungi form association with the roots and increase the plant growth, leaf, chlorophyll content as well as nutrients. Interaction of arbuscular mycorrhizal fungi, including *Glomus mosseae*, *G. fasciculatum*, *G. constrictum*, *G. intraradices* and *Acaulospora laevis* with *P. aphanidermatum* in tobacco seedbeds considerably reduced the disease incidence (Subhashini and Padmaja 2010). The endophytic fungus *Paraconiothyrium* sp. associated with ginger showed strong antagonism against *P. myriotylum* and rhizome treatment with spore suspension or metabolites (danthron) of this endophytic fungus provides effective control of rhizome rot caused by the pathogen (Anisha et al. 2018).

Soil treatment: The incorporation of *T. harzianum* in soil reduced the infection of *Pythium* species. However, the effect of *T. harzianum* was more pronounced in sterilized soil as compared to the non-sterilized soil (Devaki et al. 1992) and slightly varied with the type of substrate used (Balicevic et al. 2008). Soil application of *T. harzianum* effectively minimized the damping-off and root-rot diseases involving *P. debaryanum*, *P. aphanidermatum* and number of other phytopathogens (Xu et al. 1993, Abada 1994, Balicevic et al. 2008). Mutants of *T. viride* effectively controlled the mustard damping-off caused by *P. aphanidermatum* by more than 85% in both the sterilized and natural soils (Khare et al. 2010). Incorporation of the *Trichoderma* before sowing into soil infested with *P. spinosum* increased *Pinus massoniana* germination; after emergence the disease was reduced by sprinkling the seedbed with a spore suspension. Germination was also promoted by drenching with *Trichoderma* spore suspension (Tang and Uung 1983).

Foliar spraying: The efficacy of *Trichoderma* species showed a variation with strain to strain against different pathogens (Raut et al. 2012). *T. harzianum* and *T. viride* isolates obtained from healthy seeds, phylloplane and rhizosphere of tomato seedlings and plants showed antagonistic effects against *P. aphanidermatum* (Gulhane et al. 2005). Three foliar sprays of *T. viride* @ 4 kg/ha on ginger plants grown from rhizome treated with *T. viride* @ 10 g/l of water significantly reduced the soft rot disease incidence (Chaturvedi 2014).

Use of bacterial antagonist

Seed treatment: Some strains of *Pseudomonas fluorescens* effectively controlled different *Pythium* species and improved the crop yield. Out of 59 *P. fluorescens* isolates evaluated against cotton seedling diseases caused by *P. deliense*, only two showed good results (Erdogan et al. 2016). In several greenhouse and field trials the application of *P. fluorescens* to cucumber, capsicum, tomato and aubergine seeds effectively suppressed the infection of *Pythium* spp. and improved the crop yield (Zhang et al. 1990). It also brought a significant reduction in cotton damping-off (Aşkın and Katırcıoğlu 2009). Tomato seed treatment with *P. aeruginosa* effectively controlled damping-off disease. *Pseudomonas aeruginosa* also have the ability to persist in hydroponic system, thus providing additional protection (Buysens et al. 1995). Some species of *Pseudomonas* are equally effective as chemical fungicides. Tomato seed coating with *P. aureofaciens* at 10^8 cfu per seeds or Metalaxyl provided equal protection against *P. ultimum* (Warren and Bennett 2000). Similarly, some strains of *Bacillus* species were highly effective in reducing different *Pythium* diseases (Zaspel and Suss 1992, Kim et al. 1997, Bhai et al. 2005, Dinesh et al. 2015). They have shown to produce antifungal compounds. Pre-inoculation of ginger rhizome with *Bacillus* isolates provided effective protection against the infection by *P. myriotylum*. *Pseudomonas fluorescens* and *B. subtilis* were compatible with each other and their combined application was more effective in reducing the rhizome rot of cardamom caused by *P. vexans* and in enhancing the plant growth as compared to their individual application (Sivakumar et al. 2012).

Soil treatment: Different strains of *Pseudomonas* when incorporated in the soil showed a strong antagonistic ability against a number of pathogens including *Pythium* species (Perneel et al. 2007, Al-Hinai et al. 2010). *Pseudomonas aeruginosa* effectively inhibited the growth of *P. myriotylum* and its infection on cocoyam (Tambong and Achuo 1997, Tambong et al. 1999). However, different strains of *P. aeruginosa* differ in terms of aggressiveness to *Pythium*. The combined soil application of rhamnolipid-deficient and phenazine-deficient mutants of *P. aeruginosa* was more effective than their individual use against *P. myriotylum* on cocoyam or *P. splendens* on bean (Perneel et al. 2008). The selected rhizospheric fluorescent *Pseudomonas* isolates of alfalfa plants was found to possess strong antagonistic ability towards *P. debaryanum* (Yanes et al. 2004, 2012). Talc based formulation of *P. fluorescens,* used either as seed or soil treatment, supported better survival of the bacterium in soil (Subhashini 2012). A soil actinomycete *Streptomyces lavendulae* provided good control of tobacco damping-off caused by *P. aphanidermatum*. Spray of mixed spore suspension of the pathogen (*P. aphanidermatum*) and the antagonist (*S. lavendulae*) on tobacco seedbeds improved germination, dry weight, leaf area and nutrient content (Subhashini 2010). Species of *Pseudomonas*, *Bacillus*, *Arthrobacter*, *Brevibacillus*, *Paenibacillus* and *Rummeliibacillus* associated with composts have shown antagonistic abilities against *P. sulcatum* and many other phytopathogens (Mohamed et al. 2017).

Foliar spraying: Similar to seed and soil application, some antagonistic bacteria were also effective against *Pythium* spp., when applied as spray. *Bacillus subtilis* possessed varied antagonistic potential against different pathogens. In greenhouse cultivated tomato and cucumber, *B. subtilis* significantly controlled *P. debaryanum*, and other pathogens (Constantinescu and Sesan 2002). *B. megaterium* (AUM72) induced the activity of defense related enzymes, including phenylalanine ammonia lyase, peroxidase, polyphenol oxidase, chitinase and β-1,3-glucanase in the turmeric plant. Higher accumulation of phenolics was noticed in plants pre-treated with *B. megaterium* that provided control of rhizome rot in turmeric caused by *P. aphanidermatum* (Uthandi and Sivakumaar 2013).

Use of secondary compounds from biocontrol agents

In many cases, secondary metabolites produced by microbial antagonists were equally effective as the organism itself. The secondary metabolites of different isolates of *T. harzianum* and *T. viride* reduced the infection of different pathogens including *P. vexans* on the seedlings of different host plants. Among 13 metabolites associated with *T. harzianum*, only 6-pentyl-α-pyrone showed antifungal potential when amended to the growth medium and effectively checked the pathogen growth (Sharma and Dureja 2004). The metabolites obtained from the composted hardwood bark isolate of *T. harzianum* were found effective in enhancing plant growth and inhibiting *P. irregulare* and other pathogens (Vinale et al. 2009). Among different toxins such as gliotoxin, dimethyl gliotoxin, viridin and viridiol obtained from culture filtrate of *T. virens*, gliotoxin showed strong inhibition against *P. debaryanum* and other plant pathogens (Singh et al. 2005).

Four fungal metabolites *viz.*, arthrographol, citrinin, palitantin and patulin produced by *Penicillium* spp., associated with the *Picea glehnii* seeds showed antifungal activity against *P.*

vexans causing damping-off (Yamaji et al. 1999). *Paxillus* sp., an ectomycorrhizal fungus, produced metabolites which have antifungal potential against *P. vexans* only at pH 3-4; variation from this pH level renders the metabolites ineffective against the test pathogen (Yamaji et al. 2005). *Coniochaeta ligniaria*, an endophytic fungus from leaves of *Baeckea frutescens*, and its crude ethyl acetate extract inhibited *Phytophthora palmivora* and *Pythium aphanidermatum* (Kokaew et al. 2011). Antifungal metabolites from *Chaetomium cochliodes* extracted with EtOAc suppressed the growth of *P. aphanidermatum*, inhibited oospores formation and caused abnormalities in the host hyphae, oogonia and oospores. The mechanism of the antagonist was lysis and antibiosis (Pornsuriya et al. 2010).

Paromomycin, an aminoglycoside antibiotic produced by *Streptomyces* spp. showed inhibitory potential against *P. myriotylum* and caused lytic effects on its zoospores and hyphae (Geethu et al. 2013). The antibiotics like griseofulvin, streptomycin, aureomycin and actidione inhibited the growth of *P. irregulare* and other species and also caused various morphological abnormalities in germ tubes. Sporangium germination of *P. irregulare* was greatly affected by actidione (Vaartaja and Agnihotri 1969). Soil application of thiolutin, a sulfur-containing antibiotic prepared from *Streptomyces luteosporeus*, was found to inhibit the *P. debaryanum*, but supported the soil population of antagonists like *Trichoderma* and *Penicillium* (Deb and Dutta 1984). Pre-sowing soaking of cabbage, tomato and lettuce seeds in antibiotic solution produced by *Streptomyces* sp. completely checked the infection and enhanced the seedling germination and growth, in soil artificially infested with *P. debaryanum* (Numic et al. 1979).

Pseudomonas aeruginosa produces phenazine-1-carboxylic acid and phenazine-1-carboxamide (oxychlororaphin) that effectively inhibited the growth of *P. myriotylum* as well as reduced the severity of cocoyam root rot and ginger rhizome rot (Tambong and Hofte 2001, Jasim et al. 2013, 2014). *Pseudomonas aeruginosa* PNA1 produced antifungal metabolites phenazines and rhamnolipid-biosurfactants; both effectively minimized the development of pre-emergence bean damping-off caused by *P. splendens* when applied synergistically, but showed no antagonistic effects if applied alone (Perneel et al. 2008). Anthranilate, a compound associated with *P. aeruginosa*, is an intermediate in the tryptophan biosynthesis pathway, and effectively controlled the *Pythium* damping-off in bean, lettuce and chickpea (Anjaiah et al. 1998a, b).

Crude culture filtrates of *B. subtilis* CU12 containing cyclic dimer of 3-hydroxypropionaldehyde effectively checked the growth of many plant pathogens including *P. sulcatum* (Wise et al. 2012). Similarly, tomato damping-off caused by *P. debaryanum* was controlled by culture filtrates of *Aspergillus niveus*, *A. tamarii* and *Penicillium notatum* (Joshi and Keshwal 1969). Hexane and dichloromethane extracts of four lichen species *viz.*, *Parmelia reticulata*, *Ramalina roesleri*, *Usnea longissima* and *Stereocaulon himalayense* were more effective than their aqueous extract against a number of pathogens including *P. dabaryanum* (Mayurika et al. 2011). During *in vitro* bioassays, a secondary compound veratryl alcohol obtained from a root-inhabiting sterile fungus effectively checked the growth of *P. irregulare* and *Sclerotinia sclerotiorum* (Vinale et al. 2010). Prodigiosin obtained from *Serratia marcescens* possesses antifungal activity against a number of pathogens, including *Phytophthora capsici*, *Cochliobolus miyabeanus*, *Pythium spinosum*, *P. ultimum*, *Clavibacter michiganensis* subsp. *michiganensis* and *Erwinia carotovora* subsp. *carotovora* (Okamoto et al. 1998).

Use of plant secondary metabolites

Some plant secondary metabolites possess enormous antimicrobial potential. Reductions in number of diseases have been achieved through the use of crude plant extracts extracted with different solvents or through purified plant secondary metabolites. Soil amendment with leaf extracts of *Hyptis suaveolens*, *Murraya koenigii* and *Ocimum canum* controlled tomato damping-off caused by *P. debaryanum* and *P. aphanidermatum* (Pandey and Dubey 1997). The infection of *P. deliense* in tomato and pepper plants was significantly minimized by the ethanolic extract of garlic applied with irrigation water (Ozkaya and Ergun 2017). Cinnamon extracts effectively inhibited the growth of *P. sulcatum* and suppressed lesion development in case of cavity spot of carrot (Mvuemba et al. 2009). The extract of wild ginger species (*Zingiber zerumbet*) showed strong inhibitory potential against *P. myriotylum* causing soft rot in common ginger species (*Zingiber officinale*) and provided protection under *in vivo* conditions (Aswani et al. 2017). Similarly, botanical extracts of *Azadirchta indica, Eucalyptus globulus, Pseudarthria viscida, Eupatorium cannabinum* and *Vitex negundo* were found to inhibit number of pathogens including *P. debaryanum* and *P. aphanidermatum* (Kumar and Tripathi 1991, Deepa et al. 2004, Gomathi et al. 2011, Sadhna et al. 2012).

Essential oils showed strong inhibition potential against *Pythium* diseases. The tomato seed dressing with essential oils of *Chenopodium ambrosioides* and *Lippia alba* remarkably reduced damping-off in soil artificially infested with *P. debaryanum* and *P. aphanidermatum*; the efficacy was even higher than Agrosan, Captan and Ceresan (Kishore and Dubey 2002). Leaves of *Hyptis suaveolens, Murraya koenigii* and *Ocimum canum* or essential oils from these plants controlled damping-off disease of tomato in soil infested with *P. aphanidermatum* and *P. debaryanum* (Pandey and Dubey 1994, 1997). The essential oils of *Mentha piperita, Lavandula* sp., *Rosmarinus officinalis, Eucalyptus* sp., *Cuminum cyminum, Syzygium aromaticum, Anethum graveolens, Echinophora, Senecio amplexicaulis* and *Thymus zygis* showed inhibition against a number of plant pathogens including *P. irregulare, P. debaryanum, P. ultimum* and *P. aphanidermatum* (Perez-Sanchez et al. 2007, Kareem 2010, Arıcı et al. 2011, Singh et al. 2016).

Glucolimnanthin, a secondary plant metabolite obtained from seed meal and freshly ground seeds of meadowfoam (*Limnanthes alba*), has pesticidal potential and in the presence of the enzyme myrosinase can be converted into pesticidal compounds such as 3-methoxybenzyl isothiocyanate (ITC) and 3-methoxyphenylacetonitrile (nitrile) that are highly toxic to *P. irregulare* and *Meloidogyne hapla* (Ersahin et al. 2014). Compounds like gossypol, gossypolone and apogossypolone from cotton seed have strong inhibition potential against *P. irregulare, P. ultimum* and *Fusarium oxysporum* (Mellon et al. 2014). Garlic extract contains organosulfur compounds especially diallyl disulfide (allicin), diallyl trisulfide and ajoene, which inhibit different microbial enzymes essential for the growth of different pathogens including *P. aphanidermatum* and *P. splendens* (Tedeschi et al. 2007, Mostafa et al. 2013). Similarly, partially purified taxane extract obtained from the needles of ornamental yews (*Taxus* spp.) showed high inhibitory activity against *P. aphanidermatum* and *P. myriotylum* (Elmer et al. 1994).

Biofumigation

Biofumigation of soil is carried out through incorporating soil with different organic wastes or intercrops. It not only increases the soil nutrient availability, but also provides control of many soil-borne pathogens. Biofumigation of soil has been considered non-toxic, very effective and eco-friendly approach for the management of diseases. Besides direct suppression of phytopathogens, incorporation of organic substrate into the soil also enhances the microbial population, hence increasing their antagonistic efficacy. Soil amendment with *Brassica juncea* seed meal based substrates was found to be very effective against *Pythium* spp. especially *P. abappressorium* causing apple replant disease and *P. sulcatum* causing carrot cavity-spot disease. The emission of allyl isothiocyanate (AITC) from *Brassica* based seed meal is mainly responsible for pathogen suppression. However, the continuity in pathogen suppression even after termination of AITC in amended soil demonstrated the role of some other compounds or microbes (Breton et al. 2011, Weerakoon et al. 2012).

Volatile compounds released from crushed aerial parts of *Brassica juncea* were found effective against *P. sulcatum* and *Rhizoctonia solani*. Biofumigation of soil with mustard brought an important reduction in soil infectivity over time, as well as reducing the incidence of brown rot disease in successive carrot crops (Montfort et al. 2011). Soil biofumigation with degraded cabbage and green mustard was found highly effective for controlling *P. aphanidermatum* and produced 79% and 92% inhibition, respectively. Soil amendment with *Brassica* spp. at 20% w/w completely reduced cucumber damping-off disease (Choochuay et al. 2016). Solarization followed by biofumigation reduced damping-off disease caused by *P. aphanidermatum* and enhanced plant growth of greenhouse cucumber in Oman (Deadman et al. 2006). The addition of onion and leek byproducts in compost artificially infested with *P. ultimum* induced toxic fumigants that provided satisfactory disease control and increased the vegetative growth of cucumber (Arnault et al. 2008). It is said that at least to some extent, biofumigation may provide an alternative to highly toxic fumigant, methyl bromide (Arnault et al. 2008).

Composting

Mixtures of agri-food industry wastes (lees, rice husk, cocoa waste, wafer waste and hazelnut shells) significantly decreased the number of dead cucumber plants grown on the substrate artificially infested with the *P. ultimum* and the waste like lees and marc of the wine industry are tested against *P. ultimum* infection on cucumber (Pugliese et al. 2012, 2013). The efficacy of disease suppressive composts against *P. ultimum* was influenced by particle size of the given compost (Pugliese et al. 2013). *Pythium ultimum* infection was effectively reduced on cucumber and bean grown in greenhouse conditions by the soil amendment with cane lignin and straw lignin biomass used in the production of biodiesel (Moreno Alvarez et al. 2013). In greenhouse study, the composts prepared from a wide variety of crop wastes (green grasses, banana stems, sugarcane waste, tree leaves and branches, cocoa pods, coffee hulls, oil palm waste and tree bark) reduced the pathogen infection on cocoyam plants (Perneel et al. 2004). Organic amendments in the shape of sugarcane plant residues done four months before capsicum planting increased the microbial biomass and free living nematodes, and decreased the populations of root knot nematodes, *P. aphanidermatum* and *P. myriotylum*. It resulted in reduced levels of *Pythium* root rot severity in contrast to plastic mulched soils (Stirling and Eden 2008). Different formulations of mushroom compost, spent golden mushroom compost and paper mill sludge effectively inhibited *P. myriotylum* root rot of tomato and watermelon (Chiu and Huang 1997).

Un-decomposed plant residues of alfalfa in the soil increased the severity of replanting alfalfa seedling damping-off caused by *P. ultimum* and *R. solani*. Therefore, complete decomposition of last crop before sowing of new one may help control the damping-off problem (Bonanomi et al. 2011). The mixture of peat moss with composted biowastes has shown a high level of disease suppression to different pathogen-host combinations including *P. ultimum* and *P. irregulare* infection in cucumber (Blok et al. 2005, Scheuerell et al. 2005). A good control of *P. splendens* population in Hawaiian soils was achieved by synergistic application of calcium and lucerne meal. Lucerne meal served as the source or promoter of beneficial microbes (Kao and Ko 1986a).

Cultural control

Crop rotation, use of suitable cover, soil solarization, mulching, sanitation and pasteurization, use of healthy seeds, fertilizers and manures may affect the growth of plants and help to reduce the diseases. *Pythium sulcatum* has comparatively narrow host range and mostly infects plant species belonging to the Apiaceae family. Cavity spot of carrots caused by *P. sulcatum* was effectively controlled by crop rotation with broccoli. It not only reduced the cavity spot incidence and severity in disease hot spots, but also improved quantity and quality of the produce (Davison and McKay 2003). Cultivation of suitable cover crops not only increased the population of beneficial microbes, but also reduced the inocula of different soil-borne pathogens. Among oats, spring vetch and tansy phacelia, grown as cover crop, oats greatly reduced the population of different fungi including *P. irregulare* in the soil as well as its mulching increased the antagonistic mycoflora in the soil (Patkowska and Konopiński 2014).

Soil solarization of fields where potato, cauliflower and cucumber were grown, was highly effective in reducing the inoculum of *P. debaryanum* and other soil-borne plant pathogens. Positive effects also reflected in increased yield of these vegetables (Abu-Blan and Abu-Gharbieh 1994). Soil solarization with transparent plastic for 10 weeks remarkably reduced the populations of soil-borne plant pathogens including *P. debaryanum* associated with tomato brown root rot disease (Bourbos and Skoudridakis 1991). The adverse effects of blast disease (*P. splendens* and *Rhizoctonia lamellifera*) were reduced by shading the oil palm seedlings during the hot dry period (Rajagopalan 1974). Foot rot in saintpaulias caused by *Phytophthora nicotianae* and *Pythium spinosum* can be prevented by nursery hygiene and by ensuring that the potting soil is not too wet or has too high salt content (Dirkse 1981).

Fertilizers also play a vital role in determining the host response to pathogen invasion. In case of root rot of *Pelargonium* caused by *P. splendens* or *Phytophthora cinnamomi*, the use of fertilizer, like P alone or with N, makes the host more susceptible to these two Oomycetes (Mohamed et al. 1987). The groundnut pod rot caused by *P. myriotylum* was controlled by soil application of gypsum and GS (a mixture of gypsum, rice hull, sulfur, oyster shells, fish meal, tobacco grounds and complex fertilizer); both amendments also caused positive effects on yield (Chen and Huang 1992). Application of calcium in soil either in the form of $CaCO_3$, $CaSO_4$ or $Ca(OH)_2$ was found highly effective in managing the cucumber seedling damping-off caused by *P. splendens*. The incorporation of $CaCO_3$ and lucerne meal alone or synergistically brought significant reduction in incidence of damping-off (Kao and Ko 1986b). Similarly, sulfur-containing salts, especially calcium sulfate and ammonium sulfate showed high inhibition of *P. sulcatum*. The carrot cavity spot disease was also effectively controlled by 50 mM calcium sulfate, sodium sulfate and metabisulfite salts (Kolaei et al. 2012). Aluminium containing salts including aluminium chloride and aluminium sulfate were also effective against the same disease (Kolaei et al. 2013).

Among different organic manure, poultry manure at 20 tons ha^{-1} increased cotton yield and decreased damping-off caused by *P. ultimum, Rhizoctonia solani* and *Fusarium* spp. (Hoshiarfard and Gharanjiki 2009). Soil amendment with animal manure based vermicompost was more effective against cocoyam root rot disease caused by *P. myriotylum* than compost prepared from coffee pulp (Artavia et al. 2010). The effectiveness of the different composts against *P. irregulare* causing pre-emergence damping-off of cucumber varied with the exposing temperatures; the maximum disease reduction was observed at 20° to 24°C (Ben-Yephet and Nelson 1999). Cavity spot disease may be minimized by altering the pH of growing medium since no disease development was found in soils having pH 8.0 or above (White 1988). Sanitation is critical for avoiding the disease spread. Transfer of infected soil particles with agriculture equipment/machinery is very common. Disinfection of vehicles' tires entering from another country with 0.240% NaOCl eliminated *Pythium* contaminants by decreasing the oospores viability (Hashem 2015).

Use of resistant varieties

Plant resistance provides the most effective and reliable control of plant diseases. Cultivars of geranium and snapdragon exhibited different levels of resistance against the *P. irregulare* or other *Pythium* species in greenhouse studies. Among the 11 geranium cultivars, two were completely resistant to *P. irregulare*, whereas, among 12 snapdragon cultivars, no one was resistant, but two were less susceptible (Múnera and Hausbeck 2015). Common bean (*Phaseolus vulgaris*) is one of the most widely grown leguminous crops, which is continuously threatened by many soil-borne root pathogens including couple of *Pythium* species. In Africa, bean cultivation is subjected to the attack of a fairly large number of *Pythium* species *viz., P. arrhenomanes, P. conidiophorum, P. cucurbitacearum, P. diclinum, P. dissotocum, P. folliculosum, P. indigoferae, P. pachycaule, P. rostratifingens, P. spinosum, P. torulosum, P. ultimum, P. vexans, P. macrosporum* and *P. rostratum*. The screening of 10 commercial varieties by artificial inoculation of these all *Pythium* species revealed that four cultivars are susceptible to all species where six varieties showed a high level of resistance (Nzungize et al. 2011a). This provided an opportunity for developing high yielding *Pythium* resistant bean germplasm by transferring the *Pythium* resistant genes from already identified resistance sources into high yielding susceptible cultivars. In each successive backcross, the incorporation of resistance genes was confirmed by the presence of PYAA19$_{800}$ SCAR marker that was linked to *Pythium* root rot resistance (Nzungize et al. 2011b). Li et al. (2014) suggested that common bean resistant isolate of any *Pythium* species is enough to demonstrate the behavior of a variety against a large number of *Pythium* species. The resistance to *Pythium* has been determined by single dominant gene PYAA19$_{800}$ (Ongom et al. 2012). In periwinkle (*Catharanthus roseus*), the resistance to dieback caused by *P. aphanidermatum* is also governed by a single gene (Kulkarni and Baskaran 2003). However, in some cases pyramiding of large number of genes against different pathogens may result in yield reduction. The attempt to incorporate three anthracnose and one *Pythium* root rot resistance genes simultaneously in four susceptible, but commercial bean varieties resulted in grain reduction (Kiryowa et al. 2015). In a comprehensive varietal screening program that consisted of 194 bean cultivars, only two were the most resistant to *P. irregulare* root rot disease (Li et al. 2016). Adegbola and Hagedorn (1970) found that only one cultivar possessed resistance to blight involving five *Pythium* species. Another study revealed that out of 138 bean cultivars, one showed resistance response against *P. aphanidermatum* (Kim and Kantzes 1972). Similarly, Dickson and Abawi (1974) found one snap bean cultivar resistant to *P. ultimum* damping-off.

　　Pythium myriotylum is considered the most destructive pathogen of cocoyam (*Xanthosoma sagittifolium*), an important food source in humid and sub-humid regions of the world. *P. myriotylum* root rot has the potential to reduce cocoyam yield up to 90%. Therefore, development of resistant cocoyam germplasm has the focus of plant scientists since long. Cocoyam cultivars may be grouped into resistant, tolerant and susceptible to *P. myriotylum* infection. The plant phenolic contents also varied with the host response types; they were more in resistant cultivars than others (Temgo and Boyomo 2002). Peroxidase and pectin methylesterase may regulate the plant resistance in case of *P. myriotylum* infection (Boudjeko et al. 2005). Mostly, the aim of development of disease resistant varieties has been achieved by selection. Under a comprehensive varietal development program, 12 high yielding accessions were selected; none were immune to *Pythium* root rot (Wutoh et al. 1991). In other crops like peanuts, breeding varieties having resistance to pod rot caused by *P. myriotylum* had been the focus of the scientists (Smith et al. 1979). *Pythium myriotylum* is more virulent on ornamentals like *Caladium* and *Amaranthus*. In two different studies, out of 19 and 23 caladium cultivars, only four and three showed partial resistance, respectively, while remaining were either susceptible or highly susceptible (Deng et al. 2005a, b). Similarly, from 126 accessions of *Amaranthus*, no one was completely resistant (Sealy et al. 1988).

　　The host-*Pythium* interactions in resistant and susceptible cultivars were unique. In case of cavity spot of carrot caused by *P. violae*, in susceptible cultivar the pathogen penetrated deep inside the host tissues very quickly: within the first 24 hours of ingression, it reached up to pericycle and the phloem parenchyma and resulted in the host wall dissolution and cytoplasm clumping. On the other hand, in resistant cultivar the penetration of *P. violae* was limited up to pericycle (Guérin et al. 1998). The tissue culture developed somaclones of carrot have been found to be a good source of resistant genetic material against cavity spot disease. Their response was not uniform and varied in greenhouse and field conditions. However, in both growing conditions, the disease rating of some of somaclones was lower as compared to the four commercial cultivars (Cooper et al. 2006).

　　In peas and strawberry, differential responses of cultivars were found against *P. ultimum*. Among eight commercial strawberry cultivars screened against the different crown and root diseases pathogens individually, only one cultivar cv. Festival was most resistant to *P. ultimum* (Xiangling et al. 2012). Similarly, among different accessions, only five showed resistance reaction to pea root rot by *P. ultimum* and *F. solani* f. sp. *pisi* (Kraft and Roberts 1970). In peas, the resistance to *P. ultimum* has been mostly governed by the presence of seed coats colored by anthocyanins. The cultivars with colored seed coats exhibited more resistance than uncolored cultivars, which means they contain some antifungal compounds (Kraft 1974, Ohh et al. 1978, Stasz et al. 1980). In cowpea, amounts of oxalic acid and polygalacturonase in plant tissues regulate the host response to *P. aphanidermatum* (Koleosho et al. 1987).

Potato germplasm also possessed significant variation to *P. ultimum* infection, ranging from highly resistant to highly susceptible (Salas et al. 2003). In Japan, all tested potato cultivars and most of the sweet potato were found susceptible to *P. ultimum* (Takahashi et al. 1970). Similarly, out of 12, only three cotton cultivars were found to be highly resistant against *P. ultimum*, both in controlled and field conditions (Wang and Davis 1997).

In cereals such as maize, sorghum, wheat and barley, various efforts have been made to find out *Pythium* resistant cultivars. Out of 1550 maize accessions, 397 were highly resistant, 304 moderately resistant, 364 moderately susceptible and 485 highly susceptible to *P. inflatum* stalk rot (Xiuqin and Furong 1998). In another study, out of 287 inbred maize lines, 43 were highly resistant, 95 were highly susceptible, while remaining showed intermediate reactions against the same pathogen (Song et al. 2012). With the passage of time, the resistance to *Pythium* species may also breakdown as happened in other host-pathogen interaction like rust. Eight wheat varieties, previously resistant to *P. arrhenomanes*, became susceptible (Mojdehi and Singleton 2000). In sorghum, on the basis of controlled and field study, resistant and susceptible cultivars against *P. arrhenomanes* were identified (Forbes et al. 1987). Similarly, barley cultivars tested for four years showed varied responses to *P. graminicola* (Ho et al. 1941). For screening rose cultivars against root rot caused by *P. helicoides*, a new bioassay was developed by using modified soilless culture; this method reliably differentiated one cultivar Matsushima No. 3 as resistant and other cultivar Nakashima 91 as susceptible (Li et al. 2007). Out of 51 tobacco varieties screened by artificial inoculation with *P. debaryanum*, eight were found highly resistant, 10 resistant and 19 fairly so (Pakdi and Claridad 1971).

Among *Pythium* species, *P. aphanidermatum* is the most versatile pathogen that infects a fairly large number of host plants. Out of 26 varieties of melon (*Cucumis melo*), three of cucumber (*C. sativus*), three of squash (*Cucurbita pepo*) and 15 of watermelon (*Citrullus lanatus*) screened against *P. aphanidermatum* by artificial inoculation, most melon and cucumber cultivars were found susceptible while watermelon was moderately resistant and squash was highly resistant. Among all these cucurbits, the *C. melo* variety Gold and Silver were highly resistant (Rahimian and Banihashemi 1979). The resistance to damping-off caused by *P. aphanidermatum* was successfully achieved when commercial cucumber was grafted onto the resistant rootstocks of ridge gourd or sponge gourd cultivars (Al-Mawaali et al. 2012). In ginger, *P. aphanidermatum* causes devastating soft rot disease because of absence of resistant germplasm. Out of 22 cultivars, only 2 were moderately resistant and none were highly resistant to artificial infestation by *P. aphanidermatum* (Balagopal et al. 1974). Another study in which 134 cultivars were tested for three years to rhizome rot complex revealed that only one was resistant, eight moderately resistant, 56 moderately tolerant, 39 susceptible and 11 highly susceptible (Senapati and Ghose 2005). Out of different wild and cultivated Zingiberaceae species evaluated in the search of resistance to soft rot disease caused by *P. aphanidermatum* in cultivated ginger, *Zingiber zerumbet* was found as the most resistant cultivar which may be used as the potential donor for resistance (Kavitha and Thomas 2007). Different species and their cultivars of turf grasses showed different levels of resistance to *P. aphanidermatum*. *Poa pratensis* cv. Courtyard appeared as the most resistant, while *Agrostis stolonifera* and *Lolium perenne* cv. Inspire were the most susceptible cultivars (Yang et al. 2008).

Integrated application of antagonistic microorganisms with fungicides or compost

Complex interaction of soil-borne pathogens with their hosts in a particular ecological niche makes the resulting disease problem more difficult to control. Successful management of such problems required integration or combined application of different control measures. Application of different fungal and bacterial antagonistic microorganisms alone or in combination with different cultural methods as well as different fungicides provided better control of diseases. In some cases, integration of these control methods was found more beneficial than their individual use. Following are some examples depicting the significance of IPDM, effective against different *Pythium* spp.

Combined use of fungal and bacterial antagonists

The combined application of specific *Bacillus* isolates along with commercially available *Trichoderma* provided promising results in terms of pathogen control and enhancing plant growth (Jimtha et al. 2016). Alone application of a mixture of four *Bacillus subtilis* strains + one isolate of *B. uniflagellatus* failed to control the disease, but their combined application with *Glomus fasciculatus* moderately controlled the disease caused by *P. debaryanum* (Raabe et al. 1981). Application of a consortium of *T. harzianum* + *Glomus mosseae* + *P. fluorescens* reduced the rhizome rot of ginger caused by *P. splendens* and *F. oxysporum* f.sp. *zingiberi*. Maximum tillers, plant height and rhizome weight were also recorded with this treatment (Meenu et al. 2010). Similarly, mixture of *Burkholderia cepacia* +*T. harzianum* appeared highly effective against the same disease (Shanmugam et al. 2013). Use of *T. viride* and *Streptomyces* for the management of root rot and blackleg of geranium (*Pelargonium hortorum*) effectively reduced rate of disease caused by *P. splendens* (Bolton 1978). Turmeric rhizome rot incited by *P. aphanidermatum* was effectively controlled with seed and soil application of *T. viride* and *P. flourescens* @ 4 g kg^{-1} of seed and 2.5 kg ha^{-1} as basal and top dressing, respectively (Muthulakshmi and Saveetha 2009).

Combined use of antagonists and fungicides

Different fungal or bacterial biocontrol agents along with chemical fungicides have been tested in various combinations. Fungicidal treatment of rhizomes of ginger with Bavistin (carbendazim) + Ridomil MZ (metalaxyl + mancozeb) coupled with soil application of biocontrol agents (*T. harzianum* + *Pseudomonas* sp.) or rhizome treatment with metalaxyl + carbendazim followed by soil application of *Gliocladium virens* provided better rhizome rot control caused by *P. myriotylum* (Ram et al. 2000, Meena and Mathur 2003). Copper oxychloride and Phorate (organophosphate insecticide) not only showed good level of compatibility with *T. harzianum* but also supported its buildup at cardamom rhizosphere (Bhai and Thomas 2010). A specific strain of *T. asperellum* showed good inhibition potential against numbers of fungal pathogens including *P. debaryanum* as well as compatibility with many commercial fungicides such as azoxystrobin, mancozeb, cymoxinil + mancozeb, metalxyl + mancozeb at 100, 200 and 300 ppm. It indicates the potential of this strain in integrated disease management where it may be used along with fungicides (Manjunath et al. 2017). Application of *T. harzianum* alone or in combination with Akomin (potassium phosphonate) or copper oxychloride provided better control of rhizome rot disease of cardamom as compared to the other control agents (Vijayan and Thomas 2002, Dhanapal and Thomas 2008). Damping-off of field pea caused by *P. ultimum* or *P. irregulare* can be effectively controlled by combined application of *B. subtilis* or *B. polymyxa* + half dose of metalaxyl (Hwang et al. 1996). In winter wheat, infection of many soil-borne plant pathogens, including *Pythium* species effectively minimized seed treatment with certain rhizobacteria-fungicide combinations that resulted in increased yields (Cook et al. 2002).

Use of antagonists with compost or FYM

Composts are usually a good source of beneficial microbes including antagonistic microorganisms. The soils amended with disease suppressive composts have high population densities of antagonistic microbes like *Pseudomonas* spp., actinomycetes, and *Trichoderma* spp., which effectively controlled root rot disease in cocoyam caused by *P. myriotylum* (Adiobo et al. 2007). The damping-off of rooibos (*Aspalathus linearis*) caused by *P. irregulare* was effectively controlled when composts were amended with nonpathogenic *Pythium* species such as *P. acanthicum*, *P. cederbergense* and *Pythium* RB II, whereas composts alone were unable to reduce *P. irregulare* infection significantly (Bahramisharif et al. 2013). Combined use of VAM and charcoal compost significantly reduced damping-off in cucumber seedlings caused by *P. splendens* and *Rhizoctonia solani*. However, under field conditions, especially in areas of disease hot spot, their efficacy became less and provided protection for a limited time (Kobayashi 1990).

The addition of antagonistic cyanobacteria/bacteria in the composts not only provided effective disease control, but also caused positive impacts on seed germination and plant growth. Addition of *Anabaena oscillarioides* and *Bacillus subtilis* provided control of diseases caused by *P. debaryanum* and other soil-borne pathogens in tomato (Dukare et al. 2011). Bacterial and cyanobacterial strains alone as well as when amended in rice straw compost showed strong antagonistic potential against *P. debaryanum* and other fungal pathogens (Dukare et al. 2013).

The soil amendment with *T. viride* alone or with combination of farmyard manure effectively minimized the activity of *P. deliense*. *Trichoderma viride* significantly enhanced the seedling emergence of tomato grown in soil artificially infested with *P. deliense* as well as reduced the incidence of damping-off (Neelamegam and Govindarajalu 2002). Pre-planting application of *T. viride* + FYM or seed treatment coupled with soil treatment with *T. viride* effectively controlled rhizome rot disease of ginger caused by *P. myriotylum* and enhanced plant growth (Singh et al. 2012). The efficacy of *T. viride* was greatly enhanced when it was applied along with suitable organic matter such as neem cake or wheat bran (Neelamegam 2004).

Combined application of fungal biocontrol agents like *T. hamatum*, *T. harzianum* and *Paecilomyces lilacinus* along with resistance inducer substances such as Bion (benzo (1,2,3) thiadiazole-7-carbothioic acid S-methyl ester) and salicylic acid remarkably reduced the cotton root rot caused by *F. oxysporum* and *P. debaryanum* as well as greatly increased seed germination. Under field conditions, the best combination was *T. hamatum* + *P. lilacinus* + Bion + salicylic acid (Abo-Elyousr et al. 2009). Tomato seed treatment with methanol extract of *Aegle marmelos* 4% + *T. viride* (4 g/kg) + *P. fluorescens* (10 g/kg) provided promising control of *P. debaryanum*, causing tomato pre- and post-emergence damping-off (Yogeshwar et al. 2017). Another study indicated that soil amendments with dried leaves of *Thespesia populnea* along with *T. viride* remarkably reduced the infection of *P. deliense* in tomato plants and increased all plant growth parameters (Neelamegam 2005). The oil obtained from the leaves of *Eupatorium adenophorum* significantly inhibited the vegetative growth and biomass production of *P. myriotylum* and was compatible with fungicides; under *in vitro* conditions, both showed synergistic effects against *P. myriotylum* and reduced its growth drastically. Pre-inoculation treatment of ginger rhizome with oil also checked the *P. myriotylum* infection in artificially infested rhizome (Liu et al. 2017).

Conclusion

The review of literature regarding the management of diseases caused by *Pythium* species shows that chemicals can provide excellent control of such diseases. However, indiscriminate use of chemical fungicides can lead to environmental as well health hazards to man and animals. Alternative methods of control such as use of biocontrol agents or metabolites produced by these microorganisms, resistant varieties, different cultural methods and secondary metabolites produced by the plants are also effective against the diseases caused by this group of pathogens. It is also evident that the integration of more than one method is more useful than the individual use of different strategies. There is, therefore, need to device an integrated strategy for the management of each pathogen to get maximum protection against the disease they cause.

References

Abada, K.A. 1994. Fungi causing damping-off and root-rot on sugar-beet and their biological control with *Trichoderma harzianum*. Agric. Ecosyst. Environ. 51: 333-337.

Abbasi, P.A. and Lazarovits, G. 2006. Seed treatment with phosphonate (AG3) suppresses Pythium damping-off of cucumber seedlings. Plant Dis. 90: 459-464.

Abdelzaher, H.M.A., Gherbawy, Y.A.M.H. and El-Naghy, M.A. 2000. Damping-off disease of maize caused by *Pythium deliense* Meurs in El-Minia, Egypt and its possible control by some antagonistic soil fungi. Egyptian J. Microb. 35: 21-45.

Abo-Elyousr, K.A.M., Hashem, M. and Ali, E.H. 2009. Integrated control of cotton root rot disease by mixing fungal biocontrol agents and resistance inducers. Crop Prot. 28: 295-301.

Abu-Blan, H.A. and Abu-Gharbieh, W.I. 1994. Effect of soil solarization on winter planting of potato, cauliflower and cucumber in the central Jordan Valley. Dirasat. Series B, Pure App. Sci. 21: 203-213.

Adegbola, M.O.K. and Hagedorn, D.J. 1970. Host resistance and pathogen virulence in Pythium blight of bean. Phytopathol. 60: 1477-1479.

Adiobo, A., Oumar, O., Perneel, M., Zok, S. and Höfte, M. 2007. Variation of *Pythium*-induced cocoyam root rot severity in response to soil type. Soil Biol. Bioch. 39: 2915-2925.

Agati, J.A. 1931. Studies on the root-rot of the sugarcane seedlings in the nursery. Philipp. J. Agric. 2: 1-26.

Al-Balushi, Z.M., Agrama, H., Al-Mahmooli, I.H., Maharachchikumbura, S.S. and Al-Sadi, A.M. 2018. Development of resistance to hymexazol among *Pythium* species in cucumber greenhouses in Oman. Plant Dis. 102: 202-208.

Al-Hinai, A.H., Al-Sadi, A.M., Al-Bahry, S.N., Mothershaw, A.S., Al-Said, F.A., Al-Harthi, S.A. and Deadman, M.L. 2010. Isolation and characterization of *Pseudomonas aeruginosa* with antagonistic activity against *Pythium aphanidermatum*. J. Plant Pathol. 92: 653-660.

Al-Kiyumi, K.S. 2006. Greenhouse Cucumber Production Systems in Oman: A Study on the Effect of Cultivation Practices on Crop Diseases and Crop Yields. PhD Thesis, University of Reading, Reading, UK.

Al-Mawaali, Q.S., Al-Sadi, A.M., Khan, A.J., Al-Hasani, H.D. and Deadman, M.L. 2012. Response of cucurbit rootstocks to *Pythium aphanidermatum*. Crop Prot. 42: 64-68.

Aloj, B., D'Errico, F.P. and Ragozzino, E. 2000. New alternative methods to the control of soil-borne parasites with methyl bromide. GF 2000. Atti, Giornate Fitopatologiche, Perugia, 16-20 Aprile, 2000, Volume Primo, 535-540.

Al-Sa'di, A.M., Drenth, A., Deadman, M.L. and Aitken, E.A.B. 2008. Genetic diversity, aggressiveness and metalaxyl sensitivity of *Pythium aphanidermatum* populations infecting cucumber in Oman. Plant Pathol. 57: 45-56.

Al-Sheikh, H. and Abdelzaher, H.M.A. 2010. Isolation of *Aspergillus sulphureus*, *Penicillium islandicum* and *Paecilomyces variotii* from agricultural soil and their biological activity against *Pythium spinosum*, the damping-off organism of soybean. J. Biol. Sci. 10: 178-189.

Alzum, M., Shilnikova, V.K. and Shkalikov, V.A. 1997. Antagonistic properties of *Fusarium* sp. (AF-976). Izv. Timiryazev. S-Kh. Akad. 2: 109-113.

Anisha, C., Sachidanandan, P. and Radhakrishnan, E.K. 2018. Endophytic *Paraconiothyrium* sp. from *Zingiber officinale* Rosc displays broad-spectrum antimicrobial activity by production of danthron. Curr. Microbiol. 75: 343-352.

Anjaiah, V., Cornelis, P., Koedam, N., Höfte, M. and Tambong, J. 1998a. Biocontrol of Fusarium wilt and Pythium damping-off by the metabolites of *P. aeruginosa* PNA1 and its TN5-derivatives: Phenazines and anthranilate. Meded. Rijksuniv. Gent Fak. Landbouwkd. Toegep. Biol. Wet. 63: 1671-1678.

Anjaiah, V., Koedam, N., Nowak-Thompson, B., Loper, J., Höfte, M., Tambong, J. and Cornelis, P. 1998b. Involvement of phenazines and anthranilate in the antagonism of *Pseudomonas aeruginosa* PNA1 and Tn5 derivatives toward *Fusarium* spp. and *Pythium* spp. Mol. Plant-Microbe. Interact. 11: 847-854.

Anon. 1971. Soil treatment in the nursery. Plant Dis. Bull., New South Wales Department of Agriculture 172: 11.

Arıcı, Ş.E., Özgönen, H., Şanlı, A., Polat, M. and Yasan, G. 2011. Antimicrobial activity of essential oils against agricultural plant pathogenic fungi and bacteria. In 4ème Conférence Internationale sur les Méthodes Alternatives en Protection des Cultures. Evolution des cadres réglementaireseuropéen et français. Nouveaux moyens et strategies Innovantes, Nouveau Siècle, Lille, France, 8-10 mars 2011. Association Française de Protection des Plantes (AFPP), pp. 249-253.

Ark, P.A. and Middleton, J.T. 1949. Pythium black rot of cattleya. Phytopathol. 39: 1060-1064.

Arnault, I., Vey, F., Fleurance, C., Nabil, H. and Auger, J. 2008. Soil fumigation with *Allium* sulfur volatiles and *Allium* by-products. Cultivating the Future Based on Science: 2nd Conference of the

International Society of Organic Agriculture Research ISOFAR, Modena, Italy, June 18-20, 2008, pp. 540-543.

Arslan, G. 1979. The effects on the germination of eggplant seed of various chemicals used for controlling damping-off caused by *Pythium* spp. Cokerten hastaligi (*Pythium* spp.) ile savasimda etkili cesitli kimyasallarin patlican (*Solanum melongena*) cimlenme gucu uzerine etkisi. Tarimsal Arastirma Dergisi 1: 132-135.

Artavia, S., Uribe, L., Saborío, F., Arauz, L.F. and Castro, L. 2010. Efecto de la aplicación de abonosorgánicosen la supresión de *Pythium myriotylum* enplantas de tiquisque (*Xanthosomas agittifolium*). Agron. Costarric. 34: 17-29.

Aşkin, A. and Katircioğlu, Y. 2009. Determination of pathogens causing damping-off and their pathogenicity in tomato seedbeds in Ankara (Ayaş, Beypazarı and Nallıhan districts) province. Bitki Koruma Bülteni 48: 49-59.

Aswani, R., Saji, V. and Radhakrishnan, E.K. 2017. Biocontrol activity of the extract prepared from *Zingiber zerumbet* for the management of rhizome rot in *Zingiber officinale* caused by *Pythium myriotylum*. Arch. Phytopathol. Plant Prot. 50: 555-567.

Attaullah, H., Lal, A.A. and Sobita, S. 2015. Eco-friendly management of damping-off (*Pythium aphanidermatum*) of chilli (*Capsicum annum* L.). In. J. Agri. Sci. Res. (IJASR). 5: 1-5.

Augusto, J., Brenneman, T.B. and Csinos, A.S. 2010. Etiology of peanut pod rot in Nicaragua II: The role of *Pythium myriotylum* as defined by applications of gypsum and fungicides. Plant Health Prog. doi:10.1094/PHP-2010-0215-02-RS.

Avikainen, H., Koponen, H. and Tahvonen, R. 1993. The effect of disinfectants on fungal diseases of cucumber. Agri. Food Sci. 2: 179-188.

Bahramisharif, A., Lamprecht, S.C., Calitz, F. and McLeod, A. 2013. Suppression of *Pythium* and *Phytophthora* damping-off of rooibos by compost and a combination of compost and nonpathogenic *Pythium* taxa. Plant Dis. 97: 1605-1610.

Bala, K., Robideau, G.P., Lévesque, A., de Cock, A.W.A.M., Adad, Z.G., Lodhi, A.M., Shahzad, S., Ghaffar, A. and Coffey, M.D. 2010. *Phytopythium* Abad, de Cock, Bala, Robideau, Lodhi and Lévesque gen. nov. and *Phytopythium sindhum* Lodhi, Shahzad & Lévesque sp. nov. Persoonia 24: 136-137.

Balagopal, C., Devi, S.B., Indrasenan, G. and Wilson, K.I. 1974. Varietal reactions of ginger (*Zingiber officinale* R.) towards soft rot caused by *Pythium aphanidermatum* (Edson) Fetz. Agri. Res. J. Kerala 12: 113-116.

Balicevic, R., Paradikovic, N. and Samota, D. 2007. Control of soil parasites (*Pythium debaryanum, Rizoctonia solani*) on tomato by a biological product. Cereal Res. Commun. 35: 1001-1004.

Balicevic, R., Paradikovic, N., Cosic, J., Jurkovic, D. and Samota, D. 2008. Influence of substrate in biological control of tomato seedlings against *Rhizoctonia solani* and *Pythium debaryanum*. Cereal Res. Commun. 36: 1499-1502.

Baličević, R., Parađiković, N., Ćosić, J., Šamota, D. and Vinković, T. 2009. Efficiency estimation of biological protection in vegetable transplants production. *In*: 2nd International Scientific/Professional Conference, Agriculture in Nature and Environment Protection, Vukovar, Croatia, 4-6 June 2009, Croatian Soil Tillage Research Organization (CROSTRO), pp. 95-98.

Bayus, G.T.S., Deshpande, R.S. and Storey, I.F. 1943. Effect of seed treatment on emergence of peas. Ann. App. Biol. 30: 19-26.

Ben-Yephet, Y. and Nelson, E.B. 1999. Differential suppression of damping-off caused by *Pythium aphanidermatum, P. irregulare* and *P. myriotylum* in composts at different temperatures. Plant Dis. 83: 356-360.

Beuzenberg, M.P. 1966. Practical experiments with methyl bromide. Praktijproeven met methylbromide. Vakbladvoor de Bloemisterij 21: 1389.

Bhai, R.S., Kishore, V.K., Kumar, A., Anandaraj, M. and Eapen, S.J. 2005. Screening of rhizobacterial isolates against soft rot disease of ginger (*Zingiber officinale* Rosc.). J. Spices Aromat. Crops 14: 130-136.

Bhai, R.S. and Thomas, J. 2010. Compatibility of *Trichoderma harzianum* (Rifai) with fungicides, insecticides and fertilizers. Indian Phytopathol. 63: 145-148.

Blok, W.J., Termorshuizen, A.J., Coenen, T.G., de Wilde, V., Veeken, A.H., Köpke, U., Niggli, U., Neuhoff, D., Cornish, P., Lockeretz, W. and Willer, H. 2005. Disease suppression of potting mixes amended with composted biowaste. Proceedings of the First Scientific Conference of the International Society of Organic Agriculture Research (ISOFAR) on Researching Sustainable Systems, Adelaide, South Australia, 21-23 September, 2005, pp. 137-141.

Bolton, A.T. 1978. Effects of amending soilless growing mixtures with soil containing antagonistic organisms on root rot and blackleg of geranium (*Pelargonium hortorum*) caused by *Pythium splendens*. Can. J. Plant Sci. 58: 379-383.

Bonanomi, G., Antignani, V., Barile, E., Lanzotti, V. and Scala, F. 2011. Decomposition of *Medicago sativa* residues affects phytotoxicity, fungal growth and soil-borne pathogen diseases. J. Plant Pathol. 93: 57-69.

Boudjeko, T., Omokolo, N.A., Driouich, A. and Balangé, A.P. 2005. Peroxidase and pectin methylesterase activities in cocoyam (*Xanthosomas agittifolium* L. Schott) roots upon *Pythium myriotylum* inoculation. J. Phytopathol. 153: 409-416.

Bourbos, V.A. and Skoudridakis, M.T. 1991. Control of tomato brown root rot in the greenhouse using soil solarization. Luttecontre la pourriturebrune des racines de tomateenserre par la solarisation du sol. Bull. SROP. 14: 172-177.

Boz, Ö., Yildiz, A., Benlioğlu, K. and Benlioğlu, H.S. 2011. Methyl bromide alternatives for presowing fumigation in tobacco seedling production. Turkish J. Agri. For. 35: 73-81.

Bradley, C.A. 2008. Effect of fungicide seed treatments on stand establishment, seedling disease, and yield of soybean in North Dakota. Plant Dis. 92: 120-125.

Breton, D. and Rouxel, F. 1994. Recent findings relevant to cavity spot of carrots in France. Donneesrecentes sur le cavity-spot de la carotte en France. Acta Hort. 354: 159-170.

Breton, D., Aubree, N., Schlaunich, E., Faloya, V. and Montfort, F. 2011. Field study on the effect of vegetation cover with fumigation on two diseases of carrot: The cavity-spot and black scurf. 4ème Conférence Internationale sur les Méthodes Alternatives en Protection des Cultures. Evolution des cadres réglementaireseuropéen et français. Nouveaux moyens et strategies Innovantes, Nouveau Siècle, Lille, France, 8-10 mars 2011, pp. 322-331.

Bruna, E.M. 1946. Damping-off of tobacco seedlings in Chile. Contribucion al estudio de la 'caida' de los almacigos de tabaco en Chile. Agric. Tec. Chile 6: 109-134.

Buchholtz, W.F. 1936. Seed treatment as a control for damping off of alfalfa and other legumes. Phytopath. 26: 88.

Buchholtz, W.F. 1942. Influence of cultural factors on alfalfa seedling infection by *Pythium debaryanum* Hesse. Iowa Agric. Exp. Stn. Res. Bull. 296: 572-592.

Buysens, S., Höfte, M. and Poppe, J. 1995. Biological control of *Pythium* sp. in soil and nutrient film technique systems by *Pseudomonas aeruginosa* 7NSK2. Acta Hort. 382: 238-243.

Campbell, L. 1939. Black root of sugar beets in the Puget Sound Section of Washington. Bull. Washington State Agri. Exp. Sta. 379: 5-14.

Cebolla, V., Martínez, P.F., Gómez de Barreda, D., Tuset, J.J., Del Busto, A. and Vanacher, A. 1995. Dosage reduction of methyl bromide fumigation in the Spanish Mediterranean coast. Acta Hort. 382: 156-163.

Chaturvedi, R.C. 2014. Eco-friendly management of rhizome rot (soft rot) disease of ginger under Pasighat condition of Arunachal Pradesh. HortFlora Res. Spectrum 3: 380-382.

Chen, S.S. and Huang, J.W. 1992. Control of pod rot of peanut (*Arachis hypogaea* L.) by soil amendments II. Effects of gypsum and GS on pod rot and yield. J. Agri. For. 41: 59-64.

Cheng, Y.H., Cheng, A.H., Chen, S.S. and Tu, C.C. 1989. The outbreaks of pod rot of peanut and its control. J. Agri. Res. China. 38: 353-364.

Chiu, A.L. and Huang, J.W. 1997. Effect of composted agricultural and industrial wastes on the growth of vegetable seedlings and suppression of their root diseases. Plant Pathol. Bull. 6: 67-75.

Choochuay, K., Davong, K., Thaveechai, N. and Lertsuchatavanich, U. 2016. Soil biofumigation with *Brassica* spp. for controlling of *Pythium aphanidermatum* [Conference poster]. Agricultural Innovation for Global Value Chain, Proceedings of 54th Kasetsart University Annual Conference, 2-5 February 2016, Kasetsart University, Thailand. Vol. 1, Plants, Animals, Veterinary Medicine, Fisheries, Agricultural Extension and Home Economics, pp. 400-406.

Constantinescu, F. and Sesan, T.E. 2002. Soil-borne fungi and host plant influence on the efficacy of *Bacillus subtilis* biocontrol agents. Bull OILB/SROP. 25: 349-352.

Cook, R.J., Sitton, J.W. and Haglund, W.A. 1987. Influence of soil treatments on growth and yield of wheat and implications for control of Pythium root rot. Phytopathol. 77: 1192-1198.

Cook, P.J., Landschoot, P.J. and Schlossberg, M.J. 2009. Inhibition of *Pythium* spp. and suppression of Pythium blight of turfgrasses with phosphonate fungicides. Plant Dis. 93: 809-814.

Cook, R.J., Weller, D.M., El-Banna, A.Y., Vakoch, D. and Zhang, H. 2002. Yield responses of direct-seeded wheat to rhizobacteria and fungicide seed treatments. Plant Dis. 86: 780-784.

Cooper, C., Crowther, T., Smith, B.M., Isaac, S. and Collin, H.A. 2006. Assessment of the response of carrot somaclones to *Pythium violae*, causal agent of cavity spot. Plant Pathol. 55: 427-432.

Cotes, A.M., Lepoivre, P. and Semal, J. 1996. Correlation between hydrolytic enzyme activities measured in bean seedlings after *Trichoderma koningii* treatment combined with pregermination and the protective effect against *Pythium splendens*. Eur. J. Plant Pathol. 102: 497-506.

Cuc, G., Pop, O., Pop, D., Oltean, I. and Oros, S. 2007. Research concerning the integrated control of diseases, pests and weeds in cabbage crop. Bull. USAMV-CN. 63: 271-276.

Damicone, J.P., Olson, J.D. and Kahn, B.A. 2012. Cultivar and fungicide effects on Pythium leak of snap bean. Plant Health Prog. doi:10.1094/PHP-2012-0418-01-RS.

Dar, W.A., Bhat, J.A., Rashid, R., Rehman, S. and Bhat, Z.A. 2012. Bioefficacy of *Pseudomonas fluorescens* in management of damping-off disease in papaya (*Carica papaya* L.). Int. J. Agri. Tech. 8: 693-697.

Davison, E.M. and McKay, A.G. 2003. Host range of *Pythium sulcatum* and the effects of rotation on *Pythium* diseases of carrots. Australas. Plant Pathol. 32: 339-346.

De Cal, A., Martinez-Treceno, A., Salto, T., López-Aranda, J.M. and Melgarejo, P. 2005. Effect of chemical fumigation on soil fungal communities in Spanish strawberry nurseries. App. Soil Eco. 28: 47-56.

De Villiers, R.J. and Agenbag, G.A. 2007. Effect of chemical seed treatment, seeding rate and row width on plant populations and yield of canola (*Brassica napus* var. *oleifera*). S. Afr. J. Plant Soil. 24: 84-87.

Deadman, M., Al-Hasani, H. and Al-Sa'di, A. 2006. Solarization and biofumigation reduce *Pythium aphanidermatum* induced damping-off and enhance vegetative growth of greenhouse cucumber in Oman. J. Plant Pathol. 88: 335-337.

Deb, P.R. and Dutta, B.K. 1984. Activity of thiolutin against certain soil borne plant pathogens. Curr. Sci. (India) 53: 659-660.

Deepa, M.A., Narmatha Bai, V. and Basker, S. 2004. Antifungal properties of *Pseudarthria viscida*. Fitoterapia. 75: 581-584.

Deng, Z., Harbaugh, B.K., Kelly, R.O., Seijo, T. and McGovern, R.J. 2005a. Pythium root rot resistance in commercial caladium cultivars. Hort. Sci. 40: 549-552.

Deng, Z., Harbaugh, B.K., Kelly, R.O., Seijo, T. and McGovern, R.J. 2005b. Screening for resistance to Pythium root rot among twenty-three caladium cultivars. HortTechnology 15: 631-634.

Desaeger, J.A., Seebold, K.W. and Csinos, A.S. 2008. Effect of application timing and method on efficacy and phytotoxicity of 1, 3-D, chloropicrin and metam-sodium combinations in squash plasticulture. Pest Manag. Sci. 64: 230-238.

Deshmukh, N.J., Deokar, C.D. and Musmade, N.A. 2016. Management of wilt and root rot disease of sugarcane in nursery. Int. J. Plant Prot. 9: 489-493.

Devaki, N.S., Bhat, S., Bhat, S.G. and Manjunatha, K.R. 1992. Antagonistic activities of *Trichoderma harzianum* against *Pythium aphanidermatum* and *Pythium myriotylum* on tobacco. J. Phytopath. 136: 82-87.

Dezeeuw, D.J. 1954. Fungicide treatment of table beet and spinach seeds for the prevention of damping-off. Quart. Bull. Mich. State Univ. Agr. Exp. Sta. 37: 105-118.

Dhanapal, K. and Thomas, J. 2008. Management of rhizome rot disease of small cardamom (*Elettaria cardamomum* Maton.) using fungicides and bioagent. J. Plant. Crops 36: 466-468.

Dickson, M.H. and Abawi, G.S. 1974. Resistance to *Pythium ultimum* in white seeded beans (*Phaseolus vulgaris*). Plant Dis. Rep. 58: 774-776.

Dinakaran, D. and Ramakrishnan, G. 1996. Studies on the control of tomato damping-off with *Trichoderma viride*. Plant Dis. Res. 11: 148-150.

Dinesh, R., Anandaraj, M., Kumar, A., Bini, Y.K., Subila, K.P. and Aravind, R. 2015. Isolation, characterization, and evaluation of multi-trait plant growth promoting rhizobacteria for their growth promoting and disease suppressing effects on ginger. Microbiol. Res. 173: 34-43.

Dirkse, F.B. 1981. Footrot in saintpaulias need not be a problem. Vakblad voor de Bloemisterij 36(7): 36-37.

Dorrance, A.E., Berry, S.A. and Lipps, P.E. 2004. Characterization of *Pythium* spp. from three Ohio fields for pathogenicity on corn and soybean and metalaxyl sensitivity. Plant Health Prog. doi:10.1094/PHP-2004-0202-01-RS.

Dukare, A.S., Prasanna, R., Dubey, S.C., Nain, L., Chaudhary, V., Singh, R. and Saxena, A.K. 2011. Evaluating novel microbe amended composts as biocontrol agents in tomato. Crop Prot. 30: 436-442.

Dukare, A.S., Prasanna, R., Nain, L. and Saxena, A.K. 2013. Optimization and evaluation of microbe fortified composts as biocontrol agents against phytopathogenic fungi. J. Microb. Biotec. Food Sci. 2: 2272-2276.

Dumitras, L. and Fratilescu-Sesan, T. 1979. Aspects of the antagonism of *Trichoderma viride* Pers. to *Pythium debaryanum* Hesse. Biol. Vegetala 31: 63-67.

Elmer, W.H. 1994. Sensitivity of plant pathogenic fungi to taxane extracts from ornamental yews. Phytopathol. 84: 1179-1185.

El-Tarabily, K.A., Nassar, A.H., Hardy, G.S.J. and Sivasithamparam, K. 2009. Plant growth promotion and biological control of *Pythium aphanidermatum*, a pathogen of cucumber, by endophytic actinomycetes. J. App. Microb. 106: 13-26.

Erdoğan, O., Bölek, Y. and Göre, M.E. 2016. Biological control of cotton seedling diseases by fluorescent *Pseudomonas* spp. Tarim Bilim. Derg. 22: 398-407.

Ersahin, Y.S., Weiland, J.E., Zasada, I.A., Reed, R.L. and Stevens, J.F. 2014. Identifying rates of meadowfoam (*Limnanthes alba*) seed meal needed for suppression of *Meloidogyne hapla* and *Pythium irregulare* in soil. Plant Dis. 98: 1253-1260.

Forbes, G.A., Ziv, O. and Frederiksen, R. 1987. Resistance in sorghum to seedling disease caused by *Pythium arrhenomanes*. Plant Dis. 71: 145-148.

Gawande, R.L. and Shukla, V.N. 1976. Effect of some fungicides on control of betelvine wilt. Magazine, College of Agriculture, Nagpur 49: 7-10.

Geard, I.D. 1959. Diseases of french beans and runner beans. Tasman. J. Agric. 30: 336-345.

Geethu, C., Sumna, S., Resna, A.K. and Aswati Nair, R. 2013. Hyphal and zoospore lysis underlies the mechanistic basis for inhibitory effect of paromomycin on *Pythium myriotylum*. Fungal Genom. Biol. 3(1): 100-107.

Gerhold, N.R. 1956. Sugar Beet diseases and their control. Iowa State College J. Sci. 30: 362.

Gholve, V.M., Tatikundalwar, V.R., Suryawanshi, A.P. and Dey, U. 2014. Effect of fungicides, plant extracts/botanicals and bioagents against damping-off in brinjal. African J. Microb. Res. 8: 2835-2848.

Gomathi, S., Ambikapathy, V. and Panneerselvam, A. 2011. Antimicrobial activity of some medical plants against *Pythium debaryanum* (Hesse). J. Microb. Biotech. Res. 1: 8-13.

Grijalba, P.E. and Ridao, A. 2017. Chemical control of *Pythium* spp. in soybean seedlings. RIA, Rev. Investig. Agropecu. 43: 67-71.

Guérin, L., Benhamou, N. and Rouxel, F. 1998. Ultrastructural and cytochemical investigations of pathogen development and host reactions in susceptible and partially-resistant carrot roots infected by *Pythium violae*, the major causal agent for cavity spot. Eur. J. Plant Pathol. 104: 653-665.

Gulhane, V.G., Gaikwad, S.J., Lanje, P.W., Zade, S.R. and Kuruwanshi, V.B. 2005. Biological control of damping-off of tomato caused by *Pythium aphanidermatum* (Eds.) Fitz. J. Soils Crops 15: 118-121.

Gutierrez, W., Melton, T. and Mila, A. 2012. Pythium root rot of flue-cured tobacco seedlings produced in greenhouses: Factors associated with its occurrence and chemical control. Plant Health Prog. doi:10.1094/PHP-2012-0925-01-RS.

Haenseler, C.M. 1930. Control of seed decay and damping-off of cucumbers. Fifty-first Ann Rept. New Jersey Agric. Exper. Stat. for the year ending June 30, 1930. pp. 254-264.

Harris, A.R. and Adkins, P.G. 1999. Versatility of fungal and bacterial isolates for biological control of damping-off disease caused by *Rhizoctonia solani* and *Pythium* spp. Biol. Control 15: 10-18.

Hashem, M. and Ali, E.H. 2004. *Epicoccum nigrum* as biocontrol agent of Pythium damping-off and root-rot of cotton seedlings. Arch. Phytopathol. Plant Prot. 37: 283-297.

He, S.S., Zhang, B.X. and Ge, Q.X. 1992. On the antagonism by hyperparasite *Pythium oligandrum*. Acta Phytopathol. Sin. 22: 77-82.

Helmy, A.A., Baiuomy, M.A.M. and Hilal, A.A. 2001. First record of root rot and wilt diseases of the medicinal plant *Ruta graveolens* L. in Egypt and their control. Egyptian J. Agri. Res. 79: 21-35.

Hiltunen, L.H. and White, J.G. 2002. Cavity spot of carrot (*Daucus carota*). Ann. App. Biol. 141: 201-223.

Ho, W.C., Meredith, C.H. and Melhus, I.E. 1941. *Pythium graminicola* Subr. on barley. Iowa Agric. Exp. Stn. Res. Bull. 287: 289-314.

Hobsfall, J.G. 1932. Dusting tomato seed with copper sulfate monohydrate for combating damping-off. NY. Agric. Exp. Stn. Tech. Bull. 198: 1-34.

Hoshiarfard, M. and Gharanjiki, A.R. 2009. Effect of source and rate of incidence and severity of important diseases, yield and yield components in cotton (*Gossypium hirsutum* L.). Iranian J. Crop Sci. 11: 237-248.

Hudge, B.V. 2015. Management of damping-off disease of soybean caused by *Pythium ultimum* Trow. Int. J. Curr. Microbiol. App. Sci. 4: 799-808.

Hwang, S.F., Chang, K.F., Howard, R.J., Deneka, B.A. and Turnbull, G.D. 1996. Decrease in incidence of Pythium damping-off of field pea by seed treatment with *Bacillus* spp. and metalaxyl. Z. Pflanzenkr. Pflanzenschutz 103: 31-41.

Hwang, S.F., Gossen, B.D., Chang, K.F., Turnbull, G.D. and Howard, R.J. 2001. Effect of seed damage and metalaxyl seed treatment on Pythium seedling blight and seed yield of field pea. Can. J. Plant Sci. 81: 509-517.

Iremiren, G.O. 1987. Management of oil palm nurseries in Nigeria. Agric. Int. 39: 76-78.

Jasim, B., Rohini, S., Anisha, C., Jimtha J.C., Jyothis, M. and Radhakrishnan, E. 2013. Antifungal and plant growth promoting properties of endophytic *Pseudomonas aeruginosa* from *Zingiber officinale*. J. Pure App. Microb. 7: 1003-1009.

Jasim, B., Anisha, C., Rohini, S., Kurian, J., Jyothis, M. and Radhakrishnan, E. 2014. Phenazine carboxylic acid production and rhizome protective effect of endophytic *Pseudomonas aeruginosa* isolated from *Zingiber officinale*. World J. Microbiol. Biotechnol. 30: 1649-1654.

Jimtha, J.C., Jishma, P., Arathy, G.B., Anisha, C. and Radhakrishnan, E.K. 2016. Identification of plant growth promoting Rhizosphere *Bacillus* sp. WG4 antagonistic to *Pythium myriotylum* and its enhanced antifungal effect in association with *Trichoderma*. J. Soil Sci. Plant Nut. 16: 578-590.

Joshi, L.K. and Keshwal, R.L. 1969. Inhibition of *Pythium debaryanum* Hesse by culture filtrates of *Penicillium notatum* and two species of *Aspergillus* from soil. JNKVV Res. J. 3: 93-94.

Kao, C.W. and Ko, W.H. 1986a. Suppression of *Pythium splendens* in a Hawaiian soil by calcium and microorganisms. Phytopathol. 76: 215-220.

Kao, C.W. and Ko, W.H. 1986b. The role of calcium and microorganisms in suppression of cucumber damping-off caused by *Pythium splendens* in a Hawaiian soil. Phytopathol. 76: 221-225.

Kapoor, A.S. 2008. Biocontrol potential of *Trichoderma* spp. against important soilborne diseases of vegetable crops. Indian Phytopathol. 61: 492-498.

Kareem, T.A. 2010. Evaluation the effectiveness of five essential plant oils in the inhibition of growth of four types of pathogenic. Diyala Agri. Sci. J. 2: 220-228.

Karthikeyan, G., Sabitha, D. and Sivakumar, C.V. 2000. Biological and chemical control of *Pythium aphanidermatum-Meloidogyne incognita* disease complex in chilli and brinjal. Madras Agric. J. 87: 105-108.

Kato, M., Minamida, K., Tojo, M., Kokuryu, T., Hamaguchi, H. and Shimada, S. 2013. Association of *Pythium* and *Phytophthora* with pre-emergence seedling damping-off of soybean grown in a field converted from a paddy field in Japan. Plant Prod. Sci. 16: 95-104.

Kavitha, P.G. and Thomas, G. 2007. Evaluation of Zingiberaceae for resistance to ginger soft rot caused by *Pythium aphanidermatum* (Edson) Fitzp. Plant Gen. Res. Newsl. 152: 54.

Kean, S., Soytong, K. and To-anun, C. 2010. Application of biofungicide to control citrus root rot under field condition in Cambodia. Int. J. Agric. Tech. 6: 219-230.

Keshwal, R.L. 1981. Seed coating with fungus culture for better germination of tomato seed. Seed Res. 9: 194-195.

Khare, A., Singh, B.K. and Upadhyay, R.S. 2010. Biological control of *Pythium aphanidermatum* causing damping-off of mustard by mutants of *Trichoderma viride* 1433. Int. J. Agric. Tech. 6: 231-243.

Kim, D., Cook, R.J. and Weller, D.M. 1997. *Bacillus* sp. L324-92 for biological control of three root diseases of wheat grown with reduced tillage. Phytopathol. 87: 551-558.

Kim, S.H. and Kantzes, J.G. 1972. Species, cultivars and lines of *Phaseolus* resistant to *Pythium aphanidermatum*. Phytopathol. 62: 769 (Abstr.)

Kiryowa, M., Nkalubo, S.T., Mukankusi, C., Talwana, H., Gibson, P. and Tukamuhabwa, P. 2015. Effect of marker aided pyramiding of anthracnose and Pythium root rot resistance genes on plant agronomic characters among advanced common bean genotypes. J. Agric. Sci. 7: 98.

Kishore, N. and Dubey, N.K. 2002. Fungitoxic potency of some essential oils in management of damping-off diseases in soil infested with *Pythium aphanidermatum* and *P. debaryanum*. Indian J. For. 25: 463-468.

Klittich, C.J. 2008. Milestones in fungicide discovery: Chemistry that changed agriculture. Plant Health Prog. doi:10.1094/PHP-2008-0418-01-RV.

Kobayashi, N. 1990. Biological control of soilborne diseases with VAM fungi and charcoal compost. pp. 153-160. *In*: Komada, H., Kiritani, K. and Bag Petersen, J. (eds.). The Biological Control of Plant Diseases. Proceedings of the International Seminar on Biological Control of Plant Diseases and Virus Vectors held in Tsukuba, Japan, Sep. 17-21.

Kokaew, J., Manoch, L., Worapong, J., Chamswarng, C., Singburaudom, N., Visarathanonth, N., Piasai, O. and Strobel, G. 2011. *Coniochaeta ligniaria* an endophytic fungus from *Baeckea frutescens* and its antagonistic effects against plant pathogenic fungi. Thai J. Agric. Sci. 44: 123-131.

Kolaei, E.A., Tweddell, R.J. and Avis, T.J. 2012. Antifungal activity of sulfur-containing salts against the development of carrot cavity spot and potato dry rot. Postharvest Biol. Tech. 63: 55-59.

Kolaei, E.A., Cenatus, C., Tweddell, R.J. and Avis, T.J. 2013. Antifungal activity of aluminium-containing salts against the development of carrot cavity spot and potato dry rot. Ann. App. Bio. 163: 311-317.

Koleosho, B., Ikotun, T. and Faboya, O. 1987. The role of oxalic acid and polygalacturonase in the pathogenicity of *Pythium aphanidermatum* on different cowpea varieties. Phytoparasitica 15: 317. https://doi.org/10.1007/BF02979547

Kraft, J.M. and Roberts, D.D. 1970. Resistance in peas to Fusarium and Pythium root rot. Phytopathol. 60: 1814-1817.

Kraft, J.M. 1974. The influence of seedling exudates on the resistance of peas to Fusarium and Pythium root rot. Phytopathol. 64: 190-193.

Krasnow, C.S. and Hausbeck, M.K. 2017. Influence of pH and etridiazole on *Pythium* species. HortTechnology 27: 367-374.

Krishnamoorthy, A.S. and Bhaskaran, R. 1990. Biological control of damping-off disease of tomato caused by *Pythium indicum* Balakrishnan. J. Biol. Control 4: 52-54.

Kulkarni, R.N. and Baskaran, K. 2003. Inheritance of resistance to Pythium dieback in the medicinal plant periwinkle. Plant Breed. 122: 184-187.

Kumar, A. and Tripathi, S.C. 1991. Evaluation of the leaf juice of some higher plants for their toxicity against soil borne pathogens. Plant & Soil 132: 297-301.

Lalfakawma, C., Nath, B., Bora, L.C, Srivastava, S. and Singh, J.P. 2014. Integrated disease management of *Zingiber officinale* Rosc. rhizome rot. The Bioscan 9: 265-269.

Lévesque, C.A. and de Cock, W.A.M. 2004. Molecular phylogeny and taxonomy of the genus *Pythium*. Mycol. Res. 108: 1363-1383.

Li, L., Kageyama, K., Kinoshita, N., Yu, W. and Fukui, H. 2007. Development of bioassay for screening of resistant roses against root rot disease caused by *Pythium helicoides* Drechsler. Engei Gakkai Zasshi 76: 79-84.

Li, Y.P., You, M.P. and Barbetti, M.J. 2014. Species of *Pythium* associated with seedling root and hypocotyl disease on common bean (*Phaseolus vulgaris*) in Western Australia. Plant Dis. 98: 1241-1247.

Li, Y.P., You, M.P., Norton, S. and Barbetti, M.J. 2016. Resistance to *Pythium irregulare* root and hypocotyl disease in diverse common bean (*Phaseolus vulgaris*) varieties from 37 countries and relationships to waterlogging tolerance and other plant and seed traits. Eur. J. Plant Pathol. 146: 147-176.

Linderman, R.G., Davis, E.A. and Masters, C.J. 2008. Efficacy of chemical and biological agents to suppress *Fusarium* and *Pythium* damping-off of container-grown Douglas-fir seedlings. Plant Health Prog. doi:10.1094/PHP-2008-0317-02-RS.

Liu, X., Yan, D., Ouyang, C., Yang, D., Wang, Q., Li, Y., Guo, M. and Cao, A. 2017. Oils extracted from *Eupatorium adenophorum* leaves show potential to control *Pythium myriotylum* in commercially-grown ginger. PLoS One 12: e0176126. doi: 10.1371/journal.pone.0176126.

Maghazy, S.M.N., Abdelzaher, H.M.A., Haridy, M.S. and Moustafa, S.M.N. 2008. Biological control of damping-off disease of *Trifolium alexandrinum* L. caused by *Pythium spinosum* Sawada var. *spinosum* using some soil fungi. Arch. Phytopathol. Plant Prot. 41: 431-450.

Male, M.F. and Vawdrey, L.L. 2010. Efficacy of fungicides against damping-off in papaya seedlings caused by *Pythium aphanidermatum*. Australas. Plant Dis. Notes 5: 103-104.

Manjunath, M., Singh, A., Tripathi, A.N., Prasanna, R., Rai, A. and Singh, B. 2017. Bioprospecting the fungicides compatible *Trichoderma asperellum* isolate effective against multiple plant pathogens *in vitro*. J. Enviro. Biol. 38: 553-560.

Martinez, C., Lévesque, C.A., Bélanger, R.R. and Tweddell, R.J. 2005. Evaluation of fungicides for the control of carrot cavity spot. Pest Manage. Sci. 61: 767-771.

Mayurika, G., Sharma, P.K., Dureja, P., Rani, A. and Uniyal, P.L. 2011. Antifungal activity of extracts of the lichens *Parmelia reticulata, Ramalina roesleri, Usnea longissima* and *Stereocaulon himalayense*. Arch. Phytopathol. Plant Prot. 44: 1300-1311.

Mbarga, J.B., Martijn Ten Hoopen, G., Kuaté, J., Adiobo, A., Ngonkeu, M.E.L., Ambang, Z., Akoa, A., Tondje, P.R. and Begoude, B.A.D. 2012. *Trichoderma asperellum*: A potential biocontrol agent for *Pythium myriotylum*, causal agent of cocoyam (*Xanthosoma sagittifolium*) root rot disease in Cameroon. Crop Prot. 36: 18-22.

McDonald, M.R., Taylor, A.G., Lorbeer, J.W. and Heide, J.J. van der. 2004. Efficacy testing of onion seed treatments in the greenhouse and field. Acta Hort. 631: 87-93.

Meena, R.L. and Mathur, K. 2003. Evaluation of biocontrol agents for suppression of rhizome rot of ginger. Ann. Agri. Bio. Res. 8: 233-238.

Meenu, G., Dohroo, N.P., Gangta, V. and Shanmugam, V. 2010. Effect of microbial inoculants on rhizome disease and growth parameters of ginger. Indian Phytopathol. 63: 438-441.

Mellon, J.E., Dowd, M., Beltz, S. and Moore, G. 2014. Growth inhibitory effects of gossypol and related compounds on fungal cotton root pathogens. Lett. App. Microbiol. 59: 161-168.

Mev, A.K. and Meena, R.L. 2003. Mass multiplication of *Trichoderma harzianum* for biocontrol of rhizome rot of ginger. J. Phytological Res. 16: 89-92.

Mihajlović, M., Rekanović, E., Hrustić, J., Tanović, B., Potočnik, I., Stepanović, M. and Milijašević-Marčić, S. 2013. In vitro and in vivo toxicity of several fungicides and Timorex Gold biofungicide to *Pythuim aphanidermatum*. Pestic. Fitomed. 28: 117-123.

Mohamed, H.A., Abdel-Sattar, M.A., Hilal, A.A. and El-Shamy, S.M.R. 1987. Host nutrition and chemical control in relation to wilt and root-rot of *Pelargonium*. Agric. Res. Rev. 65: 285-296.

Mohamed, R., Groulx, E., Defilippi, S., Erak, T., Tambong, J.T., Tweddell, R.J., Tsopmo, A. and Avis, T.J. 2017. Physiological and molecular characterization of compost bacteria antagonistic to soil-borne plant pathogens. Can. J. Microbiol. 63: 411-426.

Mojdehi, H. and Singleton, L.L. 2000. Reaction of wheat varieties to infection by *Pythium arrhenomanes* or its toxic metabolite(s). J. Agr. Set. Tech. 2: 33-39.

Molin, N. 1955. Damping-off of coniferous seedlings. Fallsjuka pa groddplantor av barrtrad. Meddeland. Statens Skogsförsöksanst. 45: 12.

Montfort, F., Poggi, S., Morlière, S., Collin, F., Lemarchand, E. and Bailey, D.J. 2011. Opportunities to reduce *Rhizoctonia solani* expression on carrots by biofumigation with Indian mustard. Acta Hort. 917: 149-157.

Moreno Alvarez, M.T., Castella, G., Pugliese, M., Gullino, M.L. and Garibaldi, A. 2013. Contenimento di patogeni terricoli con scarti del processo di produzione di biodiesel di seconda generazione. Protezione Delle Colture 2: 100-100.

Morton, H.V. and Urech, P.A. 1988. History of the development of resistance to phenylamide fungicides. pp. 59-60. *In*: Delp, C.J. (ed.). Fungicide Resistance in North America. American Phytopathological Society, St. Paul, MN.

Mostafa, A.A., Al-Rahmah, A.N., Yakout, S.M. and Abd-Alrahman, S.H. 2013. Bioactivity of garlic bulb extract compared with fungicidal treatment against tomato phytopathogenic fungi. J. Pure Appl. Microbiol. 7: 1925-1932.

Múnera, J.d.C. and Hausbeck, M.K. 2015. Integrating host resistance and plant protectants to manage Pythium root rot on geranium and snapdragon. HortScience 50: 1319-1326.

Muthukumar, A., Eswaran, A., Karthikeyan, G. and Sanjeevkumar, K. 2007. Efficacy of native isolates of biocontrol agents and neem formulations for management of damping-off in chilli. J. Plant Prot. Environ. 4: 126-129.

Muthulakshmi, P. and Saveetha, K. 2009. Management of turmeric rhizome rot using eco-friendly biocontrol consortia. J. Biol. Control 23: 181-184.

Mvuemba, H.N., Green, S.E., Tsopmo, A. and Avis, T.J. 2009. Antimicrobial efficacy of cinnamon, ginger, horseradish and nutmeg extracts against spoilage pathogens. Phytoprotection 90: 65-70.

Neelamegam, R. and Govindarajalu, T. 2002. Integrated application of *Trichoderma viride* Pers: Fr. and farmyard manure to control damping-off of tomato [*Lycopersicum esculentum* Mill.]. J. Biol. Control 16: 65-69.

Neelamegam, R. 2004. Evaluation of fungal antagonists to control damping-off of tomato (*Lycopersicon esculentum* Mill.) caused by *Pythium indicum*. J. Biol. Control 18: 97-101.

Neelamegam, R. 2005. Effect of different organic matter sources and *Trichoderma viride* Pers: Fr on damping-off of tomato (*Lycopercicum esculentum* L.) var. CO-1 seedlings caused by *Pythium indicum* Bal. J. Biol. Control 19: 149-155.

Nelson, S.D., Locascio, S.J., Allen, L.H., Dickson, D.W. and Mitchell, D.J. 2002. Soil flooding and fumigant alternatives to methyl bromide in tomato and eggplant production. HortScience 37: 1057-1060.

Nolla, J.A.B. 1932. The damping-off of tobacco and its control in Puerto Rico. J. Dept. Agr. Porto Rico. 16: 203-204.

Numic, R.M., Hauzer-Alfirevic, O. and Milicevic, S. 1979. The effect of an antifungal antibiotic of the monosaccharide type on *Pythium debaryanum* Hesse. Djelovanje antifungalnog antibiotika monosaharidnog tipa na *Pythium debarianum* Hesse. Zastita Bilja. 30: 283-288.

Nzungize, J., Geps, P., Buruchara, R., Buah, S., Ragama, P., Busogoro, J.P. and Baudoin, J.P. 2011a. Pathogenic and molecular characterization of *Pythium* species inducing root rot symptoms of common bean in Rwanda. Af. J. Microbiol. Res. 5: 1169-1181.

Nzungize, J., Gepts, P., Buruchara, R., Male, A., Ragama, P., Busogoro, J.P. and Baudoin, J.P. 2011b. Introgression of Pythium root rot resistance gene into Rwandan susceptible common bean cultivars. Afr. J. Plant Sci. 5: 193-200.

Ohh, S.H., King, T.H. and Kommedah, T. 1978. Evaluating peas for resistance to damping-off and root rot caused by *Pythium ultimum*. Phytopathol. 68: 1644-1649.

Okamoto, H., Sato, Z., Sato, M., Koiso, Y., Iwasaki, S. and Isaka, M. 1998. Identification of antibiotic red pigments of *Serratia marcescens* F-1-1, a biocontrol agent of damping-off of cucumber, and antimicrobial activity against other plant pathogens. Ann. Phytopathol. Soc. Japan 64: 294-298.

Ongom, P.O., Nkalubo, S.T., Gibson, P.T., Mukankusi, C.M. and Rubaihayo, P.R. 2012. Evaluating genetic association between Fusarium and Pythium root rots resistances in the bean genotype RWR 719. Afr. Crop. Sci. J. 20: 31-39.

Oumar, D., Akumah, B.E. and Tchinda, N.D. 2015. Induction of resistance in cocoyam (*Xanthosomas agittifolium*) to *Pythium myriotylum* by corm treatments with benzothiadiazole and its effect on vegetative growth. Am. J. Exp. Agri. 5: 164-171.

Ownley, B.H., Griffin, M.R., Klingeman, W.E., Gwinn, K.D., Moulton, J.K. and Pereira, R.M. 2008. *Beauveria bassiana*: Endophytic colonization and plant disease control. J. Invertebr. Pathol. 98: 267-270.

Ozkaya, H.O. and Ergun, T. 2017. The effects of *Allium tuncelianum* extract on some important pathogens and total phenolic compounds in tomato and pepper. Pak. J. Bot. 49: 2483-2490.

Pakdi, S. and Claridad, F.B. 1971. Screening of some tobacco varieties for resistance to damping-off disease caused by *Pythium debaryanum* Hesse. Araneta J. Agr. 18: 166-186.

Pandey, V.N. and Dubey, N.K. 1994. Antifungal potential of leaves and essential oils from higher plants against soil phytopathogens. Soil Biol. Bioch. 26: 1417-1421.

Pandey, V.N. and Dubey, N.K. 1997. Antifungal potentiality of some higher plants against *Pythium* species causing damping-off in tomato. Nat. Acad. Sci. Lett. 20: 68-70.

Paradikovic, N., Baličević, R., Vinković, T., Parađiković, D. and Karlić, J. 2007. Bioloskemjerezastite u proizvodnji gerbera ipresadnicarajcice. Agron. Glas. 69: 355-364.

Patkowska, E. and Konopiński, M. 2014. Occurrence of antagonistic fungi in the soil after cover crops cultivation. Plant, Soil & Environ. 60: 204-209.

Perez-Sanchez, R., Infante, F., Gálvez, C. and Ubera, J.L. 2007. Fungitoxic activity against phytopathogenic fungi and the chemical composition of *Thymus zygis* essential oils. Food Sci. Tech. Int. 13: 341-347.

Perneel, M., Adiobo, A., Floren, C., de Maeyer, K., Vercauteren, A., Saborio, F. and Hofte, M. 2004. Ecologically sustainable management of the cocoyam root rot disease caused by *Pythium myriotylum*. Bull. Seances Acad. R. Sci. Outre-Mer 50: 103-113.

Perneel, M., D'hondt, L., Adiobo, A., de Maeyer, K. and Höfte, M. 2007. Synergy between phenazines and biosurfactants in the biological control of *Pythium* induced soil-borne diseases is a general phenomenom in fluorescent pseudomonads. Bull. OILB/SROP. 30: 115-119.

Perneel, M., D'hondt, L., de Maeyer, K., Adiobo, A., Rabaey, K. and Höfte, M. 2008. Phenazines and biosurfactants interact in the biological control of soil-borne diseases caused by *Pythium* spp. Environ. Microbiol. 10: 778-788.

Ponmurugan, P., Babu, R.G., Mathivanan, N. and Citra, C. 2013. Studies on population density of different PGPRS in turmeric rhizosphere soils for biocontrol activity. pp. 101-111. *In*: Reddy, M.S., Ilao, R.I., Faylon, P.S., Dar, W.D., Sayyed, R., Sudini, H., Kumar, K.V.K. and Armanda, A. (eds.). Recent Advances in Biofertilizers and Biofungicides (PGPR) for Sustainable Agriculture. Proceedings of 3rd Asian Conference on Plant Growth-Promoting Rhizobacteria (PGPR) and other Microbials, Manila, Philippines, 21-24 April, 2013.

Pornsuriya, C., Soytong, K., Kanokmedhakul, S. and Lin, F.C. 2010. Efficacy of antifungal metabolites from some antagonistic fungi against *Pythium aphanidermatum*. Int. J. Agric. Tech. 6: 299-308.

Pugliese, M., Gullino, M.L. and Garibaldi, A. 2012. Control of *Pythium ultimum* on cucumber using wine industry waste. Protezione delle Colture 2: 70-71.

Pugliese, M., Marenco, M., Gullino, M.L. and Garibaldi, A. 2013. Use of compost with different particle sizes for the containment of *Pythium ultimum* on cucumber. Protezione delle Colture 2: 114.

Raabe, R.D., Hurlimann, J.H. and Byrne, T.G. 1981. Chemical and biological approaches to control of Pythium root rot of poinsettia. Calif. Plant Pathol. 52: 5-8.

Rahimian, M.K. and Banihashemi, Z. 1979. A method for obtaining zoospores of *Pythium aphanidermatum* and their use in determining cucurbit seedling resistance to damping-off. Plant Dis. Rep. 63: 658-661.

Raj, K., Pandey, J.C. and Verma, B.L. 1998. Chemical control of replant disease of apple. Recent Hort. 4: 64-68.

Rajagopalan, K. 1974. Influence of irrigation and shading on the occurrence of blast disease of oil palm seedlings. J. Nig. Inst. Oil. Palm Res. 5: 23-31.

Ram, D., Kusum, M., Lodha, B.C. and Webster, J. 2000. Evaluation of resident biocontrol agents as seed treatments against ginger rhizome rot. Indian Phytopathol. 53: 450-454.

Raut, I., Constantin, M., Vasilescu, G., Jecu, L. and Şesan, T. 2012. Screening of antagonistic *Trichoderma* for biocontrol activities on phytopathogens. Sci. Bull. Series F. Biotechnologies 16: 63-66.

Sadhna, S., Williams, P. and Ekta, S. 2012. Effect of some plant extracts on the growth of *Pythium aphanidermatum* causing damping-off of tomato. New Agriculturist 23: 229-232.

Saha, S., Naskar, I., Nayak, D.K. and Sarkar, M.K. 2008. Efficacy of different seed dressing agents in the control of damping-off disease of chilli caused by *Pythium aphanidermatum*. J. Mycopathol. Res. 46: 121-123.

Salas, B., Secor, G.A., Taylor, R.J. and Gudmestad, N.C. 2003. Assessment of resistance of tubers of potato cultivars to *Phytophthora erythroseptica* and *Pythium ultimum*. Plant Dis. 87: 91-97.

Satou, M. and Fukuta, N. 2016. Control of root rot of Russell prairie gentian (*Eustoma grandiflorum*) in hydroponic culture using a mixture of azoxystrobin and metalaxyl M. Jpn. J. Phytopathol. 82: 93-100.

Scheuerell, S.J., Sullivan, D.M. and Mahaffee, W.F. 2005. Suppression of seedling damping-off caused by *Pythium ultimum, P. irregulare*, and *Rhizoctonia solani* in container media amended with a diverse range of Pacific Northwest compost sources. Phytopathol. 95: 306-315.

Schwinn, F. and Staub, T. 1987. Phenylamides and other fungicides against Oomycetes. pp. 323-346. *In*: Lyr, H. (ed.). Modern Selective Fungicides, 2nd Edn. Gustav Fisher Verlag, Jena, Germany.

Scott, K.L. 2018. Studies in the Management of *Pythium* Seed and Root Rot of Soybean: Efficacy of Fungicide Seed Treatments, Screening Germplasm for Resistance, and Comparison of Quantitative Disease Resistance Loci to Three Species of *Pythium* and *Phytophthora sojae*. Doctoral dissertation. The Ohio State University.

Sealy, R.L., Kenerley, C.M. and McWilliams, E.L. 1988. Evaluation of *Amaranthus* accessions for resistance to damping-off by *Pythium myriotylum*. Plant Dis. 72: 985-989.

Senapati, A.K. and Ghose, S. 2005. Screening of ginger varieties against rhizome rot disease complex in eastern ghat high land zone of Orissa. Indian Phytopathol. 58: 437-439.

Shaikh, R. and Kareappa, B.M. 2008. Effect of fungicides on soft rot of turmeric. Asian J. Bio. Sci. 3: 243-244.

Shamarao, J. and Jahagirdar, S. 2012. Management of damping-off of bidi tobacco in the nursery bed with fungicides. Indian Phytopathol. 65: 206-207.

Shanmugam, V., Gupta, S. and Dohroo, N.P. 2013. Selection of a compatible biocontrol strain mixture based on co-cultivation to control rhizome rot of ginger. Crop Prot. 43: 119-127.

Sharan, N. 1958. Damping-off of gul mohur (*Delonix regia* Raf.) in India. Plant Dis. Rep. 42: 1408.

Sharma, P. and Dureja, P. 2004. Evaluation of *Trichoderma harzianum* and *T. viride* isolates at BCA pathogen crop interface. J. Mycol. Plant Pathol. 34: 47-55.

Singh, H.K., Shakywar, R.C., Singh, S. and Singh, A.K. 2012. Evaluation of comparative efficacy of native isolate of *Trichoderma viride* against rhizome rots disease of ginger. J. Plant Dis. Sci. 7: 22-26.

Singh, O., Chaukiyal, S.P. and Sharma, H.P. 1985. Control against damping-off in spruce nurseries. Indian J. For. 8: 321-322.

Singh, R., Ahluwalia, V., Singh, P., Kumar, N., Prakash, S.O. and Sati, N. 2016. Antifungal and phytotoxic activity of essential oil from the root of *Senecio amplexicaulis* Kunth (Asteraceae) growing wild in high altitude-Himalayan region. Nat. Prod. Res. 30: 1875-1879.

Singh, S., Prem, D., Tanwar, R.S. and Atar, S. 2005. Production and antifungal activity of secondary metabolites of *Trichoderma virens*. Pesticide Res. J. 17: 26-29.

Sivakumar, G., Josephrajkumar, A. and Dhanya, M.K. 2012. Evaluation of bacterial antagonists for the management of rhizome rot of cardamom (*Elettaria cardamomum* Maton). J. Spices Aromat. Crops 21: 9-15.

Smith, O.D., Boswell, T.E., Morgan, R.G., Taber, R.A. and Pettit, R.E. 1979. Breeding peanuts for pod rot resistance: Associations of physical and cellular shell properties. Agronomy abstracts. 78. https://eurekamag.com/research/000/832/000832051.php

Song, Y.C., Pei, E.Q., Dan, Y.S. and Wang, T.Y. 2012. Identification and evaluation of resistance to stalk rot (*Pythium inflatum* Matthews) in important inbred lines of maize. J. Plant Genet. Res. 13: 798-802.

Stasz, T.E., Harman, G.E. and Marx, G.A. 1980. Time and site of infection of resistant and susceptible germinating pea seeds by *Pythium ultimum*. Phytopathol. 70: 730-733.

Stirling, G.R. and Eden, L.M. 2008. The impact of organic amendments, mulching and tillage on plant nutrition, *Pythium* root rot, root-knot nematode and other pests and diseases of capsicum in a subtropical environment, and implications for the development of more sustainable vegetable farming systems. Australas. Plant Pathol. 37: 123-131.

Subhashini, D.V. 2010. Biological control of damping-off (*Pythium aphanidermatum*) using *Streptomyces lavendulae* isolate 21 on tobacco seedbeds. J. Biol. Control 24: 338-342.

Subhashini, D.V. and Padmaja, K. 2010. Interaction between arbuscular mycorrhizal fungi and *Pythium aphanidermatum* in tobacco seedbeds. J. Biol. Control 24: 70-74.

Subhashini, D.V. 2012. Studies on efficacy of different delivery systems of *Pseudomonas fluorescens* for biosuppression of damping-off disease in tobacco seedbeds. J. Biol. Control 26: 84-87.

Takahashi, M., Ichitani, T., Akaji, K. and Kawase, Y. 1970. Ecologic and taxonomic studies on *Pythium* as pathogenic soil fungi: VIII. Differences in pathogenicity of several species of *Pythium*. Bull. Univ. Osaka Prefect., Ser. B. 22: 87-93.

Tambong, J.T. and Achuo, A.E. 1997. In vitro growth suppression of the cocoyam root rot pathogen *Pythium myriotylum* by *Pseudomonas aeruginosa*. Trop. Sci. 37: 183-188.

Tambong, J.T., Poppe, J. and Hofte, M. 1999. Isozyme characterisation and biological control of the cocoyam root rot pathogen, *Pythium myriotylum*. Meded. Rijksuniv. Gent. Fak. Landbouwkd. Toegep. Biol. Wet. 64: 63-68.

Tambong, J.T. and Hofte, M. 2001. Phenazines are involved in biocontrol of *Pythium myriotylum* on cocoyam by *Pseudomonas aeruginosa* PNA1. Eur. J. Plant Pathol. 107: 511-521.

Tang, W.Z. and Uung, C. 1983. Control study of *Trichoderma* sp. on damping-off of *Pinus* seedlings caused by *Pythium spinosum*. J. South China Agric. Coll. 4: 63-71.

Tashkoski, P. 2015. Application of the signum fungicide in control of *Pythium debaryanum* Hesse on tobacco seedlings. Tobacco 65: 38-46.

Tashkoski, P., Krsteska, V., Mitreski, M. and Stojanoski, P. 2016. The impact of Strobilurin fungicides in prevention of damping off disease on tobacco seedlings. *In*: 2nd International Symposium for Agriculture and Food, Skopje, Macedonia, pp. 309-313.

Taylor, R.J., Salas, B. and Gudmestad, N.C. 2004. Differences in etiology affect mefenoxam efficacy and the control of pink rot and leak tuber diseases of potato. Plant Dis. 88: 301-307.

Tedeschi, P., Maietti, A., Boggian, M., Vecchiati, G. and Brandolini, V. 2007. Fungitoxicity of lyophilized and spray-dried garlic extracts. J. Environ. Sci. Health B. 42: 795-799.

Temgo, T.J.C. and Boyomo, O. 2002. Variations in the phenolic contents of cocoyam clones in correlation to resistance to *Pythium myriotylum*. Biol. Plantarum 45: 433-436.

Thinggaard, K., Larsen, K. and Hockenhull, J. 1988. Antagonistic *Pythium* against pathogenic *Pythium* on cucumber roots. Bull. OEPP. 18: 91-94.

Thinggaard, K. 1989. Biological control of root pathogenic *Phytophthora, Pythium* and *Phomopsis* in greenhouse crops. Vaxtskyddsnotiser 53: 25-29.

Titone, P., Mocioni, M., Garibaldi, A. and Gullino, M.L. 2009. Fungicide failure to control Pythium blight on turf grass in Italy. J. Plant Dis. Prot. 116: 55-59.

Trotus, E. and Naie, M. 2008. Controlling specific pathogens and pests of hemp by chemical treatment of seeds. Lucrari Stiintifice, Seria Agronomie 51: 219-223.

Tsror, L., Hazanovsky, M., Mordechai-Lebiush, S., Ben-David, T., Dori, I. and Matan, E. 2005. Control of root rot and wilt caused by *Pythium myriotylum* in Kangaroo Paw (*Anigozanthos*). J. Phytopathol. 153: 150-154.

Urrea, K., Rupe, J.C. and Rothrock, C.S. 2013. Effect of fungicide seed treatments, cultivars, and soils on soybean stand establishment. Plant Dis. 97: 807-812.

Uthandi, B. and Sivakumaar, P.K. 2013. *Bacillus megaterium* (AUM72)-mediated induction of defense related enzymes to enhance the resistance of turmeric (*Curcuma longa* L.) to *Pythium aphanidermatum* causing rhizome rot. Agricultura 10: 1-8.

Uzuhashi, S., Tojo, M. and Kakishima, M. 2010. Phylogeny of the genus *Pythium* and description of new genera. Mycoscience 51: 337-365.

Vaartaja, O. and Agnihotri, V.P. 1969. Interaction of nutrients and four antifungal antibiotics in their effects on *Pythium* species *in vitro* and in soil. Plant & Soil 30: 49-61.

Vargas, A. 2018. Management of seedling diseases caused by Oomycetes, *Phytophthora* spp., *Phytopythium* spp. and *Pythium* spp. using seed treatment in Ohio. Doctoral dissertation, The Ohio State University.

Ved, R., Mishra, A., Singh, R., Tomar, A., Trivedi, S. and Dixit, S. 2017. *In-vitro* evaluation of chemical fungicides and bioagents against *Pythium aphanidermatum*. J. Nat. Res. Devel. 12: 11-14.

Venturini, G., Campia, P., Salomoni, D., Toffolatti, S.L., Vercesi, A., Brunelli, A. and Collina, M. 2014. Fungicide efficacy against different fungal species affecting maize seeds and seedlings. Atti, Giornate Fitopatologiche, Chianciano Terme (Siena), 18-21 marzo 2014, Volume secondo, pp. 393-400.

Vijayan, A.K. and Thomas, J. 2002. Integrated management of rhizome rot of small cardamom using *Trichoderma* sp. Proceedings of the 15th Plantation Crops Symposium Placrosym XV, Mysore, India, 10-13 December, 2002, pp. 576-578.

Vinale, F., Flematti, G., Sivasithamparam, K., Lorito, M., Marra, R., Skelton, B.W. and Ghisalberti, E.L. 2009. Harzianic acid, an antifungal and plant growth promoting metabolite from *Trichoderma harzianum*. J. Nat. Prod. 72: 2032-2035.

Vinale, F., Ghisalberti, E.L., Flematti, G., Marra, R., Lorito, M. and Sivasithamparam, K. 2010. Secondary metabolites produced by a root-inhabiting sterile fungus antagonistic towards pathogenic fungi. Lett. App. Microbiol. 50: 380-385.

Wang, H. and Davis, R.M. 1997. Susceptibility of selected cotton cultivars to seedling disease pathogens and benefits of chemical seed treatments. Plant Dis. 81: 1085-1088.

Warren, J.E. and Bennett, M.A. 2000. Bio-osmopriming tomato (*Lycopersicon esculentum* Mill.) seeds for improved seedling establishment. pp. 477-487. *In*: Black, M., Bradford, K.J. and Vazquez-Ramos, J. (eds.). Seed Biology: Advances and Applications. Proceedings of the Sixth International Workshop on Seeds, Merida, Mexico, 1999. CABI Publishing, Wallingford.

Weerakoon, D.M.N., Muditha, D., Weerakoon, N., Reardon, C.L., Paulitz, T.C., Izzo, A.D. and Mazzola, M. 2012. Long-term suppression of *Pythium abappressorium* induced by *Brassica juncea* seed meal amendment is biologically mediated. Soil Biol. Bioch. 51: 44-52.

Weiland, J.E., Leon, A.L., Edmonds, R.L., Littke, W.R., Browning, J.E., Davis, A., Beck, B.R., Miller, T.W., Cherry, M.L. and Rose, R. 2011. The effects of methyl bromide alternatives on soil and seedling pathogen populations, weeds, and seedling morphology in Oregon and Washington forest nurseries. Cana. J. For Res. 41: 1885-1896.

Weiland, J.E., Santamaria, L. and Grünwald, N.J. 2014. Sensitivity of *Pythium irregulare, P. sylvaticum,* and *P. ultimum* from forest nurseries to mefenoxam and fosetyl-Al, and control of Pythium damping-off. Plant Dis. 98: 937-942.

Wheeler, T.A., Howell, C.R., Coltton, J. and Porter, D. 2005. *Pythium* species associated with pod rot on west Texas peanuts and in vitro sensitivity of isolates to mefenoxam and azoxystrobin. Peanut Sci. 32: 9-13.

White, J.G. 1988. Studies on the biology and control of cavity spot of carrots. Ann. App. Biol. 113: 259-268.

Wise, C., Novitsky, L., Tsopmo, A. and Avis, T.J. 2012. Production and antimicrobial activity of 3-hydroxypropionaldehyde from *Bacillus subtilis* strain CU12. J. Chem. Ecol. 38: 1521-1527.

Wutoh, J.G., Tambong, J.T., Meboka, M.M. and Nzietchueng, S. 1991. Field evaluation of cocoyam (*Xanthosoma sagittifolium* (L.) Schott) for tolerance to the root rot disease caused by *Pythium myriotylum*. *In*: Symposium on Tropical Root Crops in a Developing Economy 380: 462-466.

Xiangling, F.A.N.G., Phillips, D., Verheyen, G., Hua, L.I., Sivasithamparam, K. and Barbetti, M.J. 2012. Yields and resistance of strawberry cultivars to crown and root diseases in the field, and cultivar responses to pathogens under controlled environment conditions. Phytopathol. Mediterr. 51: 69-84.

Xiuqin, S. and Furong, W. 1998. Identification for resistance of maize varieties to stalk rot. J. Shanxi Agric. Sci. 04.

Xu, T., Zhong, J.P. and Li, D.B. 1993. Antagonism of *Trichoderma harzianum* T82 and *Trichoderma* sp. NF9 against soil-borne fungus pathogens. Acta Phytopathol. Sin. 23: 63-67.

Yadav, D.L. and Joshi, K.R. 2012. Efficacy of agrochemicals against *Pythium aphanidermatum* cause damping-off in bidi tobacco. J. Plant Dis. Sci. 7: 77-80.

Yamaji, K., Fukushi, Y., Hashidoko, Y., Yoshida, T. and Tahara, S. 1999. Characterization of antifungal metabolites produced by *Penicillium* species isolated from seeds of *Picea glehnii*. J. Chem. Ecol. 25: 1643-1654.

Yamaji, K., Fukushi, Y., Hashidoko, Y., Yoshida, T. and Tahara, S. 2001. *Penicillium* fungi from *Picea glehnii* seeds protect the seedlings from damping-off. New Phytol. 152: 521-531.

Yamaji, K., Ishimoto, H., Usui, N. and Mori, S. 2005. Organic acids and water-soluble phenolics produced by *Paxillus* sp. 60/92 together show antifungal activity against *Pythium vexans* under acidic culture conditions. Mycorrhiza 15: 17-23.

Yanashkov, I., Maneva, S. and Vatchev, T. 2017. Application of fungicides for management of major root and lower stem rot diseases of wheat. Rasteniev'dniNauki/Bulgarian J. Crop Sci. 54: 3-14.

Yanes, M.L., Fernández, A., Arias, A. and Altier, N. 2004. Método para evaluar protección contra *Pythium debaryanum* y promoción del crecimientoen alfalfa de *Pseudomonas fluorescentes*. Agrociencia 8: 23-31.

Yanes, M.L., Fuente, L., Altier, N. and Arias, A. 2012. Characterization of native fluorescent *Pseudomonas* isolates associated with alfalfa roots in Uruguayan agroecosystems. Biol. Control 63: 287-295.

Yang, Y.J., Tao, B., Li, C., Bian, X.J. and Zhang, J.L. 2008. Using the toxin of *Pythium aphanidermatum* to screen turfgrasses with disease resistance. Acta Prataculturae Sin. 17: 93-98.

Yogeshwar, M., Gade, R.M. and Shitole, A.V. 2017. Evaluation of antifungal activities of extracts of *Aegle marmelos, Syzygium cumini* and *Pongamia pinnata* against *Pythium debaryanum*. Indian J. Pharm. Sci. 79: 377-384.

Zagade, S.N., Deshpande, G.D., Gawade, D.B., Atnoorkar, A.A. and Pawar, S.V. 2012. Biocontrol agents and fungicides for management of damping-off in chilli. World J. Agric. Sci. 8: 590-597.

Zaspel, I. and Suss, R. 1992. Biological control of damping-off in *Pinus sylvestris* L. with bacterial antagonists. Bull. OILB/SROP. 15: 118-120.

Zhang, B.X., Ge, Q.X., Chen, D.H., Wang, Z.Y. and He, S.S. 1990. Biological and chemical control of root diseases on vegetable seedlings in Zhejiang province, China. pp. 181-196. *In*: Hornby, D. (ed.). Biological Control of Soil-borne Plant Pathogens. C.A.B. International, Wallingford.

Management of *Pythium* spp. by Arbuscular Mycorrhizal Fungi

Sarika R. Bhalerao[1*], Pratiksha R. Gund[1], Sunita D. Bansod[2] and Mahendra Rai[3]

[1] Department of Plant Biotechnology, Vilasrao Deshmukh College of Agricultural Biotechnology, V.N.M.K.V., Latur, Maharashtra, India
[2] Ajinkya Agrobiotech, MIDC, Amravati, Maharashtra, India
[3] Department of Biotechnology, Sant Gadge Baba Amravati University, Amravati, Maharashtra, India

Introduction

Pythium spp. were earlier classified as fungi. But these are not considered to be true fungi due to their resemblance with parasitic brown algae (Dreistadt 2004). Most species are plant pathogens. Several species of *Pythium* grow as saprophytes in soil, in water, on decaying organic matters, whereas others are facultative parasites causing root rot (root tips), fruit rot (watery fruits) and damping-off of pre-emergent and post-emergent seedling of a number of plants. The Pythium root rot is considered important for managing the disease in commercial greenhouses and continually threatened the productivity of numerous kinds of crops in hydroponic system around the world including cucumber, tomato, sweet pepper, spinach, lettuce, nasturtium, arugula, rose and chrysanthemum. *Pythium ultimum* is one of the three most common *Pythium* species. *P. aphanidermatum* and *P. irregulare* are other two pathogenic species (Moorman 2014).

After colonization of plant roots, *Pythium* causes extensive damage to the root system. The infected roots develop root rot symptoms which include lack of small feeder roots, brown lesions in the roots, and loss of cortex. The pathogen may grow luxuriantly and move to other parts of the host plant. In some cases where the plant survives, *Pythium* may infect the stem and cause stem rot. The stem turns brown with extensive damage to the vascular system resulting in blocking of xylem and phloem and cuts-off water transportation to the leaves and causes death of the plant. Sometimes the infected plants are stunted as compared to a healthy plant and do not die. In general, plants with ability to rapidly produce new roots and having an extensive root system can cope with and survive *Pythium* infection, while plants with relatively weak root system rapidly succumb.

In this chapter, we aim to discuss the role of AMF in *Pythium* management and potential mechanisms by which AM fungi may contribute to bioprotection against *Pythium*.

Diseases caused by *Pythium* spp.

Not all the species of *Pythium* are plant pathogenic and cause crop losses; some species of *Pythium* cause economically important diseases in several, especially herbaceous, crops. They invade numerous plants and severity of infection is highly dependent on environmental conditions. Some

*Corresponding author: sarikasshende@gmail.com

species grow well in moist and cool conditions (such as *Pythium irregulare* and *P. ultimum*), while others require higher temperatures (such as *Pythium aphanidermatum* and *P. myriotylum*). *Pythium* diseases are divided in two types: diseases affecting below ground plant parts in contact with the soil (roots, lower stem, seeds, tubers, and fleshy fruits) and above ground plant parts (leaves, young stems, and fruits) (Agrios 2005). *Pythium* species infect plant roots, showing necrotic lesions on root tips and/or fine feeder roots and, less commonly, on tap roots. *Pythium* infections are usually restricted to the meristematic root tips, root epidermis, cortex of roots, and fruits. Severe infections may occur when the pathogen spreads deeper into the plant tissue and reaches the vascular system. They are one of the important pathogens, causing root rot of several crops in hydroponic culture (Watanabe et al. 2008). *P. debaryanum*, *P. ultimum*, *P. aphanidermatum* and *P. irregulare* cause damping-off of seedling, of which *P. debaryanum* is very common and causes damping-off of seedling of tobacco, chilli, tomato and mustard. *P. gracilis* grows as parasite on *Vaucheria* (alga). List of the pathogens along with diseases caused by them are given in Table 19.1.

Table 19.1. Important diseases caused by *Pythium* species

S. no.	Causal organism	Name of the disease
1	*Pythium aphanidermatum*	Rhizome rot of ginger, Foot rot of papaya, Damping-off of potato, *Aquilaria agallocha* Fruit-rot of cucurbits. Pythium blight of turfgrass
2	*P. butleri*	Fruit-rot of cucurbits
3	*P. debaryanum*	Damping-off of tobacco, tomato, chilli, caster, cress (*Lepidium*)
4	*P. graminicolum*	Rhizome and root rot of turmeric
5	*P. indicum*	Fruit-rot of lady's finger (*Hibiscus esculentus*)
6	*P. myriotylium*	Rhizome rot of ginger
7	*P. ultimum*	Damping-off of soybean
8	*P. vexans*	Damping-off of cardamom
9	*Pythium* species	Root rot of Lettuce

Pythium species are soil inhabitants and remain present in most wet agricultural soils for extended periods of time. *Pythium* is easily spread by water flow from irrigation, flooding, and soil movement by tractors and other equipment. The infection can occur during every crop stage throughout the entire vegetation season. Crop's root system is usually ignored by many farmers, where the first signs of this pathogen occur. The sign of *Pythium* infection often does not appear on the leaves until significant injury and/or death occur.

Bio-control of plant pathogens presents an enthralling method of increasing crop yields by suppressing or destroying pathogens, enhancing the ability of plants to resist pathogens, and/or protecting plants against pathogens. Bio-control contributes a myriad of aids such as a component of the environment, impervious to development of chemical pesticide resistance, being relatively safe and risk free, and being compatible with sustainable agriculture. Arbuscular mycorrhizal fungi (AMF) are a group of an organism that represents a key link between plants and soil mineral nutrients and also acts as bioprotectors of plants. Thus, they are generating global interest of scientific community as natural fertilizers. AMF are obligate symbionts, of phylum Glomeromycota (Schüßler et al. 2001), and form mutualistic synergies with about 80% of land plant species, including several agricultural crops. AMF benefits the host plant by providing mineral nutrients and water, in exchange for photosynthetic products (Smith and Read 2008). The AMF hyphae emerging from the root system are much thinner than roots spread and are therefore able to penetrate smaller pores and extract nutrients from soil (Smith et al. 2000, Allen 2011). Exchange of carbohydrates and mineral nutrients takes place inside the roots across the interface between the plant and the fungus. AMF hyphae colonize the root cortex and form highly branched structures inside the cells, i.e. arbuscules, which are the functional site of nutrient exchange (Balestrini et al. 2015). Thus, AMF can aggravate the plant growth retardation caused by an inadequate nutrient supply (Nouri et al. 2014).

It has recently been suggested that a plant with non-mycorrhizal association should be viewed as abnormal. There is a marked diversity among AM fungal communities, depending on diversity of plant species, soil type, and season, or a combination of these factors in natural environments (Smith and Smith 2012). Their relationship with rhizosphere flora and fauna encourages the growth and fitness of the associated plants (Azcon-Aguilar and Barea 1992, Fitter and Sanders 1992). AMF has significant role in crop production, as discordant association between the host plant and the native AMF community can lead to serious losses in crop yields. A companionable association between AMF and plant results in enhanced plant productivity, through enhanced host P nutrition (Ravnskov and Jakobsen 1995), control of plant diseases caused by soil-borne pathogens (Caron 1989a, St-Arnaud et al. 1995), and/or enhancement of plant hormonal activity (Frankenberger and Arshad 1995).

Establishment of AM association involves fundamental reprogramming and activation of hundreds of genes of the host cells (Liu et al. 2003, Güimil et al. 2005, Hohnjec et al. 2005, Fiorilli et al. 2009, Gomez et al. 2009, Guether et al. 2009, Breuillin et al. 2010, Gaude et al. 2012, Tromas et al. 2012, Hogekamp and Küster 2013, Calabrese et al. 2017). Some of these genes are expressed primarily or exclusively in cells with arbuscules. These genes are essential for intracellular accommodation of the fungus, and for harmonization of symbiotic functions.

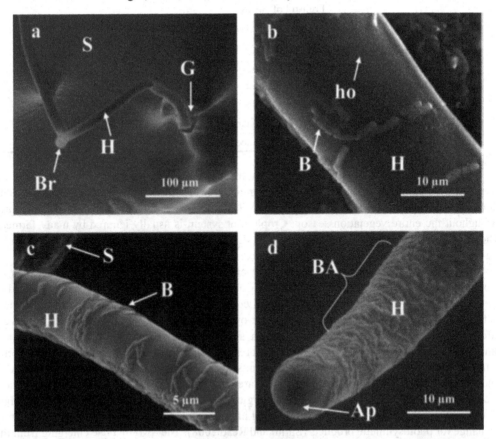

Figure 19.1. *In situ* SEM picture of *G. margarita* spore (S) and hyphae (II) colonized by PE (B), including their germination (G) and branching point (Br). Some holes (ho) and bacterial aggregates (BA) were detected on the hyphal surface, but no bacterial colonization on the apical part of hyphae (Ap). Control (no bacterial) (a); KTCIGM01 (b); KTCIGM02 (c); KTCIGM03 (d). (Source: Adopted from Andre Freire Cruz and Takaaki Ishii, 2011)

The mycorrhizosphere is the rhizospheric region subjected to modifications following AMF colonization of the host plant (Linderman 1988). The biochemical changes that occur in the plant following AMF root colonization are collectively termed the "mycorrhizosphere effect." The mycorrhizosphere effect typically results in a transient or permanent shift in the indigenous microbial community which may be helpful for the elimination or proliferation of pathogens (Meyer and Linderman 1986, Paulitz and Linderman 1989, Nemec 1994, Edwards et al. 1998). In general, these changes are mediated by modifications in host root membrane permeability that subsequently leads to modifications in root exudate composition (Ratnayake et al. 1978, Graham et al. 1981).

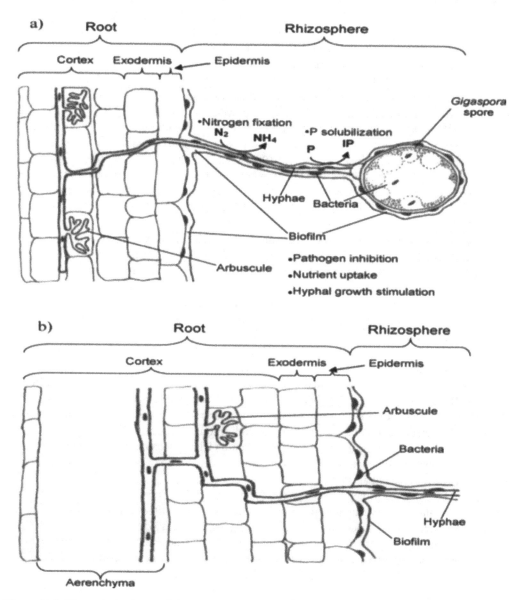

Figure 19.2. Theoretical scheme of the association between AMF and their partner bacteria within the plant-soil-microorganism system in roots under normal (a) and waterlogged conditions (b). The bacteria can be located inside and/or surrounding spores and hyphae, and they are released into roots and cells after digestion of hyphae (see arrows). IP – Insoluble phosphorus; P – solubilized phosphorus; Atmospheric nitrogen (N_2); Ammonium (NH_4). (Source: Adopted from Andre Freire Cruz and Takaaki Ishii, 2011)

Meticulous management of the mycorrhizosphere may serve as an effective, safe, and environment friendly alternative to conventional methods of plant disease control. AMF interact with bacteria during the process of plant colonization. Certain population of bacteria can live in the fungal spores, hyphae (Schuessler et al. 1994) and in the cytoplasm of AMF spores (Cruz 2004, Lumini et al. 2007), which affect nutrient biodynamics and biocontrol of soil-borne plant pathogens (Cruz and Ishii 2011).

Among the phytopathogen, fungi contribute significantly to crop damage and yield loss, followed by bacteria and viruses. The literature survey shows the potential of AMF to control various plant pathogenic fungi (Krishna and Bagyaraj 1983, Duchesne et al. 1989, Boyetchko and Tewari 1996, Kapoor et al. 1998, Kegler and Gottwald 1998, Becker et al. 1999, Bodker et al. 2002, Kasiamdari et al. 2002). Specificity of AMF for the control of crop diseases is crucial in order to mitigate any non-target effects to beneficial micro-organisms.

AMF mediated control of agronomically important pathogens has been consistently demonstrated in many cases. These include *Aphanomyces* species (Rosendahl 1985, Kjoller and Rosendahl 1996, Bødker et al. 1998, Slezack et al. 1999, 2000), *Cylindrocladium spathiphylli* Shoult, El-Gholl and Alf (Declerck et al. 2002), *Erysiphe pisi causing* powdery mildew (Liu et al. 2018), *Fusarium* species (Zambolin and Schenck 1983, Benhamou et al. 1994, Niemira et al. 1996, Jaizme-Vega et al. 1998), *Macrophomina phaseolina* (Tassi) Goid (Zambolin and Schenck 1983), *Phytophthora* species (Davies et al. 1978, Davies and Menge 1980, Vigo et al. 2000), *Pythium* species (Rosendahl and Rosendahl 1990), *Rhizoctonia* species (Zambolin and Schenck 1983, Guillon et al. 2002), *Sclerotinium* species (Krishna and Bagyaraj 1983), *Verticillium* species (Davies et al. 1979) and *Thielaviopsis basicola* (Berk and Broome) Ferraris (Baltruschat and Schöbeck 1975, Schöbeck and Dehne 1977).

AMF mediated reduction of root rot disease in cereal crops has been demonstrated by several workers (Boyetchko and Tewari 1988, Grey et al. 1989, Thompson and Wildermuth 1989, Rempel and Bernier 1990). The model systems used for AMF-mediated plant disease control is *Phytophthora* species, which cause diseases in a variety of plants (Guillemin et al. 1994, Cordier et al. 1996, Pozo et al. 1996, Trotta et al. 1996, Mark and Cassells 1999, Norman and Hooker 2000). Caron and co-workers (1985, 1986a, b) using *G. intraradices* and pathogen *F. oxysporum* f. sp. *lycopersicio*n in tomato have shown that, growth medium used, the application of P, and inoculation of the growth medium with AMF can influence disease severity. Despite proof of AMF potential in controlling plant diseases, few published reports have successfully demonstrated biological control of plant pathogens by AMF in the field (Newsham et al. 1995, Torres-Barragan et al. 1996, Bodker et al. 2002). Newsham et al. (1995) demonstrated pre-treatment of annual grass *Vulpia ciliata* var. *ambigua* with an indigenous *Glomus* species and re-introducing the grass into a natural grass population extended a favourable effect against an indigenous *F. oxysporum*. Onion pre-inoculated with *Glomus* sp. Zac-19 delayed the development of onion white rot caused by *S. cepivorum* by two weeks and protected onion plants for 11 weeks in the field after transplanting with an increased yield of 22% (Torres-Barragan et al. 1996).

Fungicides are used to cope with *Pythium* damping-off, but due to pesticide safety and the increasing popularity of organically produced crops, they have stimulated the need for further development of microbial biocontrol. Inoculation of growth media with AMF is a cost effective preventative measure for growers (Jack and Nelson 2010). Thus, AMF has great potential for control of phytopathogens particularly *Pythium* and the need for more detailed and well-planned and executed studies that will address problems of inconsistent and unreliable results.

Management of *Pythium* by AM fungi

The AMF associations benefit plant's survival, nutrition and growth due to the better utilization of soil nutrients (Odebode et al. 2001, Schwarzott et al. 2001, Gemma et al. 2005, Olawuyi et al. 2012). AMF plays a vital role in nutrients' cycling and safeguard plants against environmental and cultural stresses (Harley and Smith 1989, Odebode 2005). Moreover, the past report of reduction of the pathogenic activity of the soil-borne fungi by AMF has proved that these AMF can be used as suitable biocontrol agents for *Pythium* (Odebode et al. 1995, Olawuyi et al. 2011).

Tobacco treated with *G. fasciculatum* and *Trichoderma harzianum* together effectively controlled damping-off caused by *Pythium aphanidermatum* and black shank disease caused by *P. parasitica* var. *nicotianae* (Sreeramulu et al. 1998). In some cases, microbial mixtures act synergistically with pesticides to result in active protection against plant diseases. Changes in host root membrane permeability by modifications of root exudate composition as a result of AMF colonization (Graham et al. 1981) can impose changes in the rhizosphere microbial stability (Kaye et al. 1984, Meyer and Linderman 1986, Brejda et al. 1998, Edwards et al. 1998), which can collectively benefit host plants by creating favourable conditions for the propagation of microbial population antagonistic to pathogens such as *Phytophthora* and *Pythium* species as shown by Malajczuk and McComb (1979) for eucalyptus seedlings. The details of AMF controlling pythium infection in various crop plants is given in Table 19.2.

Table 19.2. Protective role of AMF species against *Pythium* infection in crop plants

Sr. no.	AMF	Pythium species	Crop	Reference
1	*Glomus etunicatum* and *Glomus* sp.	*Pythium ultimum*	Cucumber	Rosendahl and Rosendahl (1990)
2	*Glomus intraradices*	*P. ultimum*	*Tagetes patula* (French marigold)	St-Arnaud et al. (1994)
3	*Glomus etunicatum* and *G. monosporum*	*P. fragariae*	Strawberry roots	Norman and Hooker (2000)
4	*G. deserticola*	*P. aphanidermatum*	Cowpea (*Vigna unguiculata* L. Walp.)	Nwaga et al. (2007)
5	*Glomus mosseae* and *G. claroideum*	*P. ultimum*	White clover	Carlsen et al. (2008)
6	*Glomus fasciculatum*	*P. aphanidermatum*	*Aquilaria agallocha*	Tabin et al. (2009)
7	*Glomus mosseae, G. fasciculatum, G. constrictum, G. intraradices, Glomus* spp. (local isolate),	*P. aphanidermatum*	Tobacco (*Nicotiana tabacum* L.)	Subhashini and Padmaja (2010)
8	*Glomus intraradices, G. mosseae* and *Glomus claroideum*	*P. aphanidermatum*	Tomato	Larsen et al. (2012)
9	*Glomus mosseae* and *Glomus deserticola*	*P. aphanidermatum*	Pawpaw (*Carica papaya* L.)	Olawuyi et al. (2013)
10	*Glomus intraradices*	*P. ultimum*	Cherry tomato	Graham and Yoshimura (2014)

AM fungi may control the *Pythium* infection by means of various molecular mechanisms which are as following.

Physiological and biochemical alterations of the host

Following AMF colonization, host root tissue P levels are typically enhanced which modify the phospholipid composition and therefore the root membrane permeability results in the reduction in the leakage of net amount of sugars, carboxylic acids, and amino acids into the rhizosphere (Ratnayake et al. 1978, Graham and Menge 1982). These alterations arrest the chemotactic effect of pathogens to plant roots and discourage pathogen entry. Pathogen suppression provided by vesicular arbuscular mycorrhizae is related to improved plant progression, alterations in root morphology (Whipps 2000), competition for plant photosynthetic material, antagonism from mycorrhiza-associated bacteria (Larsen et al. 2012), and competition for access sites (Pal and Gardener 2011). Possible protective mechanisms induced by AMF include increased plant nutrient procurement, damage reimbursement and competition with pathogens for food material, induction of phytochemicals such as phenolics and expression of disease-resistance genes (Feldmann and Boyle 1998, Ruiz-Lozano et al. 1999).

Differential expression of defense-related genes in mycorrhizal plants has been the modern focus of AMF mediated biocontrol (Lambais and Mehdy 1995, Blee and Anderson 1996, Dumas-Gaudot et al. 1996, Pozo et al. 2002). According to investigators, AMF enters into host roots and induce a local, weak, and transient activation of the host protection mechanism against pathogen, which encompasses the stimulation of hydrolytic enzymes such as chitinase, chitosanase, β-glucanase and superoxide dismutase (Pozo et al. 2002). Following AMF colonization, non-mycorrhizal part of the root system appears to have alterations in the constitutive isoforms of the enzymes indicating systemic changes (Pozo et al. 2002). A high level of glucanase activity in host tissues showed positive correlation with the pathogen resistance (Graham and Graham 1991). Further investigation on the role of these glucanases will help in the development of strategies for control of pathogens using AMF.

AMF-induced changes in the expression of plant disease–related enzymes, such as peroxidase (POD), catalase (CAT) and polyphenol oxidase (PPO), and decreases in the levels of malondialdehyde (MDA) are a marker for membrane lipid peroxidation and are associated with plant disease resistance (Liu et al. 2014). MDA have been reported in perennial ryegrass (*Lolium perenne*) infected with the foliage pathogen *Bipolaris sorokiniana* (Li et al. 2018). Both POD and CAT are plant defence enzymes, which reduce reactive oxygen species and can aggravate damage sustained by plants on exposure to some biotic and abiotic stresses (Li et al. 2018).

Systemic-induced resistance

Researchers have suggested that AMF-inoculated plants may employ systemic induced resistance (SIR) as a biocontrol mechanism (Benhamou et al. 1994, Brendan et al. 1996, Trotta et al. 1996). The SIR phenomenon revealed localized and systemic resistance to the pathogen in mycorrhizal plants (Cordier et al. 1998). These effects could be the result of increased salicylic acid in *Glomus intraradices* colonized roots which activate plant protection mechanisms through the salicylic acid and jasmonate signalling pathways (Larsen et al. 2012), also known as induced systemic acquired resistance. An increase in the lignin deposition in plant cell walls following AMF colonization can restrict the spread of pathogens (Dehne and Schonbeck 1979). Some scientists have studied the role of pathogenesis related (PR) proteins in the AMF mediated disease control process (Liu et al. 1995).

Phytoalexins and phytoanticipins

AMF colonization of roots facilitate the increase in the level of total soluble plant phenolics such as isoflavonoids or flavonoids, lignin, syringic, ferulicorcoumaric acids, etc. as synthesis of phytoalexins (Harrison and Dixon 1993, Morandi 1989, 1996). Some flavonoids that are not true phytoalexins may also respond to AMF colonization of roots (Morandi and Le-Quere 1991, Harrison and Dixon 1993, Volpin et al. 1995).

Challenges and strategies for enhancing efficacy of AMF in *Pythium* control

Not many reports are available concerning the bio-control of plant diseases in field conditions using AMF; it is mainly because of two limitations: (i) production of large quantities of AMF inoculum is quite impossible due the obligate biotrophicity status of AMF, and (ii) introduced AMF and the indigenous AMF along with microbial community may have negative interactions after introduction into field. Success of AMF inoculation in disease control under field conditions often depends on interactions between AMF and indigenous microbial community and soil and environmental conditions. An element that encourages AMF efficacy as bio-control agents can further enhance their survival, competitiveness and efficacy. For example, it is known that intra-radical proliferation of AMF within roots is regulated by the host (Bever et al. 1996). Therefore, a highly mycotrophic host or host cultivar may be more favourable for AMF proliferation and reproduction than one that is less mycotrophic (Feldmann and Boyle 1998, Xavier 1999). In addition, non-conducive soil-environment combinations such as high soil P levels and soil disturbance can affect AMF colonization (Stahl et al. 1988, Gazey et al. 1992, Bever et al. 1996, Stutz and Morton 1996) and efficacy (Menge et al. 1978, Ratnayake et al. 1978, Graham et al. 1981).

Colonization of host roots by AMF is a crucial component in the AMF-mediated SIR response of host plants to phytopathogens, as the manifestation of SIR response involves a threshold level AMF presence within host roots (Cordier et al. 1996, 1998). The effect of phosphorus on AMF efficacy may be direct, wherein in optimum P levels reduces AMF activity. In contrast, soil disturbance has an indirect effect where AMF efficacy may be affected by a postponement in mycelial network commencement and use of carbon for the synthesis or repair of the external mycelial network and not for nutrient acquisition. The ability of AMF to effectively re-establish their mycorrhizal association after disturbance might partially determine their success in a disturbed site (Zak et al. 1982). It is not known to what extent soil disturbance affects the biological control activities of AMF, but it interrupts the external mycelial network resulting in a severe reduction in mycorrhizal efficiency (Stahl et al. 1988, Evans and Miller 1990).

Choice of host genotype and rotation (Johnson et al. 1992, Bever et al. 1996), levels of fertilizer application (Jasper et al. 1979, Baltruschat and Dehne 1982, Vivekanandan and Fixen 1991, McGonigle and Miller 1993, 1996), tillage (Evans and Miller 1990, Vivekanandan and Fixen 1991, McGonigle and Miller 1993), pesticide application (Manjunath and Bagyaraj 1984, Schreiner and Bethlenfalvay 1997), and the effect of associated micro-organisms (Andrade et al. 1995, Xavier and Germida 2003) are some critical factors that can affect AMF diversity in soils. For example, continuous cropping selectively enhances the proliferation of parasitic AMF, which are relatively fast growing compared to AMF, which may lead to changes in mycorrhizal biodiversity in the rhizosphere (Johnson et al. 1992). Similarly, one particular AMF host selected from native AMF pool results in the selective enrichment of certain AMF species over others (Xavier 1999).

Selection of effective AMF

Selection of AMF species for a particular activity usually depends on their survival ability, aggressive colonization of host roots, and efficacy (Bagyaraj 1984). AMF species initially isolated from test host roots were found to be advantageous for number of plant species including agricultural crops, forest tree species, and orchard crops (Reena and Bagyaraj 1990b, Vinayak and Bagyaraj 1990, Talukdar and Germida 1994). Screening procedures for selecting efficient AMF species should include selection pressure similar to that in which the AMF will be applied.

Superior application technology

Research illustrates that plants "priming" against pathogens using selective AMF species (or plant immunization) helps to protect plants by inducing a SIR response (Cordier et al. 1998). The inoculum may be applied to seeds, transplanted crops, or plantlets produced through tissue culture before being transplanted into pathogen-infested fields.

Conclusion

The global economic crisis is now forcing growers to realise the potential of sustainable agricultural systems, by reducing the input of phosphorus using AMF inocula. Applied research focused on defining the best inoculum formulation strategies (Verbruggen et al. 2013) should be encouraged. Unfortunately, solid inoculation practices are yet to be employed. The commercial sector has recognized the potential of AMF, and several firms nowadays produce and vend AMF based inocula.

The world's population is expected to exceed nine billion by 2050 (Rodriguez and Sanders 2015). Thus, global agriculture will have to set up mission of doubling food production with reduction in the usage of agrochemicals, in order to safeguard human and environmental health. The projected necessary yield increase will exceed the current worldwide capacity of food production (Rodriguez and Sanders 2015), which highlights the necessity to implement or revitalize eco-friendly technologies, such as AMF-based biofertilization and biocontrol. AMF application has not been fully accepted by farmers so far, despite its enormous potential. A literature survey shows potential of AMF in control of plant diseases. Although there are challenges in the form of AMF non-culturability, which limits bulk production for agricultural crops, there is assurance for non-direct sown crops, which is currently undervalued and underexploited. The AMF, by enhancing crop production using present resources, avoiding resistance development to chemicals, maintaining pollution and risk-free disease control, and conforming to sustainable agricultural practices, offers more than mere plant disease control. Mycorrhizosphere management is the only viable and ecosystem friendly key in the future, for plant diseases management and reducing pathogen inoculum.

The huge losses caused by the *Pythium* diseases to economically important agricultural and horticultural crops and with increasing global demand to increase productivity, use of AMF in the farm practices is one of the alternatives in order to produce disease-free and chemical-free organic food for the society. This will help in harmonizing the nature equilibrium and agricultural farming.

Reference

Agrios, G.N. 2005. Plant Pathology. 5th Edition. Academic Press. ISBN: 9780120445653

Allen, M.F. 2011. Linking water and nutrients through the vadose zone: A fungal interface between the soil and plant systems: linking water and nutrients through the vadose zone. J. Arid Land 3: 155-163. 10.3724/SP.J.1227.2011.00155

Andrade, G., Azcon, R. and Bethlenfalvay, G.J. 1995. A rhizobacterium modifies plant and soil responses to the mycorrhizal fungus *Glomus mosseae*. Appl. Soil Ecol. 2: 195-202.

Azcon-Aguilar, C. and Barea, J.M. 1992. Interactions between mycorrhizal fungi and other rhizosphere microorganisms. pp. 163-198. *In*: Allen, M.F. (ed.). Mycorrhizal Functioning: An Integrative Plant-Fungal Process. Chapman and Hall, New York.

Bagyaraj, D.J. 1984. Biological interactions with VA mycorrhizal fungi. pp. 131-153. *In*: Powell, C.L. and Bagyaraj, D.J. (eds.). VA Mycorrhiza. CRC Press, Florida.

Balestrini, R., Lumini, E., Borriello, R. and Bianciotto, V. 2015. Plant-soil biota interactions. pp. 311-338. *In*: Paul, E.A. (ed.). Soil Microbiology, Ecology and Biochemistry. Academic Press: Elsevier, London. doi: 10.1016/b978-0-12-415955-6.00011-6

Baltruschat, H. and Schöbeck, F. 1975. The influence of endotrophic mycorrhizae on the infestation of tobacco by *Thielaviopsis basicola*. Phytopathol. Z84: 172-188.

Baltruschat, H. and Dehne, H.W. 1982. Occurrence of vesicular arbuscular mycorrhiza in various cereal crop rotation and fertilizing systems [West Germany]. *In*: Meded-Fac Land bouwwet-Rijksuniv. Gent, Belgium: Het Faculteit 47: 831-839.

Becker, D.M., Bagley, S.T. and Podila, G.K. 1999. Effects of mycorrhizal associated *streptomycetes* on growth of *Laccaria bicolor*, *Cenococcum geophilum*, and *Armillaria* species and on gene expression in *Laccaria bicolor*. Mycologia 91: 33-40.

Benhamou, N., Fortin, J.A., Hamel, C., St-Arnaud, M. and Shatilla, A. 1994. Resistance responses to mycorrhizal Ri T-DNA transformed carrot roots to infection by *Fusarium oxysporum* f. sp. *chrysanthemi*. Phytopathology 84: 958-968.

Bever, J.D., Morton, J.B., Antonovics, J. and Schultz, P.A. 1996. Host dependent sporulation and species diversity of arbuscular mycorrhizal fungi in a mown grassland. J. Ecol. 84: 71-82.

Blee, K.A. and Anderson, A.J. 1996. Defence-related transcript accumulation in *Phaseolus vulgaris* L. colonized by the arbuscular mycorrhizal fungus *Glomus intraradices* Schenck and Smith. Plant Physiol. 110: 675-688

Bødker, L., Kjoller, R. and Rosendahl, S. 1998. Effect of phosphate and the arbuscular mycorrhizal fungus *Glomus intraradices* on disease severity of root rot of peas (*Pisum sativum*) caused by *Aphanomyces euteiches*. Mycorrhiza 8: 169-174.

Bodker, L., Kjoller, R., Kristensen, K. and Rosendahl, S. 2002. Interactions between indigenous arbuscular mycorrhizal fungi and *Aphanomyces euteiches* in field-grown pea. Mycorrhiza 12: 7-12.

Boyetchko, S.M. and Tewari, J.P. 1988. The effect of VA mycorrhizal fungi on infection by *Bipolaris sorokiniana* in barley. Can. J. Plant Pathol. 10: 361.

Boyetchko, S.M. and Tewari, J.P. 1996. Use of VA mycorrhizal fungi in soil-borne disease management. pp. 146-163. *In*: Utkhede, R.S. and Gupta, V.K. (eds.). Management of Soil Borne Diseases. Kalyani Publishers, New Delhi.

Brejda, J.J., Moser, L.E. and Vogel, K.P. 1998. Evaluation of switchgrass rhizosphere microflora for enhancing seedling yield and nutrient uptake. Agron. J. 90: 753-758. Brendan NA, Hamme.

Brendan, N.A., Hammerschmidt, R. and Safir, G.R. 1996. Post harvest suppression of potato dry rot (*Fusarium sambucinum*) in prenuclear minitubers by arbuscular mycorrhizal fungal inoculum. Am. Potato J. 73: 509-515.

Breuillin, F., Schramm, J., Hajirezaei, M., Ahkami, A., Favre, P. and Druege, U. 2010. Phosphate systemically inhibits development of arbuscular mycorrhiza in *Petunia hybrida* and represses genes involved in mycorrhizal functioning. Plant J. 64: 1002-1017. doi: 10.1111/j.1365-313X.2010.04385.x

Calabrese, S., Kohler, A., Niehl, A., Veneault-Fourrey, C., Boller, T. and Courty, P.E. 2017. Transcriptome analysis of the *Populus trichocarpa-Rhizophagus irregularis* mycorrhizal symbiosis: Regulation of plant and fungal transportomes under nitrogen starvation. Plant Cell Physiol. 58: 1003-1017. doi: 10.1093/pcp/pcx044

Carlsen, S.C.K., Understrup, A. and Fomsgaard, I.S. 2008. Flavonoids in roots of white clover: Interaction of arbuscular mycorrhizal fungi and a pathogenic fungus. Plant Soil 302: 33-43. https://doi.org/10.1007/s11104-007-9452-9

Caron, M., Fortin, J.A. and Richard, C. 1985. Influence of substrate on the interaction of *Glomus intraradices* and *Fusarium oxysporum* f. sp. *radicis-lycopersici* on tomatoes. Plant Soil 87: 233-239.

Caron, M., Fortin, J.A. and Richard, C. 1986a. Effect of *Glomus intraradices* on the infection by *Fusarium oxysporum* f. sp. *radicis-lycopersici* in tomatoes over a 12-week period. Can. J. Bot. 64: 552-556.

Caron, M., Fortin, J.A. and Richard, C. 1986b. Effect of preinfestation of the soil by a vesicular-arbuscular mycorrhizal fungus, *Glomus intraradices*, on *Fusarium* crown and root rot of tomatoes. Phytoprotection 67: 15-19.

Caron, M. 1989a. Potential use of mycorrhizae in control of soil-borne diseases. Can. J. Plant Pathol. 11: 177-179.

Cordier, C., Gianinazzi, S. and Gianinazzi-Pearson, V. 1996. Colonisation patterns of root tissues by *Phytophthora nicotianae* var. *parasitica* related to reduced disease in mycorrhizal tomato. Plant Soil 185: 223-232.

Cordier, C., Pozo, M.J., Barea, J.M., Gianinazzi, S. and Gianinazzi Pearson, V. 1998. Cell defence responses associated with localized and systemic resistance to *Phytophthora* induced in tomato by an arbuscular mycorrhizal fungus. Mol. Plant-Microb. Interact. 11: 1017-1028.

Cruz, A.F. 2004. Element storage in spores of *Gigaspora margarita* Becker & Hall measured by electron energy loss spectroscopy (EELS). Acta. Bot. Bras 18: 473-480.

Cruz, A.F. and Ishii, T. 2011. Arbuscular mycorrhizal fungal spores host bacteria that affect nutrient biodynamics and biocontrol of soil borne plant pathogens. Biology Open 1: 52-57. doi: http://www.biologists.com/biology-open

Davies, R.M., Menge, J.A. and Zentmayer, G.A. 1978. Influence of vesicular-arbuscular mycorrhizae on *Phytophthora* root rot of three crop plants. Phytopathology 68: 1614-1617.

Davies, R.M., Menge, J.A. and Erwin, D.C. 1979. Influence of *Glomus fasiculatum* and soil phosphorous on *Verticillium* wilt of cotton. Phytopathology 69: 453-456.

Davies, R.M. and Menge, J.A. 1980. Influence of *Glomus fasiculatus* and soil phosphorous on *Phytophthora* root rot of citrus. Phytopathology 70: 447-452.

Declerck, S., Risede, J.M., Rufyikiri, G. and Delvaux, B. 2002. Effects of arbuscular mycorrhizal fungi on severity of root rot of bananas caused by *Cylindrocladium spathiphylli*. Plant Pathol. 51: 109-115.

Dehne, H.W. and Schonbeck, F. 1979. The influence of endotrophic mycorrhiza on plant diseases. II: Phenol metabolism and lignification *Fusarium oxysporum*. Untersuchungenzum Einfluss der endotrophen Mycorrhiza auf Pflanzenkrankheiten. II. Phenol stoffwechsel und Lignifizierung. Phytopathol-Z 95: 210-216.

Dreistadt, S.H. 2004. Pests of Landscape Trees and Shrubs: An Integrated Pest Management Guide, 2nd Ed. University of California Division of Agriculture and Natural Resources. Publication 3359.

Duchesne, L.C., Peterson, R.L. and Ellis, B.E. 1989. The time-course of disease suppression and antibiosis by the ectomycorrhizal fungus *Paxillus involutus*. New Phytol. 111: 693-698.

Dumas-Gaudot, E., Slezack, S., Dassi, B., Pozo, M.J., Gianinazzi-Pearson, V. and Gianinazzi, S. 1996. Plant hydrolytic enzymes (chitinases and β-1, 3-glucanases) in root reactions to pathogenic and symbiotic microorganisms. Plant Soil 185: 211-221.

Edwards, S.G., Young, J.P.W. and Fitter, A.H. 1998. Interactions between *Pseudomonas fluorescens* biocontrol agents and *Glomus mosseae*, an arbuscular mycorrhizal fungus, within the rhizosphere. FEMS-Microbiol Lett. 166: 297-303.

Evans, D.G. and Miller, M.H. 1990. The role of the external mycelia network in the effect of soil disturbance upon vesicular arbuscular mycorrhizal colonization of maize. New Phytol. 114: 65-71.

Feldmann, F. and Boyle, C. 1998. Concurrent development of arbuscular mycorrhizal colonization and powdery mildew infection on three *Begonia hiemalis* cultivars. Z Pflanzenkr Pflanzenschutz 105: 121-129.

Fiorilli, V., Catoni, M., Miozzi, L., Novero, M., Accotto, G.P. and Lanfranco, L. 2009. Global and cell-type gene expression profiles in tomato plants colonized by an arbuscular mycorrhizal fungus. New Phytol. 184: 975-987. doi: 10.1111/j.1469-8137.2009.03031.x

Fitter, A.H. and Sanders, I.R. 1992. Interactions with the soil fauna. pp. 333-354. *In*: Allen, M.F. (ed.). Mycorrhizal Functioning: An Integrative Plant Fungal Process. Chapman and Hall, New York.

Frankenberger, W.T. and Arshad, M. 1995. Microbial biosynthesis of auxins. pp. 35-71. *In*: Frankenberger, W.T. and Arshad, M. (eds.). Phytohormones in Soil. Marcel Dekker, New York.

Gaude, N., Bortfeld, S., Duensing, N., Lohse, M. and Krajinski, F. 2012. Arbuscule-containing and non-colonized cortical cells of mycorrhizal roots undergo extensive and specific reprogramming during arbuscular mycorrhizal development. Plant J. 69: 510-528. doi: 10.1111/j.1365-313X.2011.04810.x

Gazey, C., Abbott, L.K. and Robson, A.D. 1992. The rate of development of mycorrhizas affects the onset of sporulation and production of external hyphae by two species of *Acaulospora*. Mycol. Res. 96: 643-650.

Gemma, A., Hohnjec, N., Vieweg, M.F., Puhler, A., Becker, A. and Kuster, H. 2005. Overlaps in the transcriptional profiles of *M. truncatula* roots program activated during arbuscular mycorrhizal. Plant Physiol. 137: 1283-1301.

Gomez, S.K., Javot, H., Deewatthanawong, P., Torres-Jerez, I., Tang, Y.H. and Blancaflor, E.B. 2009. *Medicago truncatula* and *Glomus intraradices* gene expression in cortical cells harbouring arbuscules in the arbuscular mycorrhizal symbiosis. BMC Plant Biol. 9: 10. doi: 10.1186/1471-2229-9-10

Graham, J.H., Leonard, R.T. and Menge, J.A. 1981. Membrane mediated decrease in root exudation responsible for inhibition of vesicular-arbuscular mycorrhiza formation. Plant Physiol. 68: 548-552.

Graham, J.H. and Menge, J.A. 1982. Influence of vesicular-arbuscular mycorrhizae and soil phosphorous on take-all disease of wheat. Phytopathology 72: 95-98.

Graham, S. and Yoshimura, D. 2014. The *Pythium* suppressive ability of *Glomus intraradices* in cherry tomamto propagation. BIO 462, 3 March 2014. https://pdfs.semanticscholars.org/33e1/e45f4c7a162c6e5c33404943ae8ad7ee58ce.pdf

Graham, T.L. and Graham, M.Y. 1991. Cellular coordination of molecular responses in plant defense. Mol. Plant-Microb. Interact. 4: 415-422.

Grey, W.E., van Leur, J.A.G., Kashour, G. and El-Naimi, M. 1989.The interaction of vesicular-arbuscular mycorrhizae and common root rot (*Cochliobolus sativus*) in barley. Rachis 8: 18-20.

Guether, M., Balestrini, R., Hannah, M., He, J., Udvardi, M. and Bonfante, P. 2009. Genome-wide reprogramming of regulatory networks, cell wall and membrane biogenesis during arbuscular-mycorrhizal symbiosis in *Lotus japonicus*. New Phytol. 182: 200-212. doi: 10.1111/j.1469-8137.2008.02725.x

Guillon, C., St-Arnould, M., Hamel, C. and Jabaji-Hare, S.H. 2002. Differential and systemic alteration of defence-related gene transcript levels in mycorrhizal bean plants with *Rhizoctonia solani*. Canad. J. Bot. 80: 305-315.

Güimil, S., Chang, H.S., Zhu, T., Sesma, A., Osbourn, A. and Roux, C. 2005. Comparative transcriptomics of rice reveals an ancient pattern of response to microbial colonization. Proc. Natl. Acad. Sci. U.S.A. 102: 8066-8070. doi: 10.1073/pnas.0502999102

Gullemin, J.P., Gianinazzi, S., Gianinazzi-Pearson, V. and Marchal, J. 1994. Control of arbuscular endomycorrhizae of *Pratylenchus brachyurus* in pineapple microplants. Agric. Sci. Fin. 3: 253-262.

Harley, J.L. and Smith, S.E. 1989. Mycorrhizal symbiosis. London: Academic Press.

Harrison, M.J. and Dixon, R.A. 1993. Isoflavonoid accumulation and expression of defense gene transcripts during the establishment of VA mycorrhizal associations in roots of *Medicago truncatula*. Mol. Plant-Microb. Interact. 6: 643-654.

Hogekamp, C. and Küster, H. 2013. A roadmap of cell-type specific gene expression during sequential stages of the arbuscular mycorrhiza symbiosis. BMC Genomics 14: 306. doi: 10.1186/1471-2164-14-306

Hohnjec, N., Vieweg, M.E., Puhler, A., Becker, A. and Küster, H. 2005. Overlaps in the transcriptional profiles of *Medicago truncatula* roots inoculated with two different *Glomus* fungi provide insights into the genetic program activated during arbuscular mycorrhiza. Plant Physiol. 137: 1283-1301. doi: 10.1104/pp.104.056572

Jack, A.L.H. and Nelson, E.B. 2010. Suppression of *Pythium* damping off with compost and vermicompost. Final report to the Organic Farming Research Foundation. http://cwmi.css.cornell.edu/organicfarmingfi nalreport.pdf

Jaizme-Vega, M.C., Herńandez, B.S. and Herńandez, J.M.H. 1998. Interaction of arbuscular mycorrhizal fungi and the soil pathogen *Fusarium oxysporum* f. sp. *cubense* on the first stages of micropropagated grandenaine banana. Acta Hort. 490: 285-295.

Jasper, D.A., Robson, A.D. and Abbott, L.K. 1979. Phosphorus and the formation of vesicular-arbuscular mycorrhizas. Soil Biol. Biochem. 11: 501-505.

Johnson, N.C., Copeland, P.J., Crookston, R.K. and Pfleger, F.L. 1992. Mycorrhizae: Possible explanation for yield decline with continuous corn and soybean. Agron. J. 84: 387-390.

Kapoor, A.R., Mukherji, K.G. and Kapoor, R. 1998. Microbial interactions in mycorrhizosphere of *Anethum graveolens* L. Phytomorphology 48: 383-389.

Kasiamdari, R.S., Smith, S.E., Smith, F.A. and Scott, E.S. 2002. Influence of the mycorrhizal fungus, *Glomus coronatum*, and soil phosphorus on infection and disease caused by binucleate *Rhizoctonia* and *Rhizoctonia solani* on mung bean (*Vigna radiata*). Plant Soil 238: 235-244.

Kaye, J.W., Pfleger, F.L. and Stewart, E.L. 1984. Interaction of *Glomus fasciculatum* and *Pythium ultimum* on greenhouse-grown poinsettia. Can. J. Bot. 62: 1575-1579.

Kegler, H. and Gottwald, J. 1998. Influence of mycorrhizas on the growth and resistance of asparagus. Arch. Phytopathol. Plant Protect. 31: 435-438.

Kjoller, R. and Rosendahl, S. 1996. The presence of the arbuscular mycorrhizal fungus *Glomus intraradices* influences enzymatic activities of the root pathogen *Aphanomyces euteiches* in pea roots. Mycorrhiza 6: 487-491.

Krishna, K.R. and Bagyaraj, D.J. 1983. Interaction between *Glomus fasciculatum* and *Sclerotium rolfsii* in peanut. Can. J. Bot. 61: 2349-2351.

Lambais, M.R. and Mehdy, M.C. 1995. Differential expression of defense-related genes in arbuscular mycorrhiza. Can. J. Bot. 73: S533-S540.

Larsen, J., Graham, J.H., Cubero, J. and Ravnskov, S. 2012. Biocontrol traits of plant growth suppressive arbuscular mycorrhizal fungi against root rot in tomato caused by *Pythium aphanidermatum*. Eur. J. Plant Pathol. 133(2): 361-369. DOI: 10.1007/s10658-011-9909-9

Li, F., Guo, Y.E., Christensen, M.J., Gao, P., Li, Y.Z. and Duan, T.Y. 2018. An arbuscular mycorrhizal fungus and *Epichloë festucae* var. *lolii* reduce *Bipolaris sorokiniana* disease incidence and improve perennial ryegrass growth. Mycorrhiza 28: 159-169.

Linderman, R.G. 1988. Mycorrhizal interactions with the rhizosphere microflora. The mycorrhizosphere effect. Phytopathology 78: 366-371.

Liu, J.Y., Blaylock, L.A., Endre, G., Cho, J., Town, C.D. and Vanden Bosch, K.A. 2003. Transcript profiling coupled with spatial expression analyses reveals genes involved in distinct developmental stages of an arbuscular mycorrhizal symbiosis. Plant Cell 15: 2106-2123. doi: 10.1105/tpc.014183

Liu, R.J., Liu, H.F., Shen, C.Y. and Chiu, W.F. 1995. Detection of pathogenesis-related proteins in cotton plants. Physiol. Mol. Plant Pathol. 47: 357-363.

Liu, X.L., Xi, X.Y., Shen, H., Liu, B. and Guo, Y. 2014. Influences of arbuscular mycorrhizal (AM) fungi inoculation on resistance of tobacco to bacterial wilt. Tob Sci. Technol. 5: 94-98.

Liu, Y., Feng, Xi, Gao, P., Li, Y., Christensen, M.J. and Duan, T. 2018. Arbuscular mycorrhiza fungi increased the susceptibility of *Astragalus adsurgens* to powdery mildew caused by *Erysiphepisi*. Mycology 9(3): 223-232.

Lumini, E., Bianciotto, V., Jargeat, P., Novero, M., Salvioli, A., Faccio, A., Becard, G. and Bonfante, P. 2007. Presymbiotic growth and sporal morphology are affected in the arbuscular mycorrhizal fungus *Gigaspora margarita* cured of its endobacteria. Cell. Microbiol. 9: 1716-1729.

Malajczuk, N. and McComb, A.J. 1979. The microflora of unsuberized roots of *Eucalyptus calophylla* R. Br. and *Eucalyptus marginata* Donn ex Sm. seedlings grown in soil suppressive and conducive to *Phytophthora cinnamomi* Rands. I. Rhizosphere bacteria, actinomycetes and fungi. Aust. J. Bot. 27: 235-254.

Manjunath, A. and Bagyaraj, D.J. 1984. Effect of fungicides on mycorrhizal colonization and growth of onion [*Agrosan, Benlate, Captan, Ceresan* and *Plantavax, Glomus fasciculatum*]. Plant Soil 80: 147-150.

Mark, G.L. and Cassells, A.C. 1999. The effect of dazomet and fosetyl-aluminium on indigenous and introduced arbuscular mycorrhizal fungi in commercial strawberry production. Plant Soil 209: 253-261.

McGonigle, T.P. and Miller, M.H. 1993. Mycorrhizal development and phosphorus absorption in maize under conventional and reduced tillage. Soil Sci. Soc. Am. J. 57: 1002-1006.

McGonigle, T.P. and Miller, M.H. 1996. Mycorrhizae, phosphorus absorption, and yield of maize in response to tillage. Soil Sci. Soc. Am. J. 60: 1856-1861.

Menge, J.A., Steirle, D., Bagyaraj, D.J. and Johnson, E.L.V. 1978. Phosphorus concentrations in plants responsible for inhibition of mycorrhizal infection. New Phytol. 80: 575-578.

Meyer, J.R. and Linderman, R.G. 1986. Response of subterranean clover to dual inoculation with vesicular-arbuscular mycorrhizal fungi and a plant growth promoting bacterium *Pseudomonas putida*. Soil Biol. Biochem. 18: 185-190.

Morandi, D. 1989. Effect of xenobiotics on endomycorrhizal infection and isoflavonoid accumulation in soybean roots. Plant Physiol. Biochem. 27: 697-701.

Morandi, D. and Le-Quere, J.L. 1991. Influence of nitrogen on accumulation of isosojagol (a newly detected coumestan in soybean) and associated isoflavonoids in roots and nodules of mycorrhizal and non-mycorrhizal soybean. New Phytol. 117: 75-79.

Morandi, D. 1996. Occurrence of phytoalexins and phenolic compounds in endomycorrhizal interactions, and their potential role in biological control. Plant Soil 185: 241-251.

Nemec, S. 1994. Soil microflora associated with pot cultures of *Glomus intraradix*-infected *Citrus reticulata*. Agric. Ecosyst. Environ. 1: 299-306.

Newsham, K.K., Fitter, A.H. and Watkinson, A.R. 1995. Arbuscular mycorrhiza protect an annual grass from root pathogenic fungi in the field. J. Ecol. 83: 991-1000.

Niemira, B.A., Hammerschmidt, R. and Safir, G.R. 1996. Postharvest suppression of potato dry rot (*Fusarium sambucinum*) in prenuclear minitubers by arbuscular mycorrhizal fungal inoculum. Am. Potato J. 73: 509-515.

Norman, J.R. and Hooker, J.E. 2000. Sporulation of *Phytophthora fragariae* shows greater stimulation by exudates of nonmycorrhizal than by mycorrhizal strawberry roots. Mycol. Res. 104: 1069-1073.

Nouri, E., Breuillin-Sessoms, F., Feller, U. and Reinhardt, D. 2014. Phosphorus and nitrogen regulate arbuscular mycorrhizal symbiosis in petunia hybrida. PLoS ONE 9: e90841. doi: 10.1371/journal. pone.0090841

Nwaga, D., Fankem, H., Essono Obougou, G., Ngo Nkot, L. and Randrianangaly Jean, S. 2007. *Pseudomonads* and symbiotic micro-organisms as biocontrol agents against fungal disease caused by *Pythium aphanidermatum*. Afr. J. Biotechnol. 6: 190-197.

Odebode, A.C., Ladoye, A.O. and Osonubi, O. 1995. Influence of Arbuscular mycorrhizal fungi on disease severity of pepper and tomato caused by *Sclerotium rolfsii*. J. Sci. Res. 2: 49-52.

Odebode, A.C., Salami, A.O. and Osonubi, O. 2001. Oxidative enzymes activities of mycorrhizal inoculated pepper plant infected with *Phytophthora infestans*. Arch. Phytopathol. Plant Prot. 33: 473-480.

Odebode, A.C. 2005. The use of Arbuscular Mycorrhiza (AM) as a source of yield increase in sustainable alley cropping system. Arch. Agron. Soil Sci. 51: 385-390.

Olawuyi, O.J., Odebode, A.C., Olakojo, S.A. and Adesoye, A.I. 2011. Host-parasite relationship of maize (*Zea mays* L.) and *Strigalutea* (Lour.) as influenced by Arbuscular mycorrhiza fungi. J. Sci. Res. 10: 186-198.

Olawuyi, O.J., Ezekiel-Adewoyin, D.T., Odebode, A.C., Aina, D.A. and Esenbamen, G.E. 2012. Effect of arbuscular mycorrhizal (*Glomus clarum*) and organomineral fertilizer on growth and yield performance of okra (*Abelmoschus esculentus*). Afr. J. Plant Sci 2: 84-88. doi: 10.5897/AJPS11.295

Olawuyi, O.J., Odebode, A.C., Oyewole, I.O., Akanmu A.O. and Afolabi, O. 2013. Effect of arbuscular mycorrhizal fungi on *Pythium aphanidermatum* causing foot rot disease on pawpaw (*Carica papaya* L.) seedlings. Archives of Phytopathology and Plant Protection pp. 1-9. DOI: 10.1080/03235408.2013.806079 http://dx.doi.org/10.1080/03235408.2013.806079.

Pal, K. and Gardener, B. 2011. Biological control of plant pathogens. The Plant Health Instructor. 10.1094/PHI-A-2006-1117-02

Paulitz, T.C. and Linderman, R.G. 1989. Interactions between fluorescent *Pseudomonads* and VA mycorrhizal fungi. New Phytol. 113: 37-45.

Pozo, M.J., Dumas-Gaudot, E., Slezack, S., Cordier, C., Asselin, A., Gianinazzi, S., Gianinazzi-Pearson, V., Azcon-Aguilar, C. and Barea, J.M. 1996. Induction of new chitinase isoforms in tomato roots during interactions with *Glomus mosseae* and/or *Phytophthora nicotianae* var. *parasitica*. Agronomie 16: 689-697.

Pozo, M.J., Cordier, C., Dumas-Gaudot, E., Gianinazzi, S., Barea, J.M. and Azcon-Aguilar, C. 2002. Localized versus systemic effect of arbuscular mycorrhizal fungi on defence responses to *Phytophthora* infection in tomato plants. J. Exp. Bot. 53: 525-534.

Ratnayake, M., Leonard, R.T. and Menge, J.A. 1978. Root exudation in relation to supply of phosphorus and its possible relevance to mycorrhiza formation. New Phytol. 81: 543-552.

Ravnskov, S. and Jakobsen, I. 1995. Functional compatibility in arbuscular mycorrhizas measured as hyphal P transport to the plant. New Phytol. 129: 611-618.

Reena, J. and Bagyaraj, D.J. 1990b. Response of *Acacia nilotica* and *Calliandra calothyrsus* to different VA mycorrhizal fungi. Arid. Soil Res. Rehabil. 4: 261-268.

Rempel, C.B. and Bernier, C.C. 1990. *Glomus intraradices* and *Cochliobolus sativus* interactions in wheat grown under two moisture regimes. Can. J. Plant Pathol. 12: 338.

Rodriguez, A. and Sanders, I.R. 2015. The role of community and population ecology in applying mycorrhizal fungi for improved food security. ISME J. 9: 1053-1061.

Rosendahl, C.N. and Rosendahl, S. 1990. The role of vesicular arbuscular mycorrhiza in controlling damping off and growth reduction in cucumber caused by *Pythium ultimum*. Symbiosis 9: 363-366.

Rosendahl, S. 1985. Interactions between the vesicular-arbuscular mycorrhizal fungus *Glomus fasiculatum* and *Aphanomyces euteiches* root rot of pea. Phytopathol. Z114: 31-40.

Ruiz-Lozano, J.M., Roussel, H., Gianinazzi, S. and Gianinazzi-Pearson, V. 1999. Defence genes are differentially induced by a mycorrhizal fungus and a rhizobium sp. in wild type and symbiosis-defective pea genotypes. MPMI12: 976-984.

Schöbeck, F. and Dehne, H.W. 1977. Damage to mycorrhizal and nonmycorrhizal cotton seedlings by *Thielaviopsis basicola*. Plant Dis. Rep. 61: 266-267.

Schreiner, R.P. and Bethlenfalvay, G.J. 1997. Mycorrhizae, biocides, and biocontrol 3: Effects of three different fungicides on developmental stages of three AM fungi. Biol. Fertil. Soil. 24: 18-26.

Schüßler, A., Schwarzott, D. and Walker, C. 2001. A new fungal phylum, the *Glomeromycota*: Phylogeny and evolution. Mycol. Res. 105(12): 1413-1421.

Schuessler, A., Mollenhauer, D., Schnepf, E. and Kluge, M. 1994. *Geosiphon pyriforme*, and endosymbiotic association of fungus and cyanobacteria: The spore structure resembles that of arbuscular mycorrhizal (AM) fungi. Bot. Acta 107: 36-45.

Schwarzott, D., Walker, C. and Schubler, A. 2001. *Glomus*, the largest genus of the Arbuscular mycorrhizal fungi (*Glomales*) is non-monophyletic. Mol-phylogeny Evol. 21: 190-197.

Slezack, S., Dumas-Gaudot, E., Rosendahl, S., Kjoller, R., Paynot, M., Negrel, J. and Gianinazzi, S. 1999. Endoproteolytic activities in pea roots inoculated with the arbuscular mycorrhizal fungus *Glomus mosseae* and/or *Aphanomyces euteiches* in relation to bioprotection. New Phytol. 142: 517-529.

Slezack, S., Dumas-Gaudot, E., Paynot, M. and Gianinazzi, S. 2000. Is a fully established arbuscular mycorrhizal symbiosis required for bioprotection of *Pisum sativum* roots against *Aphanomyces euteiches*? MPMI13: 238-241.

Smith, F.A., Jakobsen, I. and Smith, S.E. 2000. Spatial differences in acquisition of soil phosphate between two arbuscular mycorrhizal fungi in symbiosis with *Medicago truncatula*. New Phytol. 147: 357-366. doi: 10.1046/j.1469-8137.2000.00695.x

Smith, S.E. and Read, D.J. 2008. Mineral nutrition, toxic element accumulation and water relations of arbuscular mycorrhizal plants. pp. 145-148. *In*: S.E. Smith and D.J. Read (eds.). Mycorrhizal Symbiosis. 3rd Edn., Academic Press, London. ISBN-10: 0123705266.

Smith, S.E. and Smith, F.A. 2012. Fresh perspectives on the roles of arbuscular mycorrhizal fungi in plant nutrition and growth. Mycologia 104: 1-13. 10.3852/11-229

Sreeramulu, K.R., Onkarappa, T. and Swamy, H.N. 1998. Biocontrol of damping off and black shank disease in tobacco nursery. Tob Res. 24: 1-4.

Stahl, P.D., Williams, S.E. and Christensen, M. 1988. Efficacy of native vesicular-arbuscular mycorrhizal fungi after severe soil disturbance. New Phytol. 110: 347-354.

St-Arnaud, M., Fortin, J.A., Caron, M. and Hamel, C. 1994. Inhibition of *Pythium ultimum* in roots and growth substrate of mycorrhizal *Tagetes patula* colonized with *Glomus intraradices*. Can. J. of Plant Pathol. 16(3): 187-194.

St-Arnaud, M., Hamel, C., Vimard, B., Caron, M. and Fortin, J.A. 1995. Altered growth of *Fusarium oxysporum* f. sp. *chrysanthemi* in an *in vitro* dual culture system with the vesicular arbuscular mycorrhizal fungus *Glomus intraradices* growing on *Daucus carota* transformed roots. Mycorrhiza 5: 431-438.

Stutz, J.C. and Morton, J.B. 1996. Successive pot cultures reveal high species richness of arbuscular endomycorrhizal fungi in arid ecosystems. Can. J. Bot. 74: 1883-1889.

Subhashini, D.V. and Padmaja, K. 2010. Interaction between arbuscular mycorrhizal fungi and *Pythium aphanidermatum* in tobacco seedbeds. J. Biol. Control. 24(1): 70-74.

Tabin, T., Arunachalam, A., Shrivastava, K. and Arunachalam, K. 2009. Effect of arbuscular mycorrhizal fungi on damping-off disease in *Aquilaria agallocha* Roxb. seedlings. Tropical Ecology 50(2): 243-248.

Talukdar, N.C. and Germida, J.J. 1994. Growth and yield of lentil and wheat inoculated with three *Glomus* isolates from Saskatchewan soils. Can. J. Bot. 71: 1328-1335.

Thompson, J.P. and Wildermuth, G.B. 1989. Colonization of crop and pasture species with vesicular-arbuscular mycorrhizal fungi and negative correlation with root infection by *Bipolaris sorokiniana*. Can. J. Bot. 69: 687-693.

Torres-Barragan, A., Zavaleta-Mejia, E., Gonzalez-Chavez, C. and Ferrera-Cerrato, R. 1996. The use of arbuscular mycorrhizae to control onion white rot (*Sclerotium cepivorum* Berk.) under field conditions. Mycorrhiza 6: 253-258.

Tromas, A., Parizot, B., Diagne, N., Champion, A., Hocher, V. and Cissoko, M. 2012. Heart of endosymbioses: Transcriptomics reveals a conserved genetic program among arbuscular mycorrhizal, actinorhizal and legume-rhizobial symbioses. PLoS One 7: e44742. doi: 10.1371/journal.pone.0044742

Trotta, A., Varese, G.C., Gnavi, E., Fusconi, A., Sampo, S. and Berta, G. 1996. Interactions between the soil borne pathogen *Phytophthora nicotianae* var. *parasitica* and the arbuscular mycorrhizal fungus *Glomus mosseae* in tomato plants. Plant Soil 185: 199-209.

Verbruggen, E., van der Heijden, M.G.A., Rillig, M.C. and Kiers, E.T. 2013. Mycorrhizal fungal establishment in agricultural soils: Factors determining inoculation success. New Phytol. 197: 1104-1109.

Vigo, C., Norman, J.R. and Hooker, J.E. 2000. Biocontrol of the pathogen *Phytophthora parasitica* by arbuscular mycorrhizal fungi is a consequence of effects on infection loci. Plant Pathol. 49: 509-514.

Vinayak, K. and Bagyaraj, D.J. 1990. Vesicular-arbuscular mycorrhizae screened for Troyer citrange. Biol. Fertil. Soil 9: 311-314.

Vivekanandan, M. and Fixen, P.E. 1991. Cropping systems effects on mycorrhizal colonization, early growth, and phosphorus uptake of corn. Soil Sci. Soc. Am. J. 55: 136-140.

Volpin, H., Phillips, D.A., Okon, Y. and Kapulnik, Y. 1995. Suppression of an isoflavonoid phytoalexin defense response in mycorrhizal alfalfa roots. Plant Physiol. 108: 1449-1454.

Watanabe, H., Kageyama, K., Taguchi, Y., Horinouchi, H. and Hyakumachi, M. 2008. Bait method to detect *Pythium* species that grow at high temperatures in hydroponic solutions. J. Gen. Plant Pathol. 74: 417-424.

Whipps, J.M. 2000. Microbial interactions and biocontrol in the rhizosphere. J. Exp. Bot. 52: 487-511.

Xavier, L.J.C. 1999. Effects of interactions between arbuscular mycorrhizal fungi and *Rhizobium leguminosarum* on pea and lentil. PhD dissertation, Saskatoon, Saskatchewan, Canada: University of Saskatchewan.

Xavier, L.J.C. and Germida, J.J. 2003. Bacteria associated with *Glomus clarum* spores influence mycorrhizal activity. Soil Biol. Biochem. 35: 471-478.

Zak, J.C., Danielson, R.M. and Parkinson, D. 1982. Mycorrhizal fungal spore numbers and species occurrence in two amended mine spoils in Alberta, Canada. Mycologia 74: 785-792.

Zambolin, L. and Schenck, N.C. 1983. Reduction of the effects of pathogenic root rot infecting fungi on soybean by the mycorrhizal fungus *Glomus mosseae*. Phytopathology 73: 1402-1405.

Pythium Species as Biocontrol Agents

Mousa Alghuthaymi[1], Khaled Kasem[2*], Omar Atik[3] and Kamel A. Abd-Elsalam[4]

[1] Department of Biology, Science and Humanities College, Alquwayiyah, Shaqra University, Saudi Arabia
[2] Department of Plant Protection, Faculty of Agriculture, Hama University, Syria
[3] Department of Plant Protection, Aleppo Center, General Commission for Scientific Agricultural Research, Syria
[4] Plant Pathology Research Institute, Agricultural Research Center (ARC), 9 Gamaa St., 12619 Giza, Egypt

Introduction

The genus *Pythium* includes various critical species that have been placed in the Kingdom Chromista or Straminipila, unique from the Kingdom of Fungi. The protoplast of a sporangium is transmitted generally by applying an exit tube to a thin vesicle exterior the sporangium where zoospores are discriminated and produced after the rupture of the vesicle. Usually, *Pythium* species infect monocotyledonous herbaceous plants; moreover, some members of this genus also cause diseases in fish, red algae and mammals including humans (Ho 2018). Some mycoparasitic and entomopathogenic species of *Pythium* are generally used to effectively manage various other plant pathogens and harmful insects including mosquitoes whilst the others are applied to deliver useful chemical compounds for pharmacy and food industry (Ho 2018). Drechsler (1943) discussed and categorized *Pythium* species by their special spiny form oogonia; these kinds of non-pathogenic *Pythium* species mainly include *P. periplocum*, *P. acanthicum* and *P. oligandrum*.

Lifshitz et al. (1984) discovered an additional *Pythium* species, recognized as *P. nunn* and isolated from soil in Colorado, which suppressed pre-emergence damping-off disease in cucumber seedlings caused by *P. ultimum* in greenhouse tests. Simultaneously, Foley and Deacon (1985) reported that a new *Pythium* species, *P. mycoparasitic*, could parasitize a range of fungal pathogens. Some different spiny oogonial *Pythium* species with mycoparasitic activity like *P. lycopersicum* were isolated from Turkey (Karaca et al. 2008); the current species was added to the shortlist of the nonpathogenic *Pythium* species.

Mycoparasitism is a highly widespread occurrence existing normally in crop growing areas. Actually, quite a lot of interesting reviews exist concerning this subject (Adams 1990, Jefferies and Young 1994, Whipps 1997). A mycoparasite is a fungus which is able to destroy or kill other fungal pathogenic species. Whenever both fungal species grow mutually, it may overgrow and penetrate the other fungus, its mycelium might well coil surrounding the host mycelium, and it could indicate antagonism to the host fungus by producing antibiotics (Denis and Webster 1971), hazardous radicals (Kim et al. 1990) or wall structure lytic enzymes (Chet 1987). Additionally, it might enter the host mycelium, causing coagulation of the range protoplasm (Paul 1999). Mycoparasitic *Pythium* species includes *P. oligandrum* and *P. acanthicum* (Berry et al. 1993). Our consideration of the biological principles of *P. oligandrum* mechanism have significantly contributed to unravelling the complex

*Corresponding author: khaledkasem.sy@yandex.com

machinery developed by this fungal species in relation to its ability to 1) colonize the rhizosphere of various crop plants and fight for space and nutrition; 2) directly damage a few soil-borne fungi comprising ascomycetes (Bradshaw-Smith et al. 1991, Benhamou et al. 1997), basidiomycetes (Ikeda et al. 2012), fungal pathogenic oomycetes (Benhamou et al. 1999) and resting structures (i.e. e. sclerotia) (Rey et al. 2005); 3) enhance plant health via the putative creation of tryptamine (TNH2), an auxin precursor (Le Floch et al. 2003b); and 4) confer increased plant protection against fungal and bacterial diseases by the activation of the plant induced resistance (Benhamou et al. 1997, Le Floch et al. 2003a, Takenaka et al. 2006, Takenaka and Tamagake 2009).

As far as the life cycle of *P. oligandrum* is concerned, a little information is available about the connection signals that regulate the relationship of this valuable oomycete with its ecosystem and with the different host plant species. Until now, the mode of action caused by effective *P. oligandrum* strains induce the plant immune-system which have been examined and at least two types of microbe-associated molecular pattern (MAMPs) have been observed more hydrolytic enzymes (Picard et al. 2000, Takenaka et al. 2003). A more extensive evaluation performed on *P. oligandrum* using expressed sequence tag (EST) sequencing has recently presented the primary review on the molecules putatively associated with the *P. oligandrum* biochemical cross-talk (Horner et al. 2012). However, the molecular data provided remain risky and are reasonably complicated to understand the shortage of whole genome sequence details. Perhaps the sequencing of the *P. oligandrum* genome performed will offer comprehensive insight into how *P. oligandrum* modulates the plant's immune system in such a way that systemic protection of roots and shoots against a broad spectrum of pathogens is discussed.

All the useful *Pythium* species recognized so far are regarded as promising biocontrol agents due to their aggressiveness to a wide range of soil-borne pathogens. Nevertheless, *P. oligandrum*, most likely considered as a prevalent inhabitant in various soils, is the organism which has been the target of more specific reports (Ali 1985, Ribeiro and Butler 1992). It is worth referencing that the antagonism applied by *P. oligandrum* against pathogenic *Pythium* species represents a rather rare and exclusive scenario in biological control since the biocontrol agent is from the same genus of the targeted plant pathogens (Lévesque 2011). Little information is known about the fungal ecology of the species described from the soil (Foley and Deacon 1985, Ribeiro and Butler 1992, Mulligan et al. 1995, Al-Rawahi and Hancock 1997, 1998). This was partly linked to a deficiency of effective isolation techniques needed for ecological studies of these fungi (Ribeiro and Butler 1992). Some strategies that have been completely used for discovering mycoparasitic *Pythium* spp. in agricultural soil involve direct plating of soil onto a host-colonized agar, and baiting of soil with mycelia, sclerotia or resting spores of a host, then isolation from baiting onto selective cultural media (Mulligan et al. 1995). The current chapter focuses on the taxonomy and activity of some mycoparasitic *Pythium* species used as biocontrol agents against a few plant pathogenic fungi, and their particular mode of action have also been discussed.

Mycoparasitic species taxa

The genus *Pythium* is an economically essential fungal-like-organism or pseudo-fungus that has been demonstrated all over the world, but its taxonomic classification has become extremely controversial. The formation of globose oogonia in sexual reproduction has inserted them into the phylum Oomycota. Although the members of oomycota are identical to all filamentous fungi, they are generally omitted from Kingdom Fungi and placed in the Kingdom Chromista (Cavalier-Smith 1981). It was reported that several *Pythium* species with spiny oogonia have generally presented excellent mycoparasitic antifungal activity against other *Pythium* species. These mycoparasites include *P. acanthicum, P. oligandrum, P. periplocum* and *P. lycopersicum* (Fig. 20.1). In addition, two more mycoparasitic *Pythium* species with smooth oogonia such as *P. nunn* and *P. mycoparasitic* are reported to have significant mycoparasitic ability (Jones and Deacon 1995, Karaca et al. 2008).

Figure 20.1. Six known *Pythium* mycoparasitic species including *P. acanthicum, P. oligandrum, P. periplocum* and *P. lycopersicum*. In addition, two more mycoparasitic *Pythium* species with smooth oogonia such as *P. nunn* and *P. mycoparasitic*, have also been shown to have antifungal activity

Of the six known *Pythium* mycoparasitic species, only *P. oligandrum* and *P. nunn* have been investigated in detail in relation to their mechanism of parasitism and capability for the biocontrol against some plant pathogens (Laing and Deacon 1991, Abdelzaher et al. 1997).

The taxonomic groupings of the genus *Pythium* are considered as complicated since various morphological parameters have been applied during the classification of members of the genus in past (Van der Plaats-Niterink 1981); on the other hand, the large diversity between several constructions and substantial overlapping of a few of all these characters among numerous species complicates precise species identification. Depending on molecular equipment, it is significantly being employed in fungal taxonomy and disease detection. The genetic coding intended for rRNA that is generally utilized as molecular markers to describe the evolutionary history of organisms in different taxonomic levels have found the save of taxonomists and this is becoming a practical application to supplement the morphological differences between species (de Cock and Lévesque 2004). The internal transcribed spacer (ITS) region of the rRNA operon is often examined as a probable source of series variation amongst strongly related organisms due to their reduced level of conservation once investigated with those of the rRNA-coding genes. The evaluation of ITS sequences region of the rRNA has now managed to distinguish the oomycetes coming from true fungus. Although morphologically they are 'fungus-like', they will no longer be treated as "true fungi", being closer to algae than to yeast. In recent classification, the oomycetes has been treated in stramenopile family tree in eukaryotic domain alongwith brown algae. (Paul et al. 2006, Belbahri et al. 2008).

P. acanthicum

Two unique species of *Pythium* have been described based on some new morphological features such as non-proliferating sporangia consisting of sub-globose, elongated, or toruloid structures and ornamented oogonia less than 30 μm in diameter with oospores diameter less than 25 μm. Three *Pythium* species such as *P. oligandrum, P. acanthicum*, and *P. periplocum* diverge in their terminal oogonia, small oogonial spines, erotic oospores and antheridial nature (Lodhi et al. 2005).

P. lycopersicum

In Turkey, *P. lycopersicum* collected from field soil samples showed morphological similarity to *P. ornamentatum*, and the fungal culture has been deposited to the CBS culture collection. Some oomycetes are mycoparasites and come within the group having contiguous sporangia and spiny oogonia like *P. oligandrum*. Distinct from this former species, *P. lycopersicum* is a rare mycoparasite. The molecular characterization for ITS regions differentiated *P. lycopersicum* from all additionally recognized mycoparasites of the *Pythium* genus (Karaca et al. 2008). ITS region of the nuclear rRNA of *P. lycopersicum* resembled with many of the mycoparasites belonging to the genus *Pythium* such as *P. oligandrum* with the homogeneity of 97.8%. The ITS region of *P. lycopersicum* was probably the smallest with 761 bases and showed 11 main variances with *P. acanthicum*, the most closely related species (98.3% homogeneity). Morphological and molecular differences were used to describe *P. lycopersicum* as a mycoparasite for some plant pathogenic fungi such as *Botrytis cinerea* during pre-harvest diseases (Karaca et al. 2008).

P. nunn

Certain species of *Pythium*, such as *P. nunn* and *P. oligandrum*, were used to protect potted ornamentals and vegetables from plant pathogenic species of *Pythium* (Agrios 2005).

P. oligandrum

Sporangia are either terminal or intercalary, subspherical, 25–45 μ in diameter. Zoospores (20-50) are formed in thin-walled vesicles that release them on maturity. Zoospores are longitudinally grooved, reniform and biflagellate. The sexual spores, oospores are hyaline or yellow, subspherical with a thick wall that bears spiny pointed protuberances. The oospores are long-lived and highly resistant to adversarial ecological conditions. The biocontrol agent overwinters as oospores which germinate as soon as favourable conditions are presented (Cook and Baker 1983).

P. periplocum

P. periplocum is a destructive mycoparasite for a number of phytopathogenic fungi and oomycetes and consequently has the potential ability as a biological control agent (Kushwaha et al. 2017). *P. periplocum* was investigated as a biocontrol agent against damping-off of cucumbers prompted by *P. aphanidermatum* (Hockenhull et al. 1992).

Fungal host range

Some studies suggest that *P. periplocum* is an aggressive mycoparasite with a wide host range similar to congeneric broad-spectrum mycoparasites such as *P. oligandrum* and *P. acanthicum* (Ribeiro and Butler 1995). *P. oligandrum* is a soil-borne fungus belonging to the order *Oomycetes*. It is capable of living as a saprophyte but is antagonistic to many pathogenic fungi collected from different plant hosts (Table 20.1). Since 1986, over forty publications have reviewed that *P. oligandrum* is a promising biocontrol agent who works both directly and indirectly to protect plant hosts against numerous phytopathogenic species. It interacts straight with the plant pathogens by the specific or mixed mechanisms, including mycoparasitism, antibiosis, beneficial nutrient and space struggle, and/or indirectly by inducting resistance in the plant hosts (Gerbore et al. 2014). The wide range of *P. oligandrum* hosts suggests that it is significant to inhibit the excessive variety of interactions that can lower or increase the efficacy of biological control. *P. oligandrum* showed the ability to penetrate the *Lycopericon esculentum* root system without inducing extensive cell damage (Rey et al. 1998) and to activate an array of structural defense-related responses after difficulty with a Fusarium pathogen in tomatoes (Benhamou et al. 1997).

Molecular-based methods

Even though morphological features will continue to serve as the basis for molecular identity of *Phytophthora* and *Pythium* species (Grünwald et al. 2011, Ristaino 2012), there is an urgent need to study the genetic phylogeny of a large number of isolates to develop new molecular tools for recognition of the unknown isolates, mainly when a few vital morphological characters are absent. For *Pythium*, no database is developed till today on species identification depending on molecular studies. The faster detection of *Pythium* species constructed from molecular data associated with different species collected from different countries, and the morphological features of *Pythium* will be a lot more difficult and adjustable to identify *Pythium* species (Ho 2018).

Biocontrol agents

In 1858, the genus *Pythium* was defined by Pringsheim for the first time. The genus is currently known to be all-pervasive, occupying a number of environmental niches (Ho 2009). *Pythium* can be pathogenic to plants (Hendrix and Campbell 1973) and some animals (De Cock et al. 1987), or can be non-pathogenic (Van der Plaats-Niterink 1981) surviving as a saprophyte, or it might be mycoparasitic (Karaca et al. 2008). Among the mycoparasitic species, *P. oligandrum* is the most studied species and has been shown to possess great potential as a biocontrol agent (BCA). As a result, it has been exploited as a promising bio-fungicide in numerous countries (Benhamou et al. 2012, Gerbore et al. 2014).

Antimicrobial compounds

Ponchet et al. (1999) identified new protein (12 kDa) which was grouped since elicitin-like due to the similarity that marketed to the common elicitin. Picard et al. (2000) discovered an extracellular protein received from the traditional filtrate of the oomycete named oligandrin. *P. oligandrum* induces two types of elicitors, either released by the oomycete or taken out of its cell wall structure. *P. oligandrin* produce two types of elicitor out wall membrane structure (Masunaka et al. 2010).

Oligandrin

The low-molecular-weight protein called oligandrin has been effectively applied to stimulate systemic induced resistance in tobacco plants (Lherminier et al. 2003), anti-phytoplasma infection, and in tomato against some fungal pathogens like *P. parasitic*, and *B. cinerea* (Benhamou et al. 2001, Lou et al. 2011, Picard et al. 2000), and in grapevine against *B. cinerea* (Mohamed et al. 2007). This necessary protein stimulates defense responses but without activating the hypersensitive reaction (HR) connected with necrotic response (Picard et al. 2000). Mohamed et al. (2007) studied the application form of oligandrin for oospore inoculum on grapevine roots to manage *B. cinerea*. The safeguarding efficiency of pre-treated leaves reached 75%, no significant variations were detected after treatment either with the antagonistic or its elicitor (Gerbore et al. 2013). The second kind of elicitor, identified as an elicitin-like protein (Takenaka et al. 2006), refers to cell wall structure proteins (CWPs), coded POD-2 and POD-1 by Takenaka et al. (2003). Plant safety level by induced level of resistance was provided against some phytopathogenic bacteria and fungi, i.e. *Ralstonia*

Table 20.1. Application of *P. oligandrum* as a biocontrol agent for plant pathogens (Updated from Gerbore et al. 2014)

Mechanisms described	Host	Plant pathogen	Efficiency level (%)	Bioassay	References
Mycoparasitism	*Agaricus bisporus*	–	–	*In vitro*	Fletcher et al. 1990
Mycoparasitism	Peas	*Fusarium solani* f. sp. *pisi*, *Phoma medicaginis* var. *pinodella* and *Mycosphaerella pinodes*	–	*In vitro*	Bradshaw-Smith et al. 1991
Mycoparasitism	Tomato	*Rhizoctonia solani* AG-4, *Pythium ultimum*, *Pythium spinosum* and *Pythium irregulare*	≈70	–	He et al. 1992
Mycoparasitism	Wheat	*Pythium ultimum*	100	*In vitro*	Abdelzaher et al. 1997
Mycoparasitism	–	Sclerotia of *Sclerotinia sclerotiorum*	≈50	*In vitro*, field	Madsen and Neergaard 1999
Mycoparasitism	Cucumber	*Pythium ultimum*	≈37	*In vitro*	Ali-Shtayeh and Saleh 1999
Mycoparasitism	–	*Phytophthora parasitica*	–	*In vitro*	Picard et al. 2000a
Mycoparasitism	–	Sclerotia of *Botrytis cinerea* and *Sclerotinia minor*	–	*In vitro*	Rey et al. 2005
Mycoparasitism	–	*Pythium ultimum*, *Fusarium oxysporum*	–	*In vitro*	El-Katatny et al. 2006
Mycoparasitism	–	*Phytophthora parasitica*	–	*In vitro*	Horner et al. 2012
Mycoparasitism	Sugar beet, cress	*Pythium* spp.	≈26–33	Greenhouse	McQuilken et al. 1990, 1992, 1998
Mycoparasitism, antibiosis	–	*Pythium ultimum*, *Pythium aphanidermatum*, *Fusarium oxysporum*, *Rhizoctonia*, *Phytophthora megasperma*, *Verticillium albo-atrum*	–	*In vitro*	Benahmou et al. 1999
Mycoparasitism, plant growth promoting	Pepper	*Verticillium dahliae*	–	*In vitro*, green-house	Al-Rawahi and Hancock 1998
Mycoparasitism, plant growth promoting	Pepper	*Verticillium dahliae*	67	Greenhouse	Rekanovic et al. 2007
Mycoparasitism, plant growth promoting	Tomato	*Pythium dissotocum*	–	Greenhouse	Vallance et al. 2009

(Contd.)

Table 20.1. (*Contd.*)

Mechanisms described	Host	Plant pathogen	Efficiency level (%)	Bioassay	References
Induced resistance, mycoparasitism	Tomato	*Fusarium oxysporum* f. sp. *Radicis-lycopersici*	–	*In vitro*, greenhouse	Benhmou et al. 1997
Mycoparasitism, induced	Potato	*Rhizoctonia solani* AG-3	46–87	*In vitro*, field	Ikeda et al. 2012
Nutrient and/or space competition	Cress	*Pythium ultimum*	–	Field	Al-hamdani et al. 1983
Nutrient and/or space competition	Sugar beet, cress	*Pythium ultimum* and *Aphanomyces cochlioides*	–	Greenhouse	Whipps and McQuilken 2001
Nutrient and/or space competition	Sugar beet	*Aphanomyces cochlioides*	≈50	Greenhouse, field	Takenaka and Ishikawa 2013
Nutrient and/or space competition, mycoparasitism	Cotton	*Pythium ultimum*	32–66	*In vitro*	Martin and Hancock 1986
Nutrient and/or space competition, mycoparasitism	Sugar beet	*Pythium ultimum*	≈88	Greenhouse	Martin and Hancock 1987
Nutrient and/or space competition, mycoparasitism	Tomato	Pathogen communities of soil	≈15	Greenhouse	Cwalina- Ambroziak and Nowak 2012
Plant growth promotion	Rice	–	–	Greenhouse	Cother and Gilbert 1993
Plant growth promotion	Cucumber	–	–	–	Kratka et al. 1994
Plant growth promotion	Cucumber	–	–	*In vitro*	Wulff et al. 1998
Induced resistance	Tomato	*Phytophthora parasitica*	60	Greenhouse	Picard et al. 2000a, b
Induced resistance	Tomato	*Ralstonia solanacearum*	≈33	*In vitro*	Takenaka et al. 2008
Induced resistance	Sugar beet	*Aphanomyces cochlioides*	≈33	*In vitro*	Takenaka et al. 2006
Induced resistance	Tobacco	Phytoplasma	≈40	Greenhouse	Lherminier et al. 2003

(*Contd.*)

Induced resistance	Tomato	*Ralstonia solanacearum*	–	*In vitro*	Masunaka et al. 2009
Induced resistance	Sugar beet	*Cercospora beticola*	12–52	*In vitro*, field	Takenaka et al. 2009
Induced resistance	Strawberry	*Botrytis cinereal, Sphaerotheca macularis, Mycosphaerella fragariae*	43–70	Field	Meszka and Bielenin 2010
Induced resistance	Tomato	*Botrytis cinereal*	≈30	Greenhouse	Le Floch et al. 2009
Botrytis cinereal	Grapevine	*Botrytis cinerea*	75	*In vitro*	Mohamed et al. 2007
Induced resistance	Tomato	*Botrytis cinerea*	79	Greenhouse	Lou et al. 2011
Induced resistance	Tomato	*Fusarium oxysporum* f. sp. *Radicis-lycopersici*	84	Greenhouse	Benhamou and Garand 2001
Induced resistance	Arabidopsis thaliana	*Ralstonia solanacearum, Pseudomonas syringae*	87	Greenhouse	Kawamura et al. 2009
Induced resistance	Sugar beet and wheat	*Rhizoctonia solani* AG-2.2, *Fusarium graminearum*	34	*In vitro*	Takenaka et al. 2003
Induced resistance	Tomato	*Ralstonia solanacearum*	≈40	*In vitro*	Hase et al. 2006, 2008
Induced resistance, plant growth promotion	Tomato	*Botrytis cinerea*	≈50	Greenhouse	Le Floch et al. 2003a, b

solanacearum and *Pseudomonas syringae* on Arabidopsis (Kawamura et al. 2009); *Cercospora beticola, Rhizoctonia solanacearum* and *Aphanomyces cochlioides* on sugars beet (Takenaka et al. 2006, Takenaka and Tamagake 2009); and against *Fusarium graminearum* on wheat (Takenaka et al. 2003). Concerning the protection acquired following either POD-1 or POD-2 treatment of wheat plants, Takenaka et al. (2006) developed equivalent disease management of sugar beet against *A. cochlioides* with both CWPs. However, distinctness was discovered in the range of defense interrelated genes induced, five genes for POD-1 and three for POD-2. The experts confirmed that both elicitors may stimulate distinct protection reactions, if the protection level is actually noticed. Merging both CWPs and oligandrin to activate more genetics can offer a possibility to boost plant safety. Another interesting stage regarding these proteins is usually that both types of elicitors are not particular to plant varieties, which is in keeping with results acquired with *P. oligandrum* hyphae (Gerbore et al. 2013).

Induction mechanism

Induced plant resistance

Many trials described that treatment of crops with *P. oligandrum* hyphae or elicitors, oligandrin and CWPs, stimulated plant level of resistance (Picard et al. 2000, Lherminier et al. 2003, Mohamed et al. 2007, Takenaka et al. 2008, Masunaka et al. 2010). Following the entry of *P. oligandrum* inside the rhizosphere, the oomycete induces level of resistance by the neighbourhood level in the beginnings but, while demonstrated simply by Le Floch et al. (2003a), this impact can systemically extend to all or any elements of the plant. *P. oligandrum* can be used as biocontrol agent for various plant diseases caused by several plant pathogenic fungi such as *B. cinerea, P. ultimum* and *R. solani* (Le Floch et al. 2003a). This broad spectral range of pathogenic assault control is towards nonspecific plant arousal of defenses. Regarding the systemic level of resistance, Le Floch et al. (2003a) demonstrated an elevated introduction of PR proteins, i.e. PR-3b, PR-5a and a new isoform of PR-3b, in tomato plants colonized at the root system level by *P. oligandrum*, and contaminated on leaves with *B. cinerea*. Additional reports investigated that systemic resistance is usually induced simply by various additional BCAs, e.g. *Trichoderma* spp. and *F. oxysporum* (Harman et al. 2004, Veloso and Diaz 2012) on a wide range of host plants. A biocontrol agent is not particular to a host plant and may trigger standard defense reactions to control many pathogenic disorders.

Induced plant growth promotion

Plant microbes promoting crop growth will be the object of several investigations, as described in literature focused on plant growth promoting rhizobacteria and other fungal species (Hyakumachi and Kubota 2003, Mallik and Williams 2008, Lugtenberg and Kamilova 2009). This phenomenon is generally linked to the microbes' production of phytohormones and secondary metabolites by the bacteria (Helman et al. 2011, Hermosa et al. 2012). Crop growth promotion takes place after what appears to be a latent period. For example, Wulff et al. (1998) saw this sensation when *P. oligandrum* zoospores in cucumber baby plants were utilized. The initial two days, *P. oligandrum* inoculation triggered undesirable effects upon root of plants; after that, root elongation was stimulated.

Plant growth promotion was linked with the formation of axenic substances. Le Floch et al. (2003b) described that the tryptamine pathway is present in the oomycete hyphae. *P. oligandrum* can metabolise a great auxin substance, tryptamine (TNH2), from indole-3-acetaldehyde and tryptophan. TNH2 was absorbed by the main roots and secondary roots developed consequently. The tryptamine pathway is well known in various other non-pathogenic fungal species such as *Aspergillus, Penicillium* and *Rhizopus* (Frankenberger and Arshad 1995), although *P. oligandrum* varies from them since it is unable to transform TNH2 into indole-3-acetic acid (IAA).

Excitingly, the TNH2 influx can enhance IAA activity in tomato plants, especially if it is not really the endogenous progenitor of IAA in this plant (Cooney and Nonhebel 1991). Generations of auxin-like substances (TNH2) inside the rhizosphere can easily stimulate plant growth promotion of tomato plant life; thus, minor but frequent release of TNH2 by *P. oligandrum* could be good for crop development.

Pythium-microbe interaction

Applying Biological Control Agents (BCAs) appears to be a good and eco-friendly solution to control plant pathogens. Advantages and limits of BCAs have emerged through *P. oligandrum*, an oomycete which has received much interest within the last 10 years. *P. oligandrum* antagonism is usually a complicated and targeted fungus-dependent procedure. Interestingly, it generally does not appear to disrupt microflora biodiversity around the roots. Mohammed et al. (1999) indicated that *P. periplocum* and *P. oligandrum* were observed to be dynamic biocontrol agents against *P. ultimum*, the causal organism of damping-off diseases in cucumber plants. This fungal pathogen was loaded on thin films of cultural medium from water agar by the three mycoparasites, and was moderately vulnerable to *P. periplocum* while somewhat vulnerable to *P. acanthicum* and *P. oligandrum*. A direct application approach in which antagonistic mycoparasites were definitely involved in peat/sand combination infested with *P. ultimum* under room temperature circumstances (at 500 CFUg1) considerably increased seedling germination and protected seedlings from damping-off diseases. In the seed coating method, the biocontrol agents were used in two forms of or as a seed dressing, and the antagonistic results were similar to that obtained by an immediate utility.

Hyphal production of *P. oligandrum* and the prevalence of mycoparasitism were discovered on the surface of seed tubers after three days of planting: *P. oligandrum* hyphae not only colonized *R. solani* hyphae by coiling (Fig. 20.2A), but also colonized densely on the surface of *R. solani* sclerotia (Fig. 20.2B). Two weeks after planting, the hyphal development of both *P. oligandrum* and *R. solani* was not detected on seed tubers (Fig. 20.2C), whereas *R. solani* hyphae had abundantly colonized the control seed potatoes two weeks after planting (Fig. 20.2D) (Ikeda et al. 2012).

Many species of the genus *Pythium* are a causal organism for some plant species such as grass, additional fungus, and algae (Van der Plaats-Niterink 1981). Nevertheless, some species have been completely discovered to become slightly to extremely pathogenic to bugs and insects. A *Pythium* sp., caused a higher degree of death in a disciplined assortment of the tree hole mosquito *Ochlerotatus sierrensis* (Clark et al. 1966).

Saunders et al. (1988) associated *Pythium flevoense* from the wild populations of *Ochlerotatus sierrensis* in California, 42% of the collected samples from tree holes, although this kind of fungus triggered infections in a mere 14% of larvae during 21 weeks of exposure in *in vitro* bioassays. The fungal infection that contaminated the harmed larvae rather than healthy larvae shows that the fungi can be opportunistic instead of harshly entomopathogenic.

Su et al. (2001) detected *P. carolinianum* coming from Guizhou region, China, in outdoor bioassays. The researchers revealed that the disease degrees ranged from 13.3% to 100% in *Culex quinquefasciatus* larvae, and mentioned a population of *Aedes albopictus* (Skuse) was 'markedly controlled', but any infection proportions were not received. In spite of the pathogenicity of many *Pythium* species to insects, overall, they will not be deemed ideal for biocontrol of insects (Abd El-Ghany et al. 2018). Biological control agent (BCA) usually manages plant pathogens through different modes of action, some research stated that *P. oligandrum* can directly attack many fungal pathogens using several mechanisms (Benhamou et al. 1997, 1999, Picard et al. 2000, Rey et al. 2005). Regarding the fungus target, included in these are mycoparasitism and antibiosis. Some particular interactions with fungi that created protective reactions to avoid *P. oligandrum* episodes were also noticed (Gerbore et al. 2013).

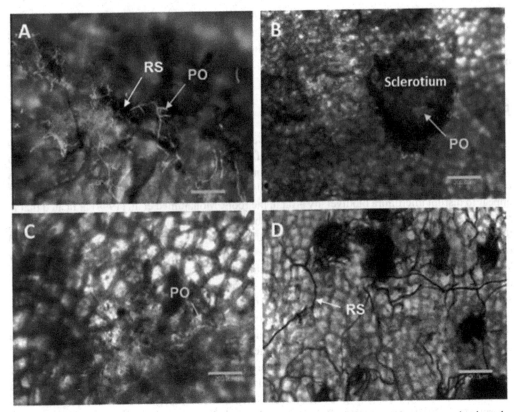

Figure 20.2. Confocal laser-scanning micrographs demonstrating colonization of potato seed tubers by *P. oligandrum* (PO) and *Rhizoctonia solani* AG-3 (RS). Seed tubers infected with sclerotia of RS were dropped in an oospore suspension of PO (A, B and C) or distilled water (D), and then simply grown in artificial soil for horticulture in plastic trays and incubated for three days (A and B) or 14 days (C and D). PO hyphae were observed by visual images of a green fluorescence antibody. The size bar represents 200 lm in all panels
(Reprinted from Ikeda et al. 2012)

Mechanisms involved in mycoparasitism of *P. oligandrum*

Mycoparasitism

P. oligandrum mycoparasitism is usually characterized by active growth along with the host hyphae and the creation of digestive support enzymes that deteriorate or break the cell wall structure. *P. oligandrum* is able to penetrate the host cells and destroy the cytoplasm of host cells. This process could be observed, for example, in the conversation with *F. oxysporum* f. sp. *radicis-lycoperici* (FORL) or perhaps *P ultimum*. These complex relationships and hydrolytic enzymes such as chitinases and cellulases are obviously included (Benhamou et al. 1997, 1999). *P. oligandrum* will be able to produce great levels of cellulolytic enzymes against *Phytophthora megasperma* (Picard et al. 2000). Recently, the type of *P. oligandrum* mycoparasitism was investigated based on the molecular methods by Horner et al. (2012). A complementary DNA library has been established, and transcripts encoding proteases, protease inhibitors, glucanases, putative elicitors and effectors have been determined during *P. oligandrum* interaction with heat-killed *P. infestans* hyphae. Some kind of protein group may take action during mycoparasitism, but additional investigations will be a need to regulate the role of every recognized necessary protein (Gerbore et al. 2013).

Antibiosis

Antibiosis is defined as the accurate connection where the target is damaged by harmful secondary metabolites created by antagonistic fungi (Haas and Défago 2005, Alabouvette et al. 2009, Gerbore et al. 2013). In case of *P. oligandrum*, this phenomenon was studied simply by Bradshaw-Jones et al. (1991) against the three main foot rot pathogens of pea: a volatile antiseptic compound produced by *P. oligandrum* inhibited the fungal growth of *Phoma medicaginis* and *Mycosphaerella pinodes,* yet this molecule is not recognized and purified. The destruction of *Phytophothra megasperma* without a physical connection with *P. oligandrum* hyphae was investigated (Benhamou et al. 1999). As stated by Campeón et al. (2008), antibiosis and mycoparasitism can be seen through the same interaction as having a fungal host.

It is hypothesized that *P. oligandrum* modulates the formation of antifungal substances based on the targeted host, resulting in two strategies: possible mycoparasitism, connected with hydrolytic digestive enzymes, or parasitism via antibiosis. The former technique seems even more regular than antibiosis. These types of outcomes show that one of the many advantages of using BCAs may be the selection of interactions they can set up with all the fungal pathogens. *P. oligandrum* appears to adjust to the fungal efforts and pathogens to destroy all of them by mycoparasitism, antibiosis, or perhaps a combined mix of both procedures; this kind of displays the multiplicity of interactions which exist in nature (Benhamou et al. 2012, Gerbore et al. 2013).

Nutrient and space competition

Alabouvette et al. (2006), suggested that a great over-all phenomenon regulating the populace aspect of microbes sharing similar ecological niche market and physical requirements if the resources are partial. Benhamou et al. (2012) described the French and Japanese investigations to identify *P. oligandrum* mycoparasitism and/or induce a degree of resistance as a major mechanism. Thus, parasitic fungi competitors designed for space and nutrients are usually mechanisms used for natural control by *P. oligandrum*.

Fungal defense reactions against *P. oligandrum* mycoparasitism

Benhamou et al. (1999) and Picard et al. (2000) demonstrated two behaviours of fungal defense response that could possibly be created on the cell wall of fungi, with one becoming localized and the additional one much more generalized to the entire hyphae. These pathogen defense reactions were, in two circumstances, not sturdy enough to prevent *P. oligandrum* invasion and penetration with the reacting web host cell phone, showing that the great ability of *P. oligandrum* to produce a wide range of cell wall structures degrading digestive support enzymes may be of primary importance. Indication provided that the host's defense reactions had been initiated by essential molecules secreted by *P. oligandrum*, but so far, no compounds have been recognized (Gerbore et al. 2013).

Microbial antagonists of *P. oligandrum*

P. oligandrum could be confronted by various fungi, several cases of cell damage to the oomycete have been observed after inoculation with *P. oligandrum* (Benhamou et al. 1999). Both microbes, pathogens and antagonists had been noted for morphological alterations, and three days following the beginning of the interaction, *R. solani* collapsed as well as the hyphae cells of clear *P. oligandrum* became cluttered. *P. oligandrum* may qualify for assault by regenerating new resting structures such as sclerotia (Foley and Deacon 1986). Madsen and Neergaard (1999) reported that, compared to the controlled treatment and survival of *Sclerotinia sclerotiorum,* sclerotia was reduced by 50 percent following *P. oligandrum* soil treatment. These total effects underlined that *P. oligandrum* features a great choice of degradative enzymes capable of penetrating into resting structures similar to sclerotia and that it may be useful for biological control approach directed at reducing sclerotia (Gerbore et al. 2013). *P. oligandrum* penetrated *B. cinerea* sclerotia through holes at the junction of rind cells that corresponded to gaps in melanin debris (Rey et al. 2005). The melanin a mixture obviously protect the rind cells against digestive support enzymes from antagonistic microorganisms (Gerbore et al. 2013). *P. oligandrum* interactions with the natural environments are incredibly complicated, interaction not only related to biotic components but also with abiotic natures (Le Floch et al. 2003a, Takenaka et al. 2008). Le Floch et al. (2007) used molecular-based methods such as quantitative real-time PCR and DNA microarray and culture-dependent methods to monitor *P. oligandrum* multitude offered contrary results. Every time a selective culture media was used, *P. oligandrum* persisted for just three months; however, relating to quantitative real-time DNA and PCR macroarray, the oomycete can persist in the root system for up to six months, suggesting that one treatment was adequate.

Induced plant resistance

Following insertion of *P. oligandrum* in the rhizosphere, the oomycete induces resistance at the area level inside the roots. However, as demonstrated by Le Floch et al. (2003a), this kind of effect is certainly systemic also, spreading to all the portions of the plant. The plants showing higher levels of resistance to more than one phytopathogenic fungi, such as *B. cinerea*, *P. ultimum* and *R. solani*, is induced by *P. oligandrum* (Le Floch et al. 2003a). This wide range of pathogenic assault control is toward a nonspecific plant stimulus of defenses. When it comes to the systemic level of resistance, Le Floch et al. (2003a) have shown an improved induction of PR proteins, i.e. PR-3b, PR-5a and a novel isoform of PR-3b, in tomato seedlings colonized by *P. oligandrum*, and infected in leaves with after that *B. cinerea*. Another study reported that systemic levels of resistance are normally caused by many other BCAs, e.g. *Trichoderma* spp. and *F. oxysporum* (Harman et al. 2005, Veloso and Diaz 2012) on a broad range of plant hosts. A biocontrol agent is, therefore, not essentially particular to a good host plant and may lead to general defense answers to managing different pathogenic outbreaks.

Conclusion and future trends

Biological control is considered as a novel and promising agricultural pest control approach, which is one of the most promising applications for sustainable agriculture. It is really a cost-effective, and green pest control strategy by utilizing living organisms to reduce infestation population of plant pests. The present chapter discusses taxonomy, biocontrol activity and mechanism of mycoparasitic *Pythium* species employed against some phytopathogens. In the *Pythium* genus, some mycoparasitic species such as *P. acanthicum*, *P. lycopersicum*, *P. mycoparasitic*, *P. nunn*, *P. oligandrum* and *P. periplocum* have been applied effectively for the management of other plant pathogens. *P. oligandrum* generates two types of elicitor, possibly released by the oomycete or perhaps purified from the cell wall. Understanding the molecular mechanisms of mycoparasitic *Pythium* species can be achieved through whole-genome sequencing. Plant associated microorganisms applied as fungal biocontrol agents may have a role in minimizing losses to some plant diseases and utilized as unique control methods to assist sustainable agriculture.

References

Abd El-Ghany, N.M., Abdel-Razek, A.S., Djelouah, K. and Moussa, A. 2018. Efficacy of bio-rational insecticides against *Tuta absoluta* (Meyrick) (Lepidoptera: Gelechiidae) on tomatoes. Biosc. Res. 15(1): 28-40.

Abdelzaher, H.M.A., Elnaghy, M.A. and Fadl-Allah, E.M. 1997. Isolation of *Pythium oligandrum* from Egyptian soil and its mycoparasitic effect on *Pythium ultimum* var. *ultimum* the damping-off organism of wheat. Mycopathologia 139: 97-106.

Adams, P.B. 1990. The potential of mycoparasites for biological control of plant diseases. Annu. Rev. Phytopathol. 72: 58-73.

Agrios, G.N. 2005. Plant Pathology. Fifth edition. Elsevier Academic Press, New York.

Alabouvette, C., Olivain, C. and Steinberg, C. 2006. Biological control of plant diseases: tThe European situation. Eur. J. Plant Pathol. 114(3): 329-341.

Alabouvette, C., Olivain, C., Migheli, Q. and Steinberg, C. 2009. Microbiological control of soil-borne phytopathogenic fungi with special emphasis on wilt-inducing *Fusarium oxysporum*. New Phytol. 184(3): 529-544.

Al-Hamdani, A.M., Lutchmeah, R.S. and Cooke, R.C. 1983. Biological control of *Pythium ultimum*-induced damping-off by treating cress seed with the mycoparasite *Pythium oligandrum*. Plant Pathol. 32(4): 449-454.

Ali, M.S.A.M. 1985. *Pythium* populations in Middle Eastern soils relative to different cropping practices. Trans. Br. Mycol. Soc. 84: 695-700.

Ali-Shtayeh, M.S. and Saleh, A.S. 1999. Isolation of *Pythium acanthicum, P. oligandrum*, and *P. periplocum* from soil and evaluation of their mycoparasitic activity and biocontrol efficacy against selected phytopathogenic Pythium species. Mycopathologia 145(3): 143-153.

Al-Rawahi, A.K. and Hancock, J.M. 1997. Rhizosphere competence of *Pythium oligandrum*. Phytopathol. 87: 951-959.

Al-Rawahi, A.K. and Hancock, J.M. 1998. Parasitism and biological control of *Verticillium daliae* by *Pythium oligandrum*. Plant Dis. 82: 1100-1106.

Belbahri, L., McLeod, A., Paul, B., Calmin, G., Moralejo, E., Spies, C.F., Botha, W.J., Clemente, A., Descals, E., Sánchez-Hernández, E. and Lefort, F. 2008. Intra-specific and within-isolate sequence variation in the ITS rRNA gene region of *Pythium mercuriale* sp. nov. (Pythiaceae). FEMS Microbiol. Lett. 284: 17-27.

Benhamou, N., Rey, P., Chérif, M., Hockenhull, J. and Tirilly, Y. 1997. Treatment with the mycoparasite *Pythium oligandrum* triggers induction of defense-related reactions in tomato roots when challenged with *Fusarium oxsporum* f. sp. *radicis-lycopersici*. Phytopathology. 87: 108-122.

Benhamou, N., Rey, P., Picard, K. and Tirilly, Y. 1999. Ultrastructural and cytochemical aspects of the interaction between the mycoparasite *Pythium oligandrum* and soilborne plant pathogens. Phytopathology 89(6): 506-517.

Benhamou, N., Bélanger, R.R., Rey, P. and Tirilly, Y. 2001. Oligandrin, the elicitin-like protein produced by the mycoparasite *Pythium oligandrum*, induces systemic resistance to *Fusarium* crown and root rot in tomato plants. Plant Physiol. Biochem. 39(7-8): 681-696.

Benhamou, N., le Floch, G., Vallance, J., Gerbore, J., Grizard, D. and Rey, P. 2012. *Pythium oligandrum*: An example of opportunistic success. Microbiol. 158: 2679-2694.

Berry, L.A., Jones, E.E. and Deacon, J.W. 1993. Interaction of the mycoparasite *Pythium oligandrum* with other *Pythium* species. Biol. Sci. Technol. 3: 247-260.

Bradshaw-Smith, R.P., Whalley, W.M. and Craig, G.D. 1991. Interactions between *Pythium oligandrum* and the fungal foot rot pathogens of peas. Mycol. Res. 95(7): 861-865.

Cavalier-Smith, T. 1981. Eukarotic kingdoms: Seven or nine? Biosystemetics 14: 461-481.

Chang, T.T. 1993. Investigation and Pathogenicity tests of *Pythium* species from rhizosphere of *Cinnamomum osmophloeum*. Plant Pathol. Bull. 2: 66-70.

Chet, I. 1987. *Trichoderma* application, mode of action and potential as a bio-control of soil-borne plant pathogenic fungi. pp. 137-160. *In*: Chet, I. (ed.). Innovative Approaches to Plant Disease Control. Wiley, New York.

Clark, T.B., Kellen, W.R., Lindegren, J.E. and Sanders, R.D. 1966. *Pythium* sp. (Phycomycetes: Pythiales) pathogenic to mosquito larvae. J. Invertebr. Pathol. 8: 351-354.

Cook, R.J. and Baker, K.F. 1983. The nature and practice of biological control of plant pathogens. The American Phytopathological Society. St. Paul

Cooney, T. and Nonhebel, H. 1991. Biosynthesis of indole-3-acetic acid in tomato shoots: Measurement, mass-spectral identification and incorporation of 2H from 2H$_2$O into indole-3-acetic acid, D- and Ltryptophan, indole-3-pyruvate and tryptamine. Planta 184(3): 368-376.

Cother, E.J. and Gilbert, R.L. 1993. Comparative pathogenicity of Pythium species associated with poor seedling establishment of rice in Southern Australia. Plant Pathol. 42(2): 151-157.

Cwalina-Ambroziak, B. and Nowak, M. 2012. The effects of biological and chemical controls on fungal communities colonising tomato (*Lycopersicon esculentum* Mill.) plants and soil. Folia Horticult. 24(1): 13-20.

Dde Cock, A.W., Mendoza, L., Padhye, A.A., Ajello, L. and Kaufman, L. 1987. *Pythium insidiosum* sp. nov., the etiologic agent of pythiosis. J. Clin. Microbiol. 25,: 344-349.

de Cock, A.W. and Lévesque, A. 2004. New species of *Pythium* and *Phytophthora*. Stud. Mycol. 50: 481-487.

Denis, C. and Webster, J. 1971. Antagonistic properties of species-groups of *Trichoderma*. 1. Production of non-volatile antibiotics. Trans. Br. Mycol. Soc. 57: 363-369.

Drechsler, C. 1943. Two species of *Pythium* occurring in southern States. Phytopathology. 33: 261-299.

El-Katatny, M.H., Abdelzaher, H.M. and Shoulkamy, M.A. 2006. Antagonistic actions of *Pythium oligandrum* and *Trichoderma harzianum* against phytopathogenic fungi (*Fusarium oxysporum* and *Pythium ultimum* var. ultimum). Archives Phytopathol. Plant Protect. 39(4): 289-301.

Fletcher, J.T., Smewin, B.J. and Obrien, A. 1990. *Pythium oligandrum* associated with a cropping disorder of *Agaricus bisporus*. Plant Pathol. 39: 603-605.

Foley, M.F. and Deacon, J.W. 1985. Isolation of *Pythium oligandrum* and other necrotrophic mycoparasites from soil. Trans. Br. Mycol. Soc. 35: 631-639.

Foley, M.F. and Deacon, J.W. 1986. Susceptibility of *Pythium* spp. and other fungi to antagonism by the mycoparasite *Pythium oligandrum*. Soil Biol. Biochem. 18(1): 91-95.

Frankenberger, W.T. and Arshad, M. 1995. Phytohormones in soil: Microbial production and function. Marcel Dekker, New York.

Gerbore, J., Benhamou, N., Vallance, J., Le Floch, G., Grizard, D., Regnault-Roger, C. and Rey, P. 2014. Biological control of plant pathogens: Advantages and limitations seen through the case study of *Pythium oligandrum*. Environ. Sci. Pollut. Res. 21: 4847-4860.

Gerbore, J., Benhamou, N., Vallance, J., Le Floch, J., Grizard, D., Regnault-Roger, C. and Rey, P. 2013. Biological control of plant pathogens: advantages and limitations seen through the case study of *Pythium oligandrum*. Environ. Sci. Pollut. Res. 21(7): 4847-4860.

Grünwald, N.J., Martin, F.N., Larsen, M.M., Sullivan, C.M., Press, C.M., Coffey, M.D., Hansen, E.M. and Parke, J.L. 2011. *Phytophthora*-ID.org: A sequence-based *Phytophthora* identification tool. Plant Dis. 95: 337-342.

Haas, D. and Defago, G. 2005. Biological control of soil-borne pathogens by Fluorescent pseudomonads. Nat. Rev. Microbiol. 4: 307-319.

Harman, G.E., Howell, C.R., Viterbo, A., Chet, I. and Lorito, M. 2004. *Trichoderma* species opportunistic, avirulent plant symbionts. Nat. Rev. Microbiol. 2(1): 43-56.

Hase, S., Shimizu, A., Nakaho, K., Takenaka, S. and Takahashi, H. 2006. Induction of transient ethylene and reduction in severity of tomato bacterial wilt by *Pythium oligandrum*. Plant Pathol. 55(4): 537-543.

Hase, S., Takahashi, S., Takenaka, S., Nakaho, K., Arie, T., Seo, S., Ohashi, Y. and Takahashi, H. 2008. Involvement of jasmonic acid signalling in bacterial wilt disease resistance induced by biocontrol agent *Pythium oligandrum* in tomato. Plant Pathol. 57(5): 870-876.

He, S.S., Zhang, B.X. and Ge, Q.X. 1992. On the antagonism by hyperparasite *Pythium oligandrum*. Acta Phytopathol. Sinica. 22: 77-82.

Helman, Y., Burdman, S. and Okon, Y. 2011. Plant growth promotion by rhizosphere bacteria through direct effects. pp. 89-103. *In*: Rosenberg, E. and Gophna, U. (eds.). Beneficial Microorganisms in Multicellular Life Forms. Springer, Berlin.

Hendrix, F.F. and Campbell, W.A. 1973. *Pythiums* as plant pathogens. Annu. Rev. Phytopathol. 11: 77-98.

Hermosa, R., Viterbo, A., Chet, I. and Monte, E. 2012. Plant-beneficial effects of *Trichoderma* and of its genes. Microbiology 158: 17-25.

Ho, H.H. 2009. The genus *Pythium* in Taiwan, China (1) – A synoptic review. Front. Biol. China 4: 15-28.

Ho, H.H. 2018. The taxonomy and biology of *Phytophthora* and *Pythium*. J. Bacteriol. Mycol. 6(1): 40-45.

Hockenhull, J., Jensen, D.F. and Yudiarti, T. 1992. The use of *Pythium periplocum* to control damping-off of cucumber seedlings caused by *Pythium aphanidermatum*. pp. 203-206. *In*: Tjamos, E.S., Papavizas, G.C. and Cook, R.J. (eds.). Biological Control of Plant Disease: Progress and Challenges for the Future. Plenum Publishing Corporation, New York.

Horner, N.R., Grenville-Briggs, L.J. and van West, P. 2012. The oomycete *Pythium oligandrum* expresses putative effectors during mycoparasitism of *Phytophthora infestans* and is amenable to transformation. Fungal Biol. 116: 24-41.

Hyakumachi, M. and Kubota, M. 2003. Fungi as plant growth promoter and disease suppressor. pp. 101-110. *In*: Arora, D.K. (ed.). Fungal Biotechnology in Agricultural, Food, and Environmental Applications. Mycology. CRC, Boca Raton.

Ikeda, S., Shimizu, A., Shimizu, M., Takahashi, H. and Takenaka, S. 2012. Biocontrol of black scurf on potato by seed tuber treatment with *Pythium oligandrum*. Biol. Cont. 60: 297-304.

Jefferies, P. and Young, T.W.K. 1994. Inter-fungal Parasitic Relationship. CAB International, Wallingford.

Jones, E.E. and Deacon, J.W. 1995. Comparative physiology and behavior of the mycoparasites *Pythium acanthophoron*, *P. oligandrum* and *P. mycoparasiticum*. Biocontrol Sci. Technol. 5: 25-39.

Karaca, G., Tepedelen, G., Belghouthi, A. and Paul, B. 2008. A new mycoparasite, *Pythium lycopersicum*, isolated in Isparta, Turkey: Morphology, molecular characteristics, and its antagonism with phytopathogenic fungi. FEMS Microbiol. Lett. 288: 163-170.

Kawamura, Y., Takenaka, S., Hase, S., Kubota, M., Ichinose, Y., Kanayama, Y., Nakaho, K., Klessig, D.F. and Takahashi, H. 2009. Enhanced defense responses in Arabidopsis induced by the cell wall protein fractions from *Pythium oligandrum* require SGT1, RAR1, NPR1 and JAR1. Plant Cell Physiol. 50(5): 924-934.

Kim, K.K., Fravel, D.R. and Papvizas, G.C. 1990. Glucose oxidase as the antifungal principle of talaron from *Talaromyces flavus*. Can. J. Microbiol. 36: 760-767.

Kratka, J., Bergmanova, E. and Kudelova, A. 1994. Effect of *Pythium oligandrum* and *Pythium ultimum* on biochemical changes in cucumber (*Cucumis sativus* L.)/Wirkung von *Pythium oligandrum* und *Pythium ultimum* auf biochemische Veränderungen in Gurkenpflanzen (*Cucumis sativus* L.). J. Plant Dis. Protect. 406-413.

Kushwaha, S.K., Vetukuri, R.R. and Grenville-Briggs, L.J. 2017. Draft genome sequence of the mycoparasitic oomycete Pythium periplocum strain CBS 532.74. Genome Announc. 5(12): e00057-17.

Laing, S.A.K. and Deacon, J.W. 1991. Video microscopy comparison of mycoparasitism by Pythium *oligandrum*, *P. nunn* and an unnamed *Pythium* species. Mycol. Res. 95: 469-479.

Le Floch, G., Rey, P., Déniel, F., Benhamou, N., Picard, K. and, Tirilly, Y. 2003a. Enhancement of development and induction of resistance in tomato plants by the antagonist, *Pythium oligandrum*. Agronomie 23: 455-460.

Le Floch, G., Rey, P., Benizri, E., Benhamou, N. and Tirilly, Y. 2003b. Impact of auxin-compounds produced by the antagonistic fungus *Pythium oligandrum* or the minor pathogen *Pythium* group F on plant growth. Plant Soil 257(2): 459-470.

Le Floch, G., Tambong, J., Vallance, J., Tirilly, Y., Levesque, A. and Rey, P. 2007. Rhizosphere persistence of three *Pythium oligandrum* strains in tomato soilless culture assessed by DNA macroarray and real-time PCR. FEMS Microbiol. Ecol. 61(2): 317-326.

Le Floch, G., Vallance, J., Benhamou, N. and Rey, P. 2009. Combining the oomycete Pythium oligandrum with two other antagonistic fungi: Root relationships and tomato grey mold biocontrol. Biol. Cont. 50(3): 288-298.

Lévesque, C.A. 2011. Fifty years of oomycetes: from consolidation to evolutionary and genomic exploration. Fungal Divers. 50: 35-46.

Lherminier, J., Benhamou, N., Larrue, J., Milat, M.L., Boudon-Padieu, E., Nicole, M. and Blein, J.P. 2003. Cytological characterization of elicitin-induced protection in tobacco plants infected by Phytophthora parasitica or phytoplasma. Phytopathology 93(10): 1308-1319.

Lifshitz, R., Stanghellini, M.E. and Baker, R. 1984. A new species of *Pythium* isolated from soil in Colorado. Mycotaxon. 20: 373-379.

Lodhi, A.M., Shahzad, S. and Ghaffar, A. 2005. A new report of *Pythium oligandrum* from Pakistan. Pak. J. Bot. 37(2): 487-491.

Lou, B.G., Wang, A.Y., Lin, C., Xu, T. and Zheng, X.D. 2011. Enhancement of defense responses by oligandrin against *Botrytis cinerea* in tomatoes. Afr. J. Biotechnol. 10(55): 442-449.

Lugtenberg, B. and Kamilova, F. 2009. Plant-growth-promoting rhizobacteria. Annu. Rev. Microbiol. 63: 541-566.

Madsen, A.M. and de Neergaard, E. 1999. Interactions between the mycoparasite *Pythium oligandrum* and sclerotia of the plant pathogen *Sclerotinia sclerotiorum*. Eur. J. Plant Pathol. 105(8): 761-768.

Mallik, M.A.B. and Williams, R.D. 2008. Plant growth promoting rhizobacteria and mycorrhizal fungi in sustainable agriculture and forestry. Sanashui, China.

Martin, F.N. and Hancock, J.G. 1986. Association of chemical and biological factors in soils suppressive to *Pythium ultimum*. Phytopathology 76(11): 1221-1231.

Martin, F.N. and Hancock, J.G. 1987. The use of *Pythium oligandrum* for biological control of preemergence damping-off caused by *P. ultimum*. Phytopathology 77(7): 1013-1020.

Masunaka, A., Nakaho, K., Sakai, M., Takahashi, H. and Takenaka, S. 2009. Visualization of *Ralstonia solanacearum* cells during biocontrol of bacterial wilt disease in tomato with *Pythium oligandrum*. J. Gen. Plant Pathol. 75(4): 281-287.

Masunaka, A., Sekiguchi, H., Takahashi, H. and Takenaka, S. 2010. Distribution and expression of elicitin-like protein genes of the biocontrol agent *Pythium oligandrum*. J. Phytopathol. 158(6): 417-426.

McQuilken, M.P., Whipps, J.M. and Cooke, R.C. 1990. Control of damping-off in cress and sugar-beet by commercial seed-coating with *Pythium oligandrum*. Plant Pathol. 39(3): 452-462.

McQuilken, M.P., Whipps, J.M. and Cooke, R.C. 1992. Use of oospore formulations of *Pythium oligandrum* for biological control of *Pythium* damping-off in cress. J. Phytopathol. 135(2): 125-134.

McQuilken, M.P., Powell, H.G., Budge, S.P. and Whipps, J.M. 1998. Effect of storage on the survival and biocontrol activity of *Pythium oligandrum* in pelleted sugar beet seed. Biocontrol Sci. Technol. 8(2): 237-241.

Meszka, B. and Bielenin, A. 2010. Polyversum WP: A new biological product against strawberry grey mould. Phytopathology 58: 13-19.

Mohamed, N., Lherminier, J., Farmer, M.J., Fromentin, J., Béno, N., Houot, V., Milat, M.L. and Blein, J.P. 2007. Defense responses in grapevine leaves against *Botrytis cinerea* induced by application of a *Pythium oligandrum* strain or its elicitin, oligandrin, to roots. Phytopathology 97(5): 611-620.

Mohammed, S.A. and Amjed, S.F.S. 1999. Isolation of *Pythium acanthicum*, *P. oligandrum*, and *P. periplocum* from soil and evaluation of their mycoparasitic activity and biocontrol efficacy against selected phytopathogenic *Pythium* species. Mycopathologia 145: 143-153.

Mulligan, D.F., Jones, E.E. and Deacon, J.W. 1995. Monitoring and manipulation of populations of *Pythium oilgandrum, Pythium mycoparasiticum* and a *Papulaspora* species in soil. Soil Biol. Biochem. 27: 1333-1343.

Patrice, R.E.Y., Le Floch, G., Benhamou, N., Salerno, M.I., Thuillier, E. and Tirilly, Y. 2005. Interactions between the mycoparasite Pythium oligandrum and two types of sclerotia of plant-pathogenic fungi. Mycol. Res. 109(7): 779-788.

Paul, B. 1999. Suppression of *Botrytis cinerea* causing the grey mould disease of grape-vine by an aggressive mycoparasite, *Pythium radiosum*. FEMS Microbiol. Lett. 176: 25-30.

Paul, B., Bala, B., Lassaad, B., Gautier, C., Sanchez-Hernandez, E. and Lefort, F. 2006. A new species of *Pythium* with ornamented oogonia: Morphology, taxonomy, internal transcribed spacer region of its ribosomal RNA and its comparison with related species. FEMS Microbiol. Lett. 254: 317-323.

Picard, K., Ponchet, M., Blein, J., Rey, P., Tirilly, Y. and Benhamou, N. 2000. Oligandrin: A proteinaceous molecule produced by the mycoparasite *Pythium oligandrum* induces resistance to *Phytophthora parasitica* infection in tomato plants. Plant Physiol. 124: 379-395.

Ponchet, M., Panabieres, F., Milat, M.L., Mikes, V., Montillet, J.L., Suty, L., Triantaphylides, C., Tirilly, Y. and Blein, J.P. 1999. Are elicitins cryptograms in plant-oomycete communications. Cell Mol. Life Sci. 56(11-12): 1020-1047.

Rekanovic, E., Milijasevic, S., Todorovic, B. and Potocnik, I. 2007. Possibilities of biological and chemical control of Verticillium wilt in pepper. Phytoparasitica 35(5): 436.

Rey, P., Benhamou, N., Wulff, J. and Tirilly, Y. 1998. Interactions between tomato (*Lycopersicon esculentum*) root tissues and the mycoparasite *Pythium oligandrum*. Physiol. Mol. Plant. Pathol. 53: 105-122.

Rey, P., Le Floch, G., Benhamou, N., Salerno, M.I., Thuillier, E. and Tirilly, Y. 2005. Interactions between the mycoparasite *Pythium oligandrum* and two types of sclerotia of plant-pathogenic fungi. Mycol. Res. 109: 779-788.

Rey, P., Gl, F., Benhamou, N. and Tirilly, Y. 2008. *Pythium oligandrum* biocontrol: Its relationships with fungi and plants. pp. 43-67. *In*: Ait Barka, E. and Clément, C. (eds.). Plant-microbe interactions. Research Signpost, Kerala.

Ribeiro, W.R.C. and Butler, E.E. 1992. Isolation of mycoparasitic species of *Pythium* with spiny oogonia from soil in California. Mycol. Res. 96: 857-862.

Ribeiro, W.R.C. and Butler, E.E. 1995. Comparison of the mycoparasites *Pythium periplocum*, *P. acanthicum* and *P. oligandrum*. Mycol. Res. 99(8): 963-969.

Ristaino, J.B.A. 2012. Lucid key to the common *Phytophthora* species. USA: APS Press.

Saunders, G.A., Washburn, J.O., Egerter, D.E. and Anderson, J.R. 1988. Pathogenicity of fungi isolated from field collected larvae of the Western treehole mosquito, *Aedes sierrensis* (Diptera: Culicidae). J. Invertebr. Pathol. 52: 360-363.

Su, X.Q., Zou, F., Guo, Q., Huang, J. and Chen, T.X. 2001. A report on a mosquito-killing fungus, *Pythium carolinianum*. Fungal Diversity. 7: 129-133.

Takenaka, S., Nishio, Z. and Nakamura, Y. 2003. Induction of defense reactions in sugar beet and wheat by treatment with cell wall protein fractions from the mycoparasite *Pythium oligandrum*. Phytopathology. 93: 1228-1232.

Takenaka, S., Nakamura, Y., Kono, T., Sekiguchi, H., Masunaka, A. and Takahashi, H. 2006. Novel elicitin-like proteins isolated from the cell wall of the biocontrol agent *Pythium oligandrum* induce defence-related genes in sugar beet. Mol. Plant Pathol. 7: 325-339.

Takenaka, S., Sekiguchi, H., Nakaho, K., Tojo, M., Masunaka, A. and Takahashi, H. 2008. Colonization of *Pythium oligandrum* in the tomato rhizosphere for biological control of bacterial wilt disease analyzed by real-time PCR and confocal laser-scanning microscopy. Phytopathology 98(2): 187-195.

Takenaka, S. and Tamagake, H. 2009. Foliar spray of a cell wall protein fraction from the biocontrol agent *Pythium oligandrum* induces defence-related genes and increases resistance against Cercospora leaf spot in sugar beet. J. Gen. Plant Pathol. 75: 340-348.

Takenaka, S. and Ishikawa, S. 2013. Biocontrol of sugar beet seedling and taproot diseases caused by *Aphanomyces cochlioides* by *Pythium oligandrum* treatments before transplanting. Japan Agric. Res. Quarterly: JARQ, 47(1): 75-83.

Vallance, J., Le Floch, G., Déniel, F., Barbier, G., Lévesque, C.A. and Rey, P. 2009. Influence of *Pythium oligandrum* biocontrol on fungal and oomycete population dynamics in the rhizosphere. Appl. Environ. Microbiol. 75(14): 4790-4800.

Van der Plaats-Niterink, A.J. 1981. Monograph of the genus *Pythium*. Stud Mycol. Baarn. 21: 1-242.

Veloso, J. and Diaz, J. 2012. *Fusarium oxysporum* Fo47 confers protection to pepper plants against *Verticillium dahliae* and *Phytophthora capsici*, and induces the expression of defence genes. Plant Pathol. 61(2): 281-288.

Whipps, J.M. 1997. Developments in the biological control of soil-borne plant pathogens. Adv. Bot. Res. 26: 1-134.

Wulff, E.G., Pham, A.T.H., Cherif, M., Rey, P., Tirilly, Y. and Hockenhull, J. 1998. Inoculation of cucumber roots with zoospores of mycoparasitic and plant pathogenic *Pythium* species: Differential zoospore accumulation, colonization ability and plant growth response. Eur. J. Plant Pathol. 104(1): 69-76.

Index

About the Editors

Mahendra Rai, Ph.D. is a Senior Professor and UGC-Basic Science Research Faculty Fellow at the Department of Biotechnology, Sant Gadge Baba Amravati University, Maharashtra, India. He was a Visiting Scientist at University of Geneva, Debrecen University, Hungary; University of Campinas, Brazil; Nicolaus Copernicus University, Poland; VSB Technical University of Ostrava, Czech Republic; and National University of Rosario, Argentina. He has published more than 400 research papers in national and international journals. In addition, he has edited/authored more than 48 books and six patents.

Kamel A. Abd-Elsalam, Ph.D. is a head researcher at Plant Pathology Research Institute, Agricultural Research Center, Giza, Egypt. Dr. Kamel's research interests include molecular plant pathology and developing eco-friendly hybrid nanomaterials for controlling plant diseases. He has published four books, and more than 120 research articles.

Dr. Kamel served as a visiting scholar in the Institute of Excellence in Fungal Research, Thailand; Institute of Microbiology, TUM, Germany; Laboratory of Phytopathology, Wageningen University, The Netherlands and Plant Protection Department, Sassari University, Italy.

Avinash P. Ingle, Ph.D. has completed his doctoral degree from Department of Biotechnology, Sant Gadge Baba Amravati University, Amravati, Maharashtra (India). He had worked as Research Scientist in the same department from 2013-2016. He has more than 65 research publications, 40 book chapters and four books to his credit. He has been awarded travel grants from different funding agencies to present his research work in international conferences held in Malaysia, China, Spain and France. Currently, Dr. Ingle is working as Post-Doctoral Fellow at Department of Biotechnology, Engineering School of Lorena – University of São Paulo, Brazil and area of his research interest includes plant pathology, nanobiotechnology and biofuel production.

Color Section

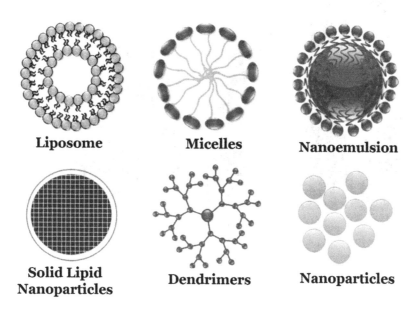

Figure 1.1. Nanomaterials for delivery of fungicides

Figure 9.3(a). Damping-off of cucumber at seedling stage

Chapter 9

Figure 9.3(b). Cottony leak of cucumber fruits

Chapter 10

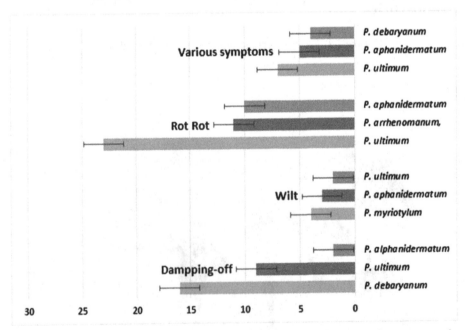

Figure 10.1. Three major *Pythium* species ranking caused various type of symptoms such as damping-off, wilt, root rot and various symptoms on different plant species

Chapter 15

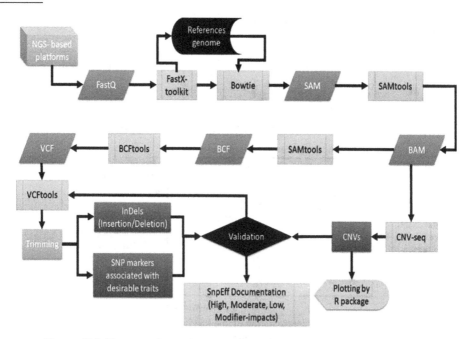

Figure 15.3. The general workflow of bioinformatics tools to NGS dataset analysis

Chapter 16

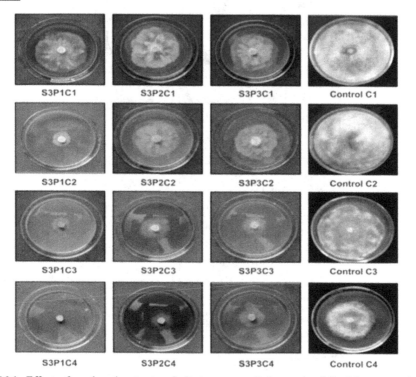

Figure 16.1. Effect of methanol extracts of plants on mycelial growth of *P. debaryanum* Solvents (S): S1-acetone, S2-ethanol, S3-methanol, S4-chloroform. Plant leaves (P): P1-*A. marmelos* leaf extract, P2-*S. cumini* leaf extract, P3-*P. pinnata* leaf extract. Concentrations (C): C1-250 µl, C2-500 µl, C3-750 µl, C4-1000 µl

Chapter 16

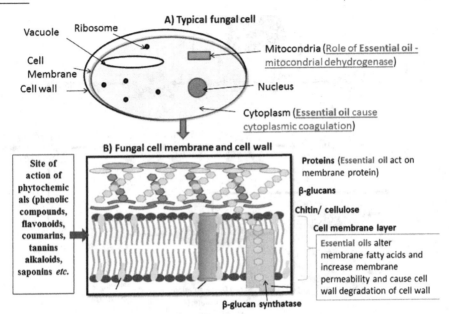

Figure 16.3. Diagrammatic representation of site and mechanism of action of antifungal phytochemicals on fungal cell

Chapter 17

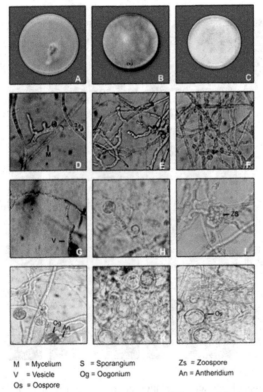

M = Mycelium	S = Sporangium	Zs = Zoospore
V = Vesicle	Og = Oogonium	An = Antheridium
Os = Oospore		

Figure 17.1. Microscopic images of *P. aphanidermatum*

Printed in the United States
by Baker & Taylor Publisher Services